FORERUNNERS OF MAMMALS

Life of the Past — *James O. Farlow, editor*

FORERUNNERS OF MAMMALS

Radiation · Histology · Biology

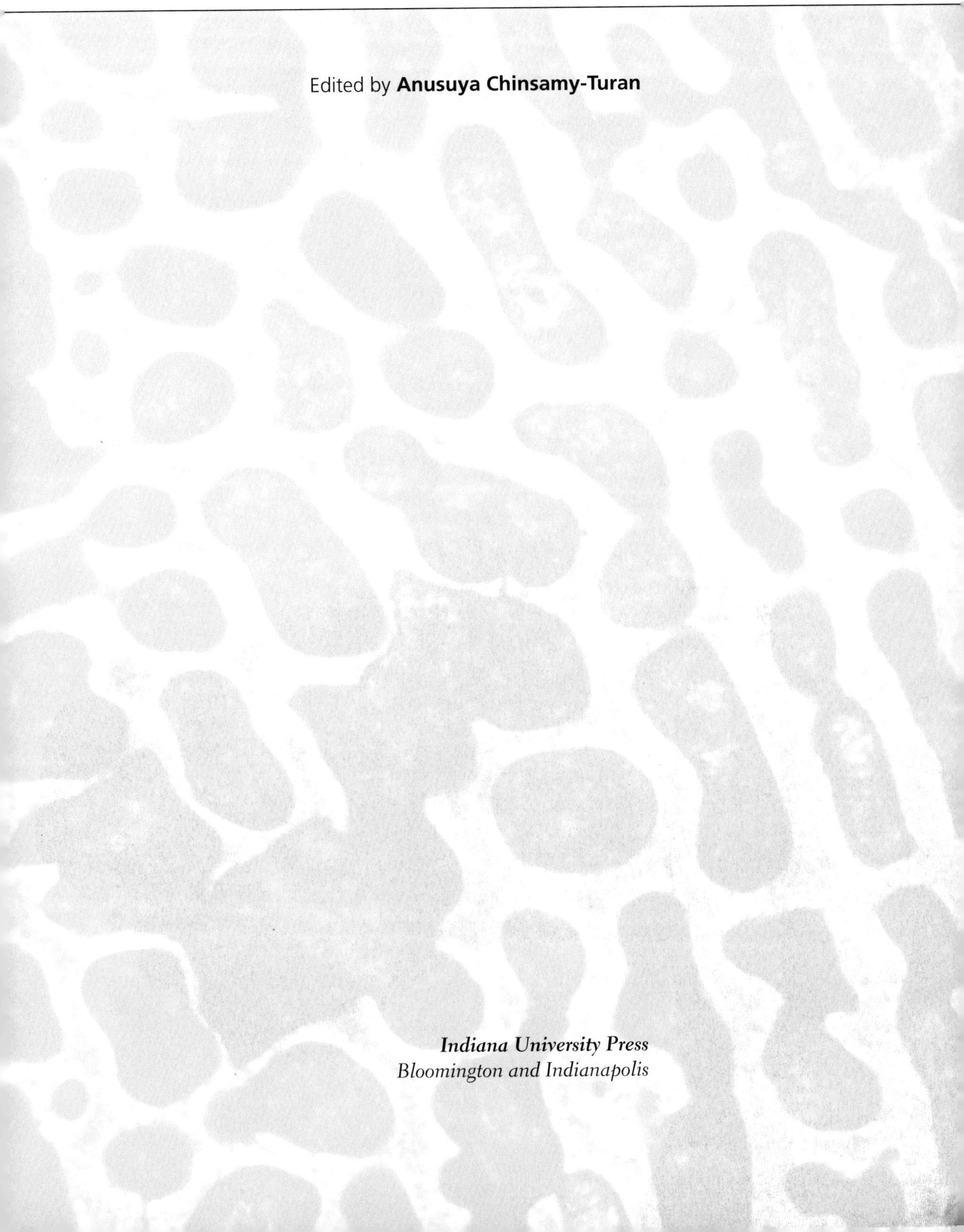

Edited by **Anusuya Chinsamy-Turan**

Indiana University Press
Bloomington and Indianapolis

This book is a publication of

Indiana University Press
601 North Morton Street
Bloomington, Indiana 47404-3797 USA

iupress.indiana.edu

Telephone orders 800-842-6796
Fax orders 812-855-7931

∞ The paper used in this publication meets the minimum requirements of the American National Standard for Information Sciences—Permanence of Paper for Printed Library Materials, ANSI Z39.48-1992.

Manufactured in the United States of America

Library of Congress
Cataloging-in-Publication Data

Forerunners of mammals : radiation, histology, biology / edited by Anusuya Chinsamy-Turan.
p. cm. — (Life of the past)
Includes bibliographical references and index.
ISBN 978-0-253-35697-0 (cloth : alk. paper)
1. Reptiles, Fossil. 2. Mammals—Evolution. 3. Bones—Histology. I. Chinsamy-Turan, Anusuya.
QE861.F67 2012
567.9'3—dc23

2011016622

1 2 3 4 5 17 16 15 14 13 12

This book is dedicated to the memory of

JAMES W. KITCHING

(1922–2003)

Gondwanan Paleontologist

An Exceptional Fossil Hunter and an Inspiration to Many

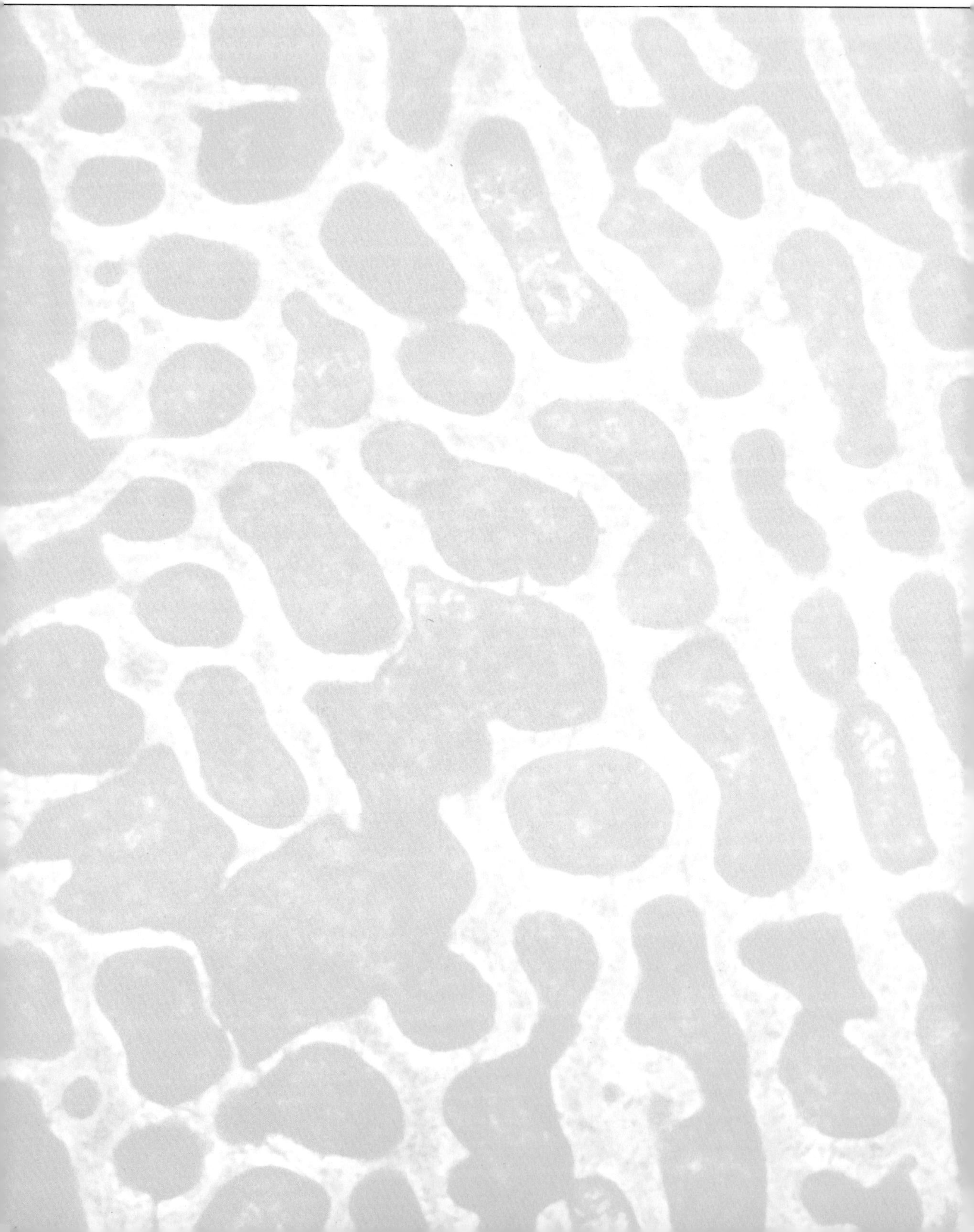

Contents

Preface

P

This book brings together a group that has over many years researched various aspects of the evolution and paleobiology of the synapsids. Many of us have collaborated in our research endeavors, and all of us have at some stage shared information and had many hearty discussions about the biology of our distant relatives.

The book comprises eleven chapters. The first two chapters provide an introduction to the predecessors of mammals and their relatives, and an assessment of the ancient world in which they radiated. The opening chapter sets the scene, providing a guide of "who" the synapsids were and how they are related to one another. In this chapter, Tom Kemp provides an up to date assessment of the radiation of the synapsids from their earliest pelycosaur members, to the diverse nonmammalian therapsids, and later to the increasingly more mammal-like cynodonts. All this is done from a global perspective.

The second chapter of this book deals more specifically with the Karoo Basin of South Africa and documents an unparalleled track record of the evolution and radiation of the therapsids. Here, Karoo paleontologists, Roger Smith and Bruce Rubidge, together with a recent PhD graduate Merrill van der Walt, provide a unique perspective of therapsid biodiversity and paleoenvironmental analysis of the Karoo Basin of South Africa. For the first time, faunal turnover in the Karoo Basin is provided through a lens of absolute numbers of genera and have permitted detailed trophic level analyses for each of the biozones.

The third chapter by yours truly sets the scene for the bone microstructure chapters that follow. The first part of this chapter takes the form of an atlas of bone microstructure that will enable a novice to identify particular types of bone tissues in synapsids. The second part examines the biological implications of particular types of bone microstructures and how these can be utilized to deduce various aspects of the biology of extinct animals.

The next seven chapters focus on particular synapsid lineages. Each of these chapters is structured to provide a phylogenetic and paleobiological context before delving into the bone microstructure of that particular group. The first in this series of chapters (chapter 4) predictably deals with the earliest members of the Synapsida, the pelycosaurs. Here, Adam Huttenlocker and Elizabeth Rega present an overview of the bone microstructure of pelycosaurian-grade synapsids from both "normal" and pathological skeletal elements.

The fifth chapter is by Sanghamitra Ray, my former postdoctoral fellow, Jennifer Botha-Brink, my former PhD student, and me. Here we review what is known about the bone microstructure of the dicynodonts, which were the dominant herbivores of the Permian. We then focus on a selection of basal to more derived dicynodont taxa from South Africa and India in the interest of deciphering growth patterns through the Dicynodontia.

The sixth chapter by Sandra Jasinoski, my current postdoctoral fellow, and myself, presents the first comprehensive assessment of dicynodont cranial microstructure. Besides documenting intercranial element variability in histology, we also assess whether functional signatures are recorded in the cranial microstructure.

The seventh chapter is by Jeremy Green, who provides an assessment of the bone and tusk microstructure of large Late Triassic dicynodonts from North America. Here he provides a comparative assessment of the bone microstructure of *Placerias*, a Kannemeyeriiform, and another large as yet unnamed dicynodont, as well as a detailed account of the growth increments recorded in the tusks.

The eighth chapter is by Sanghamitra Ray and me. Here we review the bone microstructure of some therocephalians and gorgonopsians, and we present novel data pertaining to bone damage caused by fungi in Permian-aged bones. This chapter also highlights the importance of extensive sampling of skeletal elements to appreciate histovariability within and between bones.

Chapter nine, by Jennifer Botha-Brink, Fernando Abdala, and me, reviews the bone microstructure of the nonmammaliaform cynodonts and presents fresh data on four additional taxa from South Africa and two from Brazil.

Jørn Hurum and I review the bone microstructure of early mammals from the United Kingdom and Mongolia in chapter ten, and we provide new data pertaining to the bone microstructure of extant monotremes, marsupials, and placentals.

The final chapter brings together for the first time three physiologists, John Ruben, Willem Hillenius, and Devon Quick, and paleontologist Tom Kemp, who provide an interesting perspective on when and how endothermy evolved among the synapsids.

From this book it is evident that even though the dicynodonts are relatively well-studied, many questions still exist such as their particular lifestyle and growth dynamics. The same could be said for the cynodonts, while the therocephalians and gorgonopsians are comparatively less well-studied. However, of all the therapsids, least is known about dinocephalians, which may be a direct result of the fact that their taxonomy has been so problematic. However, recent work has resolved some of these problems and plans are afoot to fill in this gap.

This unique book on the radiation, histology, and biology of the synapsids provides a documentation of the bone microstructure of the forerunners of mammals and insights into their biology, as well as highlights areas for future research.

Acknowledgments

A

Having just completed a book on dinosaur bone microstructure, the notion of another book was furthest from my mind when Jim Farlow seeded the idea of a book on "therapsid bone" in my head. Once implanted, it simply grew and has now materialized. Thanks, Jim!

I am indebted to all the contributors to this book. The successful completion of an edited volume of this nature is directly attributable to the hard work and diligence of the entire team. I count myself as incredibly fortunate to have had the opportunity to collaborate with all of you at some stage in the past, and now again on this book. Bob Sloan, Jim Farlow, Bernadette Zoss, Dan Pyle, June Silay, Karen Hallman, and others at Indiana University Press, thank you for dealing with my queries so efficiently.

I am grateful to all the collection managers from around the world who grant permission for histological analyses of fossils under their curation because they understand and appreciate the importance of studies of bone microstructure. Kholeka Sidinile, Kerwin von Willig, and especially Andrea Plos are acknowledged for technical support.

Over the years, I've had the opportunity of learning and growing with several research students, and currently, I have an exceptional group of postdoctoral fellows in my lab—Romala Govender, Sandra Jasinoski, Daniel Thomas, Yasemin Tulu, and, more recently, Aurore Canoville—as well as graduate students Nicholas Fordyce, Ian Brumfitt, and, most recently, Tobias Nasterlack. Thank you to all of you for choosing to unravel the biology of extinct animals with me.

My friends and colleagues from around the world—you know who you are—thank you for being just an email or a call away. Special thanks are necessary to Luis Rey, who has brought several synapsids to life with his vivid artwork. Peter Dodson, my friend, colleague, and mentor, I am so glad our paths crossed way back.

I am particularly indebted to my parents and sisters and their families for their support through the years. I am especially thankful to my husband, Yunus Turan, for his abiding back-up of all that I do, and to my sons, Evren and Altay, for simply bearing with me. You all inspire me more than you know.

A. C-T

My co-authors and I would like to thank and acknowledge the following reviewers for their insightful reading and constructive comments of chapters in the book:

Kenneth Angielczyk, Field Museum, Chicago, Illinois, United States

Al Bennett, University of California, Irvine, California, United States

Jennifer Botha-Brink, National Museum, Bloemfontein, South Africa

Greg Erickson, Florida State University, Tallahassee, Florida, United States

James Farlow, Indiana University—Purdue University Fort Wayne, Fort Wayne, Indiana, United States

Jeremy Green, Kent State University, Tuscarawas, Ohio, United States

Tom Hübner, Niedersächsisches Landesmuseum, Hannover, Germany

Adam Huttenlocker, University of Washington, Seattle, Washington, United States

Sandra Jasinoski, University of Cape Town, Cape Town, South Africa

Christian Kammerer, American Museum of Natural History, New York, New York, United States

Tom Kemp, Oxford Museum, Oxford, United Kingdom

Michel Laurin, Muséum National d'Histoire Naturelle, Centre National de la Recherche Scientifique, Paris, France

Sanghamitra Ray, Indian Institute of Technology, Kharagpur, India

Bruce Rubidge, Bernard Price Institute for Palaeontological Research, University of the Witwatersrand, Johannesburg, South Africa

Stuart Sumida, California State University, San Bernadino, California, United States

Allison Tumarkin-Deratzian, Temple University, Philadelphia, Pennsylvania, United States

Contributors

C

Fernando Abdala

Fernando Abdala, vertebrate paleontologist, Bernard Price Institute for Palaeontological Research, School for Geosciences, University of the Witwatersrand, Johannesburg 2050, South Africa. Fernando.abdala@wits.ac.za

Jennifer Botha-Brink

Jennifer Botha-Brink, specialist scientist and head of Karoo Vertebrate Palaeontology, has published on therapsid bone histology and the biology and ecology of vertebrates associated with the end-Permian extinction event. National Museum, Karoo Palaeontology, PO Box 266, Bloemfontein 9300, South Africa, and University of the Free State, Department of Zoology and Entomology, Bloemfontein 9300, South Africa. jbotha@nasmus.co.za

Anusuya Chinsamy-Turan

Anusuya Chinsamy-Turan, FRSSA, Professor and Fellow, has published prolifically on mineralized tissues and biology of various extinct vertebrates and is author of *The Microstructure of Dinosaur Bone: Deciphering Biology through Fine Scale Analysis* (Johns Hopkins University Press, 2005) and *Famous Dinosaurs of Africa* (Struik, 2008). University of Cape Town, Zoology Department, Private Bag X3, Rhodes Gift 7701, South Africa. Anusuya.Chinsamy-Turan@uct.ac.za

Jeremy L. Green

Jeremy L. Green, Assistant Professor of Geology, has published on growth and feeding ecology in dicynodonts, xenarthrans, and proboscideans. Kent State University at Tuscarawas, 330 University Drive NE, New Philadelphia, Ohio 44663, United States. jgreen72@kent.edu

Willem J. Hillenius

Willem J. Hillenius, Professor and Chair, College of Charleston, Department of Biology, Charleston, South Carolina 29424-0001, United States. hilleniusw@cofc.edu

Jørn H. Hurum

Jørn H. Hurum, vertebrate paleontologist, has published on Mesozoic mammals, dinosaurs, and early primates. University of Oslo, Natural History Museum, PO Box 1172, Blindern, 0318 Oslo, Norway. j.h.hurum@nhm.uio.no

Adam K. Huttenlocker

Adam K. Huttenlocker, PhD candidate, has published on pelycosaurs and therocephalians. University of Washington, Department of Biology, 24 Kincaid Hall, Box 351800, Seattle, Washington 98195-1800, United States. huttenla@u.washington.edu

Sandra C. Jasinoski

Sandra C. Jasinoski, Claude Leon Postdoctoral Fellow, has published on the cranial functional morphology of therapsids, including investigation of bone microstructure and implementation of computational analysis. University of Cape Town, Zoology Department, Private Bag X3, Rhodes Gift 7701, South Africa. sandra_jas@hotmail.com

Tom S. Kemp

Tom S. Kemp, Emeritus Research Fellow, has published on many aspects of the paleobiology of therapsids and is author of *Mammal-Like Reptiles and the Origin of Mammals* (Academic Press, 1982), *Fossils and Evolution* (Oxford University Press, 1999), and *The Origin and Evolution of Mammals* (Oxford University Press, 2005). University of Oxford, St John's College, Oxford OX1 3JP, England, United Kingdom. tom.kemp@sjc.ox.ac.uk

Devon E. Quick

Devon E. Quick, Zoology lecturer, physiologist, Oregon State University, Corvallis, Oregon 97331-2914, United States. quickd@science.oregonstate.edu

Sanghamitra Ray

Sanghamitra Ray, Assistant Professor, has published on paleobiology of dicynodonts and Indian Gondwana stratigraphy and sedimentation. Indian Institute of Technology, Department of Geology and Geophysics, Kharagpur 721302, India. sray@gg.iitkgp.ernet.in

Elizabeth Rega

Elizabeth Rega, Associate Professor of Anatomy, Western University of Health Sciences, 309 E. Second St., Pomona, California 91766-1854, United States. erega@westernu.edu

John A. Ruben, Professor of Zoology, has published extensively on major selective factors linked to the origins of avian and mammalian endothermy. Oregon State University, Corvallis, Oregon 97331, United States. rubenj@bcc.orst.edu

John A. Ruben

Bruce Rubidge, FRSSA, FGSSA, Professor and Director of Bernard Price Institute, has published extensively on Permian-Jurassic biostratigraphy, biogeography, and biodiversity changes, Karoo Basin development, and tetrapod systematics, particularly therapsids, and has a special interest in the origins of basal therapsids. Bernard Price Institute for Palaeontological Research, School for Geosciences, University of the Witwatersrand, Johannesburg 2050, South Africa. Bruce.rubidge@wits.ac.za

Bruce Rubidge

Roger Smith, FRSSA, FGSSA, sedimentologist and vertebrate paleontologist, has published extensively under the general title of "Palaeoecology of Gondwana." His research is mainly field-based and integrates paleontological and sedimentological data into paleoenvironmental reconstructions of ancient landscapes—especially in dramatic changes that took place in the Karoo Basin during the End-Permian mass extinction. Curator of Karoo Palaeontology, Iziko South African Museum, PO Box 81, Cape Town 8000, South Africa. rsmith@iziko.org.za

Roger Smith

Merrill van der Walt, vertebrate paleontologist, has a special interest in geographic information system and biodiversity trend analysis. She is the manager of education at Origins Centre, University of the Witwatersrand, Johannesburg 2050, South Africa. merrill@origins.org.za

Merrill van der Walt

FORERUNNERS OF MAMMALS

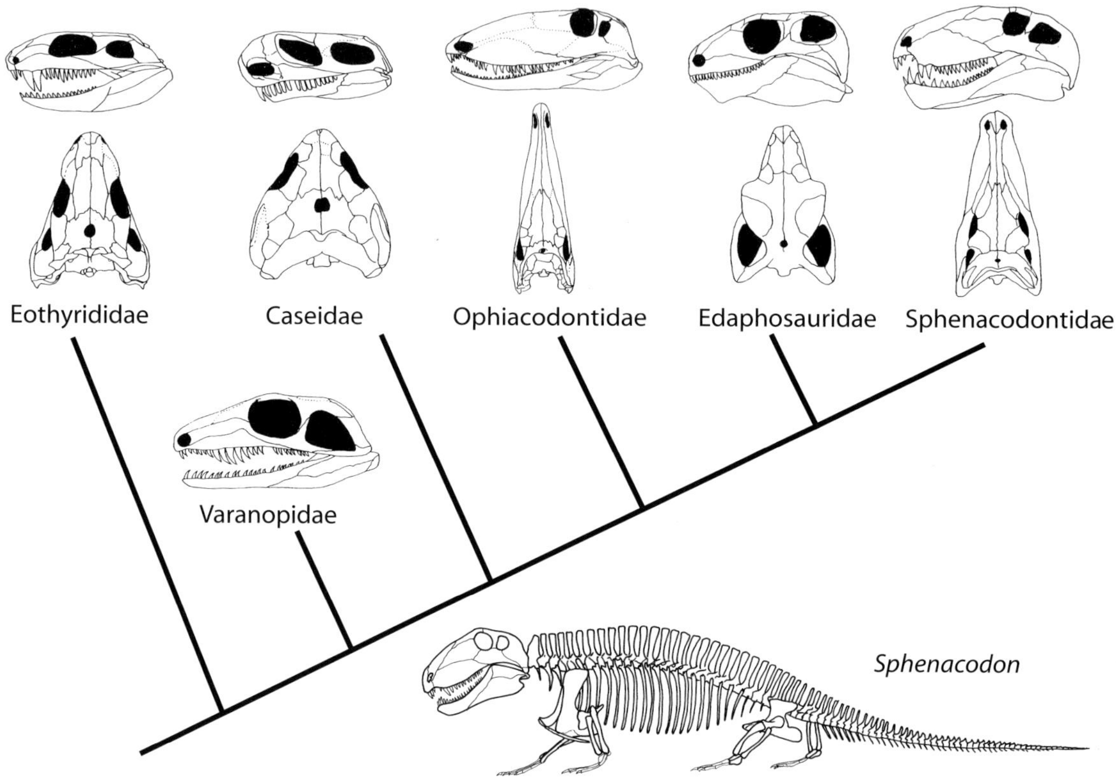

1.1. Phylogeny of the pelycosaurian-grade synapsids (reproduced by permission of Oxford University Press from "Interrelationships of the Synapsida" by Kemp [1988]), with a reconstruction of the skeleton of the sphenacodontid pelycosaur *Sphenacodon* (redrawn from Romer and Price [1940]).

The Origin and Radiation of Therapsids

1

Tom S. Kemp

Introduction

The earliest fossils of amniotes—the clade that now consists of the reptiles birds and mammals—occur in 320 million-year-old rocks of the Late Carboniferous (sensu Laurin 2004; Laurin and Reisz 1995; Voigt and Ganzelewski 2010). They are characterized by several modifications that indicate an increased independence of freestanding bodies of water. For example, the aquatic sensory system of lateral lines is no longer present, the skull and jaws are strengthened for the biting action that terrestrial animals tend to rely on, and the limbs and girdles are robust. Whether they possessed the single, most characteristic of all amniote characters, an amniotic egg capable of developing on dry land, is unknown but seems probable (Packard and Seymour 1997). Surprisingly perhaps, the divergence between the two major living amniote sister groups was in existence from the very beginning of the fossil record of amniotes (Laurin and Reisz 1995). The Sauropsida consists of the modern reptiles and birds, plus the great range of dinosaurs, pterosaurs, and aquatic reptiles that dominated the land, sea, and air of the Mesozoic. The second group, the Synapsida, consists nowadays only of the mammals with their high-energy, active lifestyle and extraordinary ability to adapt successfully to so many terrestrial, subterranean, arboreal, aquatic, and aerial habitats in such a wide range of environmental conditions.

The connection between the mammals and their remote Carboniferous relatives is one of the most remarkable parts of the fossil record. These animals, referred to as nonmammalian synapsids, or more affectionately if a trifle misleadingly as "mammal-like reptiles," are the stem-group mammals, and they exhibit a range of different combinations of basal amniote and mammalian characters. Most of these nonmammalian synapsids are included in Therapsida, the monophyletic taxon that originated within a primitive, paraphyletic group called "pelycosaurs," at least by the Middle Permian when the first undisputed therapsids are found. The therapsids are of fundamental importance in the history of terrestrial life on Earth for three reasons: first, because they dominated the terrestrial vertebrate scene from the Middle through the Late Permian and remained one of the most significant groups of the Triassic; second, because they established for the first time a terrestrial ecosystem based on very large numbers of fully terrestrial herbivorous tetrapods as the primary consumers, plus related carnivores as the major secondary consumers, a pattern to be repeated later by the dinosaurs of the Mesozoic and the mammals of the Tertiary; and third, because it was the therapsids that commenced the evolution of an elevated energy budget, and therefore a lifestyle in which relative independence of environmental fluctuations in temperature allowed continuous high levels of activity. Progressive development of this revolutionary biological strategy can be followed all the way to its expression in the fully endothermic mammals (Kemp 2006b, 2007b).

The most recent overall review of the evolution of the synapsids is that of Kemp (2005). Reisz (1986) reviewed the pelycosaurs to genus level. Rubidge and Sidor (2001), and somewhat egregiously Ivakhnenko (2003), reviewed the major therapsid subtaxa.

The Origin and Early Radiation of the Synapsida: Grade "Pelycosauria"

Synapsida is a monophyletic taxon recognized by a number of characters, including most distinctively the eponymous synapsid temporal fenestra in the skull roof, lying behind the orbit, and primitively bounded by the postparietal, parietal, squamosal, quadratojugal and jugal bones of the skull. The earliest record of synapsids consists of footprints of early Late Carboniferous age found in Germany, which predates the oldest actual fossil synapsids by 5–10 Ma (Voigt and Ganzelewski 2010). The latter consists of the humerus and other fragments described as *Protoclepsydrops* by Carroll (1964), which are dated as Westphalian B of the Late Carboniferous of Nova Scotia, and the slightly younger Westphalian D ophiacodontid pelycosaur *Archaeothyris* from the same region, which is represented by a skull lacking the mandible, and by the vertebral column and parts of the limbs (Reisz 1972).

The Late Carboniferous and Early Permian witnessed a radiation of nontherapsid synapsids, usefully referred to as the paraphyletic group "Pelycosauria." They retained many unmodified amniote features such as a more or less homodont dentition, short and heavily built sprawling limbs, and a long massive tail (Fig. 1.1).

The great majority are known from North America, although there are also representatives in Europe and, at the very end of their temporal range, rare representatives of the family Varanopidae in South Africa (Modesto et al. 2001; Botha-Brink and Modesto 2009). In their classic review, Romer and Price (1940) divided all pelycosaurs into three major subtaxa, the primitive, mostly long-snouted piscivorous Ophiacodontia, the herbivorous Edaphosauria, and the carnivorous Sphenacodontia. Reisz (1986) reduced these three taxa to family status, and recognized the more basal position of certain forms (Fig. 1.1), an arrangement largely supported by subsequent studies (e.g., Berman et al. 1995; Reisz, Godfrey, and Scott 2009; see chapter 4 of this book). The Early Permian *Eothyris* and *Oedaleops* are each known from a single, small skull and constitute a clade Eothyrididae that had retained the primitively broad form, large supratemporal bone in the temporal region, and carnivorous dentition (Reisz, Godfrey, and Scott 2009). The family Caseidae is related to the eothyridids and like them had a broad skull roof (Maddin, Sidor, and Reisz 2008). However, caseids were highly specialized for an herbivorous diet, exhibiting shortening of the skull and an enlarged temporal fenestra indicating powerful adductor musculature. The marginal dentition consists of closely packed leaf-like teeth and there is also a well-developed palatal dentition against which a tough tongue is presumed to have worked. Although low on the phylogenetic tree, caseids are actually the last major group to appear in the fossil record and were also one of the two last representatives of pelycosaurian-grade animals to survive, being still present in the Middle Permian of Russia.

All other pelycosaurs are included in the monophyletic clade Eupelycosauria, distinguished by narrowing of the skull and a reduction of the supratemporal bone. Varanopids were relatively small, carnivorous forms, and there are three derived eupelycosaurian groups, each specialized for a different basic mode of life. The Ophiacodontidae (which include *Archaeothyris*) are long-snouted forms with a marginal row of small, sharp teeth suitable for piscivory. The Edaphosauridae have a number of features for herbivory that are convergent on the caseids, such as short jaws, large temporal fenestra, and a battery of palatal teeth that in this case was opposed by a similar battery on the medial side of each mandible. The edaphosaurid postcranial skeleton is distinguished by enormously elongated neural spines bearing short crosspieces or tubercles (see Plate A). The presumed soft tissue sail that these supported could hardly fail to have had the potential to enhance ectothermic temperature regulation of these animals (Bennett 1996), although others have argued that they were primarily related to sexual selection (Tomkins et al. 2010). The final taxon is the carnivorous Sphenacodontidae, most famously *Dimetrodon* (see Plate A; Fig. 4.5C), with its caniniform teeth and a body up to three meters in length. It too possesses elongated neural spines, in this case lacking the crosspieces and no doubt evolved independently of those of edaphosaurids. *Sphenacodon* possesses much less elongated neural spines, and *Secodontosaurus* is a curiously long-snouted, piscivorous form, although Reisz, Berman, and Scott (1992) suggested that such a sail-bearing tetrapod was more likely a specialized insectivore or carnivore taking its food from crevices.

As far as is presently known, pelycosaurs were restricted during the Late Carboniferous and Early Permian to the equatorial latitudes of central Pangaea, a warm, permanently humid biome, which suggests that during this period they were unable to survive excessively arid or cool conditions (Kemp 2006b). The very few caseids and varanopids that lingered into the Middle Permian occur in higher latitudes and may perhaps have evolved seasonal torpor strategies to survive the less equable conditions they met. There was also, paradoxically, a far higher proportion of carnivorous to herbivorous pelycosaurs than expected in a terrestrial tetrapod community, even among ectotherms with low metabolic rates. This can be accounted for by the hypothesis that the ecosystem was still dependent to a considerable extent on freshwater aquatic productivity (Milner 1993; Berman, Sumida, and Lombard 1997; Kemp 2006b; Olson 1975, 1986). The presence of abundant fish-eating ophiacodontids and perhaps *Secodontosaurus* as intermediates in the energy flow from freshwater to dry land is certainly consistent with this view.

The Origin and Interrelationships of Therapsids

The origin of the therapsids by the Middle Permian was a revolution in the history of terrestrial life, because they show clear morphological evidence of higher activity levels and, therefore, presumably higher energy budgets (reviewed in Kemp [2005]; see chapter 11 of this book). The evolutionary changes in the skull mainly concern the mechanics of feeding. Enlargement of the temporal fenestra and a correlated enlargement of the coronoid region of the mandible indicate enlarged, more strongly attached adductor mandibuli musculature. This is associated with a forward shift of the jaw articulation, which shortens the jaw and so increases the bite force, and a general strengthening of the posterior part of the skull. Clearly this stronger bite was associated with the considerably enlarged canine teeth and well-developed, often interdigitating incisors. There was an equally significant transition in the postcranial skeleton, from the stout-limbs and sprawling gait of the pelycosaurs to longer and more gracile limbs and increased mobility at the shoulder and hip joints of therapsids (see Fig. 1.5D). The new gait appears to have endowed therapsids with a great deal more agility as well as speed (Gebauer 2007; Kemp 1978), and by raising the body clear of the ground, it increased the potential ventilation capacity of the lungs (Carrier 1987). The bone histology of therapsids also offers evidence for a more active, higher energy mode of life, as discussed in later chapters of this book. Much of the therapsid bone is fibrolamellar, which indicates a relatively high rate of bone deposition and growth (Botha and Chinsamy 2004, 2005; Ray, Botha, and Chinsamy 2004; Ray and Chinsamy 2004), although how exactly this relates to the evolution of temperature physiology remains debatable. For example, there is a good deal of variation even among related genera in the extent to which growth was continuous or interrupted during the unfavorable season, when food would have been in short supply.

It is surely no coincidence that therapsids make their first appearance in temperate paleolatitudes where seasonality was pronounced, suggesting that there had already been a significant degree of evolution of the temperature and osmotic regulatory strategies that were to culminate in those of the mammals (Kemp 2006a, 2006b). Apart from one or two briefly lingering forms, the pelycosaurs disappeared at this time, leaving the therapsids as the dominant terrestrial tetrapods of the Late Permian. There was also a significant shift toward a higher proportion of herbivores to carnivores (Nicolas and Rubidge 2009; Sidor and Smith 2007; see chapter 2 of this book), indicating the development of a community structure analogous to that of today's mammals, with virtually complete dependence on land plants as the primary producers and herbivorous tetrapods as the primary consumers.

The phylogenetic relationships of Therapsida are indicated by several features shared only with the sphenacodontid pelycosaurs, of which the most memorable is a reflected lamina of the angular of the lower jaw. This is little more than a notch in the ventral edge of the back of the jaw in sphenacodontids, but has become a large, thin lateral sheet in therapsids, whose function continues to be unclear. One view is that it represents the attachment of ventral jaw musculature, the other that it was involved in airborne sound reception, as an incipient stage in the origin of mammalian hearing (Allin 1975). Other similarities are the enlarged caniniform teeth and the powerful build of the occiput.

Several subtaxa of therapsids appeared approximately simultaneously around 265–270 Ma in Guadalupian aged rocks of South Africa, Russia, and China (Abdala, Rubidge, and van den Heever 2008; Fröbisch 2009). Together they constitute a very strongly supported monophyletic taxon, characterized by many features of the skull and dentition related to feeding, such as the enlarged canines and enlarged temporal fenestra, and a major reorganization of the locomotory system involving longer, more gracile limbs and a greater range of movements at their joints. Little is yet known of more basal therapsids. *Tetraceratops* was described long ago on the basis of a single poorly preserved skull lacking lower jaws and dating from the Early Permian of Texas; Romer and Price (1940) interpreted it as possibly an eothyridid pelycosaur. However, when Laurin and Reisz (1996) restudied the specimen, they concluded that it possessed a number of therapsid characters such as the form of the temporal fenestra and the manner of attachment of the braincase. Others, however, remain skeptical that the specimen is adequate for such an interpretation (Liu, Rubidge, and Li 2009; Rubidge and Sidor 2001), and resolution of this potentially very important issue must await further material of the species. Apart from the possibility of *Tetraceratops*, Liu, Rubidge, and Li (2009) have recently interpreted a fragmentary snout lacking jaws, from the Middle Permian Dashankou Formation of China, as the most basal therapsid known. Named *Raranimus*, it appears to combine the sphenacodontid characters of two functional canines, the shape of the alveolar margin of the premaxilla, and the shape of the vomer with an otherwise therapsid-like structure.

There also are a number of poorly known early therapsids from the Middle Permian Guadalupian of Russia whose phylogenetic relationships are not well understood (Kemp 2005). Among them is the formerly much discussed lower jaw of *Phthinosaurus* that Ivakhnenko (2003, 1996) classified along with a few other fragmentary specimens as the Rhopalodonta, regarding them as a primitive herbivorous group. The very incomplete skull of *Phthinosuchus*, another widely discussed specimen, has been interpreted by Ivakhnenko (2003) as a member of a carnivorous group related to the gorgonopsians. The family Nikkasauridae was established by Ivakhnenko (2000b) for a number of small, primitively built skulls bearing a long series of small, similar-sized teeth, the posterior ones of which bear anterior and posterior accessory cusps. *Niaftasuchus* (Ivakhnenko 1990), known from a single skull lacking the lower jaw, is another very small form, but with a very different dentition; it has three well-developed incisors, four smaller precanines, a larger canine, and a series of laterally compressed postcanines that decrease in size backward along the tooth row.

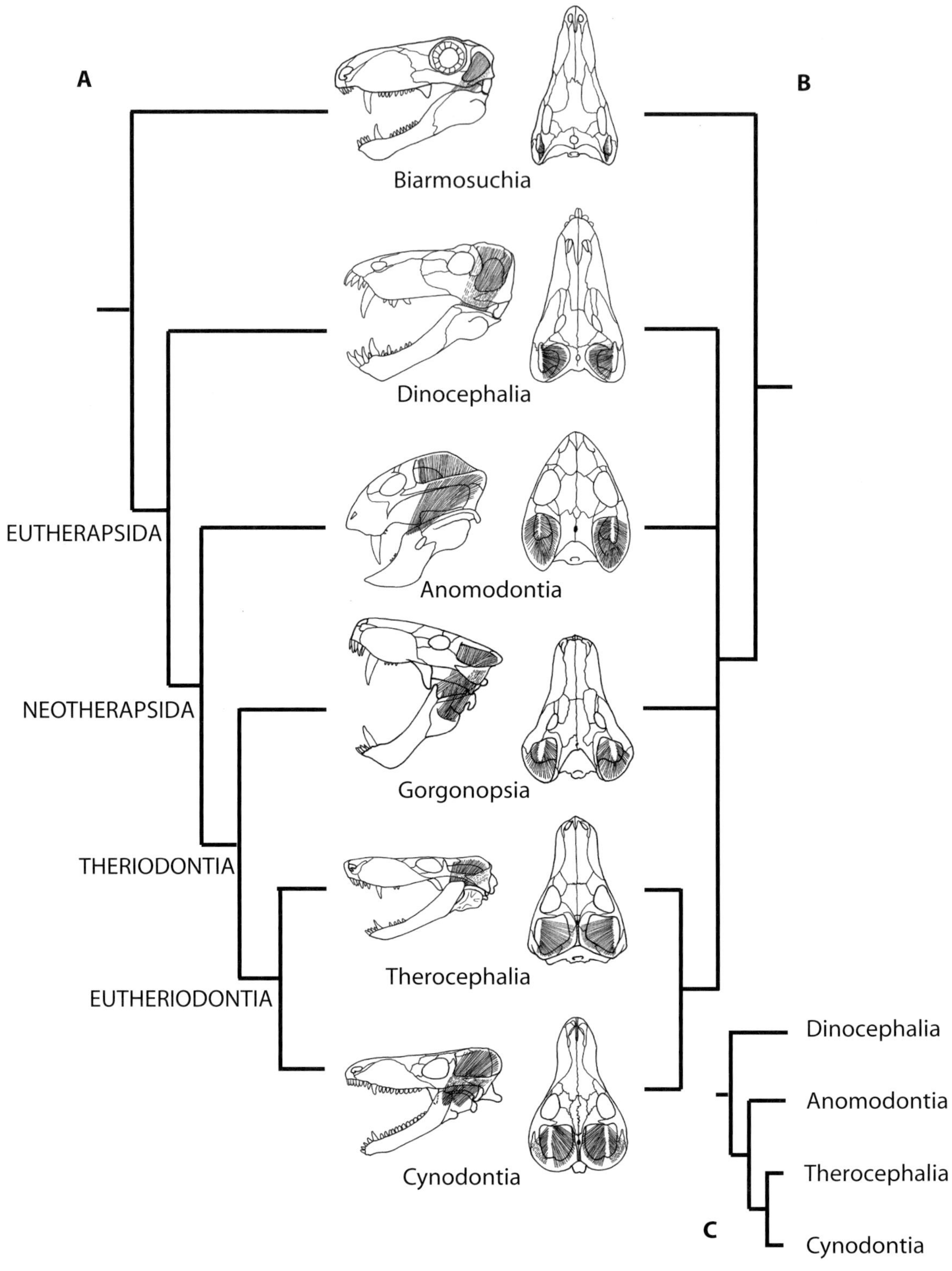
A
B
Biarmosuchia
Dinocephalia
EUTHERAPSIDA
Anomodontia
NEOTHERAPSIDA
Gorgonopsia
THERIODONTIA
Therocephalia
EUTHERIODONTIA
Cynodontia
Dinocephalia
Anomodontia
Therocephalia
C
Cynodontia

Apart from these poorly understood taxa, there are six well-established major therapsid subtaxa (Fig. 1.2), of which four are universally accepted as monophyletic, based on numerous unambiguous synapomorphies. One, Biarmosuchia, has retained a number of plesiomorphic characters and has been considered possibly paraphyletic. Another, Therocephalia, is also regarded by some but not all authors as paraphyletic on the grounds that it contains the ancestry of Cynodontia.

In marked contrast to the indisputable monophyly of Therapsida and the large measure of agreement over its major constituent subtaxa, there is considerable disagreement about the interrelationships among these subtaxa (Fig. 1.2). In their classic work, Watson and Romer (1956) recognized two groups: Anomodontia (sl) for the primarily herbivorous anomodonts and dinocephalians, and Theriodontia for the primarily carnivorous taxa of gorgonopsids, therocephalians, and cynodonts. Parts of this arrangement are still to be found in a number of subsequent classifications. King (1988) retained Anomodontia in this sense although virtually all other authors have excluded the Dinocephalia and use Anomodontia in the restricted sense of dicynodonts and their basal relatives. Theriodontia has had a more robust history and is still recognized in the widely used classification (Fig. 1.2A) based originally on Hopson and Barghusen (1986), for example, Rubidge and Sidor (2001) and Abdala, Rubidge, and van den Heever (2008). Other authors, however, such as Gauthier, Kluge, and Rowe (1988); Modesto, Rubidge, and Welman (1999); and Huttenlocker (2009) rejected the monophyly of Theriodontia in favor of a sister group relationship between Anomodontia and Therocephalia plus Cynodontia (Eutherocephalia) to the exclusion of Gorgonopsia (Fig. 1.2C).

Kemp (2009b) recently concluded that the problem of the interrelationships of the major therapsid subtaxa arises from the fact that the divergences at the base of the therapsids were so closely spaced in time and, at such a low taxonomic level, that the fossil morphology is unable even in principle to resolve a sequence of dichotomies (Fig. 1.2B). There was in effect a polytomy of four lineages,—Dinocephalia, Gorgonopsia, Anomodontia, and Eutheriodontia—that diverged virtually simultaneously from a biarmosuchian-like hypothetical ancestor, and each lineage rapidly evolved its own unique character complex independently of the others. Certainly the anatomical and functional differences among the four, coupled with the lack of unambiguous synapomorphies between any two of them, imply an evolutionary pattern in which the ancestor had dispersed into a radically new kind of habitat, namely a highly seasonal one, and proceeded to realize a potential for very rapid diversification into a variety of new ecotypes.

1.2. The interrelationships of the major therapsid subtaxa. (A) Rubidge and Sidor's (2001) fully resolved phylogeny. (B) Kemp's (2009a) concept of a fourfold polytomy. (C) Phylogeny according to Gauthier, Kluge, and Rowe (1988) and Modesto, Rubidge, and Welman (1999). (Thumbnails reproduced by permission from Kemp [2009a].)

Biarmosuchia

Biarmosuchians retain a number of primitive, sphenacodontian-like characters such as a convex dorsal margin of the skull and a relatively small

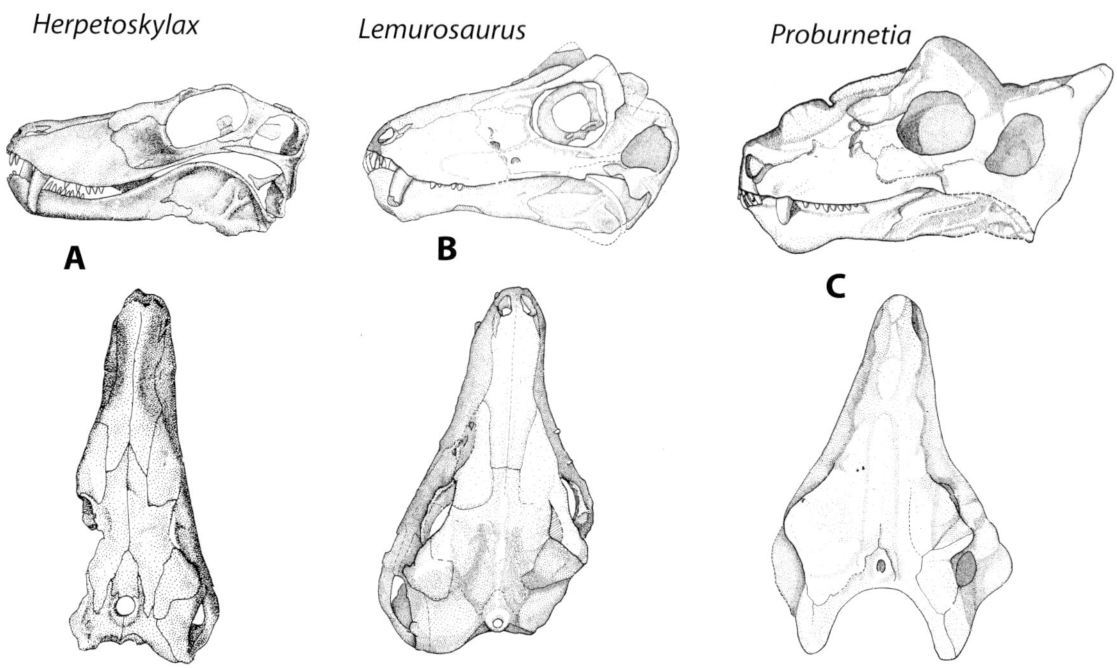

1.3. Biarmosuchia. (A) *Herpetoskylax* (reproduced from Sidor and Rubidge [2006]). (B) *Lemurosaurus* (reproduced from Sidor and Welman [2003]. © 2003 *Journal of Vertebrate Paleontology*. Reproduced by permission of Taylor & Francis Group.) (C) *Proburnetia* (reproduced from Rubidge and Sidor [2002]. © 2002 *Journal of Vertebrate Paleontology*. Reproduced by permission of Taylor & Francis Group.)

temporal fenestra. When the clade was first proposed by Hopson and Barghusen (1986), they found no unambiguous synapomorphies and that certain genera possess minor, possibly convergent characters found in other therapsid taxa such as gorgonopsians, indicating the possibility that Biarmosuchia is paraphyletic. However, recent intensive study including several new taxa has revealed a number of probable synapomorphies of the skull, leading to the conclusion that it is in fact a monophyletic taxon at the base of the therapsid phylogeny (Liu, Rubidge, and Li 2009; Rubidge, Sidor, and Modesto 2006; Sidor and Rubidge 2006; Rubidge and Kitching 2003)

Biarmosuchus occurs in the Early Kazanian of Russia (Ivakhnenko 1999, 2003) and is the most morphologically conservative and, therefore, the most ancestral-like therapsid known (Fig. 1.2). Its skull length is around 200 mm and its raptorial nature is indicated by the large upper and lower canines. The upper and lower incisor teeth may intermesh, although there is some doubt about this (Sigogneau and Tchudinov 1972). There is a set of scleral ossicles around the eye that are also found in another biarmosuchian, *Lemurosaurus*, and certain other pelycosaurs and therapsids, indicating that it is probably a plesiomorphic therapsid character. *Hipposaurus* from the *Tapinocephalus* Assemblage Zone (AZ) of South Africa is quite similar to *Biarmosuchus*, and a relatively basal member (Rubidge, Sidor, and Modesto 2006) while another relatively unspecialized South African form, *Herpetoskylax* (Fig. 1.3A), occurs in the later *Cistecephalus* AZ (Sidor and Rubidge 2006).

Burnetiamorphs form a distinctive monophyletic group within Biarmosuchia. They retain the basically primitive anatomy of the skull,

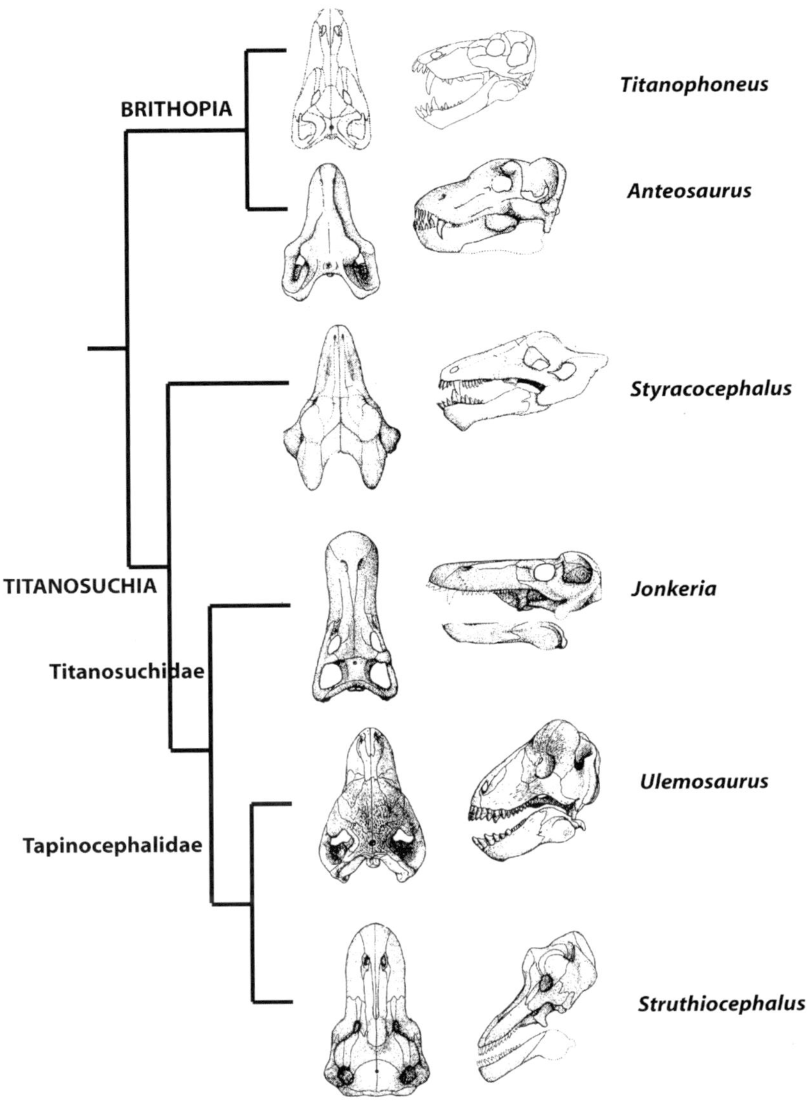

1.4. Phylogeny of Dinocephalia (reproduced by permission of Oxford University Press from *Origin and Evolution of Mammals* by Kemp [2005]).

such as convex shape and small temporal fenestra, but are characterized by pachyostosis of the cranial bones and the development of bony bosses above and below the orbit, and on the squamosal, presumably for behavioral rather than mechanical reasons. *Lemurosaurus* (Fig. 1.3B) represents a less specialized form, compared to the full expression of the pachyostosis, for example in *Proburnetia* (Fig. 1.3C). The earliest, such as *Bullacephalus* (Rubidge and Kitching 2003), date from the *Tapinocephalus* AZ (see Plate 1B), and the group persists right through to the end of the Permian, with *Burnetia* occurring in the *Dicynodon* AZ (see Plate 1B) (Smith, Rubidge, and Sidor 2006). Burnetiamorphs are also found in Russia, where *Proburnetia* is extremely similar to the South African *Burnetia* (Smith, Rubidge, and Sidor 2006).

Dinocephalia

The dinocephalians (Fig. 1.4) are large, heavily built animals with a strong tendency toward pachyostosis of the skull, which reaches its maximum in the 50-mm thick dorsal bones of such forms as *Moschops* and *Tapinocephalus* (see Plate C). The incisor teeth are distinguished by small lingual heels and interdigitation between the uppers and lowers. The group appears in the *Eodicynodon* AZ (see Plate 1B) of South Africa and was abundant in the early parts of the overlying *Tapinocephalus* AZ (see Plate 1B and Plates B and D). However, they have completely disappeared from the record by the end of that zone. Very similar contemporary dinocephalians occur in Russia, suggesting relatively free dispersal between northern and southern hemispheres of the Middle Permian, and they have also been described from the Xidagou Formation of China (Li, Rubidge, and Cheng 1996), the Madumabisa Mudstone Formation of Zimbabwe (Lepper, Raath, and Rubidge 2000), and as isolated teeth from Brazil (Langer 2000).

Relatively little work has been done on the interrelationships of dinocephalians since Boonstra (1963; Boonstra 1971), apart from Rubidge and van den Heever's (1997) cladistic analysis (Fig. 1.4). A good part of the reason for this is the extremely hard nature of the *Tapinocephalus* AZ matrix in which the majority of these very large, mostly fragmentary fossils, are preserved. The brithopians, such as *Titanophoneus* from Russia (Orlov 1958), the early South African *Australosyodon* (Rubidge 1994), and the Chinese anteosaurids (Li, Rubidge, and Cheng 1996), are the least derived forms. They are specialized carnivores with large canines and a reduced number and size of postcanine teeth. *Anteosaurus* was huge, with a skull length of 800 mm. The sister group of Brithopia is Titanosuchia (sensu Kemp 1982), which consists of forms that had evolved a herbivorous mode of life. The canines were reduced, the heels on the incisor and by now also the postcanine teeth were more prominent, and the number of postcanines had increased up to about twenty in the most derived forms. The most basal member is the peculiar *Styracocephalus*, which combines the heavy pachyostosis and reduced temporal fenestra of typical titanosuchians with retention of small canines and palatal teeth, the presence of bony bosses above the orbits, and possibly horn-bearing protuberances at the back of the skull.

Tapinocaninus from the *Eodicynodon* AZ (see Plates B and C) and several genera such as *Moschops*, *Tapinocephalus*, and the closely related *Ulemosaurus* of Russia constitute the most derived family, Tapinocephalidae (Atayman, Rubidge, and Abdala 2009). Their cranial bones were so thickened that the temporal fenestra was partially occluded.

One problematic form is *Estemmenosuchus* (Plate E). This is a Russian genus which shares with the dinocephalians large size, intermeshing teeth, shortened jaws, and moderately thickened cranial bones. However, it lacks heeled incisor teeth and the temporal fenestra is expanded posteriorly as well as dorso-ventrally in a fashion that apparently differs

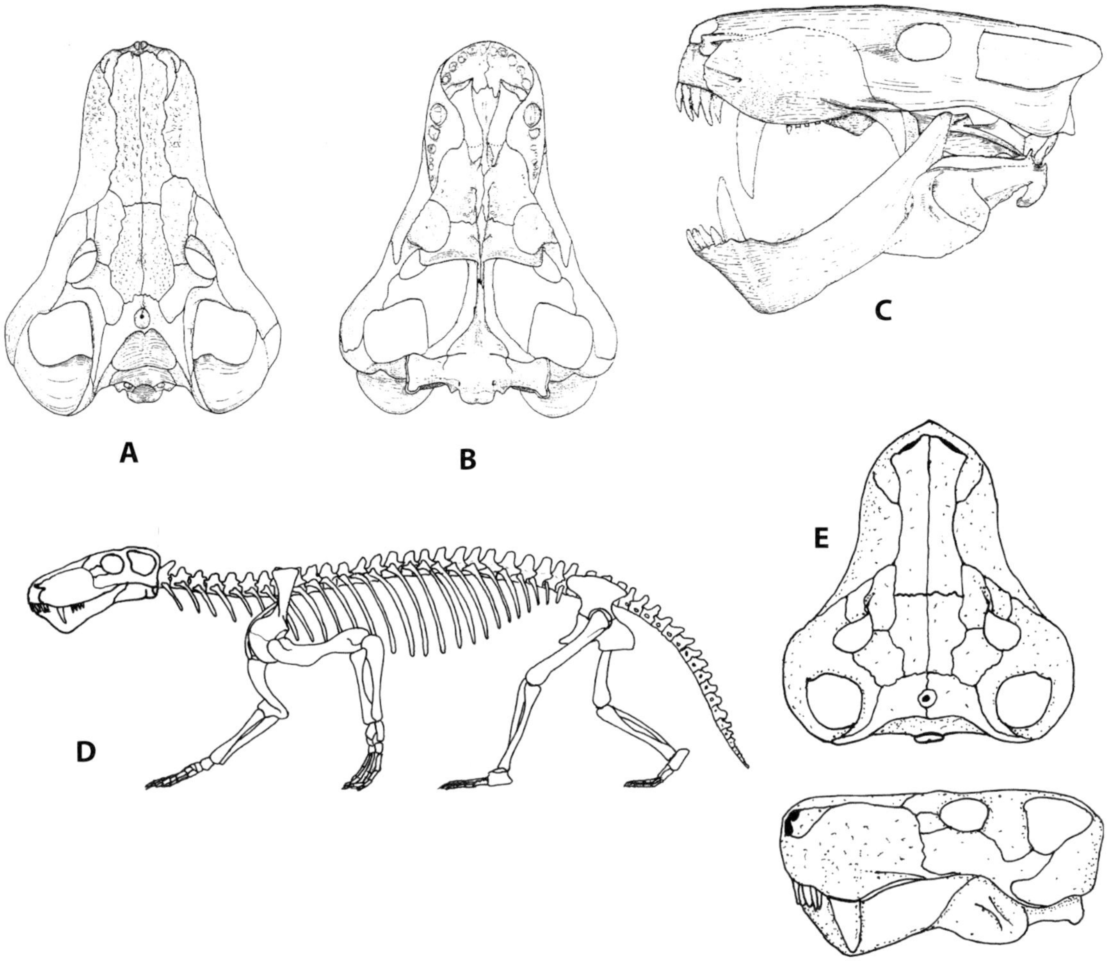

1.5. Gorgonopsia. (A and B) Dorsal and ventral views of gorgonopsian skull (Kemp 1969). (C) Lateral view of gorgonopsian skull (redrawn from Kemp [1969]). (D) Skeleton of *Lycaenops* (redrawn from Colbert [1948]). (E) Dorsal and lateral views of the skull of the rubidgeine *Clelandina* (redrawn from Gebauer [2007]).

from the general dinocephalian pattern. Rubidge and van den Heever's (1997) cladogram places it as a titanosuchian, but Ivakhnenko (2000a) believes that it evolved adaptations as a large herbivore independently of the Dinocephalia.

Gorgonopsia

Gorgonopsians (Fig. 1.5) are a group of carnivorous therapsids that evolved a unique jaw mechanism for dealing with large prey that included enormous canines, a jaw hinge allowing a gape of more than 90°, and an expansion of the temporal fenestra both posteriorly and laterally (Gebauer 2007; Kemp 1969). The earliest reported specimens occur in the *Eodicynodon* AZ, at the base of the Beaufort Group of the South African Karoo (see Plate 1B), but these are fragmentary and of indeterminate identification (Rubidge 1990b; Abdala, Rubidge, and van den Heever 2008). They became increasingly dominant medium to large carnivores during the Late Permian, but no gorgonopsian is known to have survived the Permo-Triassic extinction. Specimens are known from other parts of Africa, including north Africa (Smiley et al. 2008) and from Russia, but neither Asian nor American ones have been found.

Phylogenetic relationships within Gorgonopsia have proved difficult to determine because it is a highly conservative group, with only relatively trivial differences between genera. Sigogneau (1970; Sigogneau-Russell 1989) published a detailed but noncladistic classification, and concluded that, apart from *Watongia* from North America, which is now believed to be a varanopid pelycosaur (Reisz, Dilkes, and Berman 1998), all genera can be included in a single family Gorgonopsidae (*sic*). She divided this family into three subfamilies, the great majority of the genera being in the Gorgonopsinae (Fig. 1.5A–D). The large, long-snouted Russian form *Inostrancevia* had its own subfamily, Inostranceviinae, and she created a subfamily Rubidgeinae for several genera with a broadened posterior part of the skull and a deep, ventrally extended zygomatic arch (Fig. 1.5E). More recently, Gebauer (2007) has revised the gorgonopsians, reducing the twenty-five genera of Sigogneau to fifteen. Her cladistic analysis reached largely the same conclusion as Sigogneau in that almost all gorgonopsians fall into the single family Gorgonopsidae. The only significant difference is that she found the Gorgonopsinae to be paraphyletic, because both *Inostrancevia* and the rubidgeines are nested within it.

Anomodontia

The great majority of anomodontians are members of the highly specialized herbivorous group Dicynodontia, which were the most abundant and diverse of all the major therapsid subtaxa. Evolution of a turtle-like keratinous beak that partially or completely replaced the dentition, propaliny of the lower jaw, and massive adductor jaw musculature ensured not only a highly effective herbivorous feeding mechanism, but the opportunity for wide variation among the taxa that is presumably associated with different dietary specializations.

There is also a series of basal anomodonts (Fig. 1.6) which have only incipiently evolved the characteristic dicynodont specializations, and a full set of marginal teeth is still present. The adductor jaw musculature exhibits varying degrees of enlargement, as indicated by the extent of the characteristic dorsal bowing of the zygomatic arch. *Biseridens*, from the Xidagou Formation of China, is the most basal anomodont known (Liu, Rubidge, and Li 2009) for, although it possesses the other anomodonts characters, it lacks a mandibular fenestra. Prior to the description of this genus, the South African *Anomocephalus* was regarded as the most basal anomodont (Fröbisch 2007; Liu, Rubidge, and Li 2009; Modesto, Rubidge, and Welman 1999). It does have a mandibular fenestra, but the zygomatic arch is only weakly bowed, and the skull appears to have lacked a caniniform tooth or process that is characteristic of all other anomodonts (Modesto et al. 2003; Modesto and Rubidge 2000). Cisneros et al. (2011) have described a highly specialized basal anomodont, *Tiarajudens*,

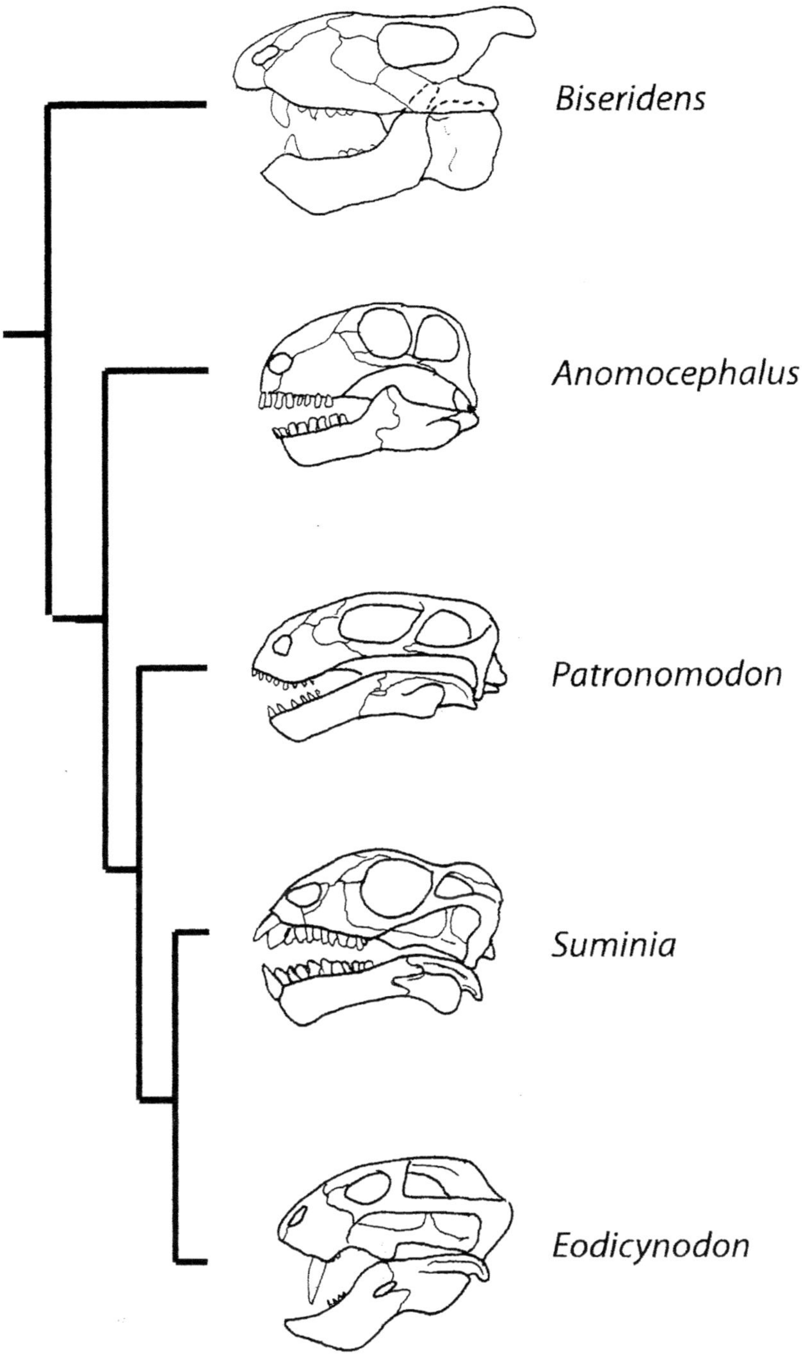

1.6. Phylogeny of basal anomodonts and the basal dicynodont *Eodicynodon*.

on the basis of a very poorly preserved skull from the Middle Permian of Brazil. It possesses an extraordinarily long, slender, laterally compressed upper canine tooth, behind which is a grinding surface formed from a row of thirteen expanded palatal teeth that presumably occluded with a similar battery on the missing lower jaw. Cisneros et al. (2011) regard *Tiarajudens* as related to *Anomocephalus* with the two together forming

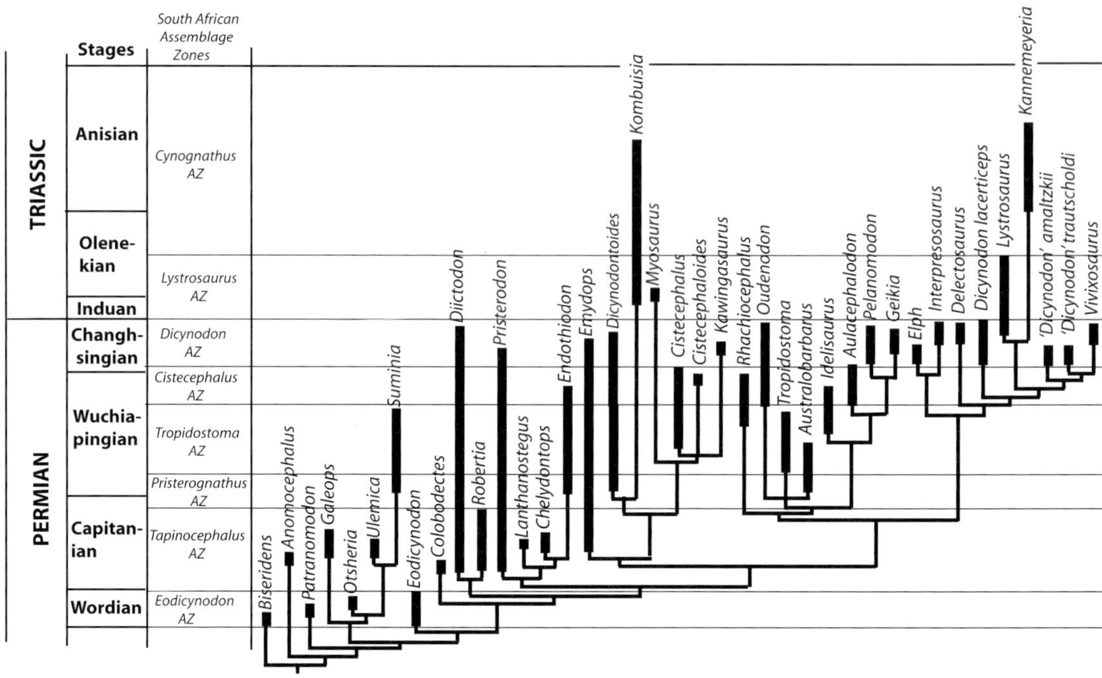

1.7. Phylogeny of anomodonts (redrawn and modified from Fröbisch [2007]).

a basal anomodont clade Anomocephaloidea that radiated in Gondwana. Although until better material is found, this conclusion is tentative. However, it is widely accepted that there is a monophyletic clade of Russian basal anomodonts, termed Venyukovoidea and consisting of *Otsheria, Ulemica,* and *Suminia,* although there is disagreement about its exact relationships among the basal anomodonts subsequent to *Anomocephalus.* Modesto and Rubidge (2000) and Fröbisch (2007), followed in the classification of Kammerer and Angielczyk (2009), find the relatively well-known South African *Patranomodon* to be more basal than the venjokovoids (Fig. 1.7). However Liu, Rubidge, and Li (2009), in the only analysis so far to include *Biseridens,* find that *Patranomodon* rather than the Russian clade is closer to the dicynodonts. This matter is unlikely to be resolved until better preserved material of the critical forms is found.

All other anomodonts are members of the Dicynodontia, recognized by their more extreme adaptations for an efficient herbivorous mode of life (Fig. 1.8). None have incisor teeth, and all relied largely or entirely on the keratinous covering of the jaws, as indicated by both the form and the natures of the bony surfaces of the jaw. The adductor jaw musculature was massively enlarged to operate what must have been a very effective triturating mechanism.

Eodicynodon (Fig. 1.6, Plate B) is the earliest and the most plesiomorphic dicynodont. It was a small animal, with a skull length of approximately 100 mm, and a broad intertemporal roof. Its dentition consists of a pair of upper tusks plus a few small, laterally and medially positioned postcanine teeth. From such a morphotype, the extensive Permian radiation of dicynodonts can be traced (Fig. 1.7), which produced over thirty currently recognized genera (Kammerer and Angielczyk 2009; Fröbisch 2008). The size range is considerable; the smallest Permian species, such as *Emydops* and *Pristerodon*, had a skull length under 100 mm and the largest, such as *Platycyclops* and *Rhachiocephalus*, over 600 mm. Much of the variation among dicynodonts concerns the morphology of the feeding structures, indicating specialization for different diets, as might be expected of such a diverse, wide-ranging, and successful terrestrial herbivorous group (Cox 1998; King 1990). The dentition is occasionally enhanced, but in most genera it is further reduced or altogether lost. The morphology of the presumed horny beak varied considerably, to judge by the form of the premaxilla, maxilla, and dentary bones, with some genera possessing sharp, shearing blades and others broad triturating surfaces. Pylaecephalids, such as *Diictodon* and *Robertia* (Kammerer and Angielczyk 2009), for example, possessed a notch in front of the caniniform process of the maxilla suggesting a tough stem-cutting device; *Dicynodontoides* (*Kingoria*) appears to have had a combination of opposed blunt lateral areas on the maxillae and dentary plus a sharp edge on the anterior tip of the upturned dentary. *Endothiodon* was a large form that had evidently redeveloped a more extensive posterior dentition, but with both upper and lower teeth biting against horny pads on the opposing bony surfaces rather than against one another. The breadth of the intertemporal roof also varies and its reduction to a narrow sagittal crest in some, for example *Endothiodon* and *Dicynodon*, indicates even more extreme modification of the adductor mandibuli musculature (King, Oelofsen, and Rubidge 1989).

Locomotory adaptations among Permian dicynodonts tend to be more conservative than the feeding structure adaptations. However, there is a tendency in several genera to develop burrowing habits, with stout limbs and strongly built claws, which is most extreme in *Cistecephalus*, which had a very strong, pointed skull and mole-like limbs. Actual burrows are known for several genera, notably *Diictodon* (Plate F), in which specimens are preserved (Smith 1987a). Recently Fröbisch and Reisz (2009) have described arboreal adaptations of the limbs and vertebral column of the basal anomodonts *Suminia*, which makes it the earliest record of a tetrapod following this mode of life. As yet, there is no known anomodont possessing obvious semi-aquatic adaptations.

Dicynodon (Fig. 1.8A) is a genus of fairly large, mostly tusked forms, with a narrow intertemporal region and a tendency to shorten the palate. It occurs until the *Dicynodon* AZ (see Plate 1B), which represents the very end of the Permian, and like almost all dicynodonts and other contemporary therapsids, it disappeared at the Permo-Triassic extinction event (Fig. 1.7; see chapter 2 of this book). However, *Dicynodon* is closely related to *Lystrosaurus* (Fig. 1.8B), in which the skull has undergone even further shortening coupled with a marked downturn of the snout. *Lystrosaurus* is one of the very few genera that did survive the Permo-Triassic extinction, reappearing in the Early Triassic at which time it had become the most abundant tetrapod genus, with a worldwide distribution (Grine

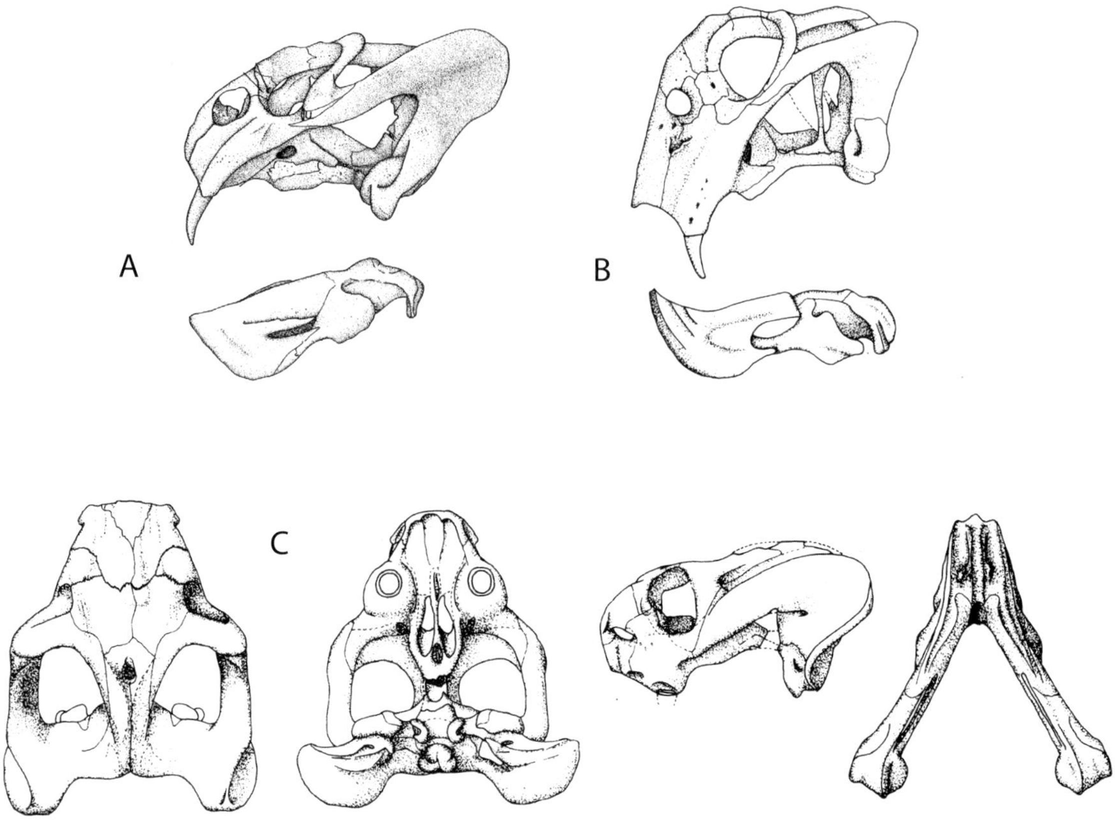

1.8. (A) Lateral view of skull and lower jaw of *Dicynodon*. (B) Lateral view of the skull and lower jaw of *Lystrosaurus*. (C) Skull and lower jaw of *Tetragonias*. (Reproduced from King [1988].)

et al. 2006) (Plates J, K). There has been much speculation about how *Lystrosaurus* managed so successfully to survive the end-Permian event. The powerfully built limbs and downturned snout indicate an ability to dig for perennial tubers and similar food (King and Cluver 1991), while the presence of burrows (Modesto and Botha-Brink 2010) suggests an ability to aestivate during an extreme arid season.

Three other lineages of dicynodonts are also known to have crossed the Permo-Triassic boundary. Two of these were relatively insignificant, being represented only as rare forms in the Early-Middle Triassic *Cynognathus* AZ (see Plate 1B) of South Africa. *Kombuisia* (Fröbisch 2007) is a kingoriid that has also been discovered in similarly dated rocks in Antarctica (Fröbisch, Angielczyk, and Sidor 2010). *Myosaurus* is a small relative of that family. However, the most successful Triassic dicynodonts were the kannemeyeriiforms, although they never achieved the same level of diversity that the Permian dicynodonts had enjoyed (see Plate N). They are the sister group of *Lystrosaurus* and were generally large-bodied herbivores (King 1988), characterized by a narrow parietal crest, wide interorbital roof, and details of the palatal structure (Fig. 1.8C). The interrelationships of the kannemeyeriiforms are in a state of flux at present, with different authors offering alternative hypotheses (Damiani et al. 2007; Maisch 2001; Surkov, Kalandadze, and Benton 2005; Vega-Dias,

Maisch, and Schultz 2004), in large part because of incomplete knowledge of many of the taxa (Kammerer and Angielczyk 2009). The cladogram of Vega-Dias, Maisch, and Schultz (2004), for example, gives an unresolved polytomy of several families, plus a generally more progressive monophyletic Stahleckeriidae. The one clear evolutionary trend they noted was a tendency to reduce and finally, in the stahleckeriids, to lose all trace of the paired tusks.

Having enjoyed a worldwide distribution in the Middle Triassic, kannemeyeriiforms declined during the Late Triassic, and finally disappeared in the Norian, with one intriguing exception. A single cranial fragment and tusk of undoubtedly dicynodont and probably kannemeyeriiform nature has been described from Early Cretaceous shallow, epicontinental marine deposits of Australia (Thulborn and Turner 2003). There is no reason to doubt either the identification or the date of this specimen, apart from the fact that it extends the range of dicynodonts, indeed of nonmammalian therapsids, by an astonishing 80 million years.

Therocephalia

The majority of the Therocephalia were carnivorous, and while the early ones have a superficial resemblance to gorgonopsians due to large canines and reduced postcanines, they are actually very different in skull structure and arrangement of the adductor mandibuli musculature. They also have a progressive postcranial skeleton, with a very slender pectoral girdle, enlarged ilium, long, slender limbs, and much reduced tail (Kemp 1986; Fourie and Rubidge 2009).

The oldest therocephalians are known from the *Eodicynodon* AZ of South Africa (Abdala, Rubidge, and van den Heever 2008), and they were the most abundant medium to large carnivores through the succeeding *Tapinocephalus* AZ, a role increasingly taken over thereafter by gorgonopsians (Fig. 1.9). About eight therocephalian lineages are known to have survived the Permo-Triassic extinction (Huttenlocker, Sidor, and Smith 2011), being found in the *Lystrosaurus* AZ, while one family, the specialized, secondarily herbivorous bauriids occurs in the Early to Middle Triassic *Cynognathus* AZ of South Africa and the Omingonde Formation of Namibia (Heever, Abdala, and Rubidge 2009). Permian therocephalians have been found in Late Permian rocks of Russia (Tatarinov 1963) and China (Sun 1991) and Triassic members in these same regions and also in Antarctica (Colbert and Kitching 1981).

The *Tapinocephalus* AZ forms tend to be medium to large in size, with relatively few and small postcanine teeth (Fig. 1.9A). The majority of the later genera were small to medium sized and had a more prominent postcanine dentition. However, there were some spectacular exceptions to this. The whaitsiids, such as *Theriognathus* (see Plate H), were relatively large and while they possessed well-developed incisors and canines, the postcanines were evidently replaced by keratinous tooth plates, perhaps as an adaptation for an omnivorous diet (Kemp 1982). *Moschorhinus*

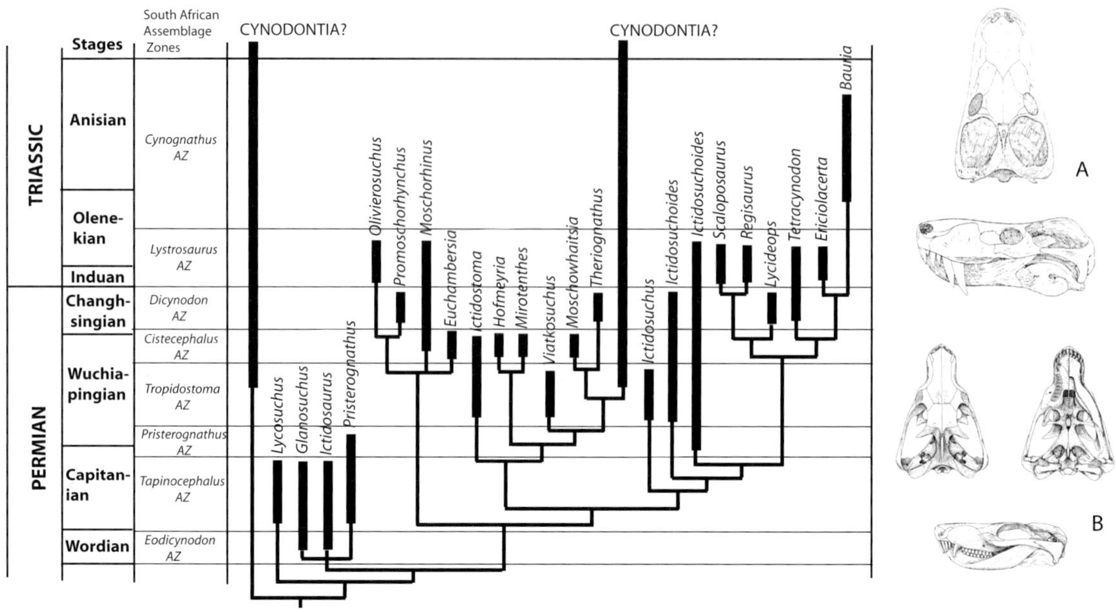

1.9. Phylogeny of Therocephalia (redrawn and modified from Huttenlocker [2009]). (A) *Lycosuchus* (reproduced from Broom [1932]). (B) *Bauria* (reproduced by permission of Brink [1963]).

was another fairly large form, with a very broad, low snout and insignificant postcanine dentition. *Euchambersia* is one of the most peculiar of all therapsids. The postorbital bar and zygomatic arch are weak; the postcanine teeth are absent; and there is a deep groove in the canine associated with a huge recess behind it, which suggests a snake-like venom gland and injecting system (Mendrez 1975). The bauriids (Fig. 1.9B) of the Middle Triassic are another extraordinary group of therocephalians. These, the last therocephalians to survive, were specialized herbivores, with broadly expanded postcanine teeth meeting in occlusion much as in the contemporary gomphodont cynodonts. The bauriids, like the cynodonts, have a complete secondary palate, indicating the possibility that bauriids had evolved a high metabolic rate and endothermy convergently with the cynodonts.

There have been two recent cladistic analyses of Therocephalia—Abdala (2007) and Huttenlocker (2009), respectively—and they agree in large measure (Fig. 1.9). The most basal form is the short-snouted *Lycosuchus*. The other early, *Tapinocephalus* AZ genera constitute the monophyletic group Scylacosauridae, which is in turn the sister group of Eutherocephalia, consisting of all other therocephalians.

It is currently disputed whether the Therocephalia are strictly monophyletic, or whether the cynodonts are nested within them (Fig. 1.9). Kemp (1972) first pointed out a number of characters shared only by whaitsiid therocephalians and cynodonts, including the absence of the suborbital fenestrae of the palate found in all other therocephalians, and the extreme broadening of the epipterygoid. In their full cladistic analyses, Abdala (2007) and Botha, Abdala, and Smith (2007) also found whaitsiids and cynodonts to be sister groups, but Huttenlocker's (2009) analysis, in contrast, generates a monophyletic Therocephalia, sister to cynodonts.

Cynodontia

The cynodonts have received intense attention because, paraphyletically speaking, they are the ancestral taxon of mammals. As such, they include a series of grades between therocephalian and mammalian, and together they illustrate much of the sequence of acquisition of mammalian characters, and by inference the origin of mammalian functional anatomy and physiology. All cynodonts have such mammalian characters as the fully differentiated dentition, sagittal crest and zygomatic arches defining an enormous temporal fenestra, and the progressively expressed enlargement of the dentary and reduction of the postdentary bones and quadrate that eventually led to the mammalian mandible and middle ear ossicles. The postcranial skeleton also became extremely mammal-like in the more derived members of the group (see Fig. 1.11D).

Cynodonts were the last of the major therapsid subtaxa to appear in the fossil record, the earliest being a single, well-preserved skull and fragmentary postcranial bones of *Charassognathus* from the *Tropidostoma* AZ of South Africa (Botha, Abdala, and Smith 2007). It is a representative of the most plesiomorphic grade, Procynosuchia (Fig. 1.10A), of which the best-known is the otter-like *Procynosuchus* itself (Kemp 1979, 1980b), which occurs in Europe (Sues and Boy 1988) as well as southern Africa (Fig. 1.10A). *Dvinia* (Tatarinov 1968) is a Russian procynosuchian possessing a specialized, herbivorous postcanine dentition. Plesiomorphic features of procynosuchians include the relatively small temporal fenestra and coronoid process of the dentary, along with relatively large postdentary bones, and an incomplete secondary palate.

The epicynodontians (Fig. 1.10B) are the next most progressive grade, in which the temporal fenestra and coronoid process are more enlarged, and the postdentary bones are more reduced. They include the Late Permian *Cynosaurus* and constitute an important part of the Early Triassic, *Lystrosaurus* AZ, fauna, with *Galesaurus* (see Plate M, top), *Thrinaxodon* (see Plate M, bottom), *Progalesaurus* (Sidor and Smith 2004), and the broad-snouted *Platycraniellus* (Abdala 2007) present.

All other cynodonts belong to the Eucynodontia (Fig. 1.10C), in which the temporal fenestra achieved maximum size and the dentary completely dominates the postdentary bones. The postcranial skeleton also has numerous mammalian characters, such as a reduced coracoid in the shoulder girdle and a long anterior extension of the ilium (Jenkins 1971; Kemp 1980a; Oliveira, Soares, and Schultz 2010). Eucynodontians underwent a radiation during the Middle and Late Triassic, culminating in very mammaliaform-like taxa, and indeed in mammals themselves. The two most comprehensive cladistic analyses of eucynodontians are by Hopson and Kitching (2001) and Abdala (2007), and they agree that there was a major dichotomy into a largely, though not exclusively, herbivorous

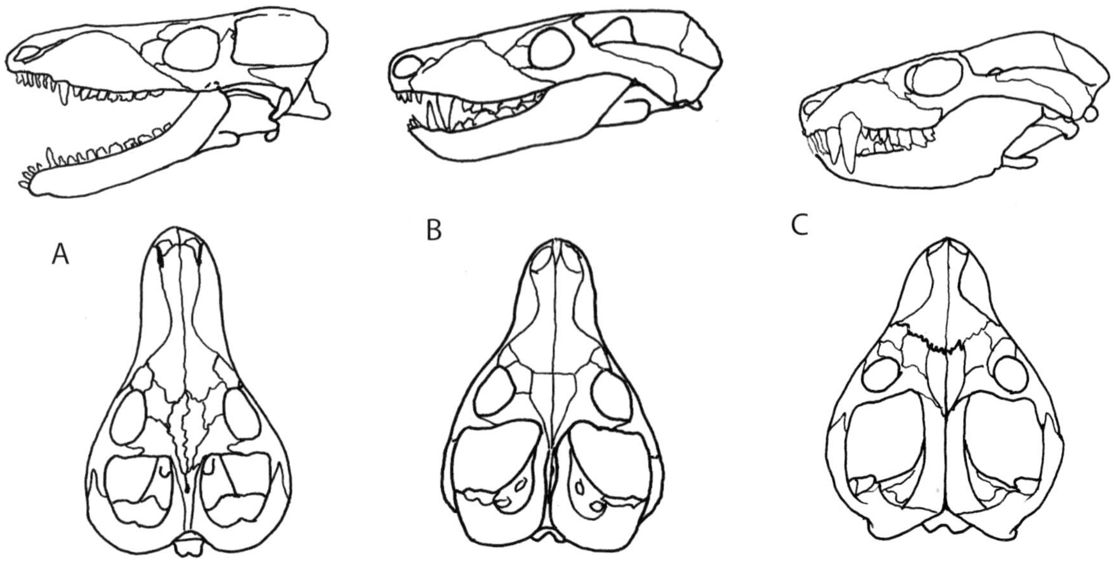
A
B
C

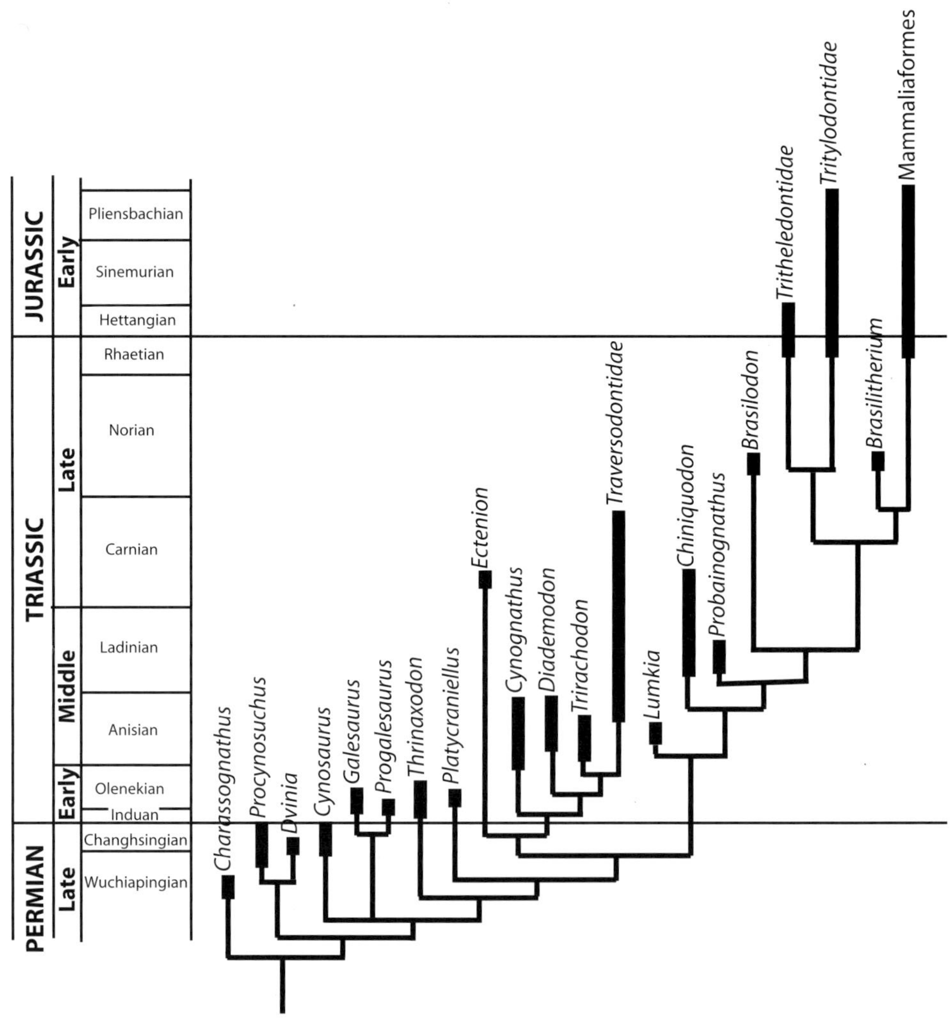
JURASSIC
Early
Pliensbachian
Sinemurian
Hettangian
TRIASSIC
Late
Rhaetian
Norian
Carnian
Middle
Ladinian
Anisian
Early
Olenekian
Induan
PERMIAN
Late
Changhsingian
Wuchiapingian
Charassognathus
Procynosuchus
Dvinia
Cynosaurus
Galesaurus
Progalesaurus
Thrinaxodon
Platycraniellus
Ectenion
Cynognathus
Diademodon
Trirachodon
Traversodontidae
Lumkia
Chiniquodon
Probainognathus
Brasilodon
Tritheledontidae
Tritylodontidae
Brasilitherium
Mammaliaformes

clade Cynognathia and a largely carnivorous clade Probainognathia that includes Mammaliaformes. *Cynognathus* (see Plate O), at the base of the Cynognathia is a large, carnivorous eucynodont with strongly sectorial postcanine teeth. For the rest, cynognathians are a diverse clade of principally herbivorous cynodonts, referred to informally as "gomphodonts," in which the postcanine teeth were expanded transversely across the jaw, so that the uppers and lowers met in true tooth occlusion. *Diademodon* was a contemporary of *Cynognathus* in the Early to Middle Triassic. It possessed three kinds of postcanine teeth, three or four simple conical anteriors, nine gomphodont central ones, and three or four cynognathid-like sectorial posterior ones. The Traversodontidae was a more derived family, in which all the postcanine teeth are gomphodont, here characterized by a strong transverse crest and various patterns of accessory cuspules (Abdala and Ribeiro 2003). They have a wide distribution, with genera known from South America, southern Africa including Madagascar (Kammerer et al. 2008), India (Chatterjee 1982), North America (Sues and Hopson 2010), and Europe (Hopson and Sues 2006), and they constituted an important component of the tetrapod radiation of the Middle and early Late Triassic (Abdala and Ribeiro 2003; Kammerer et al. 2008; Reichel, Schultz, and Soares 2009; Sues and Hopson 2010). Like the kannemeyeriiform dicynodonts, traversodontid eucynodonts declined in diversity during the Late Triassic, with only one genus remaining by the close of the Norian (Abdala and Ribeiro 2003).

The second major group of eucynodonts were the Probainognathia, most of which possess sectorial postcanine teeth, and an elongated secondary palate and upper tooth row (Abdala 2007; Hopson and Kitching 2001). The Chiniquodontidae are characterized by the particularly strongly sectorial postcanine teeth: *Lumkuia* (Fig. 1.10C) from the *Cynognathus* AZ of South Africa is a basal member (Hopson and Kitching 2001) while *Chiniquodon* (Abdala and Giannini 2002a; Romer 1969) is a well-known Middle Triassic South American representative. *Probainognathus* is a relatively small form, with a skull length of approximately 70 mm, and is celebrated for its incipient secondary contact between the dentary and the squamosal (Romer 1970; Crompton 1972a). The most completely preserved chiniquodontid is the recently described *Trucidocynodon* from Brazil (Oliveira, Soares, and Schultz 2010).

The remaining probainognathians consist of a series of small, Late Triassic to Early Jurassic forms with sharp teeth bearing various patterns of accessory cusps. They had a simple version of dentary-squamosal contact, prismatic enamel, loss of the postorbital bar, and an extremely mammal-like postcranial skeleton (Gow 2001; Martinelli et al. 2005). A number of these genera constitute the clade Tritheledontidae (Fig. 1.11A), which is closely related to Mammaliaformes (Martinelli et al. 2005; Sidor and Hancox 2006; Martinelli and Rougier 2007). In addition to the tritheledontids there are two small Brazilian genera, *Brasilodon* and *Brasilitherium* (Fig. 1.11B) which, on the basis of the cusp pattern of the molar teeth appear to be even more closely related to mammaliaforms

1.10. Cynodontia. (A) The procynosuchian. *Procynosuchus* skull (redrawn from Kemp [1979]). (B) The epicynodontian *Thrinaxodon* (redrawn from Parrington [1946]). (C) The eucynodontian *Lumkuia* (redrawn from Hopson and Kitching [2001]). (Bottom) A phylogeny of Cynodontia (redrawn and modified from Abdala [2007]).

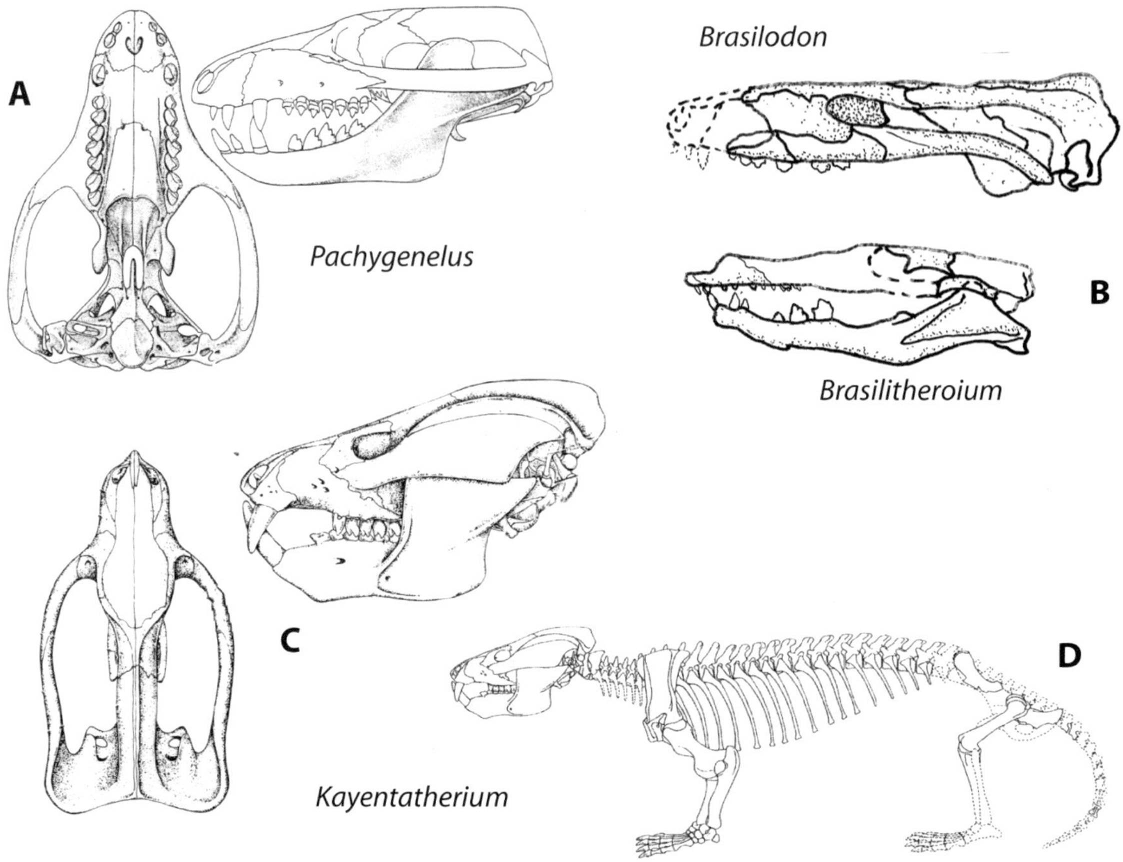

1.11. Mammaliamorphs. (A) The tritheledontid *Pachygenelus* (reproduced from Bonaparte et al. [2003]). (B) *Brasilodon* (redrawn from Bonaparte et al. [2003]). (C) Skull of the tritylodontid *Kayentatherium* (redrawn from Carroll [1988]). (D) Skeleton of *Kayentatherium* (redrawn from Sues and Jenkins [2006]).

(Bonaparte et al. 2003; Bonaparte, Martinelli, and Schultz 2005; Martinelli and Rougier 2007). The detailed interrelationships to one another and to mammaliaforms such as *Morganucodon* are yet to be confidently determined. What is clear is that during the Late Triassic there was a complex radiation of several lineages of small, insectivorous, extremely mammal-like animals, one of which includes the mammaliaforms and ultimately the mammals.

The greatest remaining problem of eucynodont interrelationships concerns the tritylodontids (Figs. 1.11C and 1.11D). These are highly derived nonmammalian cynodonts with a number of mammaliaform characters and a very mammalian postcranial skeleton, but which are also highly specialized for a rodent-like herbivorous or omnivorous diet. The postcanine teeth are greatly expanded and bear rows of sharp cusps, three rows on the uppers and two on the lowers, which meet in broad occlusion involving a bilateral propalinal action of the mandible. In this mode of action of the mandible and the broad, gomphodont-like postcanine teeth, tritylodontids resemble traversodontid cynognathians (Crompton 1972b), and several cladistic analyses nest them within Traversodontidae (Hopson and Kitching 2001; Bonaparte, Marinelli, and Schultz 2005; Sidor and Hopson 1998), which was defended most recently by Sues and Jenkins

(2006). However, tritylodontids also possess a number of similarities to mammals that are not known in traversodontids, including several characters of the braincase and jaw articulation, loss of the postorbital bone, and multirooted teeth (Kemp 1983). Thus, several other recent cladistic analyses have found tritylodontids to be mammaliamorphs within Probainognathia (Abdala 2007; Kielan-Jaworowska, Cifelli, and Luo 2004; Martinez, May, and Forster 1996; Oliveira, Soares, and Schultz 2010), as portrayed in Fig. 1.10. Despite the best part of thirty years of airing (cf. Kemp 1983; Sues 1985b; Sues and Jenkins 2006; Kemp 2007a), the question of what the tritylodontids are has still not been satisfactorily resolved.

A Summary of Synapsid Evolution

The radiation of synapsids is an extraordinarily well-documented complex phylogenetic pattern that was involved in a series of episodes of fundamental importance to understanding how the modern mode of terrestrial tetrapod communities first arose and diversified, including the origin of endothermy and the associated high energy budget, fine physiological regulation in the face of wide ambient conditions, and adaptable behavior. The main stages of synapsid evolution can be summarized briefly thus:

1. Pelycosaur-grade synapsids originated as one of the amniote lineages that constituted part of the explosive radiation of tetrapods in the Carboniferous. During the Late Carboniferous and Early Permian, pelycosaurs radiated into a variety of ecotypes, including herbivores, carnivores, and piscivores, and they became the most abundant taxon of terrestrial tetrapods. Most were medium to large in body size, but small forms are also known. During this time, they appear to have been largely restricted to humid, nonseasonal tropical climates and much of the primary consumption in the food web was still provided by freshwater fish and amphibians.
2. Therapsids originated in the Middle Permian. They show indications of elevated metabolic rates and appear to have evolved to be physiologically more independent of ambient fluctuations in temperature and aridity levels, since they are found in more seasonal, higher paleolatitude biomes. Here a radiation into many ecotypes led to complete domination of the Upper Permian terrestrial tetrapod habitat. Of particular note were the highly specialized, herbivorous dicynodonts, which evolved great diversity and large populations, and as a taxon occupied the role played in the later Mesozoic by ornithopod dinosaurs and in the Tertiary by ungulates, rodents, and other herbivorous mammals. Equally specialized carnivores, notably gorgonopsians preyed upon them.
3. The end-Permian mass extinction removed the vast majority of Late Permian lineages of therapsids, but, nevertheless, there were a number of survivors, notably *Lystrosaurus*, the most successful herbivorous tetrapod genus of all time.

4. These surviving lineages also included the kannemeyeriiform dicynodonts and diverse cynodonts, which both underwent more modest radiations during the Triassic, only to decline gradually to extinction by the end of that period.
5. The eucynodonts were most remarkable for the evolution of increasingly mammalian characters associated with enhanced feeding, locomotory and sensory systems, implying increasingly high metabolic rates and an approach to fully mammalian endothermy. One of the more highly derived of the eucynodont lineages underwent a process of miniaturization of the body, and then radiated into a series of mostly small, insectivorous or omnivorous taxa, from within which emerged the mammaliaforms—effectively the first mammals.
6. The only synapsids to have survived significantly beyond the Triassic were mammaliaforms. For the rest of the Mesozoic, mammals radiated extensively and were probably abundant, but with the exception of one or two cat- or dog-sized carnivores, they remained small. Whether this was due to competitive exclusion by dinosaurs, or because of some as-yet unclear physiological or ecological constraint on body size at this time remains debated (see Kemp [2005] for a review of opinions, and chapter 10 of this book).

Concluding Remarks: Future Studies

By the beginning of the twenty-first century, the general pattern of synapsid evolution was well understood, along with the major features of the paleoenvironmental and paleogeographic setting in which it was played out. The first decade, however, witnessed developments in evolutionary research that are promising a revolutionary new level of understanding of synapsid evolution, and indeed of major evolutionary radiations and transition in general. A combination of new fossils, new techniques for extracting far more information from the fossils and from the paleoenvironment, and the development of new, more realistic concepts for interpreting this knowledge are starting to combine to generate radical new insights (Kemp 2010).

New Fossils. The most significant new synapsid fossils have been mentioned already, notably the range of biarmosuchians at the base of the therapsid phylogeny, the sequence of increasingly plesiomorphic basal anomodonts, and the several taxa of small, extremely mammal-like eucynodonts from the Late Triassic associated with the emergence of mammals at this time.

New Techniques. Several techniques that were developed for entirely different purposes, have begun to be applied to paleontological study. Foremost among these is computed tomography scanning (Ketcham and Carson 2001). For some years now, standard hospital scanners have been used to reveal the gross anatomical structure of therapsids that are still partially encased in matrix, and more modern methods such as a synchrotron X-ray source (Sutton 2008) and neutron radiography (De Beer et al. 2008) have been introduced, which are capable of far better density discrimination and the spatial resolution necessary to reveal fine microstructural detail.

The prime function of the skeleton is to resist stresses and transmit strains, and application of the engineering technique of finite element analysis to determine the probable stress patterns within skulls and post-cranial skeletons imposed by the musculature is a hugely significant step toward interpreting functional morphology and evolutionary transformations (Jasinoski, Rayfield, and Chinsamy 2009, 2010a, 2010b; Rayfield 2007). Another recent development is the application of computer simulation to study the stance and gait of tetrapods.

As for paleobiological inference, stable isotope analysis is being applied to throw light on new aspects of biology. For example, from the carbon and oxygen isotope content of tooth enamel, Botha, Lee-Thorp, and Chinsamy (2005) were able to propose that the eucynodont *Diademodon* fed in shadier, damper areas than the contemporary *Cynognathus* did.

Beyond question, the application of isotope and also trace element analysis will be able to provide a great deal more information in the future about the paleoecology of synapsids.

New Concepts. The molecular revolution in biology has profoundly affected all branches of the subject, and paleobiology is proving to be no exception. As developmental geneticists move painstakingly forward in their search for the underlying molecular basis of morphology and morphological differences, so it is hoped that the transitions revealed by the fossil record may come to be explicable in terms of genetically controlled, specifiable alterations to the developmental processes. In the case of the synapsids, for example, a clear appreciation of how the angular, articular, and quadrate bones of the cynodont mandibular arch transformed into the ectotympanic, malleus, and incus of the mammal will depend on knowing the details of the activity and expression of the genes governing formation of the mandibular arch in vertebrate embryos, knowledge which is perhaps not so far away (Depew and Simpson 2006; Depew et al. 2005).

For many years, phylogenetic studies of fossils have used standard cladistic methodology for discovering relationships and inferring evolutionary patterns. The strength of this approach lies in the objectivity and, at least in principle, the repeatability it brings to analyses. When a cladogram is strongly supported by well-defined characters, there is little reason to doubt that it is a good estimate of relationships. However, the limitation of cladistics results from an over-simple, not very realistic model of characters evolution that it assumes: namely that morphological characters are objectively definable, discrete, independently evolving parts of the organism. Cladistics has also traditionally rejected the use of extrinsic information from geochronology, paleogeography, and paleoecology in assessing evolutionary patterns. It follows logically from the methodology that the only phylogenetic pattern to be sought is a fully resolved, dichotomous branching among the taxa studied, based solely on the distribution of character states.

A realistic view of evolutionary transitions, however, accepts that characters are artificial abstracts of the whole, are functionally integrated so they cannot evolve independently of one another, and that transitions occur temporally from ancestral to descendant states as a consequence of adaptations to habitats and of paleogeographic events. As information accrues about the pattern of phenotypic change and its functional significance, so it is starting to be possible to treat a sequence of inferred ancestors and descendants as evolving functionally integrated systems, rather than evolving collections of atomized characters (Kemp 2009b). Add to this all the new information about the environmental setting in which such transitions occurred (chapter 2 of this book), then a far more profound understanding should emerge of the major evolutionary events that synapsids underwent, such as the emergence of pelycosaurs in the Carboniferous, the explosive Middle Permian radiation of therapsids, the end-Permian mass-extinction, and the origin of mammals as the culmination of a long lineage of premammalian synapsids (Kemp 2007b). There is no other part of the fossil record with greater potential for revealing the causes of megaevolution.

Acknowledgments

I would like to thank the President and Fellows of St John's College, Oxford University, for electing me to an Emeritus Research Fellowship and for continuing to support my research. I would also like to thank Anusuya Chinsamy-Turan for inviting me to contribute this chapter. I am grateful to Fernando Abdala, Ken Angielczyk, Anusuya Chinsamy-Turan, Adam Huttenlocker, and Bruce Rubidge for their many helpful comments on the draft manuscript.

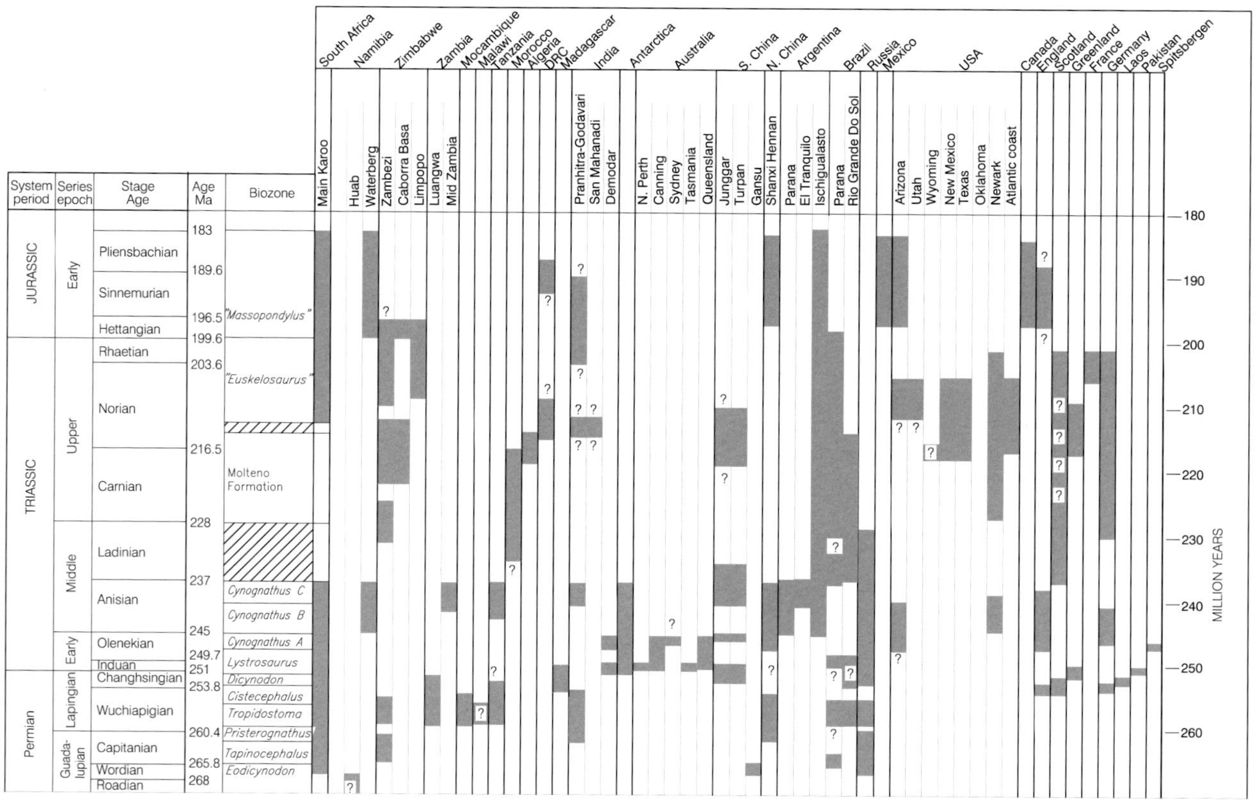

2.1. Global biostratigraphic correlation chart for Karoo-aged basins based mainly on therapsids (modified after Rubidge [2005]).

Therapsid Biodiversity Patterns and Paleoenvironments of the Karoo Basin, South Africa

2

Roger Smith, Bruce Rubidge, and Merrill van der Walt

Introduction

Therapsids lived in Pangaean times and their fossils have been found on every continent (see Plate 1A); however, there is no doubt that the most abundant and diverse collecting grounds are in the main Karoo Basin of South Africa. This 10-km-thick sedimentary succession accumulated in a large intracratonic, retro-arc, foreland basin (Johnson 1991; Catuneanu et al. 2005) in front of the rising Cape Fold Belt portion of the Gondwanide Mountain range that fringed the southern margin of Gondwana. Today, rocks of the Karoo Supergroup have a spatial distribution of some 300,000 km^2 (Smith 1990b), which is more than one-half the land surface of South Africa. They were deposited under environments ranging from glacial through intracontinental shallow marine, fluvio-lacustrine, fluvial, and aeolian and provide an almost continuous 110-My record of continental sedimentation from the Late Carboniferous (300 Ma) to the Early Jurassic (190 Ma) (Smith 1995).

Fossils from the Karoo, especially the therapsids, provide the best record of continental biodiversity from the Middle Permian to the Early Jurassic and have been crucial to studies of the global Permo-Triassic mass extinction, as well as giving insight into other extinction events. Recent research has demonstrated the importance of Karoo fossils for global stratigraphic correlation and for the conceptualization of basin development models (e.g., Catuneanu, Hancox, and Rubidge 1998; Hancox 1998; Neveling 2002; Rubidge 2005). The discovery of datable ash beds that can be linked to biozone-defining fossils has considerably enhanced research possibilities by providing radiometric ages for vertebrate biozones (Rubidge et al. 2010), and thus opening up a way to ascertain rates of evolution in fossil tetrapod lineages, as well as the timing and duration of extinction events. The recent development of a geographic information system–based database of all the fossil tetrapods from the Beaufort Group, which are housed in South African museum collections, has facilitated research on tetrapod biodiversity and biogeographic changes during the Permian and Triassic (Nicolas 2007; Nicolas and Rubidge 2009) and has endless possibilities for future research.

History of Stratigraphy of Karoo Basin

Andrew Geddes Bain was the first to draw attention to the rocks of the Karoo and named these strata the Karoo "Series" (Bain 1856). Later, Jones (1867 cited in North 1878) introduced the nomenclature Beaufort "Beds" for sedimentary rocks within the lower part of the Karoo Series around Beaufort West. In 1873, Dunn proposed the term Stormberg "Beds" for the uppermost part of the succession and then subdivided the sequence into the "Coal Measures, Red Beds, Cave Sandstone and Volcanic Beds"(Dunn 1878). Rogers (1903) lithostratigraphically subdivided what he termed the Karoo "System," into the Dwyka, Ecca, Beaufort, and Stormberg groups, and in 1905, he and Broom (1906b) further subdivided the Beaufort Group into a Lower, Middle, and Upper Series. This stratigraphic subdivision remained until Johnson (1976) erected a new lithostratigraphic subdivision for the Beaufort Group in the Eastern Cape. He established the Adelaide Subgroup to broadly correspond to the Lower Beaufort Series of Rogers (1905) and Broom (1906b) and the Tarkastad Subgroup to their Middle and Upper Series. Johnson's nomenclature was accepted by the Karoo Working Group of the South African Council for Stratigraphy (SACS 1980).

For some 80 My, from the Early Permian to Early Jurassic, the alluvial plains of the intracontinental Karoo Basin were continually adjusting to base level movement caused by subduction-driven uplift along the southern margin of the basin and resultant reciprocal flexure in the foreland sag as well as experiencing periods of rapid climate change. The constantly changing base level had a major influence over the drainage networks, topography, groundwater levels, vegetation types, and the terrestrial and aquatic communities that lived there. For most of this time, Gondwanan tetrapod communities were dominated by therapsids, although after the End-Permian mass extinction, therapsid dominance was progressively challenged by archosauromorphs. Continuous environmental change kept these communities in a state of flux. Migrations, invasions, extinctions, and radiations were happening all the time and form the basis of the current biostratigraphic scheme.

Biozonation of the Karoo Supergroup

The extended and almost continuous depositional history of the Karoo Supergroup affords a unique opportunity to study changes in ecological representation from the Middle Permian (Wordian) to Early Jurassic (Pliensbachian). A therapsid-based biostratigraphic scheme (Kitching 1977; Kitching and Raath 1984; Keyser and Smith 1978; Rubidge 1995a) has facilitated division of the vertebrate-bearing sequences of the Karoo into "time slices" and enabled determination of ecological representation of vertebrates for each time slice (biozone) as well as the representation of vertebrate trophic levels (herbivore, carnivore, or omnivore). Using biozones as time slices for the purposes of this chapter has necessitated two assumptions: that the biozone time slices have served to separate temporally successive communities; and that communities in a particular biozone were largely uniform across the geographic extent of the main

Karoo Basin. Because of the time extensive nature of the Karoo succession, the biozonation scheme has global application for correlation of continental deposits (cf. Rubidge 2005) (Fig. 2.1).

History of Karoo Vertebrate Biozonation

The oldest record of vertebrate fossils from the Karoo Supergroup is in a letter by C. H. Crisbrook, published in the *South African Quarterly Journal* in 1831 under the title "Organic Remains in the Karroo" (MacRae 1999), however the locality of these fossils is unknown. In 1838, near Fort Beaufort, Andrew Geddes Bain made the first fully documented discovery of a Karoo fossil tetrapod—a dicynodont (Bain 1845; Owen 1845). Because of their abundance in the rocks of the Karoo Supergroup and particularly the Beaufort Group, fossil tetrapods have long been used for the biostratigraphic subdivision of the succession. Wyley (1859) introduced the term "Reptilian Beds" for Karoo strata near Beaufort West, Seeley (1892) was the first to subdivide the rocks of the Karoo into biozones. Broom (1906a, 1906c, 1906d, 1907, 1909) proposed a six-fold biostratigraphic zonation for the Beaufort Group, which, with slight modification (Hotton and Kitching 1963), was accepted by most for the next sixty years (Watson 1913, 1914; Von Huene 1925; Du Toit 1954).

Subsequent collecting in the Karoo Basin has allowed for further revision of the Beaufort biostratigraphy (Kitching 1970, 1977, Keyser and Smith 1978; Keyser 1979; SACS 1980; Rubidge 1995a) and includes the Stormberg Elliot and Clarens formations as well (Kitching and Raath 1984). The current ten-fold biostratigraphic subdivision of the Karoo Supergroup has facilitated the correlation of these strata with equivalent sequences elsewhere in the world, which in turn has allowed for more precise faunal dating of the succession (Rubidge 2005) (Plate 1B, Fig. 2.2). The following summary of the biozonation scheme provides an account of the major environmental changes in the Karoo Basin and how the terrestrial vertebrate populations responded.

Eodicynodon AZ (Wordian)

PALEOENVIRONMENT

Lateral amalgamation of numerous deltas that prograded into the landlocked Ecca Sea along the active southern margin provided the earliest habitable subaerial lowland plains in the main Karoo Basin (Plate 2). The first tetrapods to colonize this narrow strip of new land arrived in the Middle Permian (Wordian) some 265 My ago (Mya). These were relatively primitive herbivorous therapsids, *Eodicynodon* and *Tapinocaninus* (see Plate B), along with flesh-eating therocephalians, the basal anteosaurid dinocephalian *Australosyodon*, and semi-aquatic rhinesuchid amphibians. Water tables were still relatively high, the paleosols display hydromorphic gley textures and under the orographic effect of

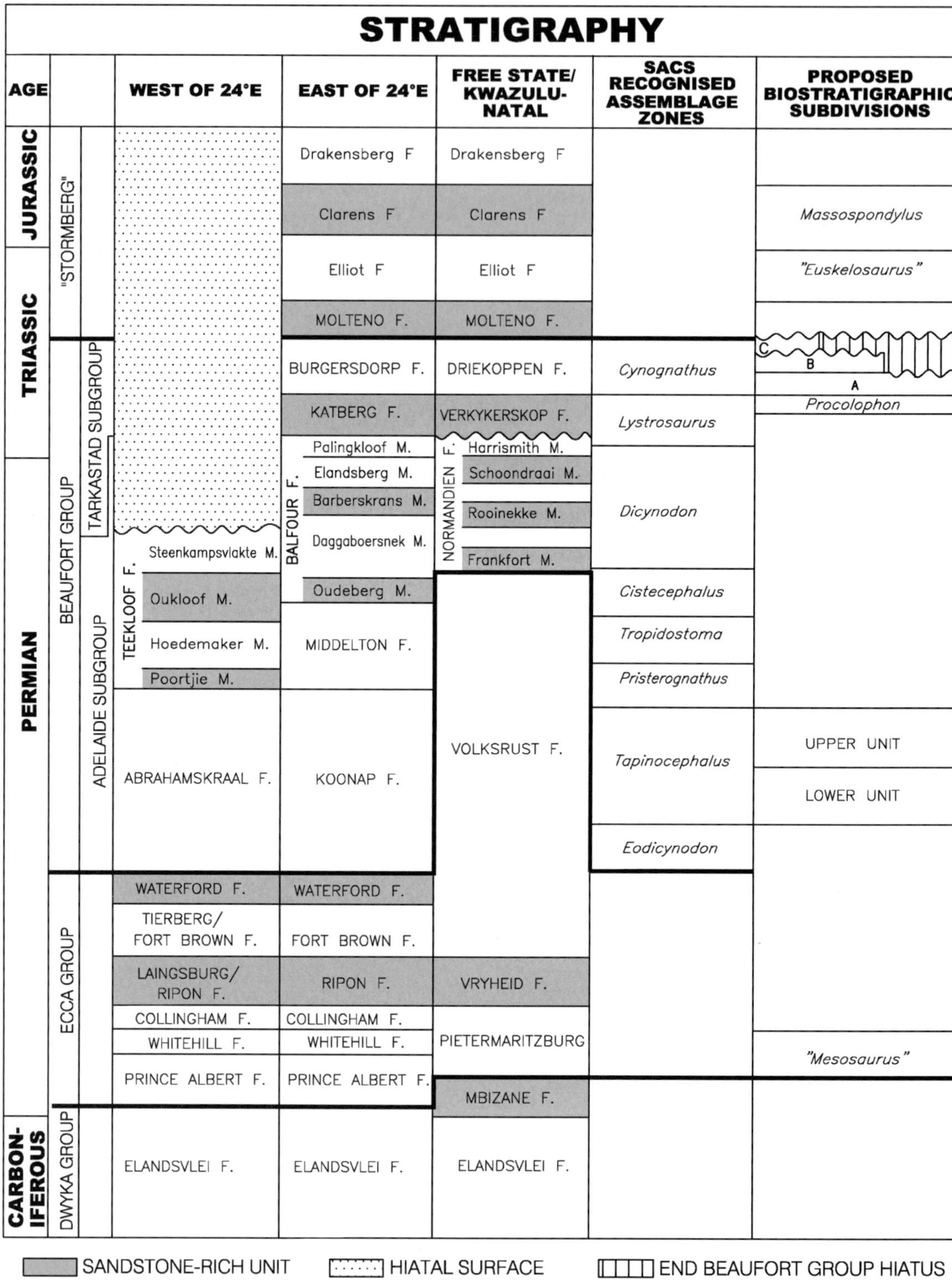
STRATIGRAPHY
AGE
WEST OF 24°E
EAST OF 24°E
FREE STATE/ KWAZULU-NATAL
SACS RECOGNISED ASSEMBLAGE ZONES
PROPOSED BIOSTRATIGRAPHIC SUBDIVISIONS
JURASSIC
TRIASSIC
PERMIAN
CARBON-IFEROUS
"STORMBERG"
BEAUFORT GROUP
TARKASTAD SUBGROUP
ADELAIDE SUBGROUP
ECCA GROUP
DWYKA GROUP
Drakensberg F
Drakensberg F
Massospondylus
Clarens F
Clarens F
Elliot F
Elliot F
"Euskelosaurus"
MOLTENO F.
MOLTENO F.
BURGERSDORP F.
DRIEKOPPEN F.
Cynognathus
C
B
A
KATBERG F.
VERKYKERSKOP F.
Lystrosaurus
Procolophon
Palingkloof M.
Elandsberg M.
Barberskrans M.
Daggaboersnek M.
Oudeberg M.
BALFOUR F.
NORMANDIEN F.
Harrismith M.
Schoondraai M.
Rooinekke M.
Frankfort M.
Dicynodon
TEEKLOOF F.
Steenkampsvlakte M.
Oukloof M.
Hoedemaker M.
Poortjie M.
MIDDELTON F.
Cistecephalus
Tropidostoma
Pristerognathus
ABRAHAMSKRAAL F.
KOONAP F.
VOLKSRUST F.
Tapinocephalus
UPPER UNIT
LOWER UNIT
Eodicynodon
WATERFORD F.
WATERFORD F.
TIERBERG/ FORT BROWN F.
FORT BROWN F.
LAINGSBURG/ RIPON F.
RIPON F.
VRYHEID F.
COLLINGHAM F.
COLLINGHAM F.
WHITEHILL F.
WHITEHILL F.
PIETERMARITZBURG
PRINCE ALBERT F.
PRINCE ALBERT F.
"Mesosaurus"
MBIZANE F.
ELANDSVLEI F.
ELANDSVLEI F.
ELANDSVLEI F.
SANDSTONE-RICH UNIT
HIATAL SURFACE
END BEAUFORT GROUP HIATUS

the Gondwanide Mountains, the climate is interpreted to have been humid temperate. Equisetalians and trees and bushes of the broad-leafed gymnosperm *Glossopteris* flourished along the branching network of low sinuosity perennial river channels. Synapsid fossils are mainly found in floodplain fines associated with horizons of near spherical calcareous nodules interpreted as a marginal lake (palustrine) habitat.

2.2. Stratigraphy of the Main Karoo Basin, South Africa. *Abbreviations:* F= formation; M= member.

TROPHIC ANALYSIS

Carnivores include therocephalians, the dinocephalian *Australosyodon*, and fragmentary remains attributed to gorgonopsians. Herbivores include the anomodonts *Eodicynodon* and *Patranomodon* and the dinocephalian *Tapinocaninus* (Table 2.1, Fig. 2.3). Of the 50 identifiable specimens from this biozone, 47 are obligate herbivores and 3 are obligate carnivores, meaning that 94 percent of the faunal assemblage of the *Eodicynodon* Assemblage Zone (AZ) is herbivorous and 6 percent carnivorous—a ratio of 16:1.

BIOSTRATIGRAPHIC CORRELATION

The *Eodicynodon* AZ tetrapod fauna comprises primitive anomodonts (Rubidge and Hopson 1990, 1996), the earliest dicynodonts (Barry 1974; Cluver and King 1983; Rubidge 1990a, 1990c), dinocephalians (Rubidge 1991, 1994), therocephalians (Rubidge, Kitching, and van den Heever 1983), and gorgonopsians (Rubidge 1988, 1995b, 1995c). In other parts of the world, Late Wordian faunas are known only from the Xidagou Formation in Gansu, North China (Li and Cheng 1995; Li, Rubidge, and Cheng 1996; Jun, Rubidge, and Jinling 2009a, 2009b), and the *Estemmenosuchus* fauna in Russia, which includes the Ocher and Mezen complexes (Modesto and Rybczynski 2000). Although no genera are common, these faunas share basal anteosaurid dinocephalians (*Archaeosyodon*, *Brithopus*, and *Chthomaloporus* from Russia; *Stenocybus* and *Sinophoneus* from China; and *Australosyodon* from South Africa) and basal anomodonts (*Otsheria* from Russia Modesto and Rybczynski 2000]; *Biseridens* from China [Battail 2000; Jun, Rubidge, and Jinling 2009a, 2009b]; and *Patranomodon* from South Africa [Rubidge and Hopson 1990, 1996; Modesto, Rubidge, and Welman 1999, Modesto and Rubidge 2000]). At this stage, dicynodonts, gorgonopsians, and therocephalians were already present in South Africa (Rubidge 1995b) but had not yet appeared in the Laurasian fauna (Battail 2000).

Tapinocephalus AZ (Capitanian)

PALEOENVIRONMENT

Continued deltaic progradation along the southern and western margin of the shrinking Ecca Sea during the mid-Permian broadened the alluvial plain to at least 100 km and extended its length around the southwestern

Table 2.1. *Eodicynodon* and *Tapinocephalus* AZ generic diversity, abundance, and trophic level

	Taxa	Genus	Total	Trophic level	Genus	Total	Trophic level
Eodicynodon AZ							
Therapsida	Dicynodontia	*Eodicynodon*	44	Herbivore			
	Dinocephalia	*Australosyodon*	1	Carnivore			
		Tapinocaninus	3	Herbivore			
	Therocephalia	*Glanosuchus*	1	Carnivore			
		Ictidosaurus	1	Carnivore			
Tapinocephalus AZ							
Amphibia		*Rhinesuchus*	11	Carnivore			
Parareptilia		*Bradysaurus*	88	Herbivore	*Nochelesaurus*	1	Herbivore
		Embrithosaurus	12	Herbivore	Pareiasauriid	116	Herbivore
		Eunotosaurus	14	Herbivore			
Pelycosauria		*Elliotsmithia*	2	Carnivore	Varanopid	1	Carnivore
Therapsida	Biarmosuchia	*Bullacephalus*	1	Carnivore	*Hipposaurus*	4	Carnivore
		Pachydectes	1				
	Anomodontia	*Anomocephalus*	1	Herbivore	*Galeops*	3	Herbivore
		Galechirus	1	Herbivore			
	Dicynodontia	*Brachyprosopus*	1	Herbivore	*Lanthanostegus*	1	Herbivore
		Chelydontops	2	Herbivore	*Pristerodon*	8	Herbivore
		Colobodectes	6	Herbivore	*Prosictodon*	1	Herbivore
		Diictodon	212	Herbivore	*Robertia*	60	Herbivore
	Dinocephalia	*Anteosaurus*	42	Carnivore	*Phocosaurus*	4	Herbivore
		Avenantia	3	Herbivore	*Riebeeckosaurus*	1	Herbivore
		Criocephalus	4	Herbivore	*Struthiocephaloides*	2	Herbivore
		Delphinognathus	2	Herbivore	*Struthiocephalus*	37	Herbivore
		Jonkeria	44	Herbivore	*Struthionops*	1	Herbivore
		Keratocephalus	4	Herbivore	*Styracocephalus*	10	Herbivore
		Mormosaurus	2	Herbivore	*Tapinocephalus*	45	Herbivore
		Moschops	30	Herbivore	*Titanosuchus*	58	Herbivore
	Gorgonopsia	*Broomisaurus*	2	Carnivore	*Gorgonops*	1	Carnivore
		Eoarctops	4	Carnivore	*Scylacognathus*	4	Carnivore
		Galesuchus	1	Carnivore			
	Therocephalia	*Alopecodon*	2	Carnivore	*Pardosuchus*	2	Carnivore
		Blattoidealestes	1	Carnivore	*Pristerognathus*	22	Carnivore
		Crapartinella	1	Carnivore	*Scylacosaurus*	2	Carnivore
		Glanosuchus	28	Carnivore	Scylacosaurid	14	Carnivore
		Ictidosaurus	1	Carnivore	*Simorhinella*	1	Carnivore
		Lycosuchus	1	Carnivore	Therocephalia indeterminate	8	Carnivore
		Lycosuchid	10	Carnivore			

2.3. Pie charts representing tetrapod higher taxon breakdown and feeding behavior for the Middle to Late Permian *Eodicynodon, Tapinocephalus, Pristerognathus,* and *Tropidostoma* AZs in the Karoo Basin.

and southern margins of the basin—a distance of approximately 2,000 km (Plate 2). Several discrete sandstone-rich packages (Cole 1992) in the Abrahamskraal Formation are interpreted as low-angled, fan-shaped channel networks made up of relatively more closely spaced low-sinuosity rivers. The rivers probably flowed perennially but were subjected to seasonal discharge peaks that overspilled onto the flanking floodplains. For most of the year, water tables were still relatively high and supported

Higher taxa breakdown

Feeding ratios

Eodicynodon Assemblage Zone

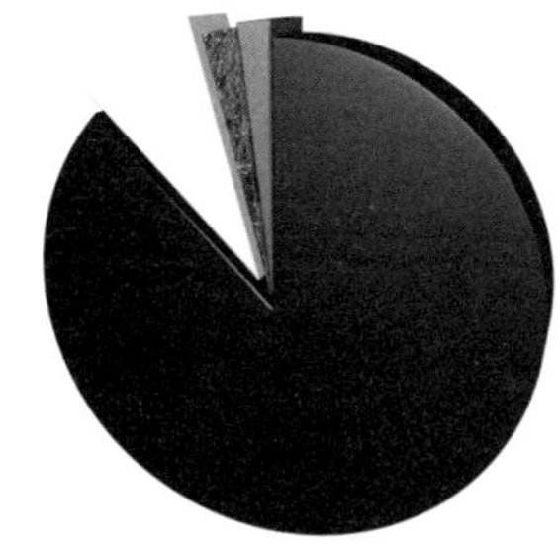

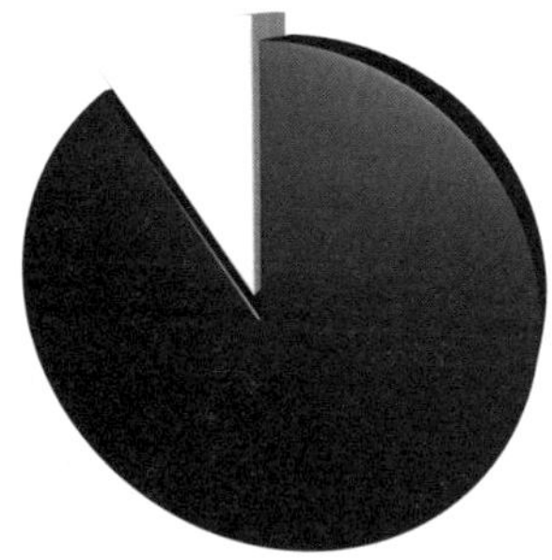

Tapinocephalus Assemblage Zone

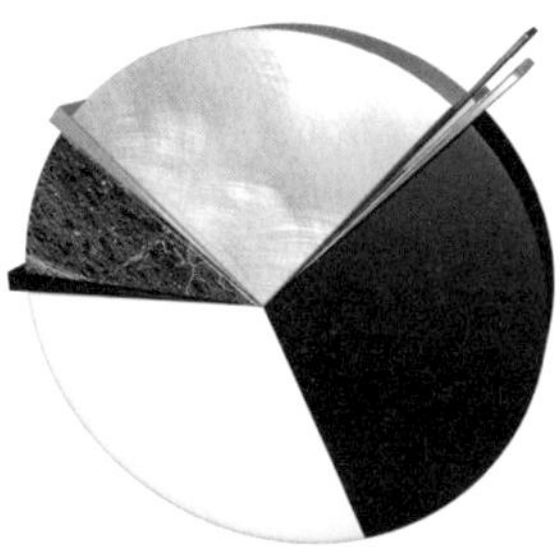

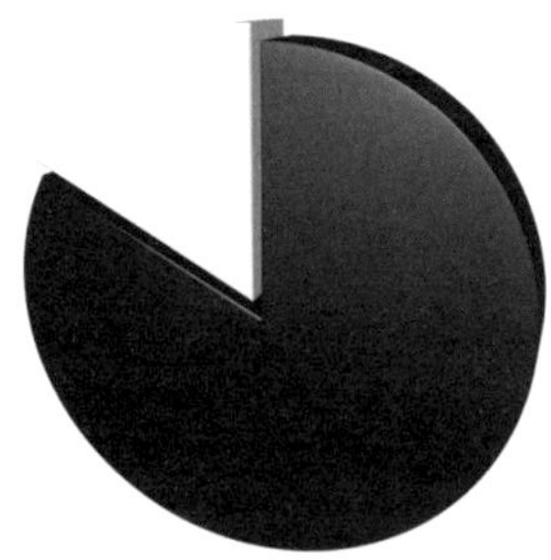

Pristerognathus Assemblage Zone

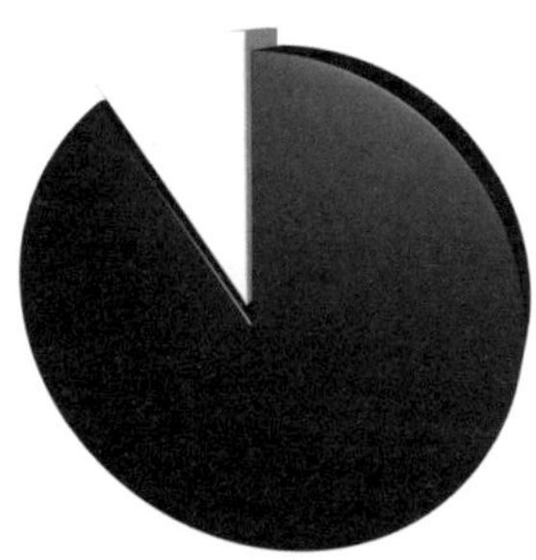

Tropidostoma Assemblage Zone

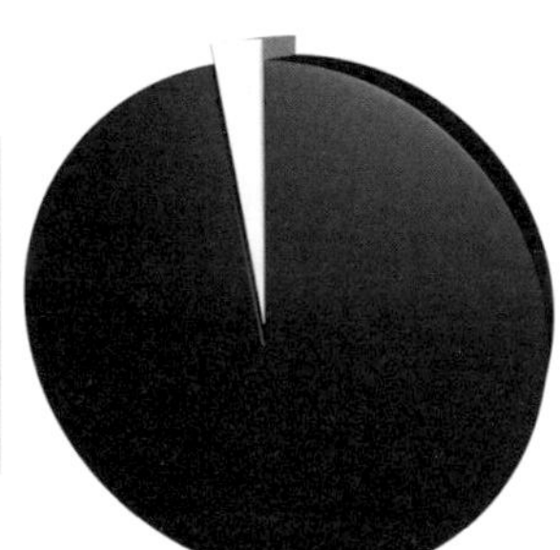

vegetation throughout the year. Dinocephalians (see Plate D) took advantage of the expanding lowland habitats and diversified into a range of large-bodied foliage grazers and some predatory carnivores. Toward the top of the *Tapinocephalus* AZ, the small but very successful long-bodied dicynodont *Diictodon* made its first appearance in the basin (Rubidge et al. 2009), perhaps exploiting the more elevated parts of the floodplains where they excavated burrows into partly consolidated alluvium (see Plates F and G). Dinocephalians disappeared at the end of the *Tapinocephalus* AZ, but it has still not been ascertained whether their demise was abrupt or gradual.

TROPHIC ANALYSIS

Among the tetrapod faunas of the *Tapinocephalus* AZ (see Plate D), all amphibians, the varanopid *Elliotsmithia,* biarmosuchians (*Bullacephalus, Hipposaurus, Pachydectes*), gorgonopsians (*Broomisaurus, Eoarctops, Galesuchus, Gorgonops, Scylacognathus*), therocephalians (*Alopecodon, Blattoidealestes, Crapartinella, Glanosuchus, Ictidosaurus, Lycosuchus, Pardosuchus, Pristerognathus, Scylacosaurus, Simorhinella*), and anteosaurid dinocephalians (*Anteosaurus*) are categorized as obligate carnivores. Herbivores include all anomodonts (*Anomacephalus, Galechirus, Galeops, Brachyprosopus, Colobodectes, Chelydontops, Diictodon, Lanthanostegus, Prosictodon, Pristerodon, Robertia*); titanosuchid (*Jonkeria, Titanosuchus*), tapinocephalid (*Avenantia, Criocephalus, Delphinognathus, Keratocephalus, Mormosaurus, Moschops, Phocosaurus, Riebeeckosaurus, Struthiocephalus* [see Plate D], *Struthiocephaloides, Struthionops, Tapinocephalus* [see Plates C and D]), and styracocephalid (*Styracocephalus*) dinocephalians; and the parareptiles *Bradysaurus, Embrithosaurus, Eunotosaurus,* and *Nochelesaurus.* The number of dinocephalian genera presented here may be inflated and is currently the subject of a doctoral study by Saniye Guven. Of the 939 identifiable specimens from this biozone, 775 are herbivorous and 164 are carnivorous, meaning that 83 percent of the faunal assemblage of the *Tapinocephalus* AZ is herbivorous and 17 percent carnivorous—a ratio of 5:1 (see Table 2.2 and Fig. 2.3).

BIOSTRATIGRAPHIC CORRELATION

A wide diversity of anapsid and synapsid taxa has been described from the *Tapinocephalus* AZ (Boonstra 1969; Keyser and Smith 1978; Nicolas and Rubidge 2009; Smith and Keyser 1995). Terrestrial tetrapod-bearing deposits of this age are also present at Mezen and Esheevo in Russia (Modesto and Rybczynski 2000), in Brazil (Cisneros et al. 2011; Langer 2000), the Zambezi Valley of Zimbabwe (K5c) (Bond 1973; Boonstra 1946; Lepper, Raath, and Rubidge 2000; Munyikwa 2001), and the most recent finds from the Ruhuhu Basin of Tanzania (Simon et al. 2010).

Mezen shares with South Africa procolophonomorphs (the Russian genus is *Nyctiphruretus*; however, in South Africa, the single nyctiphruretid specimen that has been found cannot be identified to generic level

[Cisneros and Tsuji 2009; Gow and Rubidge 1997]), biarmosuchians (*Hipposaurus*, *Bullacephalus*, and *Pachydectes* from South Africa, and *Biarmosuchus* from Russia [Rubidge and Sidor 2001, 2002; Rubidge and Kitching 2003; Rubidge, Sidor, and Modesto 2006; Sigogneau-Russell 1989; Ivakhnenko 1999]), and varanopids (*Mesenosaurus* from Russia and *Elliotsmithia* from South Africa [Dilkes and Reisz 1996; Modesto et al. 2001; Reisz 1986; Reisz, Dilkes, and Berman 1998]).

Esheevo has in common: biarmosuchians (*Biarmosuchoides*), anteosaurid dinocephalians (*Admetophoneus*, *Brithopus*, *Deuterosaurus*, *Doliosauriscus*, *Notosyodon*, *Syodon*, and *Titanophoneus* from Russia and *Anteosaurus* from South Africa), tapinocephalid dinocephalians (*Ulemosaurus* from Russia, and *Avenantia*, *Criocephalus*, *Delphinognathus*, *Keratocephalus*, *Mormosaurus*, *Moschops*, *Phocosaurus*, *Riebeeckosaurus*, *Struthiocephalus*, *Struthiocephaloides*, *Struthionops*, *Tapinocephalus*, and *Taurocephalus* from South Africa), anomodonts (*Ulemica* and *Venjukovia* from Russia, and *Galeops*, *Galechirus*, and *Anomocephalus* from South Africa), and scylacosaurid therocephalians (*Porosteognathus* from Russia and *Pristerognathus* and *Scylacosaurus* from South Africa [Battail 2000; Battail and Surkov 2000; Boonstra 1969; Modesto and Rybczynski 2000; Rubidge 1995c]).

In common with South Africa, Brazil has dinocephalians (Langer 2000) and basal anomodonts (Cisneros et al. 2011), and Zimbabwe has anteosaurid and tapinocephalid dinocephalians (Boonstra 1946; Lepper, Raath, and Rubidge 2000; Munyikwa 2001). Only the tapinocephalids have been identified to generic level and include *Criocephalus* (also from South Africa) and a new primitive undescribed form (Munyikwa 2001). Tanzania has yielded teeth of a tapinocephalid but the material is not complete enough to determine genus (Simon et al. 2010). Present in South Africa, but not known from other countries at this time, are dicynodonts, gorgonopsians, and the parareptile *Eunotosaurus*.

Pristerognathus AZ (Capitanian)

PALEOENVIRONMENT

Around 260 Mya, renewed uplift in the Gondwanides resulted in a multitude of northeasterly flowing sand-dominated braided streams crossing the southern Karoo alluvial plains toward a large intracontinental lake (Rubidge 2005) (Plate 2). These channels are preserved in the rock record as the arenaceous Poortjie Member. Another indication of renewed tectonism at this time is the presence of numerous laterally continuous thin sheets of silicified mudrocks ("cherts" of Roussouw and De Villiers 1952) that are anomalously rich in volcanic ash (Ho Tun 1979). The ashfall events do not appear to have affected the small herbivores but may have been detrimental to the broad-leafed foliage and consequently the large browsers. The small-bodied dicynodonts were unaffected and apparently flourished in the absence of the dinocephalians, judging by the increase in relative abundance of *Diictodon* during this time, as well as a diversification of dicynodonts with the first appearance of *Dicynodontoides*, *Emydops*, and *Endothiodon*. This increase in small-bodied herbivores appears to have triggered a synchronous radiation of medium-sized predatory scylacosaurid therocephalians.

TROPHIC ANALYSIS

Carnivores of the *Pristerognathus* assemblage include amphibians (*Rhinesuchus*), gorgonopsians (*Gorgonops* and *Scylacognathus*), and therocephalians (*Glanosuchus*, *Ictidosuchus*, *Ictidosaurus*, *Lycosuchus*, *Pristerognathus*, and *Scylacosaurus*), while the herbivores include parareptiles (*Bradysaurus*, *Embrithosaurus*, *Eunotosaurus*) and dicynodonts (*Diictodon*, *Dicynodontoides*, *Emydops*, *Endothiodon*, *Pristerodon*, and *Robertia*) (Table 2.2, Fig. 2.3).

Of the 1,081 identifiable specimens from this biozone, 978 are herbivorous and 103 are carnivorous, meaning that 90 percent of the faunal assemblage of the *Pristerognathus* AZ is herbivorous and 10 percent carnivorous—a ratio of 9:1. The *Pristerognathus* AZ is a relatively newly established biozone and currently only three museum collections document records from it, and thus this ratio may not be accurate (Nicolas and Rubidge 2009). Other museum collections may have older, less precise locality information of specimens collected in rocks which are now assigned to the *Pristerognathus* AZ.

BIOSTRATIGRAPHIC CORRELATION

The parareptile *Eunotosaurus* occurs in the South Africa Karoo at this time but is not known from other countries. Because this biozone is based largely on the presence of a relative abundance of scylacosaurid therocephalians and the dicynodont *Diictodon*, coupled with the absence of dinocephalians, it is difficult to correlate it with assemblages in other sedimentary basins, but it has been correlated with the Russian Zone II and III successions (Smith and Keyser 1995).

Tropidostoma AZ (Wuchiapigian)

PALEOENVIRONMENT

The *Tropidostoma* assemblage fauna of the southern Karoo Basin lived on flat, very low gradient alluvial plains some 500-km wide which were virtually featureless except for several low ridges snaking toward the central lake (Smith 1987b,1993b) (Plate 2). Each ridge contained a single meandering river. The climate in the basin was becoming drier and discharge down the rivers was highly seasonal but still perennial. Seasonal overbank floods deposited layered silts on the river banks causing the levee ridges to build up. Every 30,000 years, or so, the levees were breached by overbank flow eroding a deep crevasse which diverted the main flow and left the

Table 2.2. *Pristerognathus* and *Tropidostoma* AZ generic diversity, abundance, and trophic level

	Taxa	Genus	Total	Trophic level	Genus	Total	Trophic level
Pristerognathus AZ							
Amphibia		*Rhinesuchus*	6	Carnivore			
Parareptilia		*Bradysaurus*	8	Herbivore	*Eunotosaurus*	20	Herbivore
		Embrithosaurus	1	Herbivore	Pareiasauriid	3	Herbivore
Therapsida	Dicynodontia	*Diictodon*	865	Herbivore	*Dicynodontoides*	1	Herbivore
		Emydops	12	Herbivore	*Pristerodon*	55	Herbivore
		Endothiodon	6	Herbivore	*Robertia*	7	Herbivore
	Gorgonopsia	*Gorgonops*	14	Carnivore	*Scylacognathus*	1	Carnivore
	Therocephalia	*Glanosuchus*	3	Carnivore	*Pristerognathus*	46	Carnivore
		Ictidosaurus	1	Carnivore	*Scylacosaurus*	1	Carnivore
		Ictidosuchus	1	Carnivore	Therocephalia indeterminate	27	Carnivore
		Lycosuchus	2	Carnivore			
Tropidostoma AZ							
Amphibia		*Rhinesuchus*	4	Carnivore			
Parareptilia		*Pareiasaurus*	16	Herbivore	*Saurorictus*	1	Carnivore
Archosauromorpha		*Youngina*	4	Carnivore			
Therapsida	Cynodontia	*Charassognathus*	1	Carnivore	*Procynosuchus*	2	Carnivore
	Dicynodontia	*Cistecephalus*	7	Herbivore	*Oudenodon*	2	Herbivore
		Dicynodontoides	3	Herbivore	*Palemydops*	2	Herbivore
		Diictodon	2099	Herbivore	*Pristerodon*	206	Herbivore
		Emydops	82	Herbivore	*Rhachiocephalus*	15	Herbivore
		Endothiodon	72	Herbivore	*Tropidostoma*	124	Herbivore
	Gorgonopsia	*Aelurognathus*	1	Carnivore	*Gorgonops*	36	Carnivore
		Aloposaurus	2	Carnivore	*Lycaenops*	7	Carnivore
		Cyonosaurus	1	Carnivore	*Scymnognathus*	3	Carnivore
		Aelurosaurus	1	Carnivore			
	Therocephalia	*Hofmeyria*	3	Carnivore	*Ictidosuchus*	3	Carnivore
		Ictidosuchoides	16	Carnivore	*Lycideops*	1	Carnivore
		Ictidosuchops	1	Carnivore			

abandoned river channel as a series of ponds (Smith 1993a). The dicynodont radiation continued with the first appearance of *Cistecephalus*, *Tropidostoma*, *Oudenodon*, and *Rhachiocephalus*. *Diictodon* still dominated the ground level grubbing and foraging niche, with *Tropidostoma* and *Rhachiocephalus* becoming the dominant medium- and large-bodied herbivores. Gorgonopsians became more abundant both in numbers and in genera (Sigogneau 1970) in response to the herbivore radiation and were larger than their main carnivorous competitors, the therocephalians. A new group of carnivores, the cynodonts, made their first appearance with *Procynosuchus* and the rare and diminutive insectivore *Charassognathus gracilis* (Botha, Abdala, and Smith 2007).

TROPHIC ANALYSIS

Carnivores are represented in this zone by temnospondyl amphibians, mainly by *Rhinesuchus*, archosauromorphs (*Youngina*), and the theriodont therapsids, including gorgonopsians (*Aelurognathus*, *Aloposaurus*, *Cyonosaurus*, *Aelurosaurus*, *Gorgonops*, *Lycaenops*, and *Scymnognathus*), and therocephalians (*Hofmeyria*, *Ictidosuchoides* [see Plate L], *Ictidosuchops*, *Ictidosuchus*, and *Lycideops*). Herbivores include all dicynodonts (*Cistecephalus*, *Diictodon*, *Dicynodontoides*, *Emydops*, *Endothiodon*, *Oudenodon*, *Palemydops*, *Pristerodon*, *Rhachiocephalus*, and *Tropidostoma*), as well as the parareptiles *Pareiasaurus* and *Saurorictus*. The obligate insectivores from this zone are a small cynodont *Charassognathus*, the tiny archosauromorph *Youngina*, and the captorhinid *Saurorictus*.

Of the 2,715 identified specimens from this biozone, 2,628 are herbivorous and 87 are carnivorous, meaning that 97 percent of the faunal assemblage of the *Tropidostoma* AZ is herbivorous and 3 percent carnivorous—a ratio of 32:1 (Nicolas and Rubidge 2009; Sidor and Smith 2007) (Table 2.2, Fig. 2.3).

BIOSTRATIGRAPHIC CORRELATION

Correlative faunas of this age are known from Zimbabwe's Madumabisa Mudstone Formation (Bond 1973), Zambia's Madumabisa Mudstone Formation (Lee, Gow, and Kitching 1997), Malawi's Chiweta Beds (Haughton 1926; Jacobs et al. 2005), Mozambique's Mkanga Formation (Latimer, Gow, and Rubidge 1995), India's Kundaram Formation of the Pranhita-Godavari Valley (Ray 1999, 2000), the Morro Pelado Member of the Rio do Rasto Formation in southern Brazil (Barbarena and Araujo 1975; Battail 2000), and the fauna of the *Honania-Taihanshania* AZ of the upper Shihezi Formation, and the Hongyianchi Formation of Xinjiang Province in North China (Li and Cheng 1995).

Cistecephalus AZ (Wuchiapigian)

PALEOENVIRONMENT

During the Late Wuchiapigian, the southern Karoo Basin was again subjected to a change in fluvial style from a few large meandering rivers to more numerous low sinuosity streams (Plate 2). This is manifest in the sandstone packages of the Oukloof and Oudeberg members of the Teekloof and Balfour formations (Turner 1979). The increase in percentage of lithic fragments and volume of sand-sized clastics is an indication that this change in fluvial style was due to increased discharge and flow velocity. The floodplain paleosols of the *Cistecephalus* zone are relatively well aerated, show some rubification, and have calcareous B horizons indicating seasonal vertical migration of the water table (Smith 1990a). Large dicynodonts proliferated with genera such as *Oudenodon*, *Aulacephalodon*, *Dinanomodon*, and *Odontocyclops*. Gorgonopsians and therocephalians

Table 2.3. *Cistecephalus* AZ generic diversity, abundance, and trophic level

	Taxa	Genus	Total	Trophic level	Genus	Total	Trophic level
Cistecephalus AZ							
Amphibia		Rhinesuchus	13	Carnivore			
Parareptilia		*Anthodon*	8	Herbivore	*Owenetta*	11	Carnivore
		Milleretta	3	Carnivore	*Pareiasaurus*	37	Herbivore
		Nanoparia	5	Herbivore			
Archosauromorpha		*Acanthotoposaurus*	1	Carnivore	*Youngina*	4	Carnivore
Therapsida	Biarmosuchia	*Lemurosaurus*	2	Carnivore	Paraburnetia	1	Carnivore
		Lycaenodon	2	Carnivore	*Rubidgina*	1	Carnivore
	Cynodontia	Cynosaurus	2	Omnivore	*Procynosuchus*	4	Carnivore
	Dicynodontia	*Aulacephalodon*	156	Herbivore	*Endothiodon*	13	Herbivore
		Cistecephalus	326	Herbivore	*Kingoria*	27	Herbivore
		Dicynodon	244	Herbivore	*Kitchinganomodon*	10	Herbivore
		Dicynodontoides	13	Herbivore	*Oudenodon*	519	Herbivore
		Digalodon	1	Herbivore	*Palemydops*	1	Herbivore
		Diictodon	1909	Herbivore	*Pelanomodon*	2	Herbivore
		Dinanomodon	4	Herbivore	*Pristerodon*	98	Herbivore
		Emydops	89	Herbivore	*Rhachiocephalus*	56	Herbivore
	Gorgonopsia	*Aelurognathus*	10	Carnivore	*Gorgonops*	42	Carnivore
		Aelurosaurus	4	Carnivore	*Lycaenops*	14	Carnivore
		Aloposaurus	2	Carnivore	*Leontocephalus*	1	Carnivore
		Arctognathus	6	Carnivore	*Pardocephalus*	1	Carnivore
		Arctops	9	Carnivore	*Prorubidgea*	10	Carnivore
		Cerdorhinus	2	Carnivore	*Rubidgea*	6	Carnivore
		Clelandina	2	Carnivore	*Scylacops*	3	Carnivore
		Cyonosaurus	22	Carnivore	*Scymnognathus*	1	Carnivore
		Dinogorgon	8	Carnivore	*Sycosaurus*	3	Carnivore
	Therocephalia	*Euchambersia*	1	Carnivore	*Ictidosuchops*	18	Carnivore
		Hofmeyria	4	Carnivore	*Macroscelesaurus*	1	Carnivore
		Homodontosaurus	4	Carnivore	*Polycynodon*	1	Carnivore
		Ictidostoma	1	Carnivore	*Proalopecopsis*	1	Carnivore
		Ictidosuchoides	15	Carnivore			

still dominated the carnivore guild and *Procynosuchus*, the second small cynodont to inhabit the Karoo Basin, increased in number.

TROPHIC ANALYSIS

In the *Cistecephalus* zone fossil assemblage, carnivores are represented by amphibians (*Rhinesuchus*), biarmosuchians (*Lemurosaurus*, *Lycaenodon*, *Paraburnetia*, *Rubidgina*), several gorgonopsians (see Table 2.3), some therocephalians (see Table 2.3), and the cynodont *Procynosuchus*. Herbivores include parareptiles (*Anthodon*, *Nanoparia*, *Pareiasaurus*) and several dicynodonts. The cynodont *Cynosaurus* is considered an omnivore and probable insectivores include archosauromorphs *Acanthotoposaurus* and *Youngina* and the parareptiles *Milleretta* and *Owenetta* (Table 2.3, Fig. 2.4).

Higher taxa breakdown

Feeding ratios

Cistecephalus Assemblage Zone

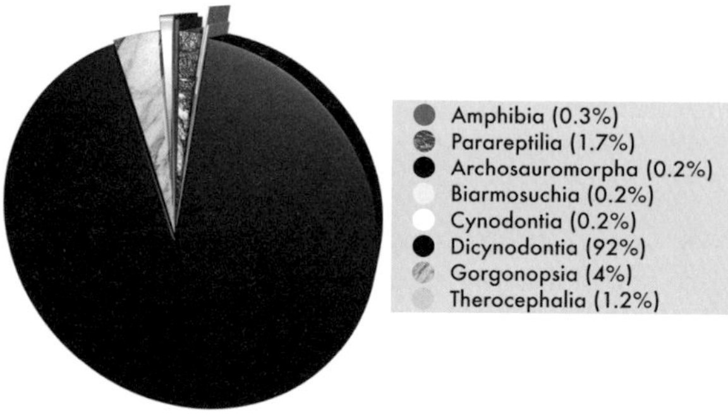

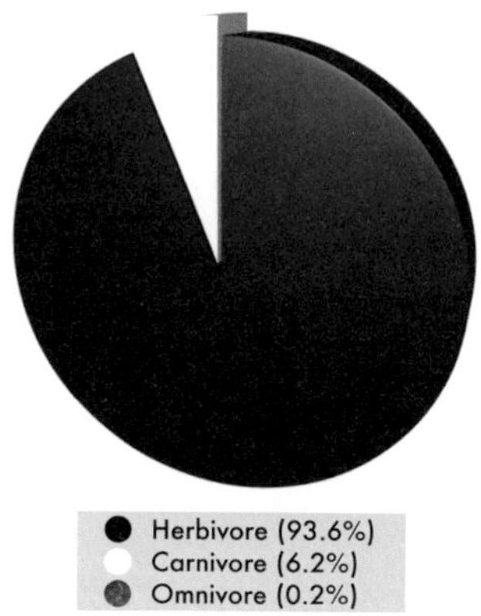

Dicynodon Assemblage Zone

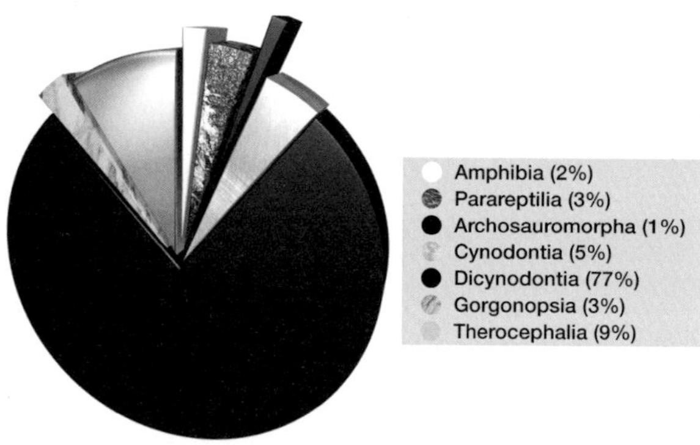

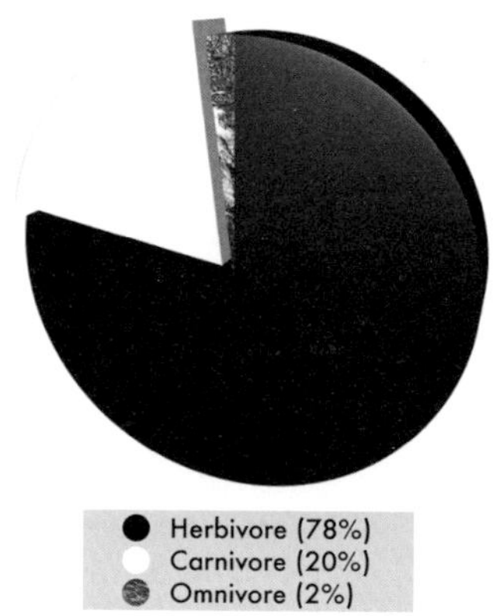

Lystrosaurus Assemblage Zone

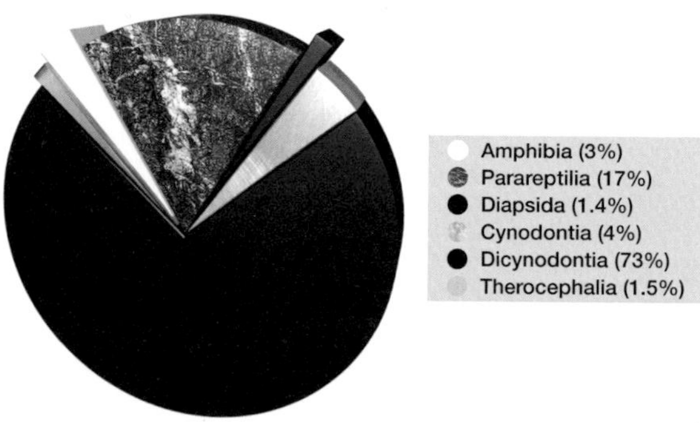

Of the 3,755 specimens identified from this biozone, 3,518 are herbivorous, 235 are carnivorous (including insectivores), and 2 omnivorous, meaning that 93.6 percent of the faunal assemblage of the *Cistecephalus* AZ is herbivorous, 6.2 percent carnivorous, and 0.2 omnivorous—a ratio of 16:1 (number of omnivores are negligible).

2.4. Pie charts representing tetrapod higher taxon and feeding behavior for the Late Permian and Early Triassic *Cistecephalus, Dicynodon,* and *Lystrosaurus* AZs in the Karoo Basin.

BIOSTRATIGRAPHIC CORRELATION

Correlatable faunas of this age are known from Zimbabwe (Madumabisa Mudstone Formation) (Bond 1973), Zambia (Madumabisa Mudstone Formation, Chikonta Group) (Lee, Gow, and Kitching 1997), Malawi (Chiweta Beds) (Haughton 1926; L. Jacobs, pers. comm. 2003), Tanzania (lower Usili Formation) (Simon et al. 2010), Mozambique (Latimer, Gow, and Rubidge 1995), India (Kundaram Formation of the Pranhita-Godavari Valley) (Ray 1999, 2000), southern Brazil (Morro Pelado Member of the Rio do Rasto Formation) (Barbarena and Araujo 1975; Battail 2000), the fauna of the *Honania-Taihanshania* AZ of the upper Shihezi Formation in North China, and the Hongyianchi Formation of Xinjiang Province (Li and Cheng 1995).

Dicynodon AZ (Wuchiapigian and Changhsingian)

PALEOENVIRONMENT

By the latest Permian, climatic conditions in the southern Karoo Basin appear to have been continental warm temperate becoming subtropical (Ward et al. 2005). For the first time, the alluvial plains were continuous all round the edge of the basin (Rubidge 2005) (Plate 2). The wide floodplains were once again drained by large meandering rivers flowing toward the center of the basin. Water tables were relatively high and the fossil record shows that in some places waterlogged patches turned to quagmire, trapping some of the larger dicynodonts such as *Dicynodon, Dinanomodon,* and *Lystrosaurus maccaigi,* as well as scavenging gorgonopsians and the saber-toothed therocephalian *Moschorhinus.*

Throughout the basin, the uppermost *Dicynodon* AZ shows a marked drop in diversity at all trophic levels marking the beginning of the End-Permian mass extinction (Smith 1995; Smith and Ward 2001). The tetrapod communities were severely disrupted by rapid climatic warming probably brought on by volcanic eruptions in northern Pangaea (White 2002). Global warming caused the already sparse ground cover on the Karoo plains to die off and with it the disappearance of herbivores such as *Diictodon, Pristerodon,* and *Oudenodon.* This was followed by the riparian forests and the broad-leafed glossopterid plants, which led to the demise of all the large-bodied dicynodonts and the few remaining gorgonopsians.

Carnivores of the *Dicynodon* AZ fossil assemblage include amphibians (*Rhinesuchus*, *Uranocentrodon*), biarmosuchians (*Burnetia*, *Lemurosaurus*, *Rubidgina*), several gorgonopsians and therocephalians (see Plate H), as well as the cynodont *Procynosuchus*. (Table 2.4, Fig. 2.4).

Dicynodon AZ herbivores are the parareptiles (*Anthodon*, *Pareiasaurus*) and a variety of dicynodonts (see Table 2.4). A notable feature is the marked decrease in the number of dicynodonts as compared to the underlying *Cistecephalus* AZ (Tables 2.3 and 2.4).

Probable insectivores include archosauromorphs (*Saurosternon*, *Youngina*) and parareptiles (*Milleretta*, *Millerosaurus*, *Nanomilleretta*, *Owenetta*, *Spondylolestes*), and probable omnivores include the cynodonts *Cynosaurus* and *Nanictosaurus*.

Various representatives of parareptiles, cynodonts, diapsids, and archosauromorphs occupied all three ecological feeding niches (carnivore, herbivore, and omnivore) so that it is not possible to apply a trophic level to any unidentified members of these taxa. Accordingly these records could not be used in calculating the ecological representation ratios for this biozone. Of the 1,070 identified specimens from this biozone, 837 are herbivorous, 219 are carnivorous, and 14 are omnivorous, meaning that 78 percent of the faunal assemblage of the *Dicynodon* AZ is herbivorous, 20 percent carnivorous, and 2 percent omnivorous, a ratio of 40:10:1. To facilitate comparison with the other biozones, the ratio of herbivores to carnivores without considering the influence of omnivores is 4:1 (Nicolas and Rubidge 2009).

BIOSTRATIGRAPHIC CORRELATION

The *Dicynodon* AZ (Kitching 1995; Smith 1995; Smith and Ward 2001) has faunal correlates with Zambia (Madumabisa Mudstone Formation), Tanzania (Kawinga Formation) (Battail 2000), and the uppermost strata of the Lower Sakamena Formation of Madagascar (Mazin and King 1991; Smith 2000). Dicynodonts, gorgonopsians, and pareiasaurs have been present in southern Africa since the Wordian, but they appear as newcomers in the Laurasian record in the Changhsingian (Battail 2000). Millerettids (a group of basal parareptiles) appear to be endemic to Gondwana, and southern Africa in particular (Modesto 2000).

Lystrosaurus AZ (Induan–Olenekian)

PALEOENVIRONMENT

Documentation of the fluvial facies and paleosols confirms that dry and hot conditions prevailed in the Karoo Basin for a prolonged period (Smith 1995) such that only those synapsids with a pre-adaptation to living under drought conditions were able to survive the Earliest Triassic. In response to the onset of semi-arid conditions, the rivers became braided as a result

Table 2.4. *Dicynodon* AZ generic diversity, abundance, and trophic level

	Taxa	Genus	Total	Trophic level	Genus	Total	Trophic level
Dicynodon AZ							
Amphibia		*Rhinesuchus*	11	Carnivore	*Uranocentrodon*	9	Carnivore
Parareptilia		*Anthodon*	3	Herbivore	*Owenetta*	3	Carnivore
		Milleretta	14	Carnivore	*Pareiasaurus*	10	Herbivore
		Millerosaurus	3	Carnivore	*Spondylolestes*	1	?
		Nanomilleretta	2	Carnivore			
Archosauromorpha		*Saurosternon*	1	Carnivore	*Youngina*	12	Carnivore
Therapsida	Biarmosuchia	*Lemurosaurus*	1	Carnivore	*Rubidgina*	1	Carnivore
	Cynodontia	*Cynosaurus*	10	Omnivore	*Procynosuchus*	34	Carnivore
		Nanictosaurus	4	Omnivore			
	Dicynodontia	*Aulacephalodon*	16	Herbivore	*Kitchinganomodon*	2	Herbivore
		Compsodon	3	Herbivore	*Oudenodon*	146	Herbivore
		Dicynodon	441	Herbivore	*Pelanomodon*	10	Herbivore
		Diictodon	161	Herbivore	*Pristerodon*	23	Herbivore
		Dinanomodon	6	Herbivore	*Rhachiocephalus*	4	Herbivore
		Emydops	10	Herbivore	*Propelanomodon*	1	Herbivore
		Newtonella	1	Herbivore			
	Gorgonopsia	*Aelurognathus*	2	Carnivore	*Dinogorgon*	1	Carnivore
		Aloposaurus	1	Carnivore	*Gorgonops*	1	Carnivore
		Clelandina	1	Carnivore	*Lycaenops*	3	Carnivore
		Aelurosaurus	3	Carnivore	*Prorubidgea*	3	Carnivore
		Cyonosaurus	10	Carnivore	*Rubidgea*	5	Carnivore
	Therocephalia	*Akidnognathus*	1	Carnivore	*Nanictidops*	1	Carnivore
		Cerdops	1	Carnivore	*Notaelurodon*	1	Carnivore
		Cerdosuchoides	1	Carnivore	*Pelictosuchus*	1	Carnivore
		Homodontosaurus	3	Carnivore	*Promoschorhynchus*	5	Carnivore
		Ictidochampsa	1	Carnivore	*Scaloporhinus*	1	Carnivore
		Ictidosuchoides	8	Carnivore	*Scaloposuchus*	1	Carnivore
		Ictidosuchops	9	Carnivore	*Tetracynodon*	2	Carnivore
		Moschorhinus	6	Carnivore	*Theriognathus*	56	Carnivore

of the lack of vegetation which would have stabilized the banks (Ward, Montgomery, and Smith 2000). Drought conditions are also indicated in the fossil record by the occurrence of bonebeds composed of up to seven subadult *Lystrosaurus* carcasses lying together (Smith and Botha 2005) and the prevalence of burrowing behavior and shelter sharing (Abdala, Cisneros, and Smith 2006). Biostratigraphic analysis of five hundred in situ vertebrate fossils in the southern Karoo Basin confirms that out of a total of forty-one therapsid genera in the mid *Dicynodon* AZ times, only three genera survived the End-Permian extinction. These were the medium-sized dicynodont *Lystrosaurus curvatus*, the cynodont ancestor to *Progalesaurus*, and the small therocephalian *Promoschorhynchus* (Smith and Botha 2005; Huttenlocker, Sidor, and Smith 2011). Applying phylogenetic ghost lineage correction to the known ranges adds more taxa to the lineages that must have continued through the End-Permian extinction event including the ancestors of the archosauromorphs *Proterosuchus* and

Prolacerta, the parareptiles *Owenetta* and *Saurópareion*, and the Triassic kannemeyeriids (Botha and Smith 2006).

TROPHIC ANALYSIS

The *Lystrosaurus* assemblage carnivores include several amphibians (see Table 2.5, Fig. 2.4), parareptiles (*Owenetta, Paliguana*), archosauromorphs (*Prolacerta, Proterosuchus*), eosuchians (*Noteosuchus*), as well as, some therocephalians and cynodonts (see Table 2.5). The herbivores are parareptiles (*Procolophon, Sauropareion*) and dicynodonts (*Lystrosaurus, Myosaurus*).

Various representatives of archosauromorphs, eosuchians, diapsids, cynodonts, and parareptiles occupied all three ecological feeding niches (carnivore, herbivore, and omnivore) so that it is not possible to apply a trophic level to any unidentified members of these taxa. Accordingly, these records could not be used in the calculation of ecological representation ratios for this biozone.

Of the 3,446 identifiable specimens from this biozone, 3,098 are herbivorous and 348 are carnivorous; 90 percent of the faunal assemblage of the *Lystrosaurus* AZ is herbivorous and 10 percent carnivorous. The ratio of herbivores to carnivores for this biozone is 9:1. A noticeable feature of the fauna of the *Lystrosaurus* AZ is that while there are many specimens of *Lystrosaurus*, and to a lesser extent *Procolophon*, there were many more small carnivores (therocephalians, cynodonts, and archosaurs) and fewer large carnivores (only *Proterosuchus* and *Moschorhinus*) in comparison to the numerous large carnivores of the underlying *Dicynodon* AZ.

BIOSTRATIGRAPHIC CORRELATION

The *Lystrosaurus* AZ has recently been split into two subzones (Neveling 2002, 2004; Shishkin, Rubidge, and Hancox 1995).

Lower *Lystrosaurus* AZ (Induan 251–249.7 Ma)

The fauna of this zone has been summarized by various authors (Keyser and Smith 1978; Kitching 1977; Groenewald and Kitching 1995) and has received more detailed attention recently (Neveling 2002, 2004, Smith 1995; Smith and Ward 2001). Tetrapod-bearing Induan aged rocks are widely distributed in both Gondwanan and Laurasian continents, being present in Antarctica, Australia, Greenland, India, China, Mongolia, and Russia. *Lystrosaurus* has been considered to have a narrow stratigraphic range and to be restricted to the very base of the Induan (Battail 1993). Neveling (2002, 2004), however, considers the range of this genus to extend from the Changhsingian until well into the Olenekian. The genera *Lystrosaurus* and *Rhytidosteus*, the lydekkerinid and lonchorhynchine amphibians, and the proterosuchid archosaurs have a Pangaea-wide distribution.

Table 2.5. *Lystrosaurus* and *Cynognathus* AZ generic diversity, abundance, and trophic level

	Taxa	Genus	Total	Trophic level	Genus	Total	Trophic level
Lystrosaurus AZ							
Amphibia		Broomistega	1	Carnivore	Micropholis	20	Carnivore
		Eolydekkerina	1	Carnivore	*Pneumatostega*	2	Carnivore
		Lydekkerina	79	Carnivore	*Putterillia*	3	Carnivore
		Micrictodon	1	Carnivore			
Parareptilia		*Owenetta*	5	Carnivore	*Procolophon*	580	Herbivore
		Paliguana	6	Carnivore	*Sauropareion*	2	Herbivore
Archosauromorpha		*Noteosuchus*	7	Carnivore	*Proterosuchus*	24	Carnivore
		Prolacerta	11	Carnivore			
Therapsida	Cynodontia	*Cynosaurus*	8	Carnivore	*Progalesaurus*	1	Carnivore
		Galesaurus	20	Carnivore	*Thrinaxodon*	105	Carnivore
		Platycraniellus	1	Carnivore			
	Dicynodontia	*Lystrosaurus*	2506	Herbivore	*Myosaurus*	10	Herbivore
	Therocephalia	*Ericiolacerta*	1	Carnivore	*Tetracynodon*	2	Carnivore
		Moschorhinus	34	Carnivore	*Regisaurus*	1	Carnivore
		Scaloposaurus	13	Carnivore	*Zorillodontops*	1	Carnivore
Cynognathus Subzone A							
Amphibia		*Bathygnathus*	1	Carnivore	*Parotosuchus*	1	Carnivore
		Kestrosaurus	45	Carnivore	*Trematosuchus*	3	Carnivore
Archosauromorpha		*Garjainia?*	37	Carnivore	*Palacrodon*	6	Omnivore
Therapsida	Cynodontia	*Cynognathus*	1	Carnivore	*Langbergia*	12	Omnivore
	Therocephalia	*Bauria*	1	Herbivore	*Sesamodon*	1	Carnivore
Cynognathus Subzone B							
Amphibia		*Batrachosuchus*	15	Carnivore	*Microposaurus*	3	Carnivore
		Cyclotosaurus	1	Carnivore	*Xenotosuchus*	24	Carnivore
		Jammerbergia	1	Carnivore			
Parareptilia		*Myocephalus*	1	Herbivore	*Thelegnathus*	12	Herbivore
Archosauromorpha		*Erythrosuchus*	44	Carnivore	*Howesia*	3	Carnivore
		Euparkeria	30	Carnivore	*Mesosuchus*	1	Carnivore
Therapsida	Cynodontia	*Cistecynodon*	1	Carnivore	*Lumkuia*	2	Carnivore
		Cynognathus	107	Carnivore	*Trirachodon*	75	Omnivore
		Diademodon	114	Herbivore	*Sysphinctostoma*	1	Carnivore
	Dicynodontia	Kannemeyeria	191	Herbivore	*Kombuisia*	2	Herbivore
	Therocephalia	*Bauria*	12	Herbivore	*Melinodon*	1	Herbivore
		Microgomphodon	2	Herbivore			
Cynognathus Subzone C							
Amphibia		*Paracyclotosaurus*	3	Carnivore			
Archosauromorpha		Unidentified	1	Carnivore			
Therapsida	Cynodontia	*Cricodon*	3	Omnivore	Diademodon	1	Herbivore
		Cynognathus	1	Carnivore	*Trirachodon*	1	Carnivore
	Dicynodontia	*Angonisaurus*	3	Herbivore	*Dicynodont indet.*	2	Herbivore
		Shansiodon	1	Herbivore			

2.5. Pie charts representing tetrapod higher taxon and feeding behavior for the *Cynognathus* A, B, and C subzones and the undifferentiated Stormberg Group (Elliot and Clarens formations) in the Karoo Basin.

Upper *Lystrosaurus* AZ (Olenekian 249.7–245 Ma)

In South Africa, the upper part of the *Lystrosaurus* AZ, that is, the uppermost part of the stratigraphic range of *Lystrosaurus* and the *Procolophon* "Abundance Zone," has produced a trematosaurid mandible (Damiani et al. 2000, Neveling, Rubidge, and Hancox 1999), which is very similar to *Trematosaurus*. This genus is known from the Upper Olenekian (Spathian) Middle Buntsandstein of Germany as well as from the *Parotosuchus* fauna of Russia (Damiani et al. 2000; Shishkin and Welman 1994; Shishkin, Rubidge, and Kitching 1996, Shishkin et al. 2000). This suggests chronostratigraphically that the *Lystrosaurus* AZ extends into the base of the Upper Olenekian (Neveling 2002, 2004).

Cynognathus AZ (Upper Olenekian–Anisian)

PALEOEVIRONMENT

The pervasive maroon colored, clay-rich floodplain fines, a relatively abundant paleocaliche and a meandering river fluvial style are characteristic of the *Cynognathus* AZ strata. These are interpreted as having been influenced by a warm, seasonally wet monsoonal climate and a generally low water table in the Karoo Basin during Mid Triassic times that is similar in many ways to the modern Indo-Gangetic alluvial plain (Hancox 1998, Fig. 6).Thin crevasse splay sands were rapidly deposited by flood events onto the floodplain surface, commonly infilling polygonal mudcracks and rarely the underground burrow complexes of cynodonts (i.e., *Trirachodon;* Groenewald, Welman, and MacEachern 2001) and parareptiles (*Thelegnathus;* Modesto and Damiani 2003). Large-bodied kannemeyeriid dicynodonts grazed on the new *Dicroidium*-dominated flora and carnivorous early archosaur precursors, *Euparkeria* and *Erythrosuchus*, made their first appearance in the basin (see Plate N). Amphibians continued a radiation that began in *Lystrosaurus* times by becoming larger, and in the case of *Xenotosuchus*, gigantic (Damiani 2008).

TROPHIC ANALYSIS

Cynognathus subzone A (Table 2.5, Fig. 2.5) carnivores include amphibians *(Bathygnathus, Kestrosaurus, Parotosuchus, Trematosuchus)*, archosauromorphs (cf. *Garjainia*), an eosuchian (*Palacrodon*), and cynodonts (*Cynognathus*). Herbivores include a cynodont with teeth similar to those of a haramyid (Abdala, Hancox, and Neveling 2005) and therocephalians *Sesamodon* and *Bauria*. The only omnivore is the cynodont *Langbergia*. Of the 108 recorded specimens from this biozone, 89 are carnivorous, 1 is herbivorous, and 18 are omnivorous. An anomaly is the shortage of herbivores, with *Langbergia* and *Palacrodon* included as omnivores. This means that 82 percent are carnivorous, 7 percent omnivorous, and 1 percent herbivorous, giving a herbivore:carnivore:omnivore ratio of 1:80:20.

Higher taxa breakdown

Feeding ratios

Cynognathus Subzone A

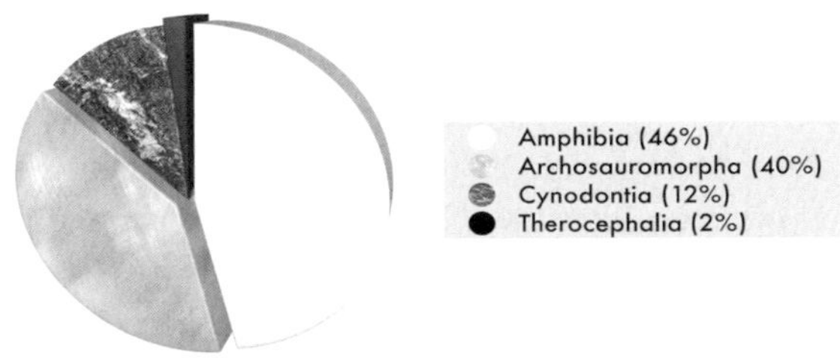

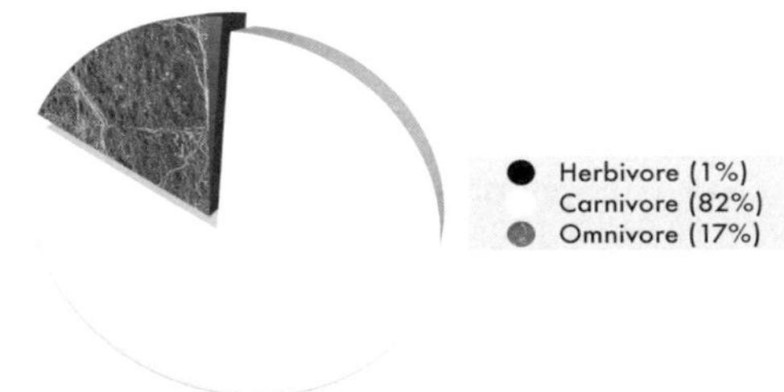

Cynognathus Subzone B

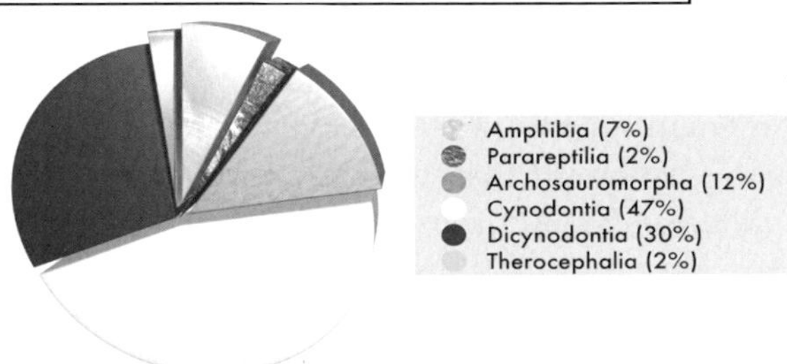

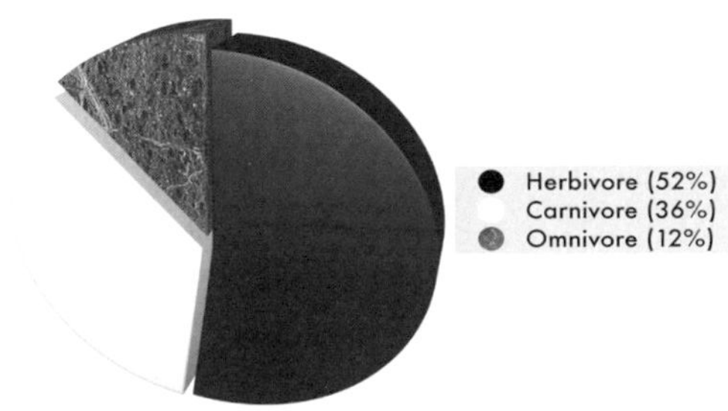

Cynognathus Subzone C

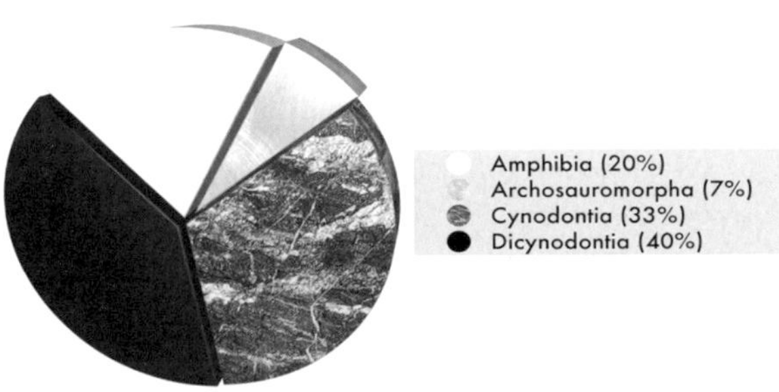

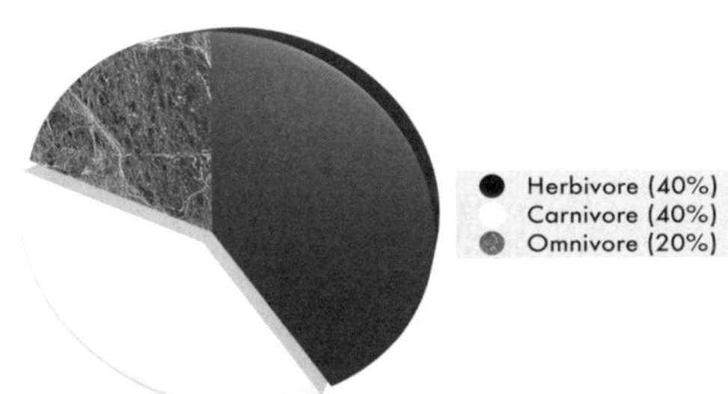

"Stormberg" Group

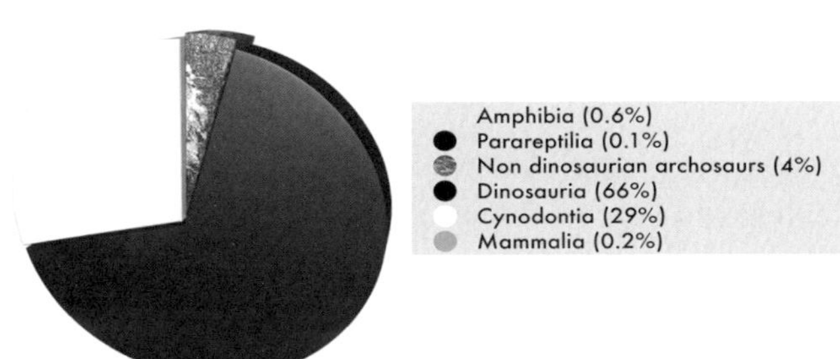

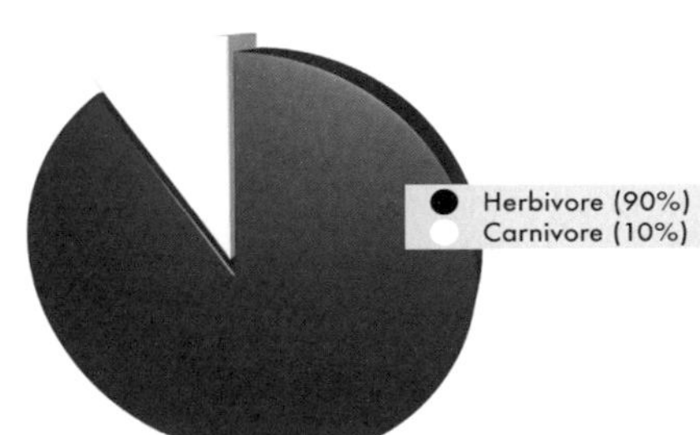

This could possibly mean that many of the carnivores were feeding on fish (P. J. Hancox, pers. comm.).

Cynognathus subzone B (Fig. 2.5) carnivores include amphibians (*Batrachosuchus, Cyclotososaurus, Jammerbergia, Microposaurus, Xenotosuchus*), archosauromorphs (*Erythrosuchus, Euparkeria, Howesia, Mesosuchus*), and cynodonts (*Cistecynodon, Cynognathus, Lumkuia*). Herbivores include dicynodonts (*Kannemeyeria, Kombuisia*), therocephalians (*Bauria, Melinodon, Microgomphodon*), cynodonts (*Diademodon*), and parareptiles (*Thelegnathus, Myocephalus*). The only omnivore is the cynodont *Trirachodon*. Of the 643 specimens recorded from this subzone, 335 are herbivorous, 233 carnivorous, and 75 omnivorous, meaning that 52 percent of the faunal assemblage of the *Cynognathus* subzone B is herbivorous, 36 percent carnivorous, and 12 percent omnivorous. The ratio of herbivores to carnivores to omnivores for this biozone is approximately 5:4:1. The ratio of herbivores to carnivores (without considering the influence of omnivores for comparative purposes) is approximately 1.4:1 and while there were more herbivores than in the underlying Subzone A, it would imply that the carnivores of this subzone were also dependent on fish for their diet.

Cynognathus subzone C (Fig. 2.5) carnivores include the large amphibian *Paracyclotosaurus*, the cynodont *Cynognathus*, and an indeterminate archosauromorph known only from isolated teeth. Herbivores are restricted to the cynodont *Diademodon*, and the dicynodonts *Shansiodon* and *Angonisaurus*. The cynodont *Trirachodon* is the only omnivore. Of the sixteen specimens catalogued from this biozone, seven are herbivorous, six carnivorous, and three omnivorous, meaning that 40 percent of the faunal assemblage of the *Cynognathus* subzone C is herbivorous, 40 percent is carnivorous, and 20 percent are omnivores. The ratio of herbivores to carnivores to omnivores for this biozone is approximately 2:2:1; however, the sample size from this subzone is too small for this analysis to be realistic.

BIOSTRATIGRAPHIC CORRELATION

Lower *Cynognathus* AZ (A subzone) (Olenekian)

Recent research has shown the lowermost subzone A of the *Cynognathus* AZ to be Olenekian, whereas the Middle and Upper parts of the biozone are Anisian (Shishkin, Rubidge, and Hancox 1995; Hancox 1998, 2000). Subzone A yields *Kestrosaurus* and *Trematosuchus*, which are close relatives of *Parotosuchus* and *Trematosaurus* from the Middle Buntsandstein in the Triassic basins of Germany, and the *Parotosuchus* fauna in the Cis-Urals of Russia (Ochev and Shishkin 1989; Shishkin and Ochev 1993; Shishkin et al. 2000). They are all considered Late Olenekian in age based on correlation with ammonite-bearing rocks of the Cis-Caspian depression and western Kazakhastan (Shishkin et al. 2000).

Middle *Cynognathus* Assemblage Zone (B subzone) (Early Anisian 245–>237 Ma)

Subzone B of the *Cynognathus* AZ is characterized by the combined presence of *Xenotosuchus africanus*, *Kannemeyeria simocephalus*, and *Erythrosuchus africanus* (Hancox et al. 1995; Shishkin, Rubidge, and Hancox 1995). *Kannemeyeria* is known from the Omingonde Formation of Namibia, the Lifua Member of the Manda Beds (K8) of Tanzania (Cruickshank 1965; Renaut and Hancox 2001), and a maxillary fragment assigned to the Kannemeyeriidae from the Fremouw Formation of Antarctica (Hammer 1995). If the synonymy of the genus with *Shaanbeikannemeyeria* from China (King 1988; Renaut 2001) and *Uralokannemeyeria* (Renaut 2001) from Russia is correct, then the *Cynognathus* B Zone correlates with parts of the Hessahngou and the Lower Ermaying formations of China and the Donguz Suite of Russia. The presence of *Kannemeyeria* in South America has been refuted (Renaut and Hancox 2001). Other components of the *Cynognathus* B subzone substantiate some of these correlations. *Cynognathus* is known from the Omingonde Formation of Namibia (Keyser 1973), the Fremouw Formation of Antarctica (Hammer 1995), and the Puesto Viejo Formation of Mendoza Province, Argentina (Abdala 1996). *Diademodon* is known from the Manda Beds (K8) of Tanzania and the lower Ntawere Formation of Zambia (Hancox 1988); the diademodontid *Titanogomphodon* is from the Omingonde Formation of Namibia (Keyser 1973); and fragments of a large diademodontid are known from the Fremouw Formation of Antarctica (Hammer 1995). Grine, Hahn, and Gow (1978) consider *Titanogomphodon* to be congeneric with *Diademodon*.

Uppermost *Cynognathus* AZ (subzone C) (Late Anisian)

This assemblage (Hancox 1998; Hancox and Rubidge 1994, 1996, 1997), is based on the presence of the dicynodonts *Angonisaurus* and *Shansiodon* and the possible presence of *Dolichuranus* (Hancox 1988), the cynodont *Cricodon* (Abdala, Hancox, and Neveling 2005), and the amphibian *Paracyclotosaurus* (Hancox et al. 2000). The presence of *Angonisaurus* allows correlation with the Lifua Member (Manda Beds, K8) of Tanzania (Hancox 2000). *Shansiodon* is known from the Upper Ermaying Formation of China and is very similar to the undescribed shansiodontid from the *Cynognathus* C subzone (Hancox 1998). The presence of the shansiodontid *Vinceria* in the Rio Mendoza Formation (Mendoza Province, Argentina), and in the Upper Puesto Viejo Formation (Renaut and Hancox 2001) further suggests a correlation with subzone C (Hancox 1998). *Dolichuranus* is known from the C subzone (Hancox 1998), the Omingonde Formation of Namibia, and the Ntawere of Zambia. Apart from the Karoo Basin, *Paracyclotosaurus* is also known from the Yerrapalli Formation of India, the Wianamata Group of Australia, and the Ntawere Formation of Zambia (Damiani et al. 2000).

"*Euskelosaurus*" Range Zone (Norian–Rhaetian)

The fossil faunas from the Elliot and overlying Clarens formations have been studied by several authors (Haughton 1924; Kitching and Raath 1984; Galton and van Heerden 1985; van Heerden 1979; Yates 2003; Yates and Kitching 2003). These formations have been divided into a lower "*Euskelosaurus*" Range Zone and an upper *Massospondylus* Range Zone (Kitching and Raath 1984). Even though this biozonation scheme is not officially recognized by the South African Committee for Stratigraphy and the genus "*Euskelosaurus*" is now considered nomen nudum (Yates 2003), the *Massospondylus* Range Zone is a valid and workable biostratigraphic unit, and the lower biostratigraphic unit of the Elliot Formation ("*Euskelosaurus* Range Zone") is also a workable biostratigraphic unit but still needs to be named.

PALEOENVIRONMENT

Following a Mid-Triassic uplift event on the eastern margin, a large braided river complex prograded to the center of the basin depositing the vertically stacked tabular bodies of coarse-grained pebbly graywackes of the Molteno Formation (Plate 3). This was succeeded by a period in the Late Triassic when the central Karoo plains were again dominated by aggrading floodplains. By now, the mean annual temperatures had increased and the diminishing synapsid populations were being superseded by dinosaurs. Long-necked sauropodomorphs such as *Antetonitrus* (Yates and Kitching 2003) and *Aardonyx* (Yates et al. 2009) browsed on the tree ferns and cycads in early Elliot Formation times and were succeeded by *Massospondylus* and the small theropod *Coelophysis* (see Plate P) (Bristowe and Raath 2004). Large rauisuchians were the apical predators of the time and rare small turkey-sized ornithischian dinosaurs (Butler, Smith, and Norman 2007) competed with the derived cynodont *Tritylodon* for insects, hatchlings, and other small prey.

An important statistic to point out is that of the 625 specimens collected from the Stormberg Group, only 174 are therapsids and 3 mammals. This means that therapsids compose only 27 percent of the tetrapod fauna of the Elliot and Clarens formations, compared with 95 percent and 85 percent for the *Lystrosaurus* and *Cynognathus* assemblage zones, respectively.

TROPHIC ANALYSIS

The taxonomy of the sauropodomorph dinosaur faunas from the Elliot and Clarens formations has not yet been fully resolved and many museum specimens, which are largely unprepared, have been identified only as an indeterminate sauropodomorph. In addition, because the proposed biozonation is relatively recent (Kitching and Raath 1984), very few of the fossils in museum collections have been recorded with biozone information. As a result, this trophic analysis is for the entire fauna of the Elliot

Table 2.6. Elliot Formation generic diversity, abundance, and trophic level

	Taxa	Genus	Total	Trophic level	Genus	Total	Trophic level
Amphibia		*Chigutisaurus*	3	Carnivore	*Siderops*	1	Carnivore
Parareptilia	Chelonia	*Australochelys*	1	Herbivore			
Reptilia	Nondinosaurian archosaurs	*Aetosaurus*	2	Carnivore			
		Basutodon	2	Carnivore	*Orthosuchus*	2	Carnivore
		Baroqueosuchus	2	Carnivore	*Pedeticosaurus*	1	Carnivore
		Clarensia	1	Carnivore	*Protosuchus*	5	Carnivore
		Litargosuchus	1	Carnivore	*Sphenosuchus*	1	Carnivore
		Notochampsa	6	Carnivore			
	Dinosauria	Therapoda	11	Carnivore			
		Sauropodomorpha	376	Herbivore			
		Ornithischia	18	Herbivore			
Therapsida	Cynodontia	*Diarthrognathus*	5	Carnivore	*Scalenodontoides*	3	Herbivore
		Diademodontidae	1	Herbivore	*Trithelodon*	2	Carnivore
		Elliotherium	1	Carnivore	*Tritylodon*	150	Herbivore
		Pachygenelus	15	Carnivore			
Mammaliaformes	Mammalia	*Erythrotherium*	1	Carnivore	*Megazostrodon*	2	Carnivore

and Clarens formations. We accept that this is not ideal but is the best that can be achieved at the moment considering the lack of stratigraphic data on the collections database.

Carnivores include amphibians (*Chigutisaurus*, *Siderops*), non-dinosaurian archosaurs (*Aetosaurus*, *Basutodon*, *Baroqueosuchus*, *Clarencia*, *Litargosuchus*, *Notochampsa*, *Orthosuchus*, *Pedeticosaurus*, *Protosuchus*, *Sphenosuchus*), theropod dinosaurs, and several small possibly insectivorous therapsids, which include cynodonts (*Diarthrognathus*, *Elliotherium*, *Pachygenelus*, *Trithelodon*) and mammaliaformes (*Erythrotherium*, *Megazostrodon*). Herbivores include a chelonian (*Australochelys*), a diverse fauna of sauropodomorph and ornithischian dinosaurs, and cynodont therapsids (*Scalenodontoides*, *Tritylodon*) (Table 2.6, Fig. 2.5).

Of the 613 specimens recorded from the Elliot and Clarens formations, 549 were herbivorous and 64 were carnivorous, meaning that 90 percent of the faunal assemblage is herbivorous and 10 percent carnivorous. The ratio of herbivores to carnivores for this biozone is 9:1.

BIOSTRATIGRAPHIC CORRELATION

Ongoing taxonomic revision of the sauropodomorph dinosaurs from the lower Elliot Formation has recognized the presence of the sauropodomorphs *Antetonitrus* (Yates and Kitching 2003), *Blikanasaurus*, *Melanorosaurus*, *Plateosaurus*, and *Aardonyx*, which is closely related to *Riojasaurus incertus* from the Los Colorados Formation of Argentina (Yates 2003; Yates et al. 2009). The basal sauropod *Antetonitrus* from the Lower Elliot Formation is very similar to *Lessemsaurus* also from the Los Colorados Formation (Yates 2003).

Based on the similarity of the lower Elliot fauna to that of the German Knollenmergel and the Los Colorados Formation of Argentina, the lower Elliot ("*Euskelosaurus*" Range Zone) is considered Norian (Hancox 2000). The contact between the "*Euskelosaurus*" and *Massospondylus* range zones, which is marked by the *Tritylodon* Acme Zone (Kitching and Raath 1984), is a reworked paleosol horizon and subaerial unconformity (Smith and Kitching 1997; Bordy, Hancox, and Rubidge 2004) and is considered to be the Triassic-Jurassic boundary (Olson and Galton 1984) although this has recently been disputed (Smith, Marsicano, and Wilson 2009).

Massospondylus Range Zone (Hettangian, Sinnemurian, and Pliensbachian)

PALEOENVIRONMENT

The Triassic-Jurassic boundary in the Karoo succession is marked by a distinctive change in the color of the floodplain fines from dark reddish brown in the Lower Elliot to light red in the Upper Elliot Formation (Smith and Kitching 1997; Bordy, Hancox, and Rubidge 2004). Thick beds of massive siltstone with articulated *Massospondylus* skeletons and, in places, clutches of dinosaur eggs with embryos (Reisz et al. 2005) characterize the Upper Elliot sediments (see Plate P). These massive silts are interpreted as loessites deposited on a barren floodplain under semi-arid climatic conditions by strong westerly winds (Smith and Kitching 1997). Near the upper contact with the aeolian Clarens Formation, dark gray laminated claystones with abundant conchostrachan tests are typical of sediments laid down in ephemeral playa lakes (Kitching and Raath 1984). The dominant herbivore in the basin at this time was *Massospondylus* (see Plate P)—a bipedal, long-necked, long-tailed prosauropod. The therapsids were almost extinct, reduced to tiny shrew-like probably nocturnal animals such as *Pachygenelus*, and the earliest mammaliaforms had already made their appearance (*Megazostrodon* and *Erythrotherium*, and a number of similar forms from China [Luo, Crompton, and Sun 2001]).

BIOSTRATIGRAPHIC CORRELATION

The fossils of the upper Elliot and Clarens formations constitute the fauna of the Early Jurassic *Massospondylus* AZ. The fauna consists of large chigutisaurid amphibians (Warren and Damiani 1999), an australochelyid turtle *Australochelys* (Gaffney and Kitching 1994), a variety of dinosaurs including the prosauropods *Aardonyx* and *Massospondylus* (Yates et al. 2009), a basal theropod *Coelophysis* (see Plate P) (Bristowe and Raath 2004, Raath 1977), and a variety of basal ornithischians including *Lesothosaurus* and *Heterodontosaurus*. There are also crocodilomorphs *Sphenosuchus, Pedeticosaurus, Protosuchus* (Gow 2000a; Gow and Kitching 1988), *Litargosuchus* (Clark and Sues 2002), *Baroqueosuchus* (Busbey and Gow 1984), *Notochampsa*, and *Orthosuchus* (Nash

1968, 1975); cynodonts *Tritylodon*, *Pachygenelus* and *Trithelodon*; and mammaliaforms *Megazostrodon* and *Erythrotherium* (Kitching and Raath 1984; Smith and Kitching 1997).

Although there are few South African Jurassic genera which are common to other Pangaean sequences, there are broad-scale similarities at the family level. Correlatable faunas of this age are known from Australia (Upper Evergreen Formation) (Schoch and Milner 2000; Warren and Huchinson 1983), India (Upper Maleri Formation, Pranhita-Godavari Basin [Bandyopadhyay, Roy Chowdhury, and Sengupta 2002; Sengupta 1995] and Dharmaram Formation, Pranhita-Godavari Valley [Kutty, Jain, and Roy Chowdhury 1987]), Namibia (Etjo Formation, Waterberg Basin) (Hölzforster, Stollhofen, and Stanistreet1999), Zimbabwe (Forest Sandstone) (Bristowe 2004; Cooper 1981; Raath 1969, 1977),the United Kingdom (British fissure fills) (Clemens 1986; Evans and Kermack 1997), Canada (McCoy Brook Formation) (Shubin, Olsen, and Sues 1997; Sues, Clark, and Jenkins 1997), China (Lower Lufeng Formation) (Crompton and Luo 1993; Luo and Wu 1997), Germany (Rhaeto-Liassic) (Sues 1985b), Mexico (La Boca Formation, Huizachal Canyon) (Clark et al. 1997), and the United States (Wingate, Moenave, Kayenta, and Navajo formations of the Glen Canyon Group) (Rowe 1989; Sues, Clark, and Jenkins 1997).

Causes of Faunal Change in the Karoo Basin

The biostratigraphic zonation of the Karoo Supergroup is based on 150 years of collecting an estimated 30,000 tetrapod fossils of which at least 12,277 are therapsids (classified to generic level). Many years of taxonomic descriptions, character revisions, and the more recent cladistic analyses have improved the certainty of identifications. The advent of seismic surveys and satellite imagery brought about a similar revision and refinement of the lithostratigraphic subdivisions such that we are now able to link rapid faunal changes with lithostratigraphic events in a way that has not previously been possible. Within the Karoo biostratigraphic scheme there are some noteworthy faunal changeovers that must have been caused by environmental disturbances, the nature of which need to be probed further.

Extinction of Dinocephalians

Dinocephalians, the first tetrapods to have attained large body sizes and the most successful terrestrial tetrapods in Pangaea during the Middle Permian, are represented by a taxonomically diverse fauna of herbivorous and carnivorous forms from the *Tapinocephalus* AZ, and yet none survived into the succeeding *Pristerognathus* zone (Boonstra 1969; Keyser and Smith 1978). At this stage, it is not possible to ascertain whether all the genera went extinct at the same time or gradually over time. The proliferation of herbivorous dinocephalians in the Karoo Basin is attributed to their large body mass and their unique tooth morphology designed for both slicing and crushing tough foliage. Trackway surfaces on point bar sandstones indicate that dinocephalians lived in herds and, judging by the dietary requirements of herbivorous forms, their preferred habitat would have been the riverine forests that lined the banks of large meandering river channels. Sedimentological studies show an abrupt regional change in fluvial style from meandering to braided at the base of the Poortjie Sandstone Member which coincides with the disappearance of dinocephalians but more research is necessary to understand the nature of the faunal turnover at this time.

Dicynodont Turnover in the *Tropidostoma*, *Cistecephalus*, and *Dicynodon* Assemblage Zones

The specimen count for *Tropidostoma* zone herbivores reveals the dominance of small, long-bodied dicynodonts, mainly *Diictodon*, *Pristerodon*, and *Emydops* with only a few medium-sized forms (*Tropidostoma* and *Oudenodon*) and even fewer large-bodied taxa (*Endothiodon* and *Rhachiocephalus*). This changes significantly in the succeeding *Cistecephalus* zone where the medium-sized *Oudenodon* and large-bodied *Aulacephalodon* and *Platycyclops* become more common. By *Dicynodon* zone times, the large-bodied forms *Dicynodon*, *Pelanomodon*, and *Dinanomodon* have become the dominant herbivores to the detriment of the small-bodied *Diictodon* and *Emydops*, which have become rare elements of the fauna (Nicolas 2007). Interestingly, this transition is accompanied by a similar trend in the carnivores whereby the smaller gorgonopsian genera *Cyonosaurus*, *Aelurognathus*, and *Lycaenops* are outnumbered by the much larger *Dinogorgon*, *Prorubidgea*, *Arctops*, and *Rubidgea* in the *Dicynodon* AZ. This relationship suggests that the fossil record is recording a real ecological turnover rather than a sampling or taphonomic effect.

The sedimentological record shows nothing out of the ordinary in terms of fluvial behavior, and therefore no extraordinary extrabasinal factors were involved. It was noted earlier that paleopedogenic depth to calcic horizons in the *Cistecephalus* zone floodplains show that the floodplain water tables were lower than the preceding and succeeding zones, which would have had a detrimental effect on the shallow rooting groundcover plants such as ferns and club mosses. This is possibly the reason for the steady decline in small ground surface grazers and the increase in abundance of burrowing grubbers and tuber feeders such as *Cistecephalus*. By *Dicynodon* zone times, the water tables were once again high and groundcover returned to the floodplains; however, this time, *Diictodon* did not respond as it had done in the past, remaining quite rare as the large-bodied dicynodonts flourished.

Therapsid Survivorship and the End-Permian Mass Extinction

The End-Permian mass extinction that occurred at approximately 251 Ma (Bowring et al. 1998) is generally regarded as the most catastrophic of the five major Phanerozoic mass extinctions. An estimated 90 percent of all marine species (Erwin 1994) and 70 percent of all terrestrial vertebrate families (Maxwell and Benton 1990) disappeared within 165 ky (Bowring et al. 1998) and perhaps as little as 10 ky (Ward et al. 2005)

The Karoo Basin has a well-preserved stratigraphic record (see Plate I) of this global phenomenon, and intensive collecting in the boundary sections has highlighted an extinction event among the tetrapods, especially the therapsids (Smith 1995; Macleod et al. 2000; Gastaldo et al. 2009). Paleomagnetic/magnetostratigraphic and carbon-isotope data allow sections to be correlated across the basin. With this stratigraphy, the vertebrate fossil data show a gradual extinction in the Upper Permian punctuated by an enhanced extinction pulse at the Permian-Triassic boundary interval, particularly among the dicynodont therapsids (Ward et al. 2005). Field data demonstrate a 69 percent mass extinction of Late Permian terrestrial vertebrate genera lasting approximately 300 ky and terminating at the Permian-Triassic boundary, followed by a lesser extinction pulse (31 percent) involving all four surviving genera, approximately 160 ky later in the Early Triassic (Smith and Botha 2005). The extinction event coincides with a general rubification of floodplain mudrocks and a change in channel sandstone geometries and internal architecture interpreted as a switch in fluvial style from meandering to braided (Smith 1995). Within as little as 100,000 years, rapid global warming caused the wet floodplains of the Permian to dry up. This in turn caused the *Glossopteris* trees and bushes to die off, leaving tetrapods without an adequate food base. In the Karoo Basin, this resulted in an ecologically "stepped" extinction of small herbivorous dicynodonts and their gorgonopsian predators before the main pulse of extinction that involved medium and large herbivores and carnivores. This pattern suggests that the smaller ground foraging animals such as *Diictodon* and *Pristerodon*, feeding on the undergrowth of ferns and club mosses, disappeared before larger browsing fauna such as *Dicynodon*, *Dinanomodon*, and *Lystrosaurus maccaigi* feeding on *Glossopteris* shrubs and trees. This is in keeping with the interpreted onset of drought conditions and plant die-off (Angielczyk, Roopnarine, and Wang 2005). The fact that the medium- and large-sized *Lystrosaurus* arrived in the basin and seemingly flourished as the *Dicynodon* fauna began to wane indicates that it was somehow preadapted to survive the worsening drought conditions. The characteristic spade-shaped premaxilla with shock-resistant sutures (e.g., Jasinoski, Rayfield, and Chinsamy 2010b) was most likely the key to the survival of *Lystrosaurus* (Botha and Smith 2006). It had the ability to continue grazing throughout the drought, probably on the tough equisetalians that remained growing in the moist areas around ponds and along the increasingly ephemeral floodplain channels.

Small cynodonts and therocephalians, being possibly insectivorous or omnivorous, managed to survive the drought and combat the dry conditions by burrowing underground. When conditions returned to normal and the monsoonal type rains became more reliable, the vegetation grew back and the few surviving herbivores such as *Lystrosaurus* found that they had plenty of food and few predators. They grew in number and species diversity and slowly began to colonize all the lowland areas of Pangaea. In the early Triassic, the small insectivorous cynodonts and therocephalians quickly recovered, finding favorable niches in which to settle, and they too rapidly evolved into a number of different species.

From facies interpretation and taphonomic data, it is proposed that pronounced climatic warming and increased seasonality and storminess at the onset of an unreliable, monsoonal rainfall regime contributed to the mass extinction of terrestrial vertebrates in southern Gondwana. This is congruent with previous work which has suggested that the warming was as a result of greenhouse gases (methane and carbon dioxide) emitted from basaltic eruptions in northern Pangaea around 251 Ma combined with a latent instability in the Earth's cyclic systems brought about by the coalescence of Pangaea.

Early Triassic Therapsid Radiation and Competition from Archosauromorphs and Amphibians

The Early Triassic recovery fauna in the Karoo Basin comprises proterosuchian archosauromorphs (*Proterosuchus*), small amphibians (*Micropholis, Lydekkerina* [see Plate K]), small procolophonoids (*Owenetta kitchingorum, Procolophon*), medium-sized dicynodonts (*Lystrosaurus murrayi* [see Plate K] *and L. declivis*), and small faunivorous cynodonts (*Progalesaurus, Galesaurus* [see Plate M, top], *Thrinaxodon* [see Plates J and M, bottom]). Ghost lineage correction data applied to the phylogenies of these new Early Triassic taxa still result in an extinction event, but of lesser impact at 39 percent of genera rather than 69 percent (Botha and Smith 2006)

There is a taphonomic bias toward preferential preservation of drought accumulations in the Early Triassic Karoo Basin which may have overemphasized the abundance and the diversity of semi-aquatic and burrowing animals and underrepresented the fauna that lived independently of water bodies. Taphonomic observations of cynodont fossils suggest that they either excavated their own burrows or were habitual occupiers of underground burrows excavated by other vertebrates—this may have been a behavior that allowed them to cope with the periods of drought in the monsoonal climate of the Earliest Triassic Karoo Basin (Smith and Botha 2005).

One of the mysteries of the Karoo biostratigraphy is: "Where did the Earliest Triassic archosauromorphs come from?" Within 5 m of the top of the Permian-Triassic boundary beds, the primitive archosauromorph *Proterosuchus* makes an appearance in relatively large numbers. Without any predecessors in the Karoo Basin, and the only Late Permian candidate being *Archosaurus* from Russia (Tatarinov 1960), trans-Pangaean migration becomes the most plausible explanation to date. Up until the Permian-Triassic boundary, therapsids were the dominant group in every terrestrial niche; after the mass extinction event, they were never to regain their former dominance. Parareptiles, amphibians, and archosauromorphs recovered diversity and numbers equally as fast as the therapsids, and even long after the ecosystem had stabilized again, these opportunistic species never relinquished the advantage they gained at this time.

As discussed in the *Dicynodon* AZ section, before the extinction, there were at least five large-bodied herbivorous dicynodont taxa (*Dinanomodon, Pelanomodon, Dicynodon, Daptocephalus, L. maccaigi*), one medium-sized (*Oudenodon*), and two small (*Diictodon, Myosaurus*), but afterward, there were only one large (*L. declivus*), one medium-sized (*L. murrayi*), and one small taxon (*Myosaurus*). The monsoonal rainfall regime and drought-prone climate prevailed long enough for those animals that were preadapted to eating drought-resistant plants (*Lystrosaurus*) and those taxa capable of living or sheltering underground (*Galesaurus, Progalesaurus, Thrinaxodon, Procolophon,* and *Lystrosaurus*) to gain a selective advantage from the new environmental conditions. There were no therapsids among the immigrant taxa, which included *Proterosuchus, Prolacerta* (see Plate J), and *Paliguana,* and it is interesting to note that they, along with the new amphibians *Lydekkerina* (see Plate K), *Micropholis,* and *Rhytidosteus,* were all nonherbivores. Thus, it remains to be resolved why it took so long for large-bodied dicynodonts to reappear in the Karoo Basin. It is possible that the loss of glossopterid trees and bushes at the Permian-Triassic boundary left insufficient vegetation to sustain herds of large-bodied herbivores until the new *Dicroidium* flora was fully established in the Mid-Triassic. At this time, the new large-bodied dicynodont *Kannemeyeria* (see Plate N) migrated into the Karoo lowlands—possibly from the African rift valleys to the north.

Concluding Remarks

Although therapsid faunas are known from all the continents, the Karoo Basin from South Africa has the most geographically and time-extensive record. This has resulted in large collections being amassed in museum collections in the country. In the distant past, most paleontological research on therapsids from the Karoo centered on taxonomic and morphological aspects. Beginning in the 1990s, research relating to therapsids has incorporated geological, stratigraphic, tectonic, taphonomic, and paleoenviornmental aspects together with the morphological, taxonomic, and phylogenetic research and has resulted in a much broader understanding of changing therapsid communities from the Permian to the Jurassic. The recent development of the Beaufort Group GIS database (Nicolas and Rubidge 2009) has broadened research horizons by providing absolute numbers of individual genera which have been collected from the Karoo and has enabled speculation on changing biomass through the Karoo succession. This data has been presented here for the first time. These studies are still at an early stage but further refinement of the database by applying the most recent taxonomic advancements to all

South African Karoo fossil collections will greatly enhance understanding of continental biodiversity changes from the Middle Permian to the Jurassic, especially in response to major extinction events.

Acknowledgments

Setting up the Karoo fossil geographic information system database has been a long-term project which would not have been possible without the co-operation and enthusiasm of the curators of all the Karoo fossil collections in South Africa. We record our gratitude to Sheena Kaal (Iziko South African Museum), Billy de Klerk (Albany Museum), Robert and Marion Rubidge (Rubidge Collection), Elize Butler and Jennifer Botha-Brink (National Museum), Bernhard Zipfel and Mike Raath (BPI Palaeontology), Heidi Fourie (Transvaal Museum), and Johann Neveling and Ellen de Kock (Council for Geoscience). We are greatly indebted to Michael Day for assistance in setting up the tables; Lyn Whitfield and Diane de Toit for drafting the figures; Tom Kemp, Ken Angielczyk, and Anusuya Chinsamy-Turan for constructive and helpful reviews; and to Fernando Abdala, Ken Angielczyk, Adam Huttenlocker, Saniye Guven, and Luke Norton for updating us on the taxonomic status of different therapsid genera. Gratitude is recorded to the Palaeontological Scientific Trust, the Department of Science and Technology, and the National Research Foundation for providing funding.

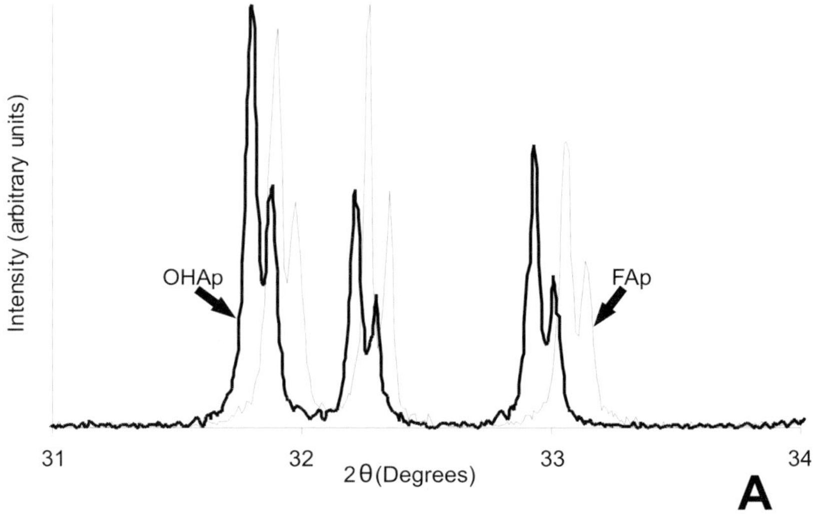

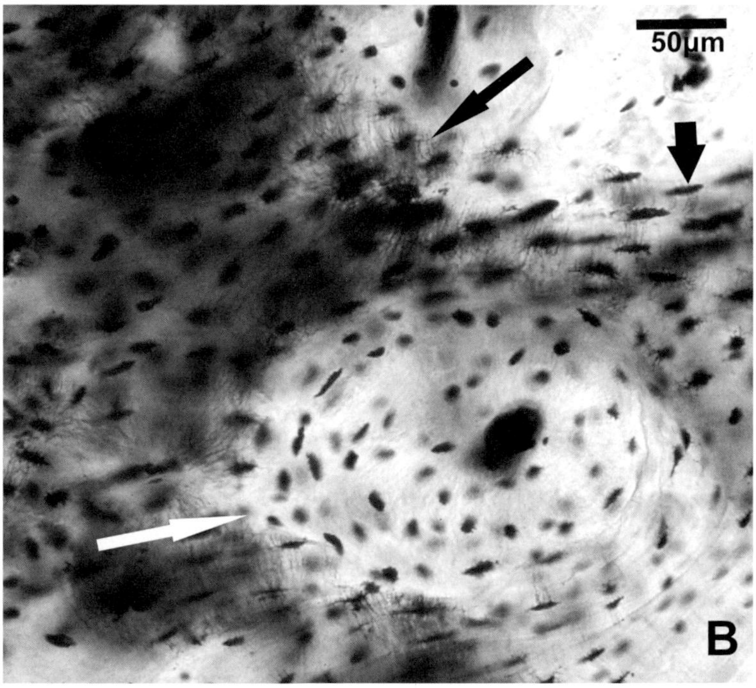

3.1. (A) X-ray diffraction spectra of hydroxyapatite (OHAp) and fluorapatite (FAp). Fluoridation of the apatite lattice (replacement of hydroxyl ions by fluoride ions) alters the mineral structure, which is detected as an increase in the 2θ position of key bands in an X-ray diffraction spectrum. (B) A thin section of a modern bone. The white arrow points to a ring of bone (the osteon) around a central dark region, which is the channel in which the blood vessels, nerves, and other connective tissue are located. The short black arrow indicates an osteocyte lacuna wherein the osteocyte (bone cell) occurs. The long black arrow points to the fine network of canaliculi that allows communication between neighboring osteocytes.

The Microstructure of Bones and Teeth of Nonmammalian Therapsids

3

Anusuya Chinsamy-Turan

Introduction

Bone is a specialized connective tissue. Except for extant jawless fishes (lampreys) and to some extent chondrichthyes (sharks), it composes the skeleton of all vertebrates. It is a composite material with an organic phase and an inorganic mineral phase consisting of carbonate hydroxyl apatite—$Ca_{10}(PO_4\ Ca_3)_6(OH)_2$. However, relatively soon after an animal dies, and becomes buried by sediment, the hydroxyl group is displaced from the bone mineral and is commonly replaced by fluorine to form carbonate fluorapatite. Thus, X-ray diffraction analysis of fossil bone often has a slightly different mineral spectrum than living bone (Fig. 3.1A).

The organic phase of bone generally comprises different types of cells (osteoblasts, the bone forming cells; osteocytes, the bone cells; and osteoclasts, the bone resorbing cells), an extracellular matrix consisting mainly of Type 1 collagen, as well as other noncollagenous proteins such as osteocalcin, osteopontin, and osteonectin; polysaccharides; and a network of blood vessels and nerves (Currey 2002).

In some vertebrates, namely, the teleost fishes and the heterostracans (the armored jawless fishes from the Ordovician) acellular bones occur. In all other vertebrates, bone cells (osteocytes) are located within small spaces in the bone matrix called lacunae (Fig. 3.1B). Neighboring osteocytes communicate with each other via cytoplasmic extensions which are located within extensions of the lacunae called canaliculi (Fig. 3.1B). During bone formation, blood vessels can be entrapped within the bone matrix as simple blood vessels, or, if the bone matrix is rapidly deposited, spaces are left around them (Fig. 3.2A). Subsequently, these channels are infilled centripetally with a more slowly formed bone tissue to form a bony ring (an osteon) around the blood vessel (Figs. 3.1B, 3.2B).

Bone formation begins with the deposition of the osteoid (mainly composed of collagen), which later becomes impregnated with hydroxyapatite to form bone. Thus, the collagen fibers and the tiny bone mineral crystals (500 Å long, 280 Å wide, and 20 Å thick), the organic and inorganic phases of bone, respectively, are intimately associated during life (Weiner and Traub 1992). When an animal dies, the organic components soon decompose, but even after millions of years, the more resilient inorganic (mineral) part of bone often undergoes only minor changes and reflects the microstructure of the bone tissue (Chinsamy-Turan 2005, and references therein).

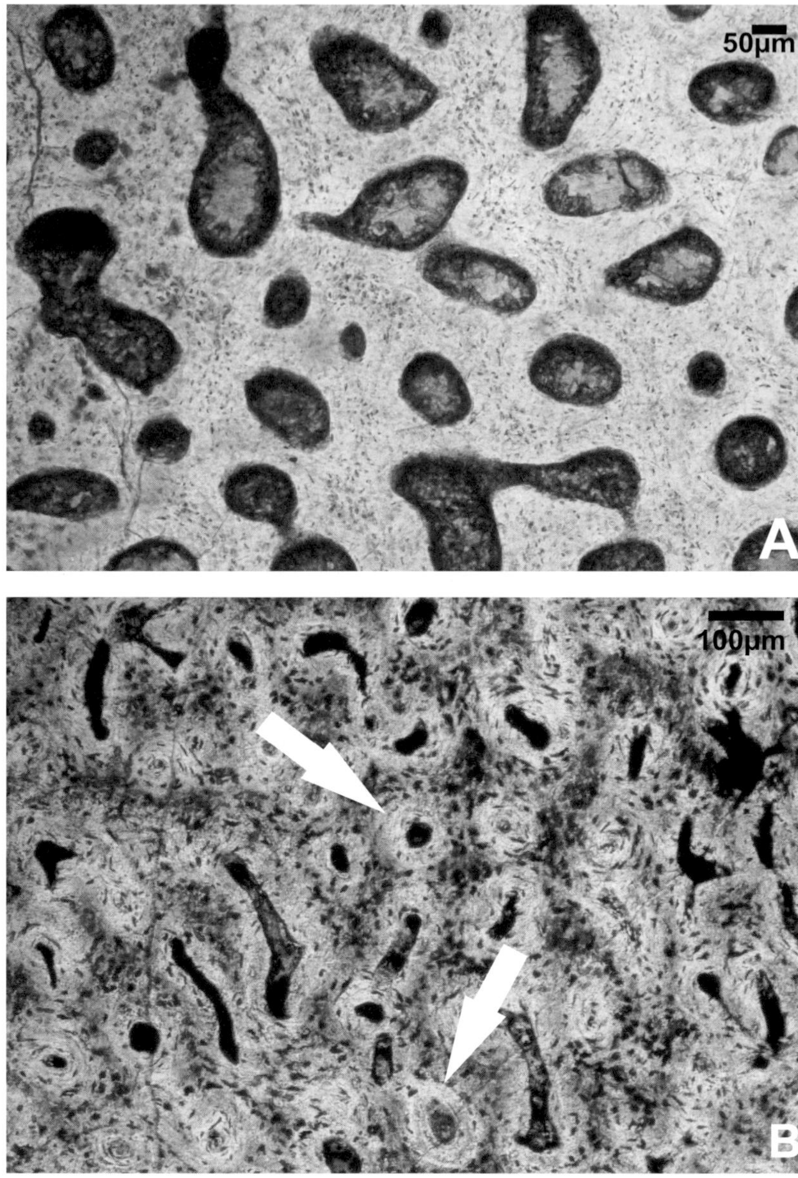

3.2. Fibrolamellar bone. (A) *Dicynodon* (SAM-PK-5576). Early stage of fibrolamellar bone formation showing the spaces that will later become infilled with a lamellar bone tissue. (B) Primary osteons (white arrows) in the fibrolamellar bone tissue in *Tropidostoma* (SAM-PK-K990).

The microstructure of fossil bones has proven to be a powerful tool to decipher biological signals recorded therein. The same basic components of bone exist across all classes of vertebrates—they are simply organized in different ways to form different types of bone tissues (e.g., Ricqlés et al. 1991). Research on bone has shown that its microstructure largely reflects the rate at which the bone was formed (Amprino 1947) and that this is affected by a number of factors, such as ontogeny, phylogeny, pathology (e.g., Ricqlés et al. 1991; Chinsamy-Turan 2005; Erickson 2005; Erickson, Makovicky et al. 2009).

This chapter serves as an atlas of bone tissues observed in nonmammalian therapsids. It also discusses some of the biological inferences that can be deduced from the occurrence of particular histological features.

Since the first step in any study of fossil bone microstructure is the preparation of thin sections (75–100 μm in thickness), this procedure is outlined.

Preparing Thin Sections of Fossil Bones

Fossil bones are generally rock-hard and indeed they are often petrographically thin sectioned in a similar way to rock samples (Chinsamy and Raath 1992). However, since fossil bones are generally brittle, the first step in the thin-sectioning process is to embed the bone in a polyester resin. The resin surrounds the bone and penetrates all the natural cavities and/or cracks. It is usually best to do this under vacuum for more porous bones, but most other bones can simply be left in the resin until the latter hardens. Once embedded, the specimen can be sectioned to remove the part of the bone that is of interest. From here onward, the process can vary depending on the kind of equipment available. Most modern petrographic equipment is fairly automated, and thin sections can be produced relatively quickly. If no special equipment is available, the specimen can be "polished" simply by using a frosted glass plate and different grits of carborundum powder (Chinsamy and Raath 1992). However, most geology departments at academic institutions will generally have some basic petrographic equipment that can be used for processing thin sections of fossil bone, and the methodology using such equipment is described. The sectioned piece of embedded bone is polished, that is, held facedown on a lap wheel using successively finer grinding paper (e.g., 400, 600, 800, and 1,000 grit). This is then followed by a final polish on a lap wheel with a velvet cloth and a solution of aluminium oxide. Once the surface of the specimen is polished to a high gloss, the polished surface is mounted onto a frosted, pre-labeled glass slide and left to cure. Thereafter, the slide, with the specimen firmly attached to it, is mounted on a vacuum chuck and sectioned using a low speed precision saw with a diamond-encrusted blade. The cut surface (now mounted on the slide) is then polished (as before) on progressively finer grit grinding paper and then on a velvet cloth, until all traces of the scratch marks are eradicated. The slide is now ready to view under a microscope—ideally a microscope with polarizing ability. The next important step is to be able to identify the bone tissues observed.

Identifying Different Bone Tissues Types: The "How and Where" of Bone

Vertebrate skeletons often have a combination of bone tissues that depend on a host of factors, such as ontogenetic age, location within the skeleton, and the actual local environment (e.g., Ricqlés et al. 1991; Chinsamy-Turan 2005; chapter 5 this book). However, beginners in studies of bone microstructure are often faced with a complex array of terms, which is further complicated by differing usages in the published literature, and by the fact that in reality tissue types are seldom discreet and often grade into one another even in a single thin section. The following is an attempt to enable a novice in bone histology to identify histological features observed in fossilized bones. Although most of the bone tissues described occur in both extinct and extant vertebrates, the focus here is on tissues documented in a variety of nonmammalian therapsids.

"How" bone is produced determines the type of bone tissue that results: If bone is produced by osteoblasts, it is referred to as osteoblastic bone, whereas if bone is produced by other types of cells such as chondrocytes or fibrocytes, it is termed metaplastic bone, which is often formed within cartilage or within tendons (Reid 1996). Most bone formed during growth is osteoblastic bone. However, the first bone formed during the early stages of development is endochondral bone, which involves the replacement of cartilaginous precursors. Later during ontogeny, endochondral bone is formed at the epiphysis, the growing ends of bones, where it forms a spongy or cancellous bone with a network of bony struts called trabeculae, which is often referred to as trabecular bone.

The origin of the newly formed bone is significant: If it is formed under the periosteum (i.e., the layer of bone-lining cells covering the external surface of bone), it is referred to as periosteal bone, whereas if it is formed under the endosteum (i.e., the layer of bone-lining cells covering the internal surface of bone), it is called endosteal bone. The bone-lining cells give rise to osteoblasts that form the osteoid. During early stages of growth, the bone that is formed is generally osteoblastic, periosteal bone, with some endosteally formed bone, and these are termed primary bone. During development and further growth, primary bone is often resorbed and replaced by secondary deposits of bone.

Organization of the Collagen Fibers

The arrangement of bundled collagen fibers in bone is referred to as the fibrillar organization (e.g., Francillon-Vieillot et al. 1990; Reid 1996). Different types of bone tissues are characterized by different arrangements of the collagen fibers, although it should be noted that in reality such categorizations are simplistic, since the organization of the fibers can be quite variable and the categories often merge into each other (e.g., Francillon-Vieillot et al. 1990; Reid 1996).

Fibrolamellar Bone

Fibrolamellar bone is characterized by a woven bone matrix wherein the collagen fibers are coarse and disorganized. Characteristically, in this type of tissue, numerous rounded or globular-shaped osteocytes are randomly distributed within the bone matrix (Plate 4; Figs. 3.1B, 3.2A, 3.2B). During the deposition of this tissue, blood vessels are entrapped within spaces in the growing bone (Fig. 3.2A). Later the spaces around the blood vessels become infilled by the inward deposition of a more slowly formed lamellar bone tissue (see next section) to form primary osteons (Plate 4, 5C, 5D; Fig. 3.2B). The arrangement of the osteons is variable and has led to the classification of different types of fibrolamellar bone tissues: laminar bone tissue, plexiform bone tissue, reticular bone tissue

(Ricqlés et al. 1991). Although this classification of different tissue types is based on the shape and arrangement of the "vascular channels," in reality, the shape of the channels or lumen within bone does not correspond to the actual shape of the vascular canal (blood vessel) therein (Starck and Chinsamy 2002). In addition, besides the blood vessels, nerves and other connective tissue also occupy the channels in the bone (Stark and Chinsamy 2002). Another point to note when using this classification of bone tissues (de Ricqlés et al. 1991) is that the various categories are based on transverse (i.e., cross) sections of bone. This is important to note since even a slightly tangential section or a longitudinal section will result in a different view of the arrangement of the vascular channels. It is therefore generally good practice to describe the plane of sectioning and/or axis of the bone before describing the orientation of the channels. For example, in cross section, the vascular channels appear circular in shape, but they run parallel to the long axis of the bone. In the illustrations provided here, unless specifically stated, sections are all in the transverse plane.

LAMINAR BONE

Laminar bone has a distinctive circumferential layering of channels, in other words, the channels are aligned parallel to the circumference of the skeletal element (e.g., Gross 1934; de Ricqlés 1975; de Ricqlés et al. 1991; Chinsamy-Turan 2005). The primary osteons in this tissue tend to be longitudinally or circumferentially arranged relative to the long axis of the bone (Fig. 3.3A).

PLEXIFORM BONE

Plexiform bone also has a circumferential arrangement of the channels in the bone, although this tissue has frequent radial connections between neighboring channels (e.g., Enlow 1962; de Ricqlés 1975; de Ricqlés et al. 1991; Chinsamy-Turan 2005) (Fig. 3.3B).

RETICULAR BONE

Reticular bone differs from laminar and plexiform tissues in having more frequent connections between neighboring channels which result in a complex three-dimensional network of channels with random organization (e.g., de Ricqlés 1975; de Ricqlés et al. 1991; Chinsamy-Turan 2005) (Figs. 3.2B, 3.3C).

RADIAL FIBROLAMELLAR BONE

Radial fibrolamellar bone is characterized by a radial arrangement of the channels (e.g., de Ricqlés 1975; Chinsamy-Turan 2005) (Fig. 3.3D).

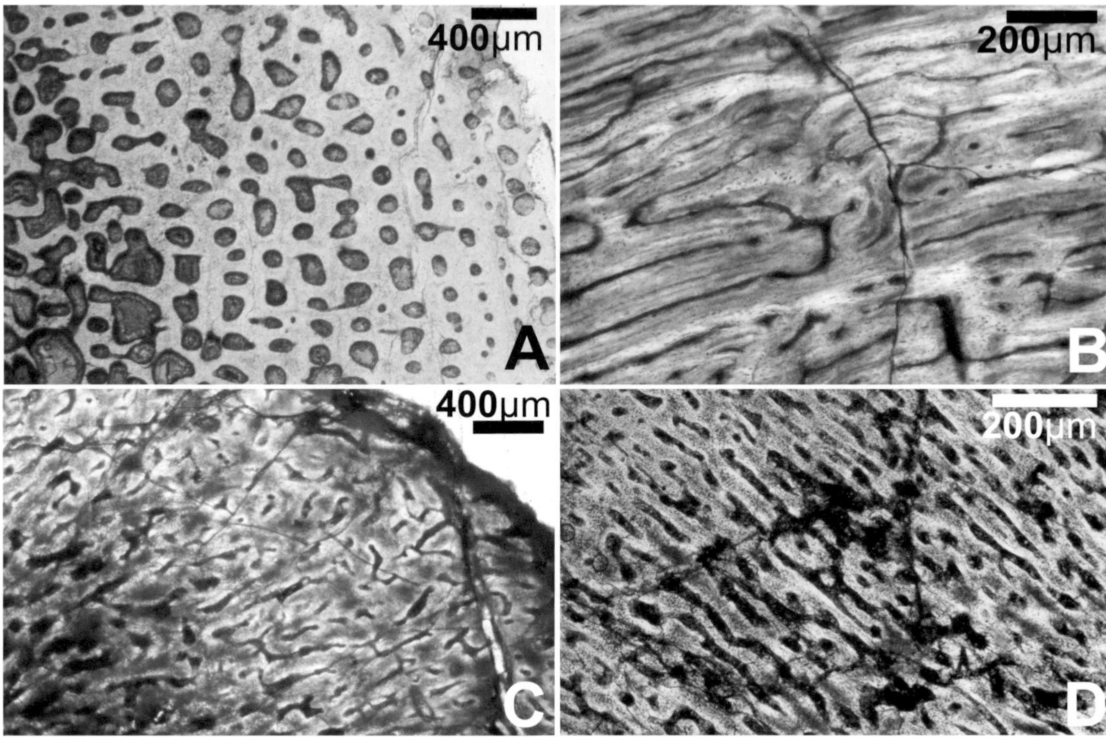

3.3. (A) Laminar fibrolamellar bone in a juvenile *Dicynodon* (SAM-PK-5576). (B) Plexiform fibrolamellar bone in a femur of *Scylacops* (SAM-PK-10188). (C) Reticular fibrolamellar bone in *Endothiodon* (SAM-PK 5065). (D) Radial fibrolamellar bone in an *Endothiodon* tibia (SAM-PK 5065C).

Lamellar Bone

Lamellar bone tissue comprises thin, successive layers of lamellae in which the collagen fibers are oriented in different directions in each successive lamellae (Francillon-Vieillot et al. 1990; Reid 1996) (Plates 5A to 5D). It is distinguishable from parallel-fibered bone since, under polarized light, lamellar bone appears as alternating dark and light bands (Plates 5B, 5D). Osteocytes in lamellar bone tend to be flattened with fewer canaliculi, and they have a more organized arrangement. Lamellar bone, when compared with fibrolamellar bone tissues, is deposited at relatively slower rates and generally entraps few blood vessels/channels while it is being deposited. Thus, blood vessels within the bone tend to be simply enclosed within the newly formed matrix of the bone (without any spaces). However, lamellar bone tissue forms the osteonal structure of primary and secondary osteons within fibrolamellar bone tissues; in other words, it is the bone tissue that forms the more slowly deposited bony ring around the channels entrapped during the rapid deposition of the fibrolamellar bone matrix (Plate 4; Fig. 3.2B).

Lamellated Bone

This bone is similar to lamellar bone but here the fine lamellations, which are formed as a result of cyclical variations in the mineral microstructure, are visible without the use of crossed polarizers (Reid 1996) (Plates 5D, 5E).

Parallel-Fibered Bone

This bone tissue is superficially similar to lamellar bone, but in this case, the collagen fibrils are organized parallel to one another, that is, all the fiber bundles run in the same direction and the bone appears anisotropic under polarized light (Plates 5G, 5H). The osteocytes within the matrix tend to be flattened and are randomly arranged, although not as irregular as in the fibrolamellar bone tissues, and not as organized as in lamellar bone (Plate 5G). Small osteons can sometimes occur in this type of bone tissue.

Growth Marks

Several different kinds of growth marks or "lines" are evident in compact bone, which often leads to a distinct layering or stratification of the cortical bone. Some of them are an expression of an animal's innate biological alternating growth rhythm (Castanet 2006), while others are formed as a result of remodeling processes such as reversals in the direction of bone deposition (e.g., Enlow 1962). The particular location and appearance of the growth marks permit the distinction between different types.

Lines of Arrested Growth

Lines of arrested growth (LAGs) are hyper-mineralized lines that indicate an arrest in appositional growth, that is, in periosteal deposition of bone (Fig. 3.4A). They often have a smooth appearance that follows the contour of the external surface of the bone. Most often, LAGs form individually but sometimes multiple (i.e., a group) LAGs occur (Fig. 3.4B). Generally, LAGs are formed annually in many poikilothermic vertebrates (see, e.g., Castanet et al. 1993; Chinsamy and Dodson 1995; Chinsamy-Turan 2005) and result in a distinctive stratification of the cortical bone into a series of growth rings that are indicative of periodic arrests in bone deposition and growth (Fig. 3.4A). Lines of arrested growth are often used for age determination, or skeletochronology (see, e.g., Castanet et al. 1993).

Rest Lines

These lines are similar to LAGs, but may be distinguished by not showing the periodicity typically expressed by LAGs (Castanet 2006). Rest lines manifest as a group of closely spaced lines near the peripheral margin of the bone wall and are formed when appositional growth has virtually ceased (see, e.g., Chinsamy 1990) (Fig. 3.4C). Thus, they form on unresorbed bone surfaces and generally have a smooth appearance. They often located in the outer circumferential layer typically seen in mammals and birds. Thus, the presence of such lines generally suggests the attainment of maximum body size (i.e., determinate growth) and the bone formed is sometimes referred to as accretionary bone (Reid 1996).

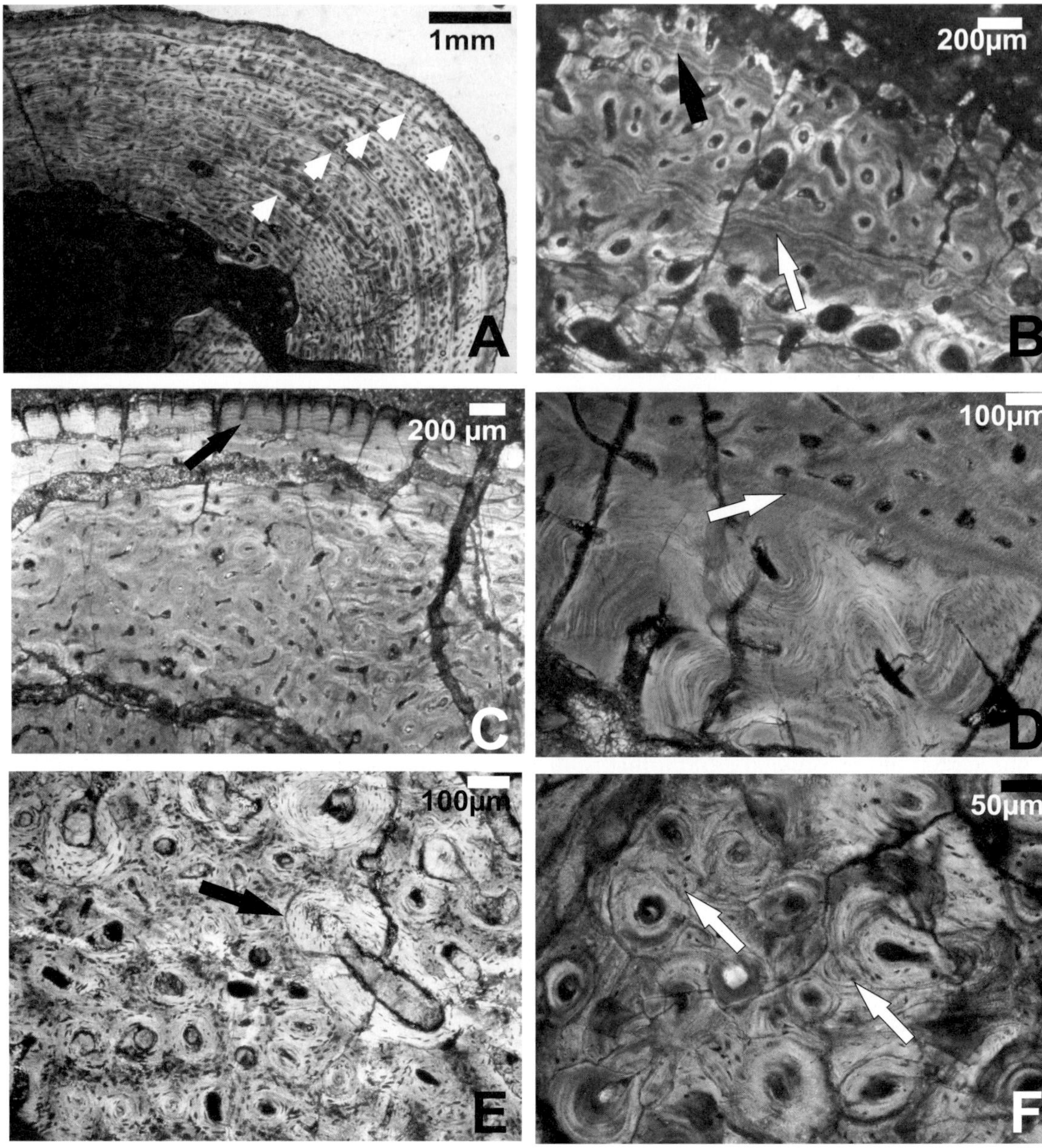

3.4. (A) *Scylacops* (SAM-PK-10188) rib. Arrows indicate periodic depositions of LAGs. (B) Multiple LAGs (arrows) interrupting the deposition of fibrolamellar bone in a scylacosaurid therocephalian (SAM-PK-11557, specimen 2A1). (C) Peripheral rest lines (arrow) in a scylacosaurid therocephalian (SAM-PK-11557, specimen 3A1). (D) *Aelurognathus* (SAM-PK-K10000) showing a tide line or erosion line (arrow) above which is primary periosteal bone tissue and underneath which is endosteally formed compacted coarse cancellous bone tissue. (E) *Tropidostoma* (SAM-PK-K990) humerus. Primary and secondary osteons are visible. An arrow indicates a cement line around secondary osteon, which marks the furthest extent of bone removal. (F) Dense Haversian bone in a *Scylacops* (SAM-PK-10188, radius). The arrow indicates remnants of secondary bone between later-formed secondary osteons.

Reversal Lines or "Tide" Lines

Such lines within bone tissues are observed on resorbed bone surfaces. These lines are often associated with remodeling or reversals in bone deposition (Fig. 3.4D). Such lines have a distinctly uneven or scalloped appearance. Often, such lines are observed at junctions between primary and secondary bone tissues, as for example, between periosteal and endosteal bone tissues around the medullary cavity (Fig. 3.4D).

Cement Lines

Technically, cement lines are reversal lines associated directly with the channels wherein blood vessels are located in bone. Secondary reconstruction (see next section) results in bone being resorbed from around the blood vessels. The outermost extent of this bone removal is marked by the presence of a cement line (Fig. 3.4E). As described for reversal lines, cement lines are also generally uneven, scalloped, and cut into pre-existing bone (which could be either primary or secondary bone).

Secondary/Haversian Reconstruction, Secondary Osteons, and Haversian Bone

Secondary reconstruction is a process that involves the resorption of bone surrounding a blood vessel and often results in erosion cavities within the compact bone (Plate 6A; Fig. 3.4E). These cavities later become infilled centripetally with lamellar bone to form secondary osteons (Plate 6A). Thus, secondary osteons are easily distinguishable from primary osteons in having a characteristic cement line which marks the furthest extent of bone removal (Plate 6; Figs. 3.4E). The extent of secondary reconstruction of primary bone tissues is variable among vertebrates. Some animals such as small birds can show virtually no secondary reconstruction (Enlow and Brown 1958), whereas others, such as the gorgonopsian *Scylacops*, show such intensive secondary reconstruction that even the tissue between neighboring secondary osteons is secondary in nature; when this occurs, it is referred to as dense Haversian bone (Plate 6B; Fig. 3.4F).

Other Tissue Types

Calcified Cartilage

Calcified cartilage has a distinctive foam-like appearance and can be easily recognized in the "articular ends" of long bones that are still growing (Fig. 3.5A). It should be noted that the articular ends of most fossil bones do not preserve the original articular surface since the latter, comprising of primarily unmineralized cartilage would have decomposed during early stages of burial (Chinsamy-Turan 2005). Thus, only the calcified cartilages of growing individuals and the more finished ossified surfaces of adult individuals survive the fossilization processes. The foam-like structures that are visible in fossil bones are the lacunae of the chondrocytes (Reid 1996). As the cartilage at the ends of the growing bone matures, it is

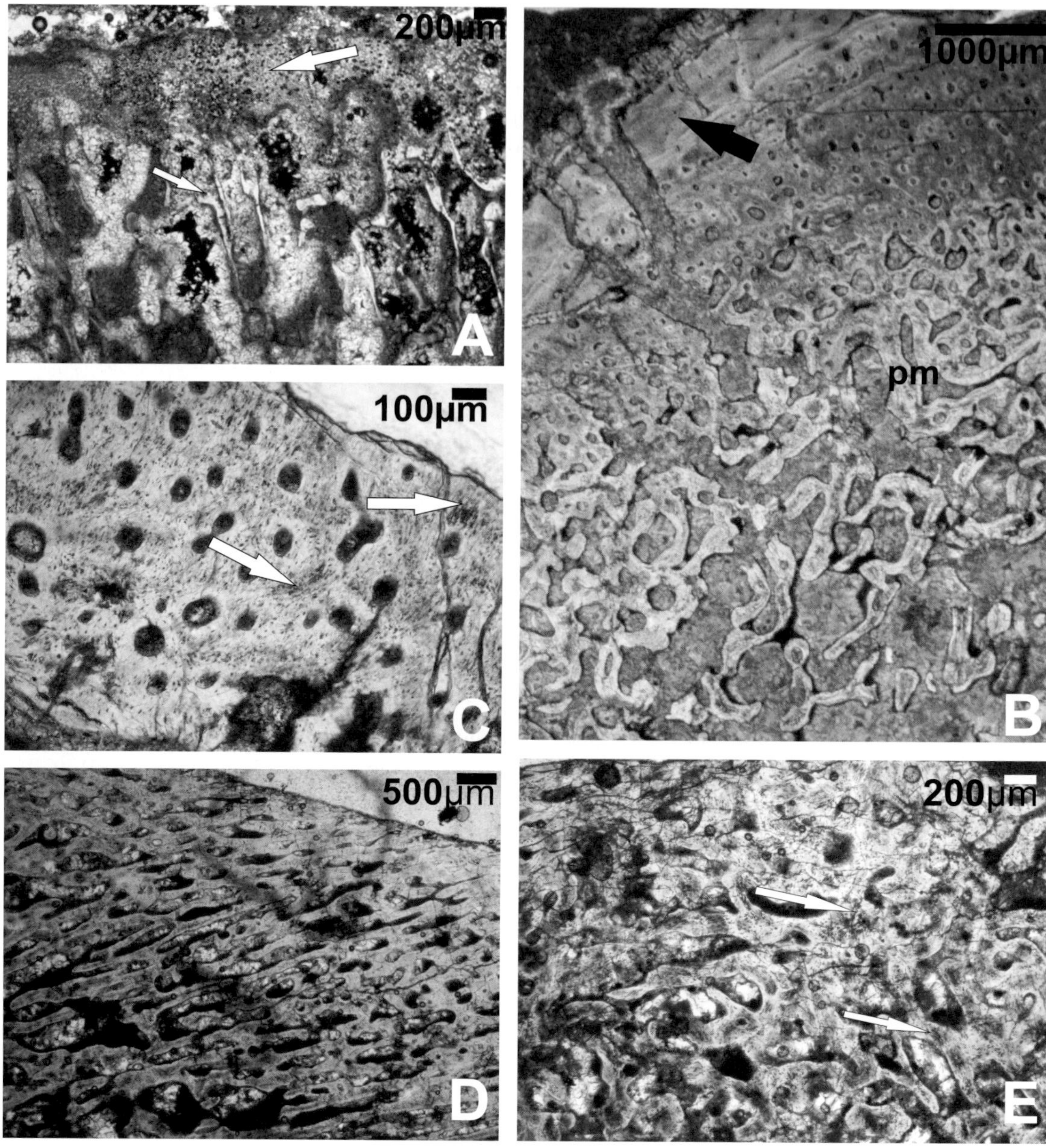

replaced by endochondral bone which forms the trabecular struts evident at the ends of long bones (Fig. 3.5A).

Secondary Cancellous (Spongy) Bone

Primary cancellous bone or trabecular bone forms as a result of endochondral ossification at the growing ends of bones (such as, the proximal or distal ends of long bones). Secondary cancellous bone, on the other hand, results from the resorption of primary bone tissue. During secondary reconstruction, bone is often redeposited and results in the formation of secondary osteons (= Haversian systems), and in some cases, dense

Haversian bone. However, occasionally resorption rates exceed the redeposition of bone, and the bone becomes increasingly porous/spongy (Fig. 3.5B). Such bone tissue is referred to cancellous or spongy bone tissue. In many individuals with thick bone compacta, the bone surrounding the medullary cavity often is the most reconstructed, and cancellous bone is often located in these regions (Fig. 3.5B).

3.5. (A) *Aelurognathus* (SAM-PK-K10000, specimen 1 B1). Longitudinal section showing calcified cartilage (big arrow) and bony trabeculae (small arrow) at the articular ends of the bone. (B) *Tropidostoma* tibia (SAM-PK-9960). Extensive bone resorption in the perimedullary (pm) region has led to the formation of secondary cancellous bone which has a distinctive spongy texture. Arrow indicates the outer circumferential layer. (C) *Endothiodon* (Sam-PK-5605) tibia. Arrows indicate Sharpey's fibers. (D) Pathological nonmammalian therapsid bone showing the spicules formed by rapid periosteal reactive growth. (E) Different region of same bone as (D), showing the woven, disorganized nature and high porosity of the tissue. Arrows indicate bundles of Sharpey's fibers.

Compacted Coarse Cancellous Bone

Compacted coarse cancellous bone is often observed in more distal or proximal sections of long bones and results when these regions become "relocated" from more endosteal positions in the younger individual into the compacta of the older individual (see, e.g., Enlow 1962; Chinsamy-Turan 2005). In young individuals, the ends of their long bones comprise endochondral cancellous bone, but as the animal matures the "ends" become incorporated into the bone wall of the older individual. In the process, the cancellous bone becomes compacted with lamellar bone deposits which fill in the cancellous spaces. The bone tissue characteristically has a swirled, convoluted appearance (Enlow 1962) (Fig. 3.4D).

Inner and Outer Circumferential Layers

Endosteal bone deposition (i.e., bone deposited under the endosteum) results in the formation of a distinctive lining of bone around the medullary cavity called the inner circumferential layer, which is often composed of lamellar, parallel-fibered, or lamellated bone (see, e.g., Enlow 1962, Chinsamy-Turan 2005). The outer circumferential lamellae (Ham 1953; Enlow and Brown 1957) refers to the periosteally formed circumferential layer of bone deposited once appositional growth has virtually completed (i.e., among extant vertebrates, it corresponds to residual growth once adult body size is attained as in mammals and birds) (Figs. 3.4C, 3.5B). Outer circumferential lamellae is synonymous with the term external fundamental system (Cormack 1987).

Sharpey's Fibers

Sharpey's fibers or fibrillary attachment processes consist of closely packed mineralized collagen fibers (de Ricqlés et al. 1991) found at attachment sites of muscles, tendons, ligaments, and possibly keratin to bone (Plate 7A; Fig. 3.5C). They are commonly found in dermal bones, such as osteoderms (e.g., Scheyer and Sander 2009), and also along the edges of sutures (see chapter 6 this book).

Pathological Bone Tissues

Although there have been a few reports of pathologic bone (and its histology) in pelycosaurs (Romer and Price 1940; Enlow and Brown 1957; Rega

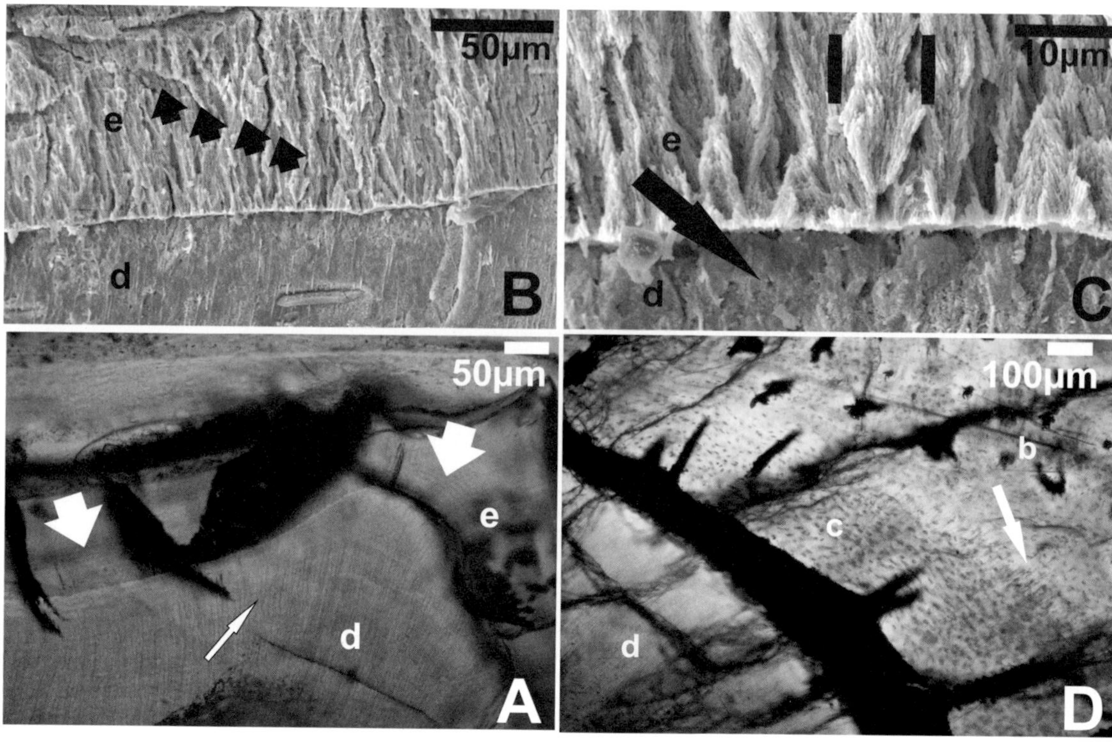

et al. 2005; chapter 4 of this book), there are even fewer reports for the nonmammalian therapsids: Enlow (1969) reported on the microstructure of a callus on a neural spine of a sphenacodontid *Dimetrodon*, while Rega et al. (2005) documented several neural spine pathologies in *Dimetrodon*. Two of the ribs of a large dicynodont (possibly *Oudenodon grandis*), nicknamed "Mamafura" (SAM-PK-K6446) at the Iziko South African Museum in Cape Town have thick calluses that suggests healed fractures (Fordyce 2010). Figures 3.5D and 3.5E show the distinctive histology associated with a callus on an unidentified bone of a nonmammalian therapsid. The histology of the bone has a number of long (almost radial) bony spicules which suggest rapid bone deposition probably caused by reactive periosteal bone growth (Chinsamy and Tumarkin-Deratzian 2009). In parts of the callus, the peripheral bone is compacted and has a distinctly disorganized woven textured appearance with a number of irregularly shaped channels, while further internally, the bone has a more porous texture (Fig. 3.5E). A large number of Sharpey's fibers are also evident.

Tooth Histology

Enamel Microstructure

Much of the research on the enamel structure of nonmammalian therapsids has involved scanning electron microscopy (SEM) (see, e.g., Grine, Vrba, and Cruickshank 1979; Sander 1997; Abdala and Malabarba 2007), although there are a few studies that have provided assessments through light microscopy. By studying thin sections under polarized light, Poole (1956) suggested that the enamel on a cynodont tooth superficially resembled mammalian enamel, while a gorgonopsid had a regular arrangement of crystallites forming "pseudo-prisms" with faint lines running parallel to the surface of the enamel. He also reported similar lines in the enamel of a *Dimetrodon* tooth, in which he described a "simulated prismatic" enamel layer. In a study of an unidentified traversodontid tooth from Tanzania, Moss (1969) also found nonprismatic enamel and incremental lines (which are also known as "Striae of Retzius" [Sander 1999]). A similar observation was made in a *Tritylodon* tooth (Fig. 3.6A).

Scanning electron microscope studies of the enamel of a postcanine tooth of the traversodontid, *Exaeretodon frenguellii* also revealed faint incremental bands in the thick enamel layer (Abdala and Malabarba 2007) (Fig. 3.6B) which seem to correspond with those observed in the enamel of some nonmammalian synapsids using light microscopy. Scanning electron microscope studies of various nonmammalian cynodonts have suggested that some, like *Diademodon* and *Thrinaxodon* (Grine, Vrba, and Cruickshank 1979), as well as *Exaeretodon frenguellii* and *Exaeretodon riograndensis* (Abdala and Malabarba 2007), have aprismatic enamel, while *Pachygenelus* (Grine, Vrba, and Cruickshank 1979) is considered to have a prismatic structure. Wood and Stern (1997) also reported prismatic enamel in *Pachygenelus*, and they documented prismatic enamel in the early mammal *Megazostrodon*. Initial SEM studies of *Tritylodon* suggested that it had aprismatic enamel (see, e.g., Grine, Vrba, and Cruickshank 1979), but more recently Kamiya et al. (2006) reported the presence of prismatic enamel in this taxon. This question of prisms or lack thereof (aprismatic or nonprismatic) among the mammaliaforms is mostly because of varying usage of the terms. In a thorough reassessment of the terminology used to describe reptilian enamel microstructure, Sander (1999) suggests that the previous references to prisms in nonmammalian synapsids is inaccurate, and that the term "prismless" be used to refer to all types of enamel that lack prisms.

Only 13 percent of all mammaliaform cynodont enamel microstructure has been previously studied (Abdala and Malabarba 2007). Of these, three different types of prismless enamel microstructures are recognized by SEM: parallel crystallites, columnar enamel, and plesiomorphic prisms (Sander 1997, 1999; Abdala and Malabarba 2007). The most common enamel type appears to be enamel with columnar divergence units in which the enamel unit is arranged in columns perpendicular to the outer enamel surface (Fig. 3.6C). According to Sander (1999) the earlier term "pseudo-prisms" (Poole 1956) probably refers to enamel having a type of columnar arrangement.

3.6. (A) *Tritylodon* tooth under light microscopy showing a band of enamel (e) that overlies a thick layer of dentine (d). Fine incremental lines are visible in the enamel (top arrows), and faint von Ebner lines are visible in the dentine (thin arrow). Dentinal fibers are evident as fine lines that run approximately perpendicular to the enamel-dentine junction. (B, C) Scanning electron microscopy micrographs of a tooth of *Exaeretodon frenguellii* showing the enamel and dentine. Arrows in the enamel layer point to the faint incremental bands. (C) Higher magnification showing the columnar divergence unit (parallel lines) of the enamel. The arrow indicates a cross section of a dentinal tubule. (D) *Tritylodon* (SAM-PK-K 407). A thin section showing a tooth located within its mandibular socket. *Abbreviations:* b = bone; c = cellular cementum. The arrow indicates Sharpey's fibers.

Dentine Microstructure

Dentine is characteristically located beneath the enamel (Figs. 3.6A to 3.6C) and extends into the pulp cavity of the tooth. Dental fibers running perpendicularly from the pulp cavity and lines of von Ebner (which indicate the periodicity of dentinogenesis) are distinct in the dentine (Fig.

3.6A). Research using tetracycline has shown that such lines are formed daily, and as such they have been utilized to assess the age of individuals (see, e.g., Erickson 1996a). Thackeray (1991) described lines of von Ebner as "growth incrementals" in tusks of *Diictodon* from the Late Permian of South Africa. He suggested that the widths between consecutive lines changed (decreased) over time in response to environmental changes near the end of the Permian. The tusk histology of the dicynodonts, *Lystrosaurus* and *Placerias hesternus*, are described in chapters 6 and 7, respectively.

Tooth Attachment Sites

Among the nonmammalian therapsids, the histology of the attachment site of the tooth in the jaws has not previously been described. Thin sections of the *Tritylodon* (SAM-PK-K407) lower jaw show the location of the tooth within a socket and that it is surrounded by a layer of cellular cementum with Sharpey's fibers (Jasinoski and Chinsamy, current research). The bone of attachment consists of primary compact bone tissue with several osteons (Fig. 3.6D).

Diagenetic Alteration of Bone Microstructure

Although fossil bones tend to generally be well preserved (Chinsamy-Turan 2005), it is sometimes possible to find bones where the microstructure is clearly degraded. There are only a few published studies describing altered bone among the therapsids (e.g., dicynodonts [Davies 1981]; *Morganucodon* [Chinsamy-Turan 2005]).

In a study of dicynodont bones from the Mid-Luangwa Valley of Zambia, Davies (1981) found that the channels in both primary and secondary bone tissues as well as within postmortem cracks in the bone were all lined with a reddish microcrystalline structure composed of the mineral heulandite. He suggested that thermal water rising under pressure along major fault planes penetrated all the pores and cracks in the fossil bones, and this resulted in the deposition of heulandite.

In a histological survey of early mammals, Chinsamy and Hurum (2006) noted that many bones of the small Early Jurassic mammal *Morganucodon* from the fissure fillings of the Pant Quarry of Wales were extensively altered. In another study of traversodontids, Chinsamy and Abdala (2008) mentioned that all the histological sections of a morphologically well-preserved tibia of the Brazilian traversodontid, *Exaeretodon riograndensis*, from the Santa Maria Formation were diagenetically altered and no histological details could be deciphered. Ray, Botha, and Chinsamy (2004) also reported that several bones of the scylacosaurid (SAM-PK-5018) exhibited diagenetic alteration (Ray, Botha, and Chinsamy 2004). In chapter 8 of this book, Chinsamy-Turan and Ray further examine the diagenetic changes and deduce that the damage observed was caused by saprophytic fungi.

Biological Interpretations from Bone

As explained earlier, the deposition of bone is influenced by a number of factors, such as ontogeny and phylogeny (see, e.g., Amprino 1947; de Ricqlés et al. 1991; Cubo et al. 2005; Chinsamy-Turan 2005; Scheyer, Klein, and Sander 2010). However, it is also evident that bone tissues reflect local conditions of growth and biomechanical adaptations (see, e.g., de Margerie, Cubo, and Castanet 2002; Starck and Chinsamy 2002). Thus, a wide range of phylogenetically distinct vertebrates can sometimes show similar types of bone tissue. A host of studies have shown how bone microstructure facilitates deductions pertaining to various aspects of the life history and biology of extinct vertebrates, such as dinosaurs (see, e.g., Sander 2000; Horner and Padian 2004; Erickson et al. 2004; Erickson et al. 2009; Chinsamy and Tumarkin-Deratzian 2009), pterosaurs (e.g., Padian, Horner, and de Ricqlès 2004; Chinsamy, Cordorniu, and Chiappe 2009), and nonmammalian reptiles (see, e.g., Botha and Chinsamy 2000, 2005; Ray, Botha, and Chinsamy 2004; Green, Schweitzer, and Lamm 2010). Some biological features that are manifested in the bone microstructure of fossil bone are discussed in the following sections.

Rates of Osteogenesis

Amprino (1947) suggested that the fibrillar organization of bone reflected the rate at which the bone was deposited, that is, that the different bone tissue types were the result of different growth rates. Therefore, bone with a woven matrix and little spatial organization, such as fibrolamellar bone, would be the result of rapid osteogenesis, while parallel-fibered, lamellar, or lamellated bone tissues with more spatially organized matrices are more slowly formed. These surmises were later confirmed by experimental studies using fluorochrome labeling (see, e.g., de Ricqlés et al. 1991; Starck and Chinsamy 2002; Castanet et al. 1993). Today we have some understanding of the types of bone tissues that result from specific depositional rates. For example, lamellar bone forms at rates of 0.05–0.3 μm per day, parallel-fibered bone forms between 0.1–0.5 μm per day, and woven-fibered bone forms at 0.4–5 μm per day (de Ricqlés et al. 1991). In addition, such studies have highlighted the fact that similar bone tissues can result from different rates of osteogenesis. Laminar fibrolamellar bone tissues, for example, can form at rates of 10–50 μm per day in young Japanese quail (Starck and Chinsamy 2002), while the same tissue is formed at 60 μm per day in mallards (Castanet et al. 1996). Such interspecific differences among modern birds caution the direct extrapolation of bone depositional rates of extant animals to the fossil record (Starck and Chinsamy 2002; Chinsamy-Turan 2005). Experimental studies have also shown that different bones in the skeleton of an individual can form exactly the same type of bone tissue at different rates (Starck and Chinsamy 2002). Thus it is evident that bone depositional rates are quite variable and depend on a number of factors including ontogeny, location in the skeleton, local growth conditions, and phylogeny.

Fibrolamellar bone (Fig. 3.3) has been documented in a variety of nonmammalian therapsids (e.g., dicynodonts [de Ricqlés 1969; Chinsamy and Rubidge 1993; Botha and Angielczyk 2007; Ray, Bandyopadhyay, and Bwahal 2009; Green, Schweitzer, and Lamm 2010]; cynodonts [de Ricqlés 1969; Botha and Chinsamy 2000, 2004, 2005; Chinsamy and Abdala 2008]; gorgonopsians [Ray, Botha, and Chinsamy 2004]). Indeed, the whole range of different types have been described, for example, laminar bone has been described in *Diictodon* (Chinsamy and Rubidge 1993) while plexiform and reticular bone types have been described in *Diademodon* and *Cynognathus* (Botha and Chinsamy 2000), respectively. It is also important to note that different types of fibrolamellar bone tissues can be present within a single bone. For example, Fig. 3.7A shows a *Diictodon* humerus having a mixture of predominantly reticular bone with a more laminar type of bone tissue peripherally, while Fig. 3.7B shows an internal region of rapidly formed fibrolamellar bone tissue and an outer more slowly formed parallel-fibered bone tissue. Quite unpredictably, even the fastest forming fibrolamellar bone, radial fibrolamellar bone, formed at rates of about 180 µm per day in Emperor penguin chicks (de Margerie et al. 2004), has been recorded in nonmammalian therapsids (e.g., *Diictodon* [Ray and Chinsamy 2004]; *Endothiodon* tibia) (Fig. 3.3D). Its presence suggests that the tissue was formed at relatively faster rates than the surrounding tissue, although we cannot be certain as to exactly how fast this rate was.

Parallel-fibered bone, lamellar bone, and lamellated bone tissues (Plate 5) have been identified in several nonmammalian therapsid bones and are commonly located near the peripheral edge of the bone wall (e.g., *Oudenodon* [Botha 2003]; *Trirachodon* [Botha and Chinsamy 2004]) or it is often found interrupting the deposition of fibrolamellar bone (e.g., in dicynodonts [Chinsamy and Rubidge 1993]; cynodonts [Botha and Chinsamy 2000]; gorgonopsians [e.g., Plate 7B]). The traversodontid *Andescynodon mendozensis* is distinctive in having most of its compacta composed of lamellar bone tissue with few simple blood vessels (Chinsamy and Abdala 2008).

Growth Dynamics

CYCLICAL BONE DEPOSITIONAL PATTERNS

Bones of extant reptiles are characterized by a stratified compacta consisting of alternating types of bone tissues. This tissue was initially described as lamellar-zonal bone (Enlow and Brown 1957; Peabody 1961; Enlow 1969; de Ricqlés 1975). Such bone is quite variable in structure, but it is often composed of parallel-fibered bone tissue with a few simple blood vessels, alternating with a narrow band of lamellar bone tissue with or without an associated LAG (Castanet et al. 1993). This has been typically described in extant lepidosaurs and crocodilians. Today, it is well recognized that cyclically formed bone does not necessarily always consist of a lamellar-zonal tissue. For example, alligators and an extinct turtle have been reported as having periodically formed fibrolamellar bone (Reid 1987; Chinsamy-Turan 2005; Tumarkin-Deratzian 2007), and several dinosaurs (both avian and nonavian) show fibrolamellar bone alternating with a more slowly formed bone tissue such as lamellar bone (with or without associated LAGs) (see, e.g., Chinsamy-Turan 2005, and references therein). Indeed, extant mammals, such as polar bears and kangaroos, are known to form bone in a cyclical manner as well (Chinsamy-Turan 2005; chapter 10 this book).

Such cycles of growth are known to be related to an innate annual biological rhythm (Castanet et al. 1993). Several studies using fluorochrome labeling and "mark-recapture" studies have demonstrated that the rhythm corresponds to an annual cycle, where wide regions termed zones (often consisting of parallel-fibered bone or fibrolamellar bone) are deposited during the favorable season (typically warmer months) at a relatively rapid rate, while narrower regions of bone, termed annuli, are formed during the unfavorable season (colder months) at a relatively slower rate (see, e.g., Hutton 1986; Castanet et al. 1993). Thus, one zone and one annulus are generally formed during a single year, although double cycles are known to occur in fish and some amphibians (Castanet et al. 1993). Cyclical growth patterns have been extensively documented among lepidosaurs (see, e.g., de Ricqlés et al. 1991; Castanet et al. 1993; Chinsamy et al. 1995) and anurans (see, e.g., Erismis and Chinsamy 2010) and were, until recently, associated with ectotherms. However, today it is recognized that cyclical patterns of growth occur among a wide variety of taxa, including a range of Mesozoic birds (see, e.g., Chinsamy, Chiappe, and Dodson 1994; Chinsamy-Turan 2005, and references therein); more derived birds such as the large, extinct, flightless moa *(Dinornis)* (Turvey, Green, and Holdaway 2005) and in the New Zealand kiwi, *Apteryx* (Bourdon et al. 2009) as well as among Mesozoic mammals (Chinsamy and Hurum 2006; chapter 10 of this book); and also in a Plio-Pleistocene bovid, *Myotragus*, from a resource-limiting environment (Köhler and Moyà-Solà 2009). When vertebrates exhibit cyclical patterns of growth, skeletochronology (Castanet et al. 1993) can be applied, thereby permitting the age of the individual to be assessed. If a growth series of individuals are represented in a sample, growth curves can be deduced using skeletochronology (see, e.g., Chinsamy 1993a; Erickson 2005; Scheyer, Klein, and Sander 2010). Such methods have been successfully applied to a variety of extinct animals (e.g., nonavian dinosaurs [Chinsamy 1993a; Chinsamy-Turan 2005; Erickson 2005]; pterosaurs [Chinsamy, Cordorniu, and Chiappe 2009]; extinct bovid [Köhler and Moyà-Solà 2009]).

Chinsamy and Rubidge (1993) documented the presence of periodically interrupted fibrolamellar bone in several dicynodont taxa (*Aulacephalodon*, *Cistecephalus*, *Dicynodon*, *Endothiodon*, *Lystrosaurus*, *Kannemeyeria*, and *Oudenodon*) from the Beaufort Group of South Africa. Many of the Dicynodontia showed LAGs interrupting the deposition

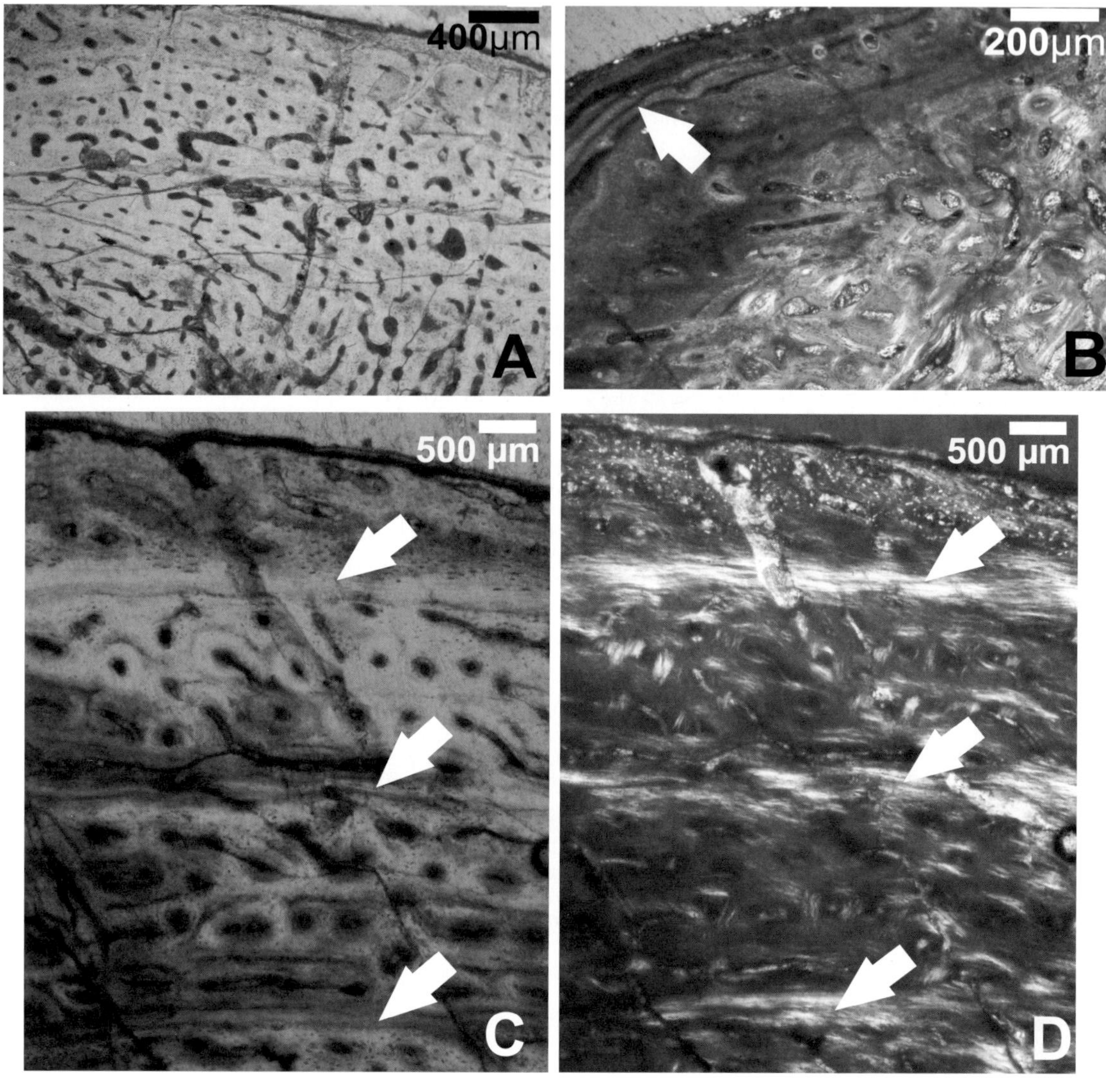

of fibrolamellar bone, but in some instead of LAGs, a distinct band, or annulus, consisting of lamellar or lamellated bone tissue interrupted the deposition of the more rapidly formed bone tissue. Several other nonmammalian therapsids were also found to have zonal bone tissues (e.g., cynodonts [Botha and Chinsamy 2000, 2004, 2005; chapter 9 of this book]; gorgonopsians; and therocephalians [[e.g., chapter 8 of this book]). More recently, zonal bone tissues consisting of periodically interrupted parallel-fibered bone have also been documented in Cretaceous eutherian mammals from the Gobi Desert (Chinsamy and Hurum 2006; chapter 10 of this book).

Given the burgeoning research in bone histology of nonmammalian therapsids, it has become increasingly apparent that the actual bone forming the zones is quite variable and depends on the rate at which the bone is formed, as well as on the ontogenetic age of the individual. For example, the cynodont *Diademodon* and the gorgonopsian *Scylacops*

(Figs. 3.7C, 3.7D) show fibrolamellar bone alternating with lamellar bone throughout their ontogeny, while the basal cynodont *Procynosuchus* typically deposits fibrolamellar bone during early ontogeny and lamellar-zonal bone during later stages (Ray, Botha, and Chinsamy 2004; chapter 9 this book).

Studies of nonmammalian therapsids have shown that cyclical growth patterns can sometimes occur during later stages of ontogeny, in other words, alternating or interrupted growth patterns can follow a period of rapid sustained growth, as for example, in the dicynodont *Diictodon*, where periodically interrupted growth patterns occur only after 70 percent of adult size has been reached (Ray and Chinsamy 2004). *Wadiasaurus* from India shows a distinct change in the nature of the zonal tissue through ontogeny, that is, initially fibrolamellar bone occurs in the zones, but at later stages of ontogeny the zonal bone tissue is parallel-fibered or lamellar (Ray, Bandyopadhyay, and Bwahal 2009; chapter 5 of this book).

Among many vertebrates that have seasonal alternating rates of bone deposition, annuli are often composed of lamellar bone, but they sometimes also consist of lamellated bone tissue. In many cases, LAGs can precede or follow an annulus, or sometimes even occur within an annulus. Thus, depending on the type of bone tissue that is located within zones or annuli information can be deduced about the relative rate of bone deposition, as well as about the relative abruptness of changes in bone depositional rates.

3.7. (A) *Diictodon* (SAM-PK-3437) humerus showing different types of fibrolamellar bone in the compacta. (B) Scylacosaurid (specimen 4 A2). The more internal bone tissue comprises fibrolamellar bone, while the external part of the bone wall consists of more parallel-fibered bone. Note the closely spaced rest lines near the peripheral edge of the bone (arrow). These lines are equivalent to the outer circumferential layer and mark the attainment of mature body size, that is, determinate growth. (C, D) *Scylacops* femur shows growth rings visible in the compact bone wall under normal (C) and polarized (D) light. Arrows indicate the layers of lamellar bone (the annulus) that interrupt the deposition of fibrolamellar bone (the zone).

SUSTAINED UNINTERRUPTED GROWTH

Many vertebrates, notably most mammals and birds, deposit bone in a sustained uninterrupted manner. The type of bone formed is generally fibrolamellar bone, but it could also be parallel-fibered bone (e.g., small birds [Enlow and Brown 1957]). The presence of this uninterrupted type of tissue indicates that the bone formed at a sustained rate irrespective of environmental cues, such as seasonality. Such uninterrupted bone tissues have been identified in the Triassic cynodont *Cynognathus*, even though the contemporaneous *Diademodon* had distinct growth cycles in its compacta (Botha and Chinsamy 2000). Interestingly, the small dicynodont *Diictodon* showed uninterrupted formation of bone tissue for most of its life (Chinsamy and Rubidge 1993) and only much later in ontogeny developed periodic interruptions (Ray and Chinsamy 2004).

Thus, the presence of such azonal bone tissues in both extant and extinct vertebrates suggests that these animals grew at consistent rates throughout the year. However, once adulthood is attained, in many instances, there is a change in the nature of bone tissue, in other words, instead of fibrolamellar bone, parallel-fibered or lamellar bone is deposited, as in for example, the nonmammaliaform cynodont *Tritylodon* (Chinsamy and Hurum 2006), and the large dicynodont *Placerias* (Green, Schweitzer, and Lamm 2010).

DETERMINATE GROWTH VERSUS INDETERMINATE GROWTH

Modern animals that stop growing once they attain adult body size are said to have determinate growth. In many of these animals, a distinct change in the nature of bone tissue occurs in the outermost regions of the compacta: a transition from the more rapidly formed fibrolamellar bone tissue, to a more slowly formed bone tissue such as lamellar or lamellated bone. This change is often accompanied by the development of closely associated LAGs termed rest lines (Figs. 3.4C, 3.5B). The bone tissue formed upon maturity is referred to as the outer circumferential layer or the external fundamental system . The presence of this type of bone tissue has been described in mammals, birds, and also among several dinosaur taxa (see, e.g., Chinsamy 1990; Reid 1996). Several nonmammalian therapsids, such as *Tritylodon* (Chinsamy and Hurum 2006) and *Placerias* (Green, Schweitzer, and Lamm, 2010) show such determinate growth patterns.

ATTAINMENT OF SEXUAL MATURITY

Once sexual maturity is attained, a distinct change in the bone microstructure is observed in many extant animals. In vertebrates that form zonal bone tissue, the amount of bone deposited in the zone following sexual maturity (and in subsequent zones) is substantially narrower than the preceding zones (see, e.g., Castanet and Baez 1991; Castanet et al. 1993).

The attainment of sexual maturity is commensurate with a distinct change in the nature of the bone tissue which manifests as a reduction in the rate of bone formation in most cases. Thus, if fibrolamellar bone was formed initially, the type of bone formed after the attainment of sexual maturity would be a more slowly formed bone tissue, such as parallel-fibered bone or lamellar bone with fewer primary osteons (e.g., *Oudenodon* [Botha 2003]).

In a study of a growth series of the small burrowing dicynodont, *Diictodon,* Ray and Chinsamy (2004) noted that once individuals were more than 70 percent of adult body size, they switched from depositing uninterrupted fibrolamellar bone to forming a distinctive periodically interrupted bone tissue. These researchers postulated that this change in the rate of bone deposition was linked to reproductive maturity being reached.

LONGITUDINAL GROWTH IN LENGTH

The extent and development of calcified cartilage in the articular ends of bones can provide a good indication of how much longitudinal growth is yet to occur. Well-finished articular surfaces consisting of a distinctive layer of bone strongly suggest that growth in length has ceased. Conversely, the presence of calcified cartilage (Fig. 3.5A) at the ends of long bones (even ones that look as though they are from large individuals) indicates that growth in length is still underway. Although cortical bone of nonmammalian therapsids has been well studied, few studies have assessed longitudinal growth (see, e.g., Ray, Botha, and Chinsamy 2004; Chinsamy and Abdala 2008). Although these studies have documented the presence of calcified cartilage at the ends of long bones, none of these studies have documented the presence of secondary ossification centers as in mammals.

Lifestyle and Biomechanical Adaptations

BONE WALL THICKNESS

It has long been recognized that bones of terrestrial and aquatic animals have different architectural (or microanatomical) characteristics in terms of amount of compact bone and the size of the medullary cavity. Wall (1983) suggested that if the bone wall of the midshaft cross section of a femur consisted of more than 30 percent of its cross-sectional diameter, the animal would more than likely be amphibious. This seemed to hold true for a variety of extant animals he analyzed. Aquatic vertebrates tend to have thick, compact bone walls (pachyostosis), which act as ballast to help them overcome buoyancy, or alternatively their bones become spongy to minimize their inertia (osteosclerosis) (see, e.g., de Buffrénil et al. 1990).

More recent works have found that many extant burrowing animals such as the naked mole rat *Heterocephalus glaber* (Chinsamy and Hurum 2006), the porcupine *Erethizon*, and several fossorial lizards (Magwene 1993) also have relatively thick bone walls. Similarly thick bone compacta have been documented in the small burrowing nonmammalian therapsids, *Thrinaxodon* (Botha and Chinsamy 2005) and *Trirachodon* (Botha and Chinsamy 2004), as well as in the traversodontid *Andescynodon* (Chinsamy and Abdala 2008), although the latter was not associated with a burrow.

Ray, Chinsamy, and Bandyopadhyay (2005) undertook an assessment of the long bone histology of the dicynodont *Lystrosaurus*. They reported that *Lystrosaurus* had a largely porous bone tissue throughout ontogeny and that most of the medullary cavity of the long bones was occupied by cancellous bone, which was used to infer an amphibious lifestyle for this large dicynodont. Interestingly, *Lystrosaurus* survived the End Permian extinction event, and Ray, Chinsamy, and Bandyopadhyay (2005) suggested that perhaps its amphibious lifestyle might have played some role in its survival.

Using a program called Bone Profiler that was specifically developed by Girondot and Laurin (2003) to analyze the compactness of bone cross sections, Kriloff et al. (2008) undertook an assessment of the microanatomy of tibiae of ninety-nine different extant and extinct taxa. They found a significant correlation between compactness of long bone microanatomy and habitat (terrestrial, amphibious, and aquatic), and using these models, they inferred terrestrial lifestyles for *Dimetrodon*, *Dicynodon*, and a gorgonopsian, while *Ophiacodon* and *Lystrosaurus* were considered to have been amphibious. Research by Canoville and Laurin (2010) also supports an amphibious lifestyle for *Lystrosaurus*.

More recently, Modesto and Botha-Brink (2010) described the skeletal remains of a juvenile *Lystrosaurus* inside a burrow cast, and Botha-Brink and Angielczyk (2010) have suggested that *Lystrosaurus*'s thick bone walls could be interpreted as an adaptation for a burrowing lifestyle or that it might reflect a plesiomorphic trait of the Dicynodontia. The latter idea was also recently proposed by Green, Schweitzer, and Lamm (2010) to explain the thick cortical bone of the large North American dicynodont *Placerias*.

Thus, it appears that thick cortical bone walls might be characteristic of both extant and extinct fossorial and amphibious animals and perhaps might reflect a phylogenetic constraint in the Dicynodontia. Testing the latter hypothesis will require more thorough investigation into the cortical bone structure of the members of the clade, while finer resolution of histological differences between extant fossorial and semi-aquatic animals is also needed.

BIOMECHANICAL LOADING AND BONE MICROSTRUCTURE

Several studies of on the bones of modern animals have suggested that repetitive loadings manifest in the matrix of the bone tissue, in the orientation of the collagen fibers, and in the extent of secondary remodeling (see, e.g., Skedros, Mason, and Bloebaum 1994; Skedros et al. 1996; Currey 2002). Other studies have shown that different types of bone tissues have different biomechanical properties, for example, Carter and Spengler (1978) experimentally showed that lamellar bone tissue with lamellae of parallel collagen fibers is biomechanically stronger than fibrolamellar bone with its coarse woven textured arrangement of collagen fibers. In addition, the orientation of the collagen fibers affects the loading strength of the tissue (both of individual osteons and whole bone), in other words, tissues with predominantly longitudinally oriented fibers are stronger in tension, while strength in compression is promoted by a transverse arrangement of the collagen fibers (Martin, Burr, and Sharkey 1998). The functional loading of a bone would therefore result in specific strain or stress in the bone, which would be accommodated by the nature of the bone microstructure. Several researchers (see, e.g., Bertram and Biewener 1988; Lanyon 1992) have suggested that regional differences in cross-sectional geometry are the result of particular loading environments, that is, bone mass and architecture respond to loads experienced. Indeed, Starck and Chinsamy (2002) reported a direct change in the lateral and medial thickness of femoral cortical bone of quail chicks (*Coturnix japonica*) in response to asymmetrical loading on their legs and that the mechanical stress also resulted in a two- to threefold increased bone deposition.

Researchers on bone have found that secondary reconstruction weakened cortical bone in terms of strength and stiffness, mainly as a result of increase in porosity and a reduction in mineralization and because of the viscoelasticity of the cement line (e.g., Martin, Burr, and Sharkey 1998). Skedros, Mason, and Bloebaum (1994) and Skedros et al. (1996) suggested that the size and shape of secondary osteons was correlated to the particular stress or strain experienced by that skeletal element; in other words, regions of compressive strain would have a higher number of small, regular-shaped secondary osteons. The extent of secondary reconstruction and the development of secondary osteons have been suggestive of the loading environment of the particular element, for example, osteon population density is higher in the cortex of bones habitually loaded in bending (Bouvier and Hylander 1981), although more recent research suggests that this hypothesis is not fully supported (Skedros, Sorenson, and Jenson 2007). Interestingly, the capacity for bone to accommodate microdamage caused during normal activity is through the development of interfaces such as cement lines of secondary osteons which are able to dissipate energy (Skedros, Sorenson, and Jenson 2007). More recent analyses suggest that the correlations between loading history and density of osteon variants are not as straightforward, and a host of other factors such as ontogenetic status, metabolism, body size, and growth-related remodeling can affect the character of secondary osteons that result (Skedros, Sorenson, and Jenson 2007).

In their analysis of limb bone tissue of a variety of modern birds, de Margerie, Cubo, and Castanet (2002) demonstrated that laminar bone tissue is more resistant to torsion and compression, whereas bone tissues with longitudinally oriented channels appeared to be better adapted to tensile loads. A host of other studies on modern mammals have suggested that collagen fiber organization and the orientation of vascular channels are correlated with cross-sectional distribution of strain during locomotion. In his study of alligators through ontogeny, Lee (2004) examined the collagen fiber orientation and distribution and organization of vascular canals in the femur and suggested that growth dynamics, rather than the mechanical environment, is the dominant influence on the histological organization of bone.

Not much research has been done in terms of the histological signals of biomechanical loading of nonmammalian therapsid bones. However, recent work by Jasinoski, Rayfield, and Chinsamy (2009, 2010a, 2010b) has identified areas in the dicynodont skull under high stress and strain during mastication using a computational methodology called finite element analysis. The results of these biomechanical studies on *Lystrosaurus* and *Oudenodon* have been compared with the bone microstructure (chapter 6 of this book) and sutural morphology (Jasinoski, Rayfield, and Chinsamy 2010a) of specific regions of the cranium to assess whether these microstructural features correlate with the predicted strain patterns.

Concluding Remarks

Over the years, our knowledge of the variety of bone microstructure has grown substantially, and new research and methodologies are constantly expanding this body of knowledge. We now have a better appreciation of the variety of intrinsic and extrinsic factors that affect the texture and nature of bone tissues, and in many ways, we are better able to recognize the biological signals within fossil bone microstructure. Studies of nonmammalian therapsid bone microstructure have substantially advanced our understanding of the biology of these extinct vertebrates and have highlighted the immense scope for more research in this field.

Acknowledgments

I am indebted to all the curators of collections of Karoo vertebrates in South Africa who have over the years so generously provided material for histological research on nonmammalian therapsid bones. Daniel Thomas provided Figure 3.1A, and Fernando Abdala kindly provided Figures 3.6B and 3.6C. Sandra Jasinoski and Daniel Thomas are acknowledged for having provided constructive comments on an earlier draft. This manuscript was peer reviewed by Greg Erickson, Tom Hübner, and Allison Tumarkin-Deratzian.

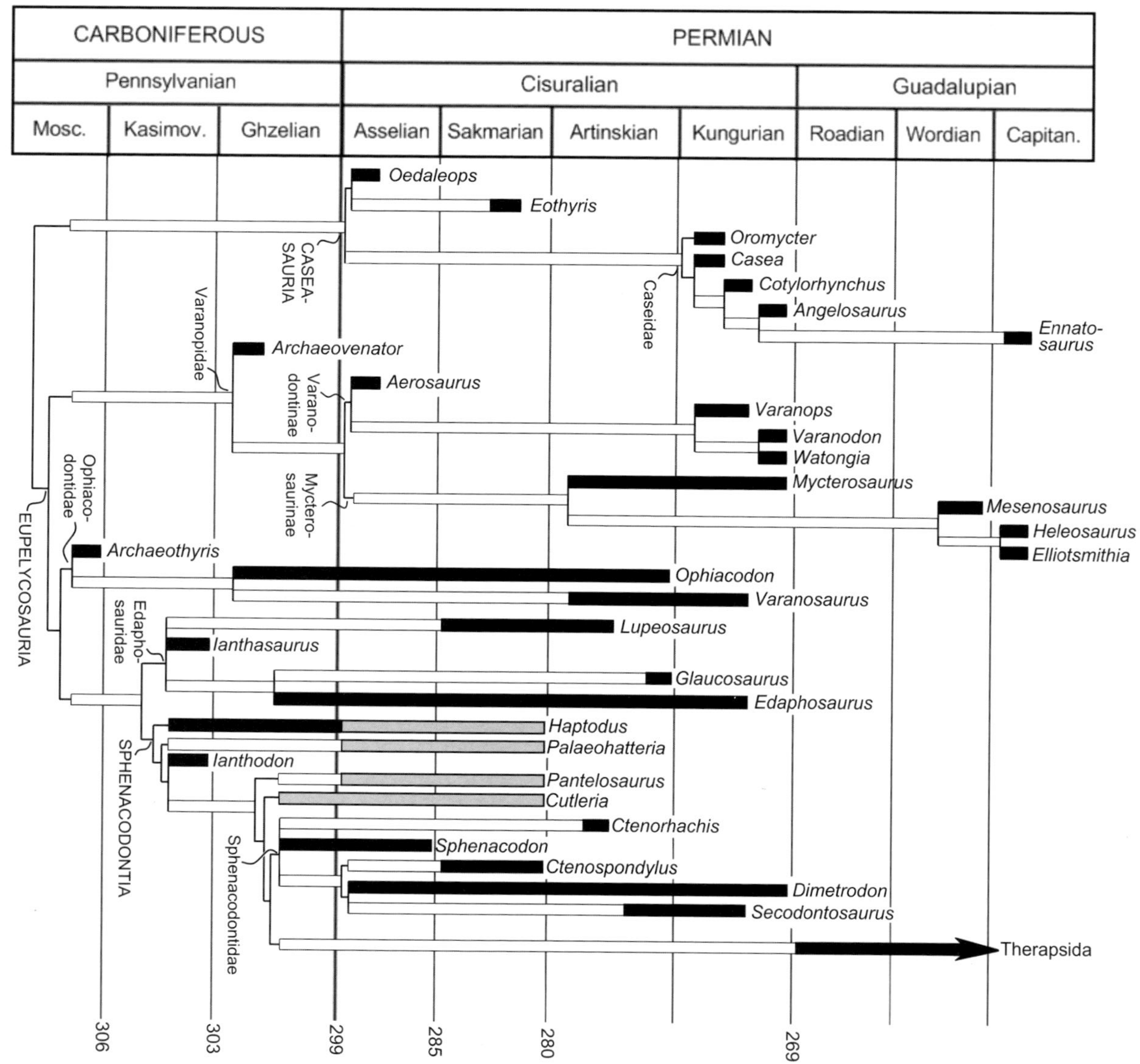
CARBONIFEROUS
PERMIAN
Pennsylvanian
Cisuralian
Guadalupian
Mosc.
Kasimov.
Ghzelian
Asselian
Sakmarian
Artinskian
Kungurian
Roadian
Wordian
Capitan.
Oedaleops
Eothyris
Oromycter
Casea
Cotylorhynchus
Angelosaurus
Ennato-
saurus
CASEA-
SAURIA
Caseidae
Archaeovenator
Varanopidae
Aerosaurus
Varano-
dontinae
Varanops
Varanodon
Watongia
Mycterosaurus
Mycterosaurinae
Mesenosaurus
Heleosaurus
Elliotsmithia
Ophiaco-
dontidae
EUPELYCOSAURIA
Archaeothyris
Ophiacodon
Varanosaurus
Edapho-
sauridae
Lupeosaurus
Ianthasaurus
Glaucosaurus
Edaphosaurus
Haptodus
Palaeohatteria
SPHENACODONTIA
Ianthodon
Pantelosaurus
Cutleria
Ctenorhachis
Sphenacodon
Sphenacodontidae
Ctenospondylus
Dimetrodon
Secodontosaurus
Therapsida
306
303
299
285
280
269

The Paleobiology and Bone Microstructure of Pelycosaurian-Grade Synapsids

4

Adam K. Huttenlocker and Elizabeth Rega

Introduction

Modern reptilian-grade amniotes (e.g., snakes, lizards, crocodilians) and the lineage that gave way to mammals (Synapsida) diverged more than 320 Mya during the Carboniferous Period. Also called pelycosaurian-grade synapsids or "pelycosaurs," basal synapsids were the dominant terrestrial vertebrates of the Late Pennsylvanian through Early Permian periods. Given their key phylogenetic position near the divergence, the adaptive evolution of their skeletal structures are especially significant, as the anamniote-amniote transition was classically characterized by a transition from a predominantly amphibious to a more terrestrial mechanical regime and lifecycle (Romer 1957, 1958; Germain and Laurin 2005).

Basal synapsids have been subject to extensive gross morphological, phylogenetic, and functional analyses which have revealed a diversity of body sizes and diets—including high fiber herbivory, as well as insectivory and apex-level carnivory—and attendant somatic and dental adaptations (e.g., Modesto and Reisz 1992; Maddin 2008). In the present chapter, we integrate existing information regarding basal synapsid gross osteology with new data regarding bone microstructure and biophysical modeling under both normal and pathological conditions. Importantly, histomorphological analysis underscores the diversity of lifestyles, growth patterns, and ecophysiological adaptations achieved early on in the synapsid lineage. For related topics omitted in the present review (in the interest of brevity), we recommend these additional readings: Kemp (1982, 2005), Hotton et al. (1986), Schultze and Trueb (1991).

4.1. Phylogeny-corrected stratigraphic ranges of basal synapsids superimposed onto the international Permo-Carboniferous global stratigraphy. Absolute dates of geologic stages (269–306) are in millions of years (Ma). Black bars represent observed ranges, white bars represent phylogeny-corrected ranges ("ghost lineages"), and gray bars represent uncertain ranges. Topology and stratigraphic ranges are based primarily on Huttenlocker (2008, unpublished), supplemented by data from Reisz, Berman, and Scott (1992), Laurin (1993), Modesto (1994), Laurin and Reisz (1995), Sidor and Hopson (1998), Kissel and Reisz (2004), and Campione and Reisz (2010).

Institutional Abbreviations

This chapter uses abbreviations for the following institutions: Carnegie Museum, Pittsburgh (CM); Field Museum, Chicago (FMNH UC); Harvard Museum of Comparative Zoology, Cambridge (MCZ); Sam Noble Oklahoma Museum of Natural History, Norman (OMNH); Royal Ontario Museum, Toronto (ROM); former University of California Vertebrate Paleontology Collections, Los Angeles (UCLA VP); University of California Museum of Paleontology, Berkeley (UCMP); Burke Museum of Natural History, Seattle (UWBM).

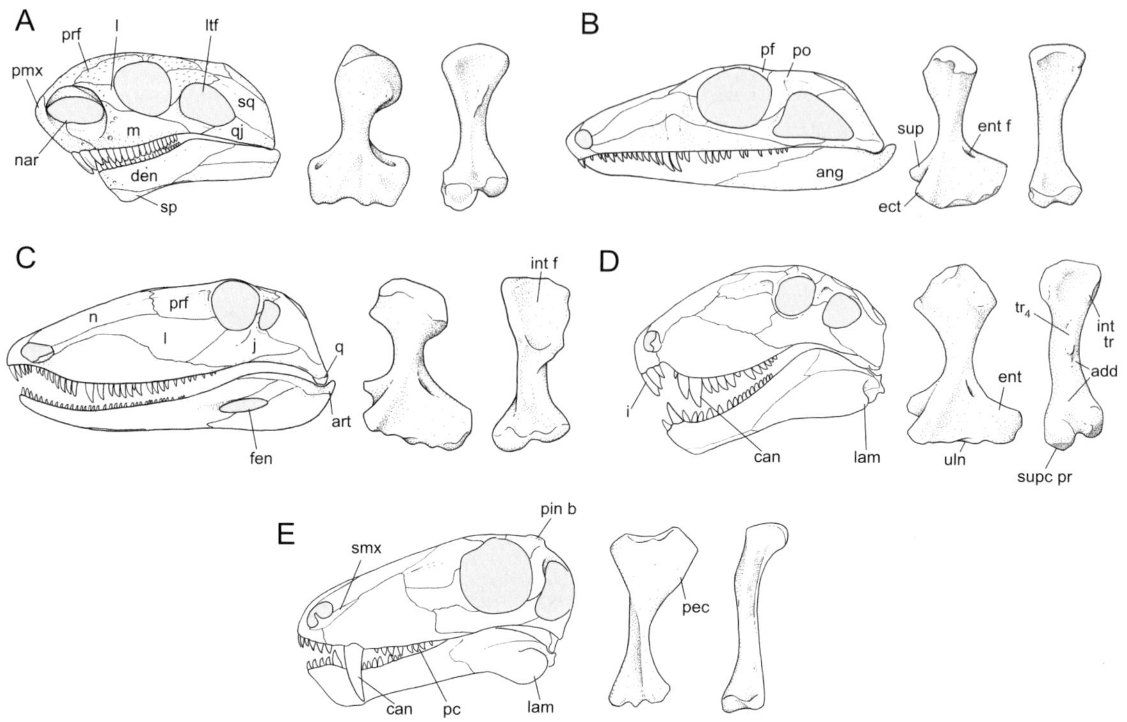

The Fossil Record of Basal Synapsids and Their Physical Environment

Basal synapsids are of significant interest to paleontologists because their record is relatively well-documented through geologic time and their evolutionary history holds clues to the origins of mammalian structure and physiology (Reisz 1986; Hopson 1991; Laurin 1993; Modesto 1994; Berman et al. 1995; Laurin and Reisz 1995; Sidor and Hopson 1998; Rubidge and Sidor 2001).The two major morphological grades of non-mammalian synapsids (introduced in chapter 1) are pelycosaurian-grade synapsids and therapsids. The pelycosaurian-grade synapsids, characterized by hyperelongated neural spines, reached the peak of their diversity during the Early Permian Period, approximately 300–269 Mya (Fig. 4.1; see Plate A). Their early diversification is best documented in Permo-Carboniferous localities of Euramerica which were arranged primarily along a paleoequatorial transect, representing subtropical to seasonally wet environments that varied greatly in local conditions, vegetation, and physiography, and indicating that early synapsids were well adapted to handle a modest array of habitats and environmental conditions (but perhaps not to the degree of their therapsid descendants; see chapters 1 and 11 of this book).

At least six well-known families of pelycosaurs were established by the Early Permian (Fig. 4.1) (chapter 1 of this book), a few of which eventually achieved distributions across North and South America, Europe, and Africa (Romer and Price 1940; Reisz 1986; Piñeiro et al. 2003). These included the caseasaurian families Eothyrididae and Caseidae, and eupelycosaurian families Varanopidae, Ophiacodontidae, Edaphosauridae, and

Sphenacodontidae. Some rare sphenacodontians from North America and Europe, the "haptodontines," have been interpreted as a paraphyletic-grade demonstrating trends toward the sphenacodontid-therapsid condition (Laurin 1993; Kissel and Reisz 2004). Their lack of hyperelongated neural spines coupled with their phylogenetic position between edaphosaurids and sphenacodontids supports the hypothesis that the dorsal sail arose more than once in early synapsids (i.e., first in edaphosaurids and later in post-haptodontine sphenacodontids). Kemp (1982, 2005) and Reisz (1986) have provided comprehensive reviews of early synapsid evolution and gross osteology. We provide only a brief summary of their diversity and evolutionary trends to supplement later discussion of their bone microstructure and its inferred paleobiological implications.

4.2. Restorations of selected synapsid skulls in left lateral view with associated humeri (dorsal view) and femora (ventral view). (A) The caseid *Cotylorhynchus;* (B) the varanopid *Varanops;* (C) the ophiacodontid *Ophiacodon;* (D) the sphenacodontid *Dimetrodon;* and (E) the basal therapsid *Biarmosuchus.* Cranial restorations are modified from Romer and Price (1940), Reisz (1986), Sigogneau-Russell (1989), and Rubidge and Sidor (2001); humerus and femur restorations are modified from Reisz (1986) and Sigogneau-Russell (1989). Skulls and limb bones are not to scale. *Abbreviations:* add = adductor ridge; ang = angular; art = articular; can = caniniform; den = dentary; ect = ectepicondyle; ent = entepicondyle; ent f = entepicondylar foramen; fen = mandibular fenestra; i = incisor; int f = intertrochanteric fossa; int tr = internal trochanter; j = jugular; l = lacrimal; lam = reflected lamina of angular; ltf = lateral temporal fenestra; m = maxilla; n = nasal; nar = external naris; pc = postcanine; pec = deltopectoral crest; pf = postfrontal; pin b = pineal boss; pmx = premaxilla; po = postorbital; prf = prefrontal; q = quadrate; qj = quadratojugal; smx = septomaxilla; sq = squamosal; sp = splenial; sup = supinator process; supc pr = supracondylar process; tr4 = fourth trochanter; uln = articular surface for ulna.

Major Features of Pelycosaurian-Grade Synapsids

Caseasauria is a monophyletic group of basal pelycosaurs that reached the zenith of its diversity during the late Early Permian and persisted into the Middle Permian (Reisz 1986; Maddin 2008; Reisz, Godfrey, and Scott 2009). Thus, it included some of the latest surviving lineages of pelycosaurian-grade synapsids (e.g., the Russian *Ennatosaurus* and *Phreatophasma*), occurring in Middle Permian deposits in Europe contemporaneous with therapsids. The eothyridids occur in lowermost Permian rocks of New Mexico, and Texas, United States, and were typified by their small size and a highly procumbent rostrum with enlarged external nares characteristic of caseasaurs as a whole (Reisz, Godfrey, and Scott 2009). The more derived caseids, generally associated with inland faunas during the latter part of the Early Permian, achieved much larger body sizes (upward of 300 kg [Romer and Price 1940]) and were characterized by their large, barrel-shaped trunk and robust limbs with a diminutive skull (e.g., *Cotylorhynchus*) (Fig. 4.2A). Interpreted as terrestrial herbivores, they elaborated the marginal dentition into teeth bearing tall pedicels crowned by apical serrations or cuspules, possibly allowing more efficient processing of fibrous plant material (Maddin 2008).

Eupelycosauria (Figs. 4.2B to 4.2E), the monophyletic sister taxon to Caseasauria, notably included the lineage that gave rise to therapsids. It consisted of four major families of pelycosaurian-grade synapsids which shared numerous synapomorphies associated with modifications of the skull, including: tall, medially compressed rostrum; broad exposure of frontal on dorsal margin of orbit; long, narrow supratemporal; and pineal foramen positioned far back on posterior skull roof (Reisz 1986). Most additionally displayed shallow lateral excavations in neural arches accommodating the insertion of interarticular muscles (except in ophiacodontids) and rearrangement of muscles associated with the hip (Romer and Price 1940; Reisz 1986).

The Varanopidae constitutes one of the longest-lived eupelycosaurian families, with a stratigraphic record spanning the upper Pennsylvanian (Ghzelian) to the upper Middle Permian (Fig. 4.1). Their biogeographic distribution was extensive as their fossils have been found in North America, Europe, South America, and southern Africa. This successful group of carnivorous synapsids was characterized by its monitor lizard–like appearance, an unusually long row of mediolaterally compressed teeth extending beyond the posterior limits of the orbit, a serrated, sharply recurved dental morphology (although some primitive forms retained a more simple, conical dentition; e.g., *Archaeovenator*), and long, slender limbs. The gracile nature of the appendicular skeleton suggested to Reisz (1986) that varanopids were active, terrestrial predators convergent to the distantly related sphenacodontids. Two derived subfamilies may have risen to dominance as common predators in North American terrestrial ecosystems during the latter part of the Early Permian: the varanodontines (e.g., *Varanops*) (Fig. 4.2B) and the mycterosaurines (e.g., *Mycterosaurus*). The latter was the longer-lived of the two lineages and includes representatives from the Middle Permian of Russia (*Mesenosaurus*) and South Africa (*Heleosaurus, Elliotsmithia*). Their osteology and systematics have been reviewed by Reisz, Dilkes, and Berman (1998), Modesto et al. (2001), Anderson and Reisz (2004), Maddin, Evans, and Reisz (2006), Reisz and Modesto (2007), and Campione and Reisz (2010).

Ophiacodontids were a comparatively more depauperate group of early eupelycosaurs typified by marked elongation of the rostrum, ascending process of the supracanine buttress on the medial surface of the maxilla, numerous "precanine" teeth, anteroposteriorly expanded neural spines (being equal or greater than the length of the centrum), and only one of two sacral ribs articulating directly with the iliac blade (Berman et al. 1995). The earliest fossils that can be assigned confidently to the family Ophiacodontidae are of the genus *Archaeothyris* from the Pennsylvanian (Moscovian) of Nova Scotia, Ohio, and the Czech Republic (Reisz 1972, 1975). These fossils represent the oldest unequivocal synapsid remains known to date. Their transported appearance and general scarcity in otherwise rich, fossiliferous coal swamp deposits suggested to Reisz (1975) that early eupelycosaurs such as ophiacodontids emerged in upland environments outside the preservational window of low-lying depositional basins. More derived forms known from the Lower Permian of Texas and Oklahoma (i.e., *Ophiacodon*) (Fig. 4.2C) apparently exhibited a secondarily aquatic or semi-aquatic habit based on their piscivore-like cranial morphology (Romer and Price 1940; Reisz 1986), stout limbs with blunt terminal phalanges (Romer and Price 1940), and aspects of the histomorphology of their long bones (de Ricqlès 1974b; Germain and Laurin 2005).

The edaphosaurids were distributed broadly across Euramerica during the Permo-Carboniferous transition and represented the earliest known tetrapod group having a dorsal sail formed by an array of hyperelongated neural spines of the cervical and dorsal vertebrae. The neural spine structure was unique in that it often bore laterally directed tubercles or crossbars, in contrast to the smooth, lightly striated lateral surface exhibited by later sphenacodontids. Thus, the family was, in general, characterized by greatly elongated neural spines that are largely subcircular in cross section; anteriorly leaning cervical and posteriorly leaning "lumbar" neural spines (Modesto and Reisz 1990a, 1990b); and laterally projecting tubercles on the presacral spines that were usually paired toward the base (although the aberrant *Lupeosaurus* appears to have lacked lateral tubercles) (Sumida 1989; Modesto and Reisz 1990a). They are generally regarded as primitively omnivorous with more derived taxa likely herbivorous, based on the possession of bulbous, isodont dentition in adults and a battery of palatal teeth. As their fossils predate those of caseids, they, along with diadectid anamniotes, are likely candidates for the first herbivorous tetrapods (Modesto and Reisz 1992). Revisions to the taxonomic composition of Edaphosauridae have identified at least four genera within the family, including *Lupeosaurus*, *Ianthasaurus*, *Glaucosaurus*, and *Edaphosaurus* (see Plate A) (Berman 1979; Reisz and Berman 1986; Sumida 1989; Modesto and Reisz 1990a, 1990b; Modesto 1994). Only recently has a general consensus of their phylogenetic relationships emerged, with most phylogenies (1) identifying progressive trends in body size evolution and ornamentation of the dorsal sail among *Edaphosaurus* species, (2) recognizing a sister taxon relationship between the genera *Glaucosaurus* and *Edaphosaurus*, and (3) recovering the genera *Lupeosaurus* and *Ianthasaurus* in basal positions (Modesto and Reisz 1992; Modesto 1994, 1995; Mazierski and Reisz 2006, 2010).

The higher-level taxon Sphenacodontia includes all haptodontine-grade synapsids, Sphenacodontidae, and Therapsida (Fig. 4.1). The "haptodontines," traditionally regarded as a subfamily of Sphenacodontidae (Romer and Price 1940), represented an early, paraphyletic grade of basal sphenacodontians that were excluded from the clade of Sphenacodontidae sensu stricto plus therapsids (Sphenacodontoidea). They were characterized by a slightly hooked snout with a sinuous alveolar margin, enlarged first incisor and subtle caniniform dentition with a prominent supracanine buttress, a well-developed reflected lamina of the angular, and an enlarged triceps process of the coracoid (Reisz, Berman, and Scott 1992; Laurin 1993).

In excluding the paraphyletic haptodontines, Sphenacodontidae has been cladistically defined as all synapsids sharing a more recent common ancestry with *Sphenacodon* and *Dimetrodon* (Fig. 4.2D) than with Therapsida (Hopson and Barghusen 1986; Reisz 1986; Laurin 1993). In addition to predatory modifications of the skull, including elaboration of the canines, sphenacodontids displayed a number of postcranial specializations indicative of a greater degree of terrestrial locomotion and energetics (many of which were shared with therapsids). These included modifications of cross-sectional geometry and second moment of area in the femur, limited by bending moments rather than torsional stresses (convergent in varanopids) (Fig. 4.3); more pronounced sigmoidal curvature of the femoral diaphysis; modestly developed adductor ridge confluent with an enlarged fourth trochanter of the femur; and three sacral ribs stabilizing the pelvic girdle instead of two or less (Romer and Price

1940; Kemp 1982). Trackways in New Mexico attributable to the ichnogenus *Dimetropus* provide evidence of narrow pes spacing and high pace angulation indicative of semi-erect posture and a lack of belly- or tail-dragging in at least some sphenacodontids (Hunt and Lucas 1998). As in edaphosaurids, the family as a whole was likely typified by elongated neural spines forming a stiff dorsal crest supporting the trunk, a trend that was taken to its greatest extreme in the sail-backed *Dimetrodon* (see Plate A) and *Secodontosaurus* (Hook and Hotton 1991; Reisz, Berman, and Scott 1992; Huttenlocker, Rega, and Sumida 2010).

Paleobiological Context: Biophysical Models and the Autecology of Basal Synapsids

Although synapsids and other early amniotes occupied a variety of body size and ecological niches during the Permo-Carboniferous transition, body size increase has been identified as a major trend characterizing the evolution and temporal succession of amniote lineages during the Permian (Romer and Price 1940; Turner and Tracy 1986; Carroll 1991; Laurin 2004). Disparate body size niches existed even in closely related species, ranging from gracile, small-bodied Permo-Carboniferous forms, less than a kilogram, to very large forms (~300 kg in the case of *Cotylorhynchus*) with broad trunks and robust limbs more common during the late Early Permian. If more than a taphonomic artifact, such patterns reflect either a passive tendency to move away from one extreme ancestral size (Stanley 1973; McShea 1994; Laurin 2004) or an active trend that was driven by some conferred selective advantage (Kingsolver and Pfennig 2004). If only probabilistic, a passive trend could have been permitted by relaxing upper bounds imposed by physiological constraints, thereby allowing the potential for new size niches to be filled (McShea 1994).

Bashkirian- (early Pennsylvanian) and Moscovian-aged amniotes, such as *Hylonomus* and *Archaeothyris*, respectively, typically exhibited presacral lengths of an estimated 10 to 25 cm and, contrary to traditional views, likely lived in well-drained basins experiencing seasonal, Permian-normative dry land conditions (Reisz 1975; Carroll 1991; Falcon-Lang 2003; Falcon-Lang et al. 2009). Some of their later relatives achieved much larger sizes during the late Early Permian and thrived in habitats characterized by increasingly arid conditions. Thus, a number of physiologic and reproductive novelties were probably necessary to relax body size limitations: drought tolerance would have relaxed constraints on aquatic reproduction, thereby allowing reproduction and/or egg-laying to take place in otherwise stressed conditions; extra-embryonic membranes facilitated gas exchange and prevented desiccation, and may have allowed extended embryo retention; and relaxed constraints on egg size allowed for more precocial neonates having larger neonatal body mass (Carroll 1991; Laurin and Girondot 1999). Laurin (2004) estimated that the common ancestor of crown-amniotes was likely larger than previously predicted (Carroll 1991) and probably would not have had a snout-vent length less than 120 mm. Whereas egg diameter correlates positively with adult body size in extant reptiles, such a strong correlation does not exist in frogs and salamanders as their egg sizes usually do not exceed 10 mm (Carroll 1991; Graham, Pianka, and McEachern 2001). Increased body size in terrestrial tetrapods may have been further influenced by increases in postnatal growth rates.

The ecophysiological and evolutionary importance of postnatal body size in synapsids was stressed by Turner and Tracy (1986), who implicated the acquisition of large body size not only in terrestriality, but in the origins and evolution of homeothermy (involuntary maintenance of a relatively constant body temperature). Having relaxed body size constraints, Permo-Carboniferous synapsids took on a range of sizes, and did so in both wetland and dry land (or seasonally dry) habitats. Seasonal variation in precipitation and temperature would have influenced foraging and daily activity in synapsids of all sizes, but more so in smaller synapsids having a high surface area-to-volume ratio.

Turner and Tracy (1986) and Tracy, Turner, and Huey (1986) noted that reptiles that behaviorally thermoregulate may preferentially sacrifice operating at their physiologically optimum activity temperature in short bouts in order to operate for longer periods at suboptimal temperatures. In the absence of observable behavioral adjustments which are difficult to detect in the fossil record, these researchers argued that testable modifications of body sizes, shapes, and accessory appendages may have conferred a selective advantage on large synapsids due to their inferred effects on thermal homeostasis. Specifically, large-bodied synapsids could have circumvented the problem of behavioral thermoregulation if their size permitted sufficient thermal inertia. The biophysical models presented by the researchers demonstrated, however, that an increase in size was necessarily associated with greater external resistance to heat transfer due to a reduced capacity for heat exchange via blood flow to the appendages; this would have been potentially problematic for active, large-bodied predators that might have otherwise operated at high activity temperatures (e.g., sphenacodontids).

Based on earlier models proposed by Bramwell and Fellgett (1973), Tracy, Turner, and Huey (1986) examined the conditions in which the dorsal sail of *Dimetrodon* might have functioned in temperature regulation—a notion first offered by Romer (1948, 1961) after observing positive allometry in the surface area of the sail. They suggested that, if the sail played any role in thermoregulation, it did not likely function to dissipate metabolic heat as the allometric coefficient of the sail ($m^{1.13}$) better fit a model of heat storage ($m^{1.0}$). These and more recent studies considering seasonal variation in daily activity (Florides et al. 2001) suggested that the evolution of body size and sail size promoted homeothermy at high body temperatures (even in smaller *Dimetrodon* species as a means of reducing the effects of daily temperature fluctuation). However, the efficacy of an ancillary heat exchanger is still unclear and alternative hypotheses have been proposed for its specific functions. For example, Haack (1986) derived a detailed biophysical model of the sail that required up to 4 hours per day for a 250-kg sail-backed synapsid to achieve its activity temperature in the early morning, as compared to 80 minutes estimated earlier by Bramwell and Fellgett (1973). Bennett (1996) suggested that, in edaphosaurids, the lateral tubercles or crossbars might have increased turbulent airflow across the sail, facilitating more rapid cooling. A more thorough understanding of the sail's growth, mechanics, and soft-tissue correlates would better substantiate some underlying assumptions of these models. The histological evidence for dorsal sail function has been reviewed by de Ricqlès (1974b); Huttenlocker, Rega, and Sumida (2010); and Huttenlocker, Mazierski, and Reisz (2011) and is summarized later in this chapter.

Review of Bone Microstructure in Pelycosaurian-Grade Synapsids

Studies of bone microstructure have been instrumental in contributing to some of the paleobiological inferences recounted herein. Most notably, numerous investigators have attempted to use bone microstructure to assess the thermal physiology of therapsids and pelycosaurian-grade synapsids based on comparative differences between the bone microstructure of extant reptiles and that of endothermic mammals and birds (Enlow and Brown 1958; Peabody 1961; Warren 1963; de Ricqlès 1974a, 1976; Bennett and Ruben 1986). De Ricqlès (1974a, 1974b, 1976) argued that highly disorganized fibrolamellar bone tissues, common to mammals and birds, were broadly distributed among (though not exclusive to) therapsids, whereas cyclic and more regularly-lamellated, or "lamellar-zonal," patterns were common to the earlier pelycosaurs as in most extant reptiles and amphibians. Notably, however, usage of the term lamellar-zonal bone in the literature often conflates concepts of matrix-structure (lamellar) and the record of growth rhythms (zonal); fibrolamellar bone may also be zonal in some large, fast-growing amniotes exhibiting multi-year growth, such as dinosaurs. Nevertheless, a dichotomy was drawn in synapsids: pelycosaurs exhibited histological organization suggestive of slow, cyclic growth, whereas therapsids attained a relatively faster, sustained growth trajectory, and were possibly "endothermic" (de Ricqlès 1974a, 1976; Bennett and Ruben 1986). This view holds that the deposition of fibrolamellar bone in therapsids should be correlated with rapid somatic growth and its associated metabolic demands, whereas its rarity in most pelycosaurian-grade synapsids was a reflection of their presumably lower metabolic costs.

Little is known about the bone microstructure of pelycosaurian-grade synapsids, as the number of genera whose osteohistological structure has been examined is vanishingly small (Table 4.1). As a result, some of the hypotheses of earlier studies presented here were limited by the small sample sizes available at the time and do not necessarily reflect the views of the authors of this chapter. The first broad surveys of bone microstructure in Paleozoic and other fossil tetrapods were provided by Enlow and Brown (1956, 1957, 1958) and synthesized later by Enlow (1969) with an emphasis on the bone microstructure of reptilian-grade

Table 4.1. List of pelycosaurian-grade synapsids that have been sampled histologically

Taxon:			Source(s):
Caseasauria			
	Caseidae		
		Ennatosaurus tecton	de Ricqlès (1974b)
Eupelycosauria			
	Varanopidae		
		Mycterosaurinae spp.	present study
		Varanops brevirostris	de Ricqlès (1974b)
		Varanops cf. *V. brevirostris*	present study
	Ophiacodontidae		
		Ophiacodon spp.	Enlow and Brown (1957); Enlow (1969); de Ricqlès (1974b); Germain and Laurin (2005)
	Edaphosauridae		
		Ianthasaurus hardestiorum	Huttenlocker, Mazierski, and Reisz (2011)
		Edaphosaurus spp.	Enlow and Brown (1957); Enlow (1969); de Ricqlès (1974b); Huttenlocker, Mazierski, and Reisz (2011)
		Lupeosaurus cf. *L. kayi*	Huttenlocker, Mazierski, and Reisz (2011)
	Sphenacodontidae		
		Dimetrodon spp.	Enlow and Brown (1957); Enlow (1969); de Ricqlès (1974b); Germain and Laurin (2005); Huttenlocker, Rega, and Sumida (2010)
		Sphenacodon spp.	Huttenlocker, Angielczyk, and Lee (2006); Huttenlocker, Rega, and Sumida (2010)

amniotes. The surveys of Peabody (1961), published posthumously, and unpublished work by Warren (1963) exposed the skeletochronological utility of growth marks in the bone of Paleozoic tetrapods, including stem amniotes and basal synapsids. De Ricqlès (1969, 1972, 1974a, 1974b, 1976) published a series of surveys that included samples from therapsids and some pelycosaurian-grade synapsids (Table 4.1). More recent work has led to the hypothesis that bone compactness profiles could shed light on habitat (e.g., aquatic versus terrestrial [Germain and Laurin 2005; Kriloff et al. 2008]) and examined the interplay between growth and mechanics of unusual skeletal structures (Huttenlocker, Rega, and Sumida 2010; Huttenlocker, Mazierski, and Reisz 2011). The present review highlights some of this work within a systematic framework.

Caseasauria

What little is known about the bone microstructure of caseasaurs comes from the seminal survey of de Ricqlès (1974b). In his survey, de Ricqlès serially sectioned and described a left femur attributable to the Middle Permian caseid *Ennatosaurus tecton*, probably from the type Moroznitsa locality (late Guadalupian) near Pinega, northwestern Russia.

The femur was a short (67 mm), robust bone with expanded proximal and distal ends and a poorly defined shaft (e.g., compare Fig. 4.2A with de Ricqlès 1974b, Fig. 10). The shaft was subtriangular in section, likely due to the extent of the intertrochanteric fossa and a sharp adductor ridge

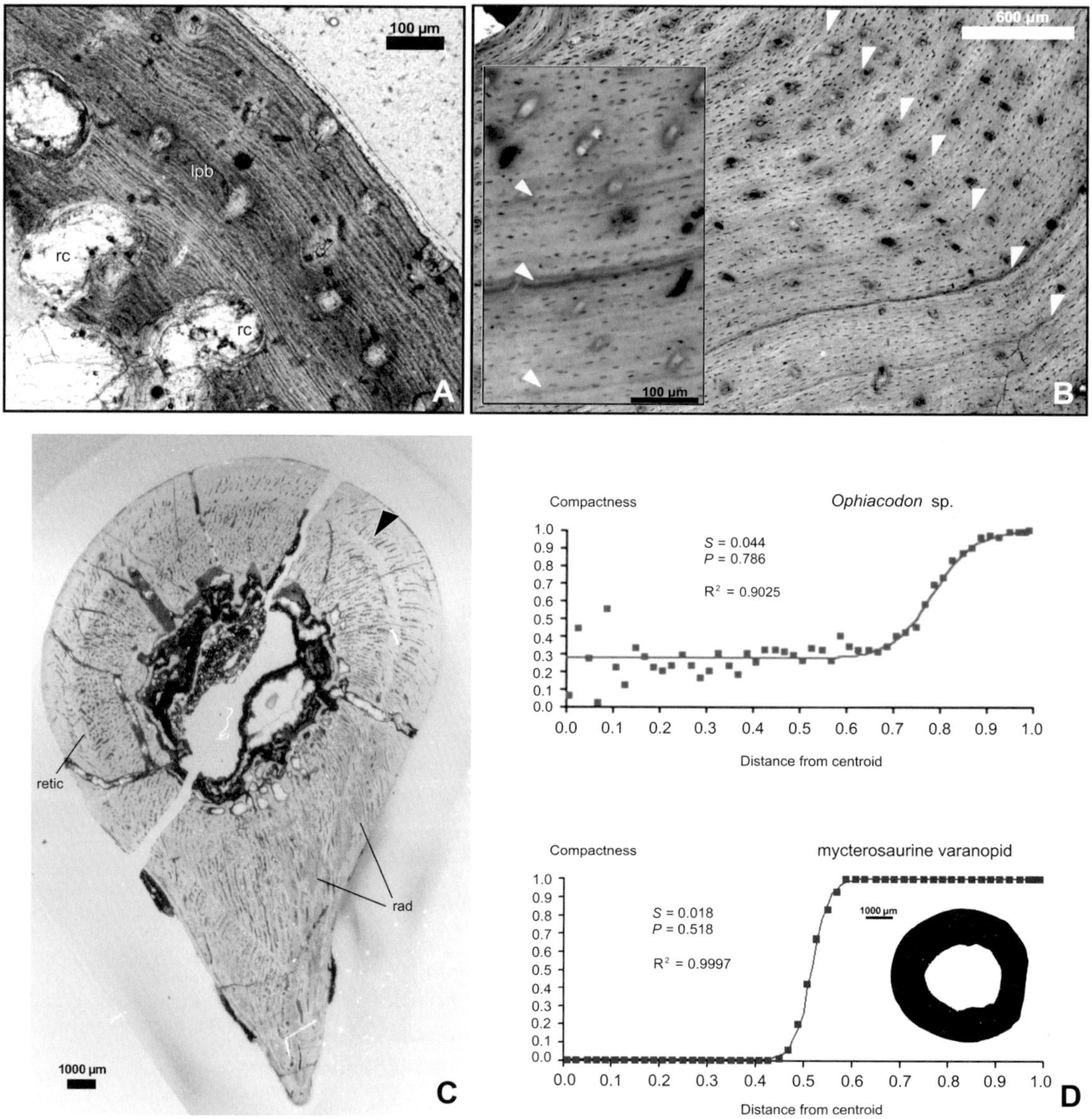

extending onto the shaft. In the anterodorsal margin of the midshaft, the cortex, largely formed of lamellar bone, was approximately 2 mm thick and moderately vascularized with longitudinal primary canals. However, in the posterior region, the cortex was nearly avascular. The slightly concave anteroventral surface exhibited an irregular lamellar structure dominated by obliquely oriented (radial) canals and numerous Sharpey's fibers, likely indicating the insertion of the adductor musculature. The inner lumen consisted of a coarse cancellous structure with enlarged "perimedullary" erosion cavities along the endosteal margins, created from resorption of lamellar bone and deposition of only a few lamellae defining the cavity borders (de Ricqlès 1974b, Plate VI, 4). There was, however, no evidence of true osteonal remodeling within the more peripheral regions of the cortex. A medullary cavity was apparently not well developed, at least not to the extent of modern terrestrial lizards, mammals, and birds. (The absence of a medullary cavity was probably somewhat common in the limb bones of Paleozoic tetrapods and has been suggested to be associated with the influence of lifestyle and mechanical usage on the overall compactness of the bone.) In addition to the small size of the bone, the general thinness of the trabecular spongiosa, erosion cavities, absence of secondary osteons, and the overall low cortical thickness suggested that the specimen represented an immature individual (de Ricqlès 1974b).

Varanopidae

Although little more histological data is available for varanopids, three separate samples are accessible for the present discussion: (1) a varanodontine, *Varanops brevirostris*, from the Clear Fork of Texas (de Ricqlès 1974b), (2) new material of *Varanops* cf. *V. brevirostris* (Fig. 4.3) from Richards Spur, Oklahoma (Kungurian, Lower Permian), and (3) multiple elements from an unidentified mycterosaurine (Figs. 4.3B, 4.3D) also from Richards Spur. De Ricqlès (1976, 129) provided anecdotal observations on the histology of the varanodontine *Watongia* (interpreted then as a gorgonopsian therapsid) stating that it "retained a completely pelycosaurian pattern of bone tissue." Despite later usage of de Ricqlès's remarks (e.g., Bennett and Ruben 1986, 209), the source of the material and the meaning of the statement were not originally made explicit.

Two elements, a poorly preserved radius and fibula attributed to the varanodontine *Varanops brevirostris*, were serially sectioned and examined by de Ricqlès (1974b). Unfortunately, diagenetic alteration rendered impossible a thorough description, and the material was therefore not figured at the time. Only gross aspects of the bone profile of the radius were resolvable, including the possession of a thin cortex (about 1 mm in thickness) preserving an abrupt transition with the spongiosa and abundant, coarse trabeculae within the lumen. Although radii were not available in the current sampling for a more precise comparison, sectioning of femoral material from Richards Spur, Oklahoma, referable to *Varanops* cf. *V. brevirostris* (Maddin, Evans, and Reisz 2006), generally corroborates these earlier observations. The section illustrated in Figure 4.3 was taken from the distal portion of the shaft of a left femur, just proximal to the metaphysis. The cortical wall is less than 1 mm in thickness and exhibits an abrupt transition with the inner lumen. Although there are numerous coarse trabeculae present along the endosteal margin, there appears to have been a well-defined medullary cavity (unlike the limb bones of *Ennatosaurus* or other sampled eupelycosaurs [Enlow and Brown 1957; Enlow 1969; de Ricqlès 1974b]). However, as in most early tetrapod bones, the inner cortex shows extensive cancellous structure with large resorption cavities giving way to coarse trabeculae, created from the sparse remaining lamellar bone at the margins of these resorption cavities and a few deposited lamellae circumscribing the cavity borders. There appears to have been no secondary reconstruction in the outer cortex. Primary

4.3. Comparison of bone microstructure in varanopids and ophiacodontids. (A) The varanodontine varanopid *Varanops* cf. *V. brevirostris* from Richards Spur, Oklahoma, United States, UWBM 89463, femur, cross section taken from distal portion of diaphysis, near metaphysis, viewed under polarized light. (B) A mycterosaurine varanopid from Richards Spur, UWBM 89464, femur, high-magnification cross section taken from mid-diaphysis, viewed under polarized light. Note poorly vascularized lamellar bone in subperiosteal layer of outer cortex, toward bottom of image (inset: high-magnification cross section taken from distal portion of the diaphysis, viewed under polarized light). (C) Photomicrograph of the radius of *Ophiacodon* (courtesy M. Laurin), showing overall low compactness with a highly cancellous lumen and porous inner cortex composed of longitudinal, reticular, and radial canals. (D) Comparison of the compactness profiles of the radius of *Ophiacodon* (top) and femur of an undescribed mycterosaurine varanopid (bottom). *S* is the reciprocal of the slope at the inflection point, representing the gradualness of the transition from the medulla to the cortex; *P* is the distance to the inflection point, representing the midpoint of the transition zone. Compactness profiles produced in Bone Profiler (Girondot and Laurin 2003). Arrows indicate LAGs. *Abbreviations:* lpb = lamellar primary bone; rad = radial canals; rc = resorption cavity; retic = reticular canals.

osteons are sparsely distributed throughout a highly lamellated matrix, and their orientation is exclusively longitudinal. No clear lines of arrested growth (LAGs) are distinguishable within the section, although a possible annulus is present within the inner cortex. The scarcity of growth marks is a likely result of the overall thinness of the bone wall.

At least two (and possibly a third) species of varanopid co-occurred at the Richards Spur locality (Reisz, Wilson, and Scott 1997; Maddin, Evans, and Reisz 2006). Hind limb elements from multiple individuals herein referred to an indeterminate mycterosaurine are distinguishable from *Varanops* by their generally smaller size, slender shaft with narrow proximal and distal ends, short intertrochanteric fossa, and indistinct internal trochanter and adductor ridge (V. Pelletier, pers. comm. 2009). Selected distal and midshaft sections are illustrated in Figure 4.3 for the present treatment.

In spite of its smaller size compared with the *Varanops* specimens, the absolute thickness of the cortical wall of the femur throughout the shaft is approximately 1.5 to 2 mm (≈21 percent its cross-sectional diameter). The cortex is highly compact with an abrupt transition to the medulla. Coarse trabecular structures are present within the medullary cavity, recording compact-to-cancellous conversion in the distal shaft, toward the metaphysis. Within the midshaft, however, resorption cavities are lacking near the endosteal margins, and the medullary cavity is instead lined by smooth, circumferentially deposited endosteal bone. There is no cancellous bone (and no trabeculae) within this region. Although the early record of growth was likely eliminated with the construction of the medullary cavity, densely packed (but unevenly distributed) longitudinal vascular canals are preserved within a lamellar matrix composing the inner cortex. Faint annuli record a regular rhythm of lamellated bone deposition. This growth period was interrupted by a distinct LAG and followed by more sparsely vascularized zones in the outer cortex, also incorporating an exclusively lamellar matrix. This transition is abrupt and is probably indicative of somewhat slowed growth (though it does not indicate cessation of growth in the individual). Albeit incomplete, this interpretation of the growth pattern is not unexpected for a pelycosaurian-grade synapsid and resembles that inferred for many extant reptiles of small body size (Castanet and Smirina 1990; Castanet 1994; Piantoni, Ibargüengoytía, and Cussac 2006a, 2006b). However, the histological profile of varanopid long bones appears to deviate from earlier expectations of pelycosaur limb bone structure (Enlow and Brown 1957; Enlow 1969) in the nature of its bone compactness profile and the possession of a well-defined medullary cavity.

Ophiacodontidae

The limb bone microstructure of *Ophiacodon* (Fig. 4.2C) has been relatively well studied compared with that of the aforementioned families. The first detailed photomicrographs were published by Enlow and Brown

(1957) and Enlow (1969). Subsequently, de Ricqlès (1974b) and, most recently, Germain and Laurin (2005) have described the cross-sectional compactness of its limb bones with reference to its inferred lifestyle (discussed later). The cross-sectional profile is illustrated in Figure 4.3D where it is compared to that of a mycterosaurine varanopid.

In contrast to some other early amniotes (e.g., the caseids and varanopids discussed earlier), Enlow and Brown noted that "circumferential lamellae in the compacta . . . are absent or poorly differentiated" (1957, 204). Instead, their sample (including long bones and ribs) revealed highly vascularized compacta with what appeared as a woven- or parallel-fibered matrix incorporating densely packed, isodiametric osteocyte lacunae typical of rapidly deposited bone. Lamellar bone tissue in these sections was restricted only to the primary osteons and not the surrounding matrix. De Ricqlès (1974b) described and figured a partial radius from the former Admiral (Nocona?) Formation of north-central Texas, which exhibited a similar histology. In general, the numerous primary osteons were arranged randomly and oriented both longitudinally and obliquely within a largely parallel-fibered matrix. In the midshaft, the cortex appeared moderately thin-walled, but exhibited no clear demarcation between the compacta and the highly cancellous lumen; in other words, the transitional profile from the endosteal lumen to the cortex was relatively gradual due to occlusion of the cancellous lumen and comparatively high porosity of the compacta. This resulting non-pathologic osteosclerosis of an occluded lumen thereby produced a greater degree of overall compactness with only a weakly developed medullary cavity.

Edaphosauridae

Early synapsid osteohistology is mainly known from studies of bone microstructure of the sail-backed Edaphosauridae and Sphenacodontidae. However, histologic sampling for each family has until recently been represented by the individual genera *Edaphosaurus* and *Dimetrodon*, respectively. Enlow (1969) and Enlow and Brown (1957) examined fragmentary material attributable to *Edaphosaurus*, including a mandible, ribs, a femur, and neural spines, and demonstrated similar histological structure among elements. The results of de Ricqlès (1974b) corroborated those earlier observations. Warren (1963) compared annular markings on the articular facets of the vertebrae of several Permian amniotes (and other tetrapods) and suggested that the putative edaphosaurid *Lupeosaurus* and the anamniote *Diadectes* exhibited prolonged, but cyclic multiyear growth to large body sizes (although the total number of annular markings could not be easily discerned in *Lupeosaurus*). Surficial skeletochronology was not supplemented with thin-sectioning of additional elements in *Lupeosaurus*.

Histological sampling by Huttenlocker, Mazierski, and Reisz (2011) has made possible rudimentary comparisons of bone microstructure between three edaphosaurid genera: *Lupeosaurus*, *Ianthasaurus*, and

4.4. Selected photomicrographs of bone microstructure in the neural spines of the edaphosaurid eupelycosaurs. (A) *Lupeosaurus* cf. *L. kayi* (UCLA VP 1651), transverse section through distal region of a mid-dorsal neural spine viewed with polarized light. Note sparse longitudinal canals within a predominantly lamellar and parallel-fibered matrix and nearly avascular outer cortex. (B) Transverse section through putative fracture callus in distal region of the same specimen as (A). Note deposition of lamellar primary bone superficial to the woven "callus" area, indicating the individual likely survived for a short time following the injury. (C) *I. hardestiorum* (ROM 59933), transverse section through distal region of mid-dorsal spine showing nearly avascular lamellar cortical bone, with large resorption cavities along endosteal margin and a well-developed central cavity (Huttenlocker, Mazierski, and Reisz 2011, Fig. 4A). (D) *Edaphosaurus sp.* (OMNH 73804), longitudinal section through distal tip of neural spine (Huttenlocker, Mazierski, and Reisz 2011, Fig. 7A). Note cancellous conversion of primary cortical bone evident in the medullary region, likely associated with the distal elongation of the spine. (E) *Edaphosaurus sp.* (OMNH 73802), longitudinal section through lateral tubercle showing reticular, woven-fibered bone indicative of rapid tubercle outgrowth, followed by lamellar bone in outer cortex (direction of periosteal surface is toward the left of image). (F) *Edaphosaurus sp.* (OMNH 73799), longitudinal section through lateral tubercle showing woven-fibered bone, obliquely oriented primary vascular canals, and large resorption cavities forming within the inner cortex. *Abbreviations:* eb = endosteal bone; lat t = lateral tubercle; lpb = lamellar primary bone; lpc = longitudinal primary canal; mc = medullary cavity; rc = resorption cavity; retic = reticular canals; wb = woven-fibered bone.

Edaphosaurus. Two vertebrae from a partial skeleton of *Lupeosaurus* cf. *L. kayi* from the former Admiral (Nocona?) Formation of Texas were sampled histologically to expose histovariability in the neural spine, reconstruct soft-tissue correlates, infer growth patterns, and to assess their similarity to those of better-known edaphosaurids. In the dorsal series, the neural spines were extremely long (at least 10 to 12 times the height of the vertebral centrum in the mid-dorsals) and subcircular in cross section with shallow anterior and posterior longitudinal grooves. Unlike *Edaphosaurus* and most of the spines in *Ianthasaurus*, the spines of *Lupeosaurus* apparently lacked lateral tubercles. The histological structure of the spines, however, shared some similarities with those of other edaphosaurids. Oblique resorption cavities and deeply penetrating Sharpey's fibers oriented at angles oblique to the periosteal surface were preserved proximally in the spine, likely demarcating the epaxial muscle-embedded region of the neural spine (Huttenlocker, Rega, and Sumida 2010, Fig. 3A). Remodeling was apparently not as extensive in this region as it was in *Dimetrodon* (discussed later). These structures apparently disappeared more distally in the spine, giving way to a more regular, finely lamellated structure with faint, localized radial structures similar to those identified in *Edaphosaurus* as Sharpey's fibers by Enlow (1969, Fig. 24). Erosion cavities were localized around the endosteal margins throughout the distal spine (as in the limb bones of many early amniotes and anamniotes) and sparse trabecular structures occupied an incipient medullary cavity, resembling the condition found in other edaphosaurids. However, the degree of compactness transitioning from the lumen to the outer cortex was still comparatively gradual. Throughout the cortex, the matrix was composed almost entirely of lamellar primary bone with very sparse, longitudinally oriented canals (Fig. 4.4A), the subperiosteal region being nearly avascular. This pattern deviated in a region of a mid-dorsal spine exhibiting a putative healing callus where reticular, woven bone scaffolding replaced the primary bone tissue in this region (Fig. 4.4B).

Two vertebrae from a purported adult skeleton of *Ianthasaurus hardestiorum* (Mazierski and Reisz 2006, 2010) from the Stanton Formation (Rock Lake Shale) of Anderson County, Kansas, were also sampled and their histological structure compared with that of *Lupeosaurus* (Huttenlocker, Mazierski, and Reisz 2011). In cross-sectional profile, the neural spines preserved throughout their length a well-defined medullary cavity (Fig. 4.4C), with the majority of bone mass distributed along the perimeter of the bone. The cavity was large and lined by smooth, circumferentially deposited endosteal bone. The inner cortex exhibited some large erosion cavities recording compact-to-cancellous conversion near the endosteal margins. The cortex was exclusively composed of nearly avascular, lamellar primary bone. Notably, the gross structure of the neural spines was more similar to that of *Edaphosaurus* than to *Lupeosaurus* in that it exhibited paired lateral tubercles. A transverse section through the distal spine of a mid-dorsal vertebra at the level of the third tubercle pair (approximately 35 mm from the spine base) demonstrated compact,

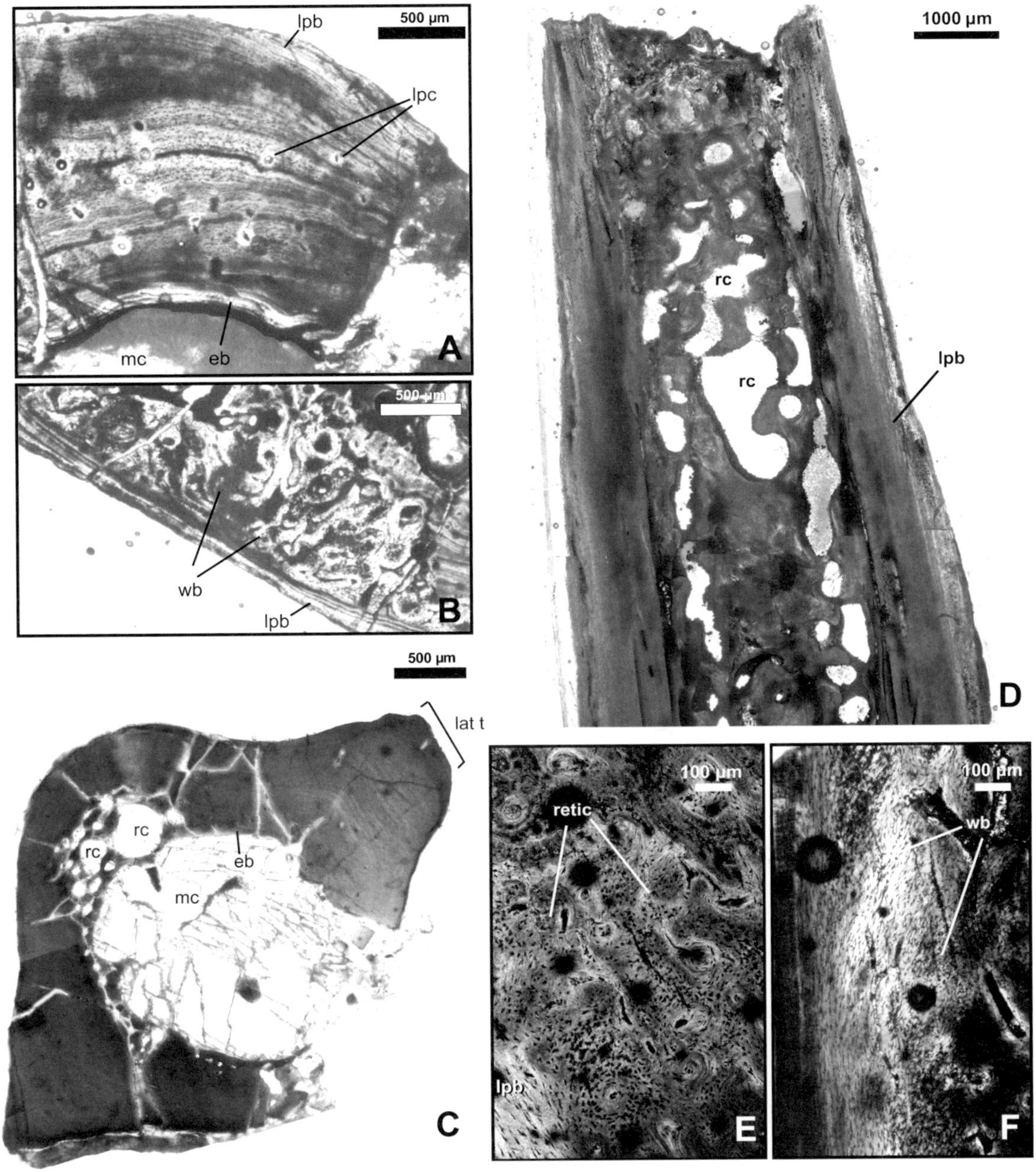

avascular lamellar bone tissue with few primary canals and a regularly lamellated matrix (Fig. 4.4C). The highly regular organization of the matrix suggested a pattern of slow, incremental growth, indicating that the tubercles did not grow with sudden or rapid onset.

The histological structure of the ribs, limb bones, and neural spines of *Edaphosaurus* exhibited large erosion cavities near the endosteal margins and a poorly vascularized outer cortex composed of lamellar bone (Enlow and Brown 1957; Enlow 1969). The ribs were comparatively more porous as is typical of amniote ribs in general (Currey 1987, 2002). The

4.5. Bone microstructure in the sphenacodontid eupelycosaur *Dimetrodon*. (A) Transverse section from distal region of dorsal neural spine of *D. giganhomogenes* (FMNH UC 1134) at low magnification, demonstrating the typical figure-8 cross-sectional shape, lamellar bone in the anterior and posterior grooves, highly vascularized lateral cortex composed of alternating lamellar and fibrolamellar bone, and a dense, fine-cancellous medullary region (Huttenlocker, Rega, and Sumida 2010, Fig. 4A). (B) Same specimen as (A) viewed at high magnification with non-polarized light, showing primary osteons within an irregularly lamellated fibrolamellar bone complex preserving distinct growth zones separated by LAGs and annuli (Huttenlocker, Rega, and Sumida 2010, Fig. 4B). (C) Articulated cast of *D. limbatus*, MCZ 1347 (Denver Museum of Nature and Science exhibition mount), showing variable distribution of neural spine height along cervical and dorsal regions, and their division into a laterally compressed, subquadrangular proximal region and rod-like distal region (Huttenlocker, Rega, and Sumida 2010, Fig. 2A). (D) Transverse section from tibia of *D. giganhomogenes* (FMNH UC 1134) showing alternating lamellar-fibrolamellar organization within the outer cortex (left of image). Arrows denote LAGs.

neural spines, however, exhibited a subperiosteal region that was more compact and nearly avascular (Huttenlocker, Mazierski, and Reisz 2011). The few vascular canals present were exclusively longitudinal. Toward the tip of the spine where distal outgrowth would have occurred, coarse trabecular structures record extensive compact-to-cancellous conversion near the endosteal region as in *Lupeosaurus* and *Ianthasaurus* (Fig. 4.4D). The concerted distal outgrowth and endosteal resorption at the tip apparently resulted in the formation of a well-defined medullary cavity (or "central canal" of de Ricqlès 1974b) lined by a smooth wall of endosteal bone as in *Ianthasaurus*. Thus, the cross-sectional profiles of the neural spines of both *Edaphosaurus* and *Ianthasaurus* take on a character similar to those of tubular bones in terrestrial amniotes (as discussed previously), in which a thickened bone wall surrounds a cavernous medulla lined by smooth, endosteal bone. Remarkably, much of the bone compactness is distributed circumferentially to the outer margins of the cortex, thereby maximizing the moment of inertia (that is, maximizing its resistance to bending in all directions) (Currey 2002). The subcircular cross-sectional geometry of the sail-bearing strut would have provided only limited resistance to torsion in all edaphosaurids (Tables 4.2, 4.3), though this may have been compensated by the plywood-like structure or material properties of the lamellated matrix.

Aside from these more general shared characteristics, the histomorphology of the lateral tubercles in *Edaphosaurus* was regarded as qualitatively different from that of *Ianthasaurus* (Huttenlocker, Mazierski, and Reisz 2011). The matrix was predominantly composed of parallel-fibered and localized woven-fibered bone deeper in the cortex (Figs. 4.4E, 4.4F), characterized by enlarged, globular osteocyte lacunae and a reticular pattern of vascularization. This pattern of organization associated with few growth zones in OMNH 73802, suggested to the authors that tubercle proliferation was rapid, occurring over a few seasons but with sudden onset (rather than incremental as in *Ianthasaurus*).

Sphenacodontidae

Isolated elements from the sphenacodontid *Dimetrodon* were first sampled by Enlow and Brown (1957) and Enlow (1969), including a spine demonstrating callus-like structure at a putative fracture site. Further histological work incorporating sphenacodontid neural spines and limb bones has been undertaken by de Ricqlès (1974b); Huttenlocker, Angielczyk, and Lee (2006); Huttenlocker, Rega, and Sumida (2010); Huttenlocker (2008); and Kriloff et al. (2008). The enamel microstructure of *Dimetrodon* was also described by Sander (1999).

As with *Edaphosaurus*, the histological structure of the neural spines in *Dimetrodon* (Figs. 4.5A, 4.5B) has received the majority of the attention in the literature (Enlow and Brown 1957; Enlow 1969; de Ricqlès 1974b; Huttenlocker, Rega, and Sumida 2010). Like *Edaphosaurus*, the neural spines were extremely long and exhibited variation in length along the

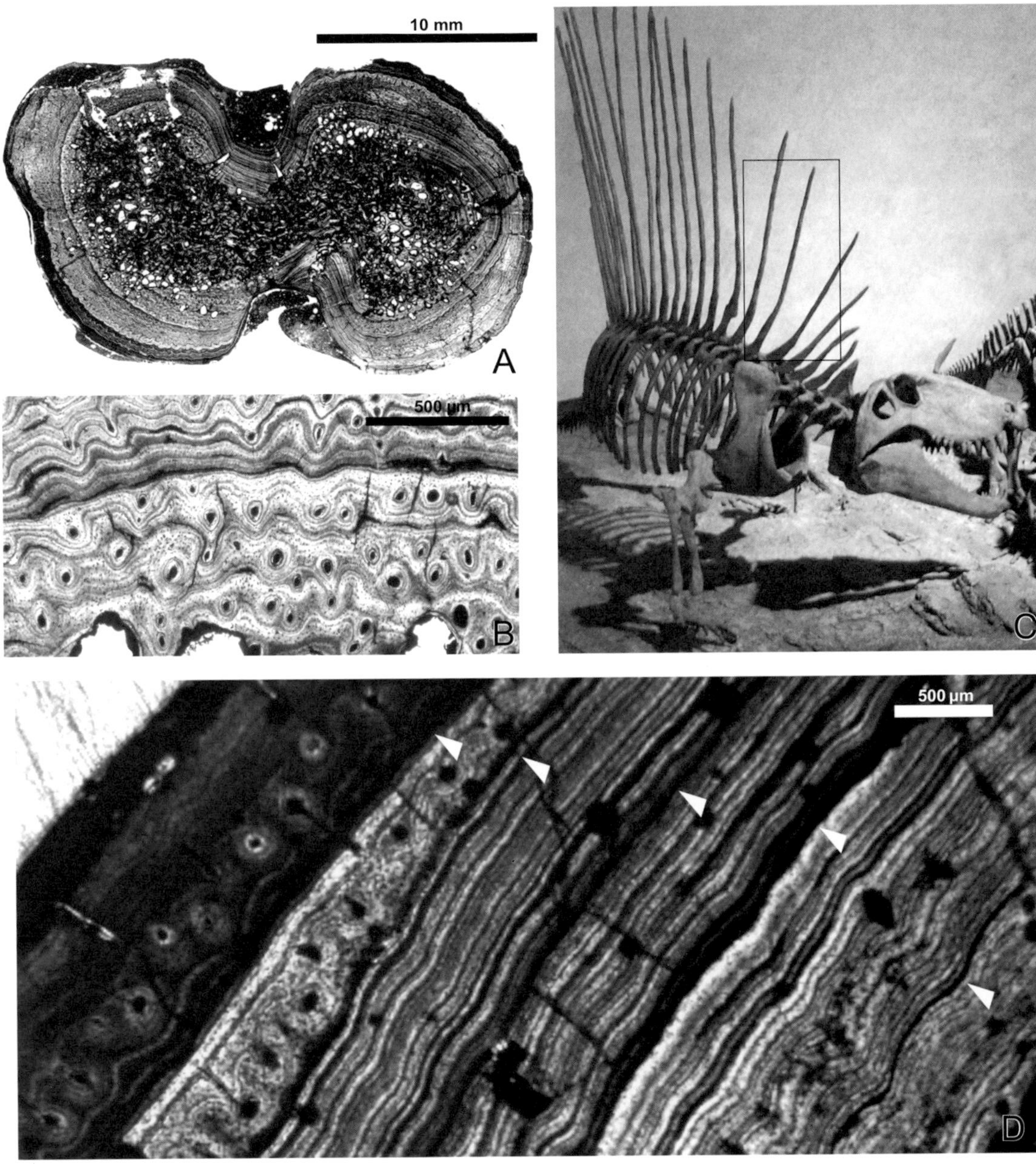

cervical and dorsal series (Fig. 4.5C). The spine was also divided into an anteroposteriorly expanded region proximally, and a distal strut-like region that presumably formed the sail membrane-bearing portion of the spine. The proximal and distal portions were separated by a shoulder-like constriction termed the "changing point" (Pivorunas 1970; Reisz, Berman, and Scott 1992; Bennett 1996). This condition has been designated "dimetrodont differentiation" (Romer and Price 1940) and also existed to some degree in *Edaphosaurus*. Taphonomic and pathological circumstances support the notion that the distal spine supported a collagenous sail membrane, including (1) regular spacing of the spines observed in articulated specimens found in situ (Romer 1927; Romer and Price 1940) and (2) the existence of fractured spines that apparently healed in place in some specimens (Enlow 1969; Rega, Wideman, and Brochu 2002). For example, in the instance of the semi-articulated type specimen of *Dimetrodon milleri* (MCZ 1365), a broken "floating" spine from the twenty-ninth presacral vertebra was presumed to have healed in place without reconnecting to the proximal part of the spine while suspended by the sail membrane (Romer and Price 1940; Pivorunas 1970). At high levels of magnification under polarized light, bundles of short Sharpey's fibers are found within the lateral cortex of the distal spine of *Dimetrodon giganhomogenes* (FMNH UC 1134), localized within individual lamellae and varying in their orientation. Though they are not as prominent or distributed as broadly, the bundles resemble those figured in the spines of *Edaphosaurus* (Enlow 1969; de Ricqlès 1974b) and may indicate the migration of the periosteum or that of an associated collagenous sail membrane.

Apart from the superficial similarities noted earlier, the cross-sectional profile of the sphenacodontid neural spine differs markedly from that of the edaphosaurids in areas of extreme vascularity and the overall compactness of bone matrix not necessarily restricted to the outer margins of the cortex (Tables 4.2, 4.3). Enlow and Brown (1957, 204) notably compared the cortical vascularity of the spine to other sampled reptiles, stating that "in no skeletal element yet examined . . . do the vascular canals number as many as they do in the spine of *Dimetrodon*." The cortical wall was thick (~1.5–2 mm), highly stratified, and composed of alternating vascular and avascular laminae exhibiting variable fiber orientations. The better vascularized laminae revealed woven- or parallel-fibered scaffolding, incorporating densely packed, globular osteocyte lacunae and high localized concentrations of circumferentially lamellated primary osteons consistent with a fibrolamellar complex (Currey 1987). Distal spines examined by de Ricqlès (1974b) and Huttenlocker, Rega, and Sumida (2010) generally exhibited a figure-8 cross section (Fig. 4.5A), revealing high vascularity in the lateral cortex, but marked avascularity in the region of anterior and posterior longitudinal grooves.

Proximal spine sections, however, are quadrangular in cross section in many large specimens, including *Dimetrodon giganhomogenes*. Micrographs reveal an inner figure-8 core inside this subrectangular section, illustrating that the figure-8 cross section was transformed over one to three growth seasons by extremely rapid growth of the horns at the corners, creating the externally apparent quadrangular shape. These horns are characterized by fibrolamellar bone of extreme vascularity with large resorption cavities circumscribed by single or double lamellae (Rega et al. 2005). This region also exhibits marked remodeling, with woven-fibered bone, oblate resorption cavities, and dense, intermingling Sharpey's fibers oriented at oblique angles. Huttenlocker, Rega, and Sumida (2010) attributed the concerted cortical remodeling and rapid primary growth to entheseal migration of epaxial musculature during normal growth

Table 4.2. Histomorphometric comparisons of neural spine data from five basal synapsid genera

Taxon		Bone compactness	Cortical porosity (%)	RBT (%)	Total cross-sectional area (mm²)	I_{min} (mm⁴)	I_{max} (mm⁴)	I_{max}/I_{min}	J (mm⁴)
Edaphosauridae									
Lupeosaurus	proximal	0.55	1.0	6.8	230	1,200	1,420	1.18	2,620
	distal	0.52	5.0	10.7	150	1,101	1,155	1.05	2,256
Ianthasaurus	distal	0.75	1.5	18.7	8.36	4.59	5.34	1.16	9.93
Edaphosaurus	proximal	0.65	1.0	8.0	214	1,878	3,761	2.00	5,639
	distal (mid)	0.77	1.2	15.5	118	1,010	1,100	1.09	2,110
	distal (tip)	0.60	2.0	14.7	36.8	102	116	1.14	218
	tubercle	0.76	2.0	—	18.7	—	—	—	—
Sphenacodontidae									
Sphenacodon	proximal	0.60	7.5	8.8	151	1,075	1,599	1.49	2,674
	distal	0.56	5.9	4.8	180	812	3,225	3.97	4,037
Dimetrodon	proximal	0.77	16.0	8.8	540	5,465	9,244		1.69
	distal	0.72	13.5	14.6	282	3,854	12,166		3.16

Source: Huttenlocker, Mazierski, and Reisz (2011).
Measurements include bone compactness, cortical porosity, relative bone wall thickness (RBT), total cross-sectional area (or "bone mass" of Cubo et al. [2005]), minimum second moment of area (I_{min}), maximum second moment of area (I_{max}), relative maximum to minimum bending rigidity (I_{max}/I_{min}), and torsional rigidity (J) of dorsal neural spines for each sampled taxon. The RBT calculations follow Chinsamy (1993a). Calculations for I_{max}/I_{min} and J follow Plochocki et al. (2008). The data are based on available proximal and distal (mid-height) sections obtained from dorsal neural spines, with additional tip and tubercle data for *Edaphosaurus*. (See Huttenlocker, Rega, and Sumida [2010] and Huttenlocker, Mazierski, and Reisz [2011] for specimen and provenance data.)

(evidenced by the deep, obliquely-oriented Sharpey's fibers in this region). Thus, the cross-sectional shape of the spine had an ontogenetic component, which should provide a cautionary note given that these differences in spine cross-sectional shape have been used to distinguish different species of *Dimetrodon*. In particular, *D. loomisi* and *D. giganhomogenes* may in fact be the same species (Sumida, Rega, and Noriega 2005). Notably, the anteroposterior grooves appear to have remained lamellar and avascular throughout ontogeny. These grooves were previously conjectured to have housed blood vessels associated with the sail membrane (Romer 1927; Romer and Price 1940; de Ricqlès 1974b). The markedly avascular nature of the grooves and the lack of emissary foramina or Volkmann's canals may cast some doubt on this hypothesis (Rega et al. 2005). Selected histomorphometric quantities of the neural spines of some sail-backed edaphosaurids and sphenacodontids are compared in Table 4.2. Qualitative comparisons are provided in Table 4.3.

The histologic structure of the long bones of *Dimetrodon* differed slightly from that of *Edaphosaurus*, but was similar in that the compacta was highly lamellated (but with localized zones adopting a fibrolamellar structure) and exhibited marked remodeling along the endosteal margins. The relatively high degree of compactness that characterized *Dimetrodon* neural spines (and sphenacodontid spines in general) was also reflected in the limb bones, owed to an occluded endosteal lumen composed of densely packed, fine-cancellous bone. The cortex was also more vascularized than in edaphosaurids, but with substantial histovariability within a given section, a result of the highly zonal nature of the

Table 4.3. Qualitative comparisons of the histological profiles of sphenacodontid and edaphosaurid neural spines

	Sphenacodontidae	Edaphosauridae
Dominant bone tissue type	fibrolamellar or parallel-fibered (lamellar in A-P groove of *Dimetrodon*)	lamellar-zonal (fibrolamellar in tubercles of *Edaphosaurus*)
"Medullary" region	occupied by dense fine-cancellous bone	cavernous with little cancellous bone or trabecular structures
Bone compactness	~0.75	~0.50–0.75
Cortical porosity	5%–15%	1%–5%
Relative bone wall thickness	5%–15% (distal) (~15% *Dimetrodon*)	10%–25% (distal) (~15–25% *Edaphosaurus*)
Resistance to torsion	greatest in distal portion of neural spine	low in distal portion of neural spine

Source: From Huttenlocker (2008).

4.6. The axial skeleton and musculature of the sphenacodontid eupelycosaur, *Sphenacodon.* (A, B) Restoration of short segmental epaxial musculature associated with the vertebral arches (Huttenlocker, Rega, and Sumida 2011, Fig. 10). (B1)–(B3) illustrate mid-dorsal vertebra in anterior, left lateral, and posterior views respectively. (C) CM 70377, photomicrograph of region of the neural spine distal to muscular attachments, viewed with non-polarized light (Huttenlocker, Rega, and Sumida 2011, Fig. 7B–7D). Note absence of Sharpey's fibers and the densely vascularized lateral cortex with large primary osteons (formed via subperiosteal absorption of blood vessels within the osteoid). (D, E) Lateral cortex at high magnification viewed with polarized light, showing large subperiosteal vascular canals in the process of becoming incorporated within the cortex. (F) Comparisons of vascularity and cross-sectional bone compactness distribution of mid-dorsal neural spines in selected eupelycosaurs (modified from Huttenlocker, Mazierski, and Reisz 2011, Fig. 9): (F1) distal spine of *Lupeosaurus* cf. *L. kayi;* (F2) distal spine of *Sphenacodon ferox;* (F3) distal spine of *Dimetrodon giganhomogenes.* Anterior is toward top of page. Scale bars ~1 mm. *Abbreviations:*

compacta; highly lamellated avascular zones frequently interrupted by LAGs and followed by zones exhibiting longitudinal primary osteons within a woven-fibered matrix (Fig. 4.5D). Kriloff et al. (2008) included a *Dimetrodon* tibia sample in testing the efficacy of inference models using tibia compactness profiles to determine lifestyle in tetrapods. Despite the high apparent compactness, their results supported a terrestrial lifestyle for *Dimetrodon*, implying that their *Dimetrodon* sample exhibited less overall compactness than those of *Ophiacodon* (which was designated as amphibious after subjected to similar analysis).

Recent sectioning of material referable to the genus *Sphenacodon*, a short-spined relative of *Dimetrodon*, exposed similarities and family-level variation in sphenacodontid bone microstructure. Huttenlocker, Rega, and Sumida (2010) examined the histology of the neural spines of *Sphenacodon*, comparing them to *Dimetrodon* and assessing soft-tissue correlates of the sail in sphenacodontids. In spite of an apparent lack of dimetrodont differentiation in *Sphenacodon*, the authors found muscle scars only within the proximal 55–60 mm of the spine in presumed mature individuals, indicating the limits of attachment of the deep, short-segmental epaxial muscles (Figs. 4.6A, 4.6B). This interpretation was corroborated by the presence of extensive erosion cavities surrounded by an irregular matrix with deeply penetrating Sharpey's fibers oriented at oblique angles in the cortex, similar to those located proximally within *Lupeosaurus* and *Edaphosaurus* spines. Distally, however, this structure gave way to a more regularly vascularized tissue having a predominantly parallel-fibered matrix lacking pronounced Sharpey's fibers (Fig. 4.6C). In this region, striations near the periosteal surface were preserved penetrating the bone and, in some instances, incorporated into the mineralized matrix as primary osteons (Figs. 4.6D, 4.6E). Although the neural spines of *Sphenacodon* were blade-like and laterally compressed in cross section, the histological differentiation between proximal and distal portions of the spine was nearly identical to that of the sail-backed *Dimetrodon.* Dorsal crests were previously hypothesized for the sphenacodontids *Ctenorhachis* and *Ctenospondylus* in the absence of histological evidence

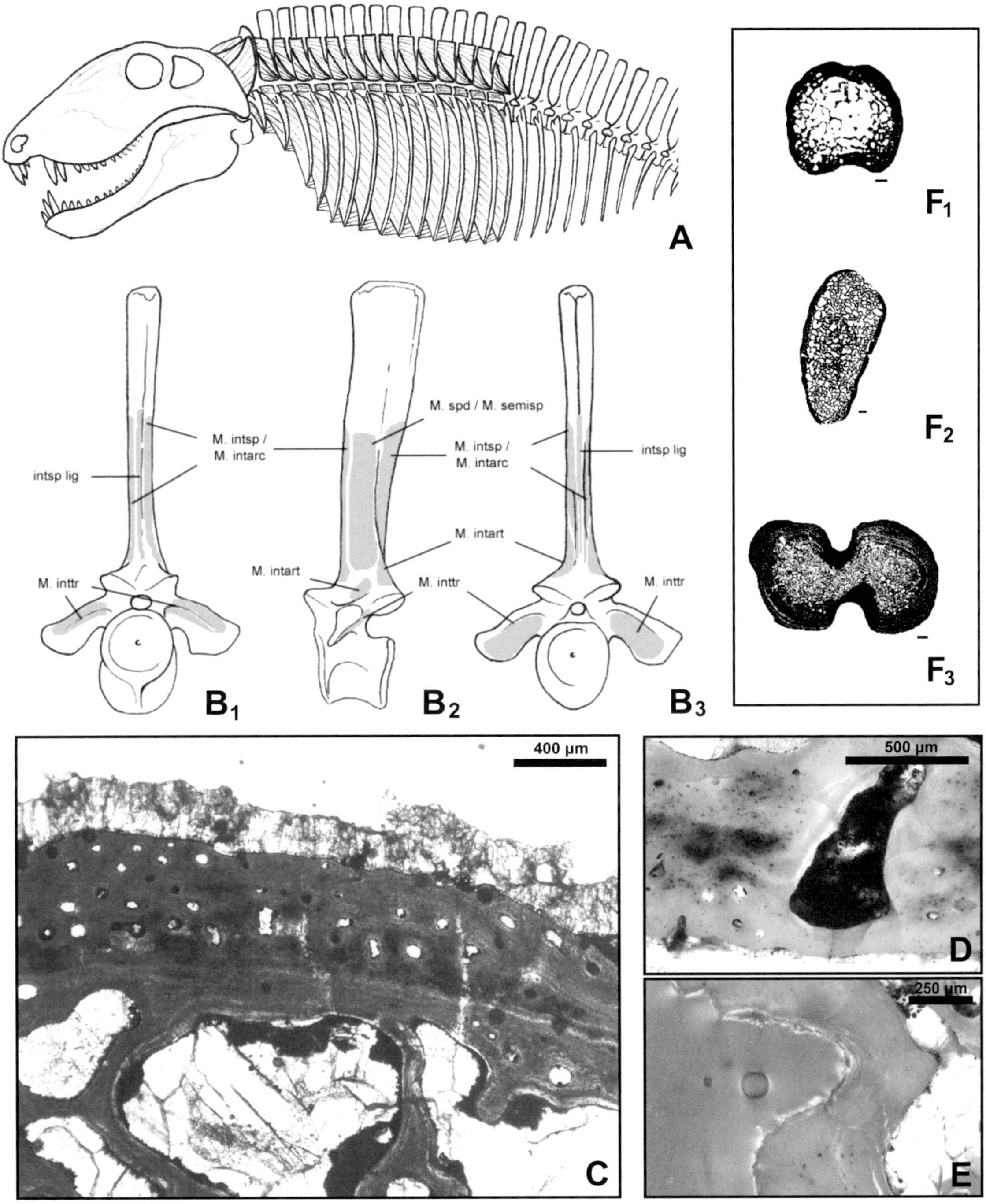

(Hook and Hotton 1991). The histological observations in *Sphenacodon* allowed inferences about the soft-tissue correlates, and, along with additional pathological evidence, permitted a restoration of *Sphenacodon* with a short, dorsal crest.

Preliminary skeletochronological analysis of the appendicular skeleton of *Sphenacodon* (Huttenlocker, Angielczyk, and Lee 2006) offered

M. intarc = interarcuate muscle; M. intart = interarticular muscle; M. intsp = interspinal muscle; M. inttr = intertransversus muscle; M. semisp = semispinalis muscle; M. spd = spinalis dorsi muscle.

4.7. Selected photomicrographs of bone microstructure in limb bones of the sphenacodontid eupelycosaur *Sphenacodon*. (A) UCMP 34209, transverse section from femur of juvenile individual, showing longitudinal and radial primary osteons within a woven-fibered bone matrix. (B) UCMP 83534, transverse section from humerus of subadult, showing transition from a woven-fibered to a predominantly parallel-fibered (but zonal) bone matrix. (C) UCMP 83535, transverse section from humerus of somatically mature adult, recording marked transition to deposition of lamellar bone, with more lenticular osteocyte lacunae, fewer vascular canals, and close spacing of successive LAGs in outer cortex (left of image). Arrows indicate LAGs. *Abbreviations:* wb = woven-fibered bone.

a glimpse into the diversity of bone tissue and growth patterns in basal synapsids, as it allowed assessments of histovariability across a complete ontogenetic series from individuals representing a single metapopulation. The researchers examined the histology of humeri and femora from multiple individuals of *Sphenacodon ferocior* from the Lower Permian Anderson Quarry, Cutler Group of New Mexico, representing five stages of ossification (Brinkman 1988). In general, the bone wall of the humerus was moderately thick (~18–19 percent of its cross-sectional diameter) when compared with that of the femur (~10–15 percent), and the cortex was moderately well vascularized in both elements throughout ontogeny. During early stages of ossification, both elements showed thick growth zones composed of densely packed radial and longitudinal primary osteons within a woven-fibered matrix (Fig. 4.7A), thereby constituting a fibrolamellar tissue complex. Lines of arrested growth and annuli were lacking and the transition between the cancellous lumen and the compacta was gradual and irregular. Larger individuals interpreted as subadults exhibited several LAGs and annuli within the cortex and a transition from a woven-fibered to a predominantly parallel-fibered matrix incorporating longitudinal and radial primary osteons (Fig. 4.7B). Erosion cavities were prevalent along the endosteal margins and the transition with the cortex was more abrupt, although the lumen was still highly occluded with fine-cancellous bone (as in *Dimetrodon*). The largest individuals, interpreted as somatically mature adults, exhibited a matrix dominated by parallel-fibered bone (with lamellar bone deposited in the outermost layers of some individuals), more closely spaced LAGs, and fewer vascular canals in the outer cortex (Fig. 4.7C). This trajectory was suggestive of a pattern of initially rapid, but asymptotic growth in *Sphenacodon*; a pattern that did not corroborate earlier predictions of growth based on *Dimetrodon*. A complete skeletochronological analysis of *Dimetrodon*, however, has not yet been attempted.

Bone Response and Pathologies in Basal Synapsid Bone

Abnormal surface lesions, having the roughened appearance of healed fracture calluses, are frequently encountered in the axial skeletons of basal synapsids and other fossil tetrapods, particularly the ribs and neural spines of sail-bearing tetrapods (e.g., the ctenosauriscid archosaur *Arizonasaurus* [Nesbitt 2003, 2005]; the sphenacodontid *Dimetrodon* [Romer and Price 1940; Enlow 1969; Rega, Wideman, and Brochu 2002; Rega et al. 2005]). (See, for example, sixth through ninth presacral vertebrae in Fig. 4.5C.) Enlow (1969) sectioned a neural spine demonstrating callus-like structure at a putative fracture site near the changing point from the proximal (epaxial muscle embedded) to distal (sail-bearing) region. Woven bone was present, applied to the external surface of the bone, but as no clear fracture plane was evident, the lesion resembled more a subperiosteal microslippage accumulation than a true fracture callus.

Plate A. This scene by Luis Rey depicts the carnivorous *Dimetrodon* (left) attacking its herbivorous relative, *Edaphosaurus* (right). These taxa are examples of the pelycosaurs, or more accurately, pelycosaurian-grade synapsids or basal synapsids which are characterized by hyperelongated neural spines that formed a "dorsal sail." They were the dominant terrestrial vertebrates of the Late Carboniferous through to the Early Permian.

Plate B. A reconstruction by Gerhard Marx of the landscape of the basal *Eodicynodon* AZ of the Beaufort Group, Karoo Basin, South Africa. The scene shows the general features of the landscape about 265 Mya. Two types of nonmammalian therapsids are depicted: a large, hippo-sized dinocephalian, *Tapinocaninus* and three small basal dicynodonts called *Eodicynodon.* The dominant plants at the time were *Glossopteris* and sphenophytes, or horsetails, and lycopods and ferns were also present. *Image courtesy of Billy De Klerk and the Albany Museum, Grahamstown.*

Plate C. (Top) John Nyaphuli with *Tapinocaninus* (NMQR 2987), the most basal tapinocephalid dinocephalian from South Africa, which he found on Modderdrift Farm (Prince Albert District, South Africa). (Bottom) A close-up of the thick-walled skull (median length of approximately 50 cm) of *Tapinocaninus.* The specimen is currently housed in the National Museum, Bloemfontein, South Africa. *Top and bottom images are courtesy of Bruce Rubidge and Saniye Guven, respectively.*

Plate D. This painting by Gerhard Marx shows the general features of the Karoo landscape about 260 Mya and depicts some of the animals and plants recovered from the *Tapinocephalus* AZ, Beaufort Group, Karoo Basin, South Africa. There are two dinocephalians depicted in the scene: At the back, are four individuals of *Tapinocephalus,* two of which are engaged in head-butting, while at center and to the right are two large individuals of the herbivorous *Struthiocephalus.* Note the prominent "boss" on the forehead of the animal in front. At left foreground is a small carnivorous therocephalian called *Lycosuchus,* and at right foreground are two small herbivorous dicynodonts called *Robertia.* The squat animal in the front (and two more toward the back on the right) with prominent body ornamentation are basal anapsid reptiles called *Bradysaurus.* Forests of *Glossopteris,* conifers, and ginkgos grew along the river banks. Other floral elements would have included ferns and sphenophytes (horsetails) growing near the water's edge. *Image courtesy of Billy De Klerk and the Albany Museum, Grahamstown.*

Plate E. A reconstruction of *Estemmenosuchus mirabilis,* a large Mid-Permian dinocephalian from the Ocher Complex of Russia which is correlated to the *Eodicynodon* AZ of South Africa and the Xidagou Formation of China. The skull of the animal is about 30-cm long with distinctive bony "horn-like" projections. *Artist, Luis Rey.*

Plate F. (Top) A reconstruction by Cedric Hunter of *Diictodon,* a small dicynodont commonly found in the Late Permian of the South African Karoo Basin. Several individuals have been recovered in the interpreted terminal chambers of distinctive spiral-shaped burrow casts. *Image courtesy of Roger Smith and Iziko South African Museum*. (Bottom) Two complete skeletons of tightly curled *Diictodon* individuals (SAM-PK-K 5477) within a burrow chamber. Arrows point to the top of each of the skull. Notice the distinctive canine tusk visible on the sides of the upper maxilla on both skulls. The specimen was found by Paul October on the farm Dunedin in the Beaufort West District, South Africa, in 1978. The fossil was prepared by Annelise Crean. The specimen is housed in Iziko South African Museum. *Image courtesy of Sandra Jasinoski.*

Plate G. (Top) Five tiny *Diictodon* juveniles (SAM-PK-K1650) found by Michael Cluver, in 1971, on the farm Leeukloof (Eldorado) in the Beaufort West District of the Karoo (*Trop-idostoma* AZ), South Africa. The fossils are interpreted as evidence of juvenile aggregation and possibly a single brood. The arrow points to the narial region of the skull. The specimen was prepared by Ione Rudner. *Image courtesy of Sandra Jasinoski*. (Bottom) Cedric Hunter's reconstruction of these siblings. *Image courtesy of Iziko South African Museum.*

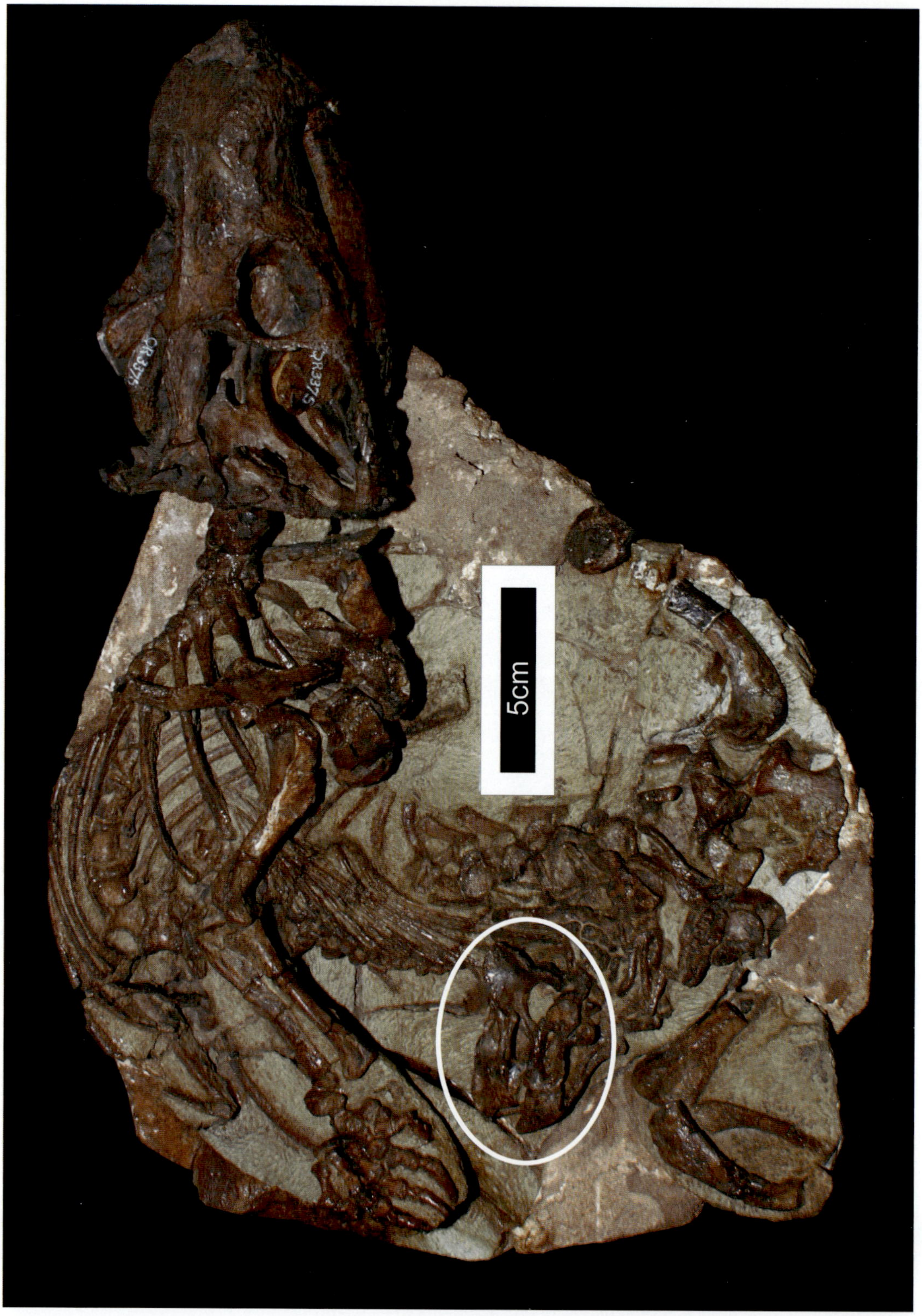

Plate H. The therocephalian *Theriognathus* (NMQR 3375) with a juvenile dicynodont (probably *Oudenodon,* circled) in the belly region. The specimen was recovered from the *Dicynodon* AZ by John Nyaphuli and was also prepared by him. *Image courtesy of Billy De Klerk and the Bloemfontein National Museum.*

Plate I. Exposed Permo-Triassic boundary sequences on the farm Bethel 763 in the Bethulie District of the Karoo Basin, South Africa. Here, the last phase of the End-Permian mass extinction event (arrow) lies just above a distinctive bed of maroon laminated mudrocks. *Image courtesy of Roger Smith and Iziko South African Museum.*

Plate J. A reconstruction of the landscape of the *Lystrosaurus* AZ, shortly after the catastrophic End-Permian extinction event. The landscape shows the recovery of the fauna and flora (as preserved in the fossil record). Following this extinction, the meandering rivers of the Lower Beaufort Group were replaced by multi-channeled, braided river systems that deposited sand rather than the silts and muds of the rivers in earlier times. The *Lystrosaurus* AZ fauna is dominated by the dicynodont *Lystrosaurus,* one of the few survivors of the End-Permian extinction. At lower right, are two small procolophonids ahead of a three *Thrinaxodon* cynodonts. At left in the foreground, a single amphibian, *Lydekkerina,* emerges from shallow water and approaches the proto-lizard, *Prolacerta* in the center. There is a sparse record of plant fossil for the post-extinction period across Gondwana. However, it appears that the flora changed from being *Glossopteris*-dominated to the beginnings of the *Dicroidium* flora typical of the Triassic Period. Sphenophytes and ferns were abundant, and early cycads are also evident in the fossil record. *Image courtesy of Billy De Klerk and the Albany Museum, Grahamstown.*

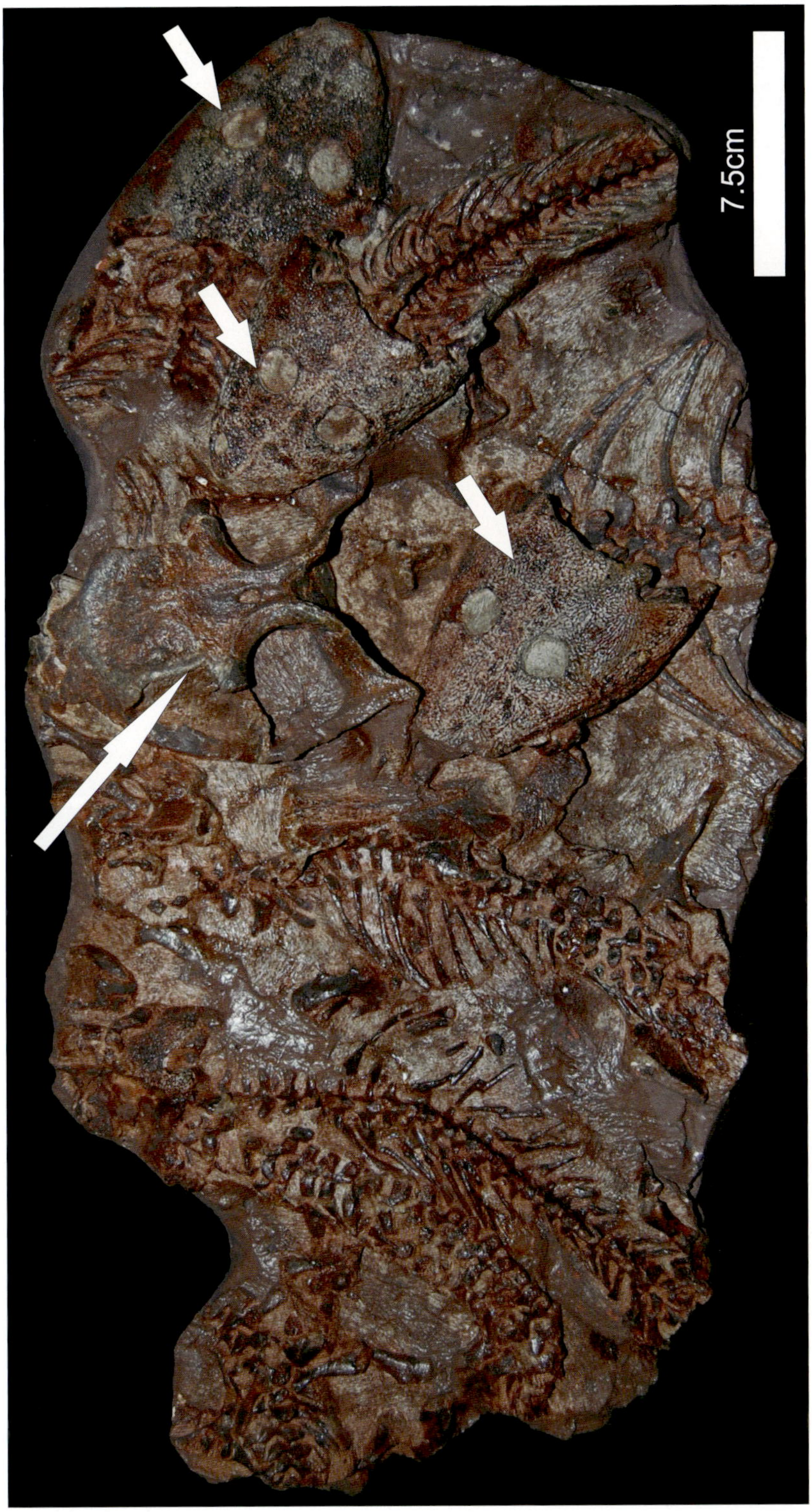

Plate K. Specimen BP/1/3900 showing a skull and skeleton of the Early Triassic dicynodont *Lystrosaurus murrayi* (the thin arrow points to the dorsal part of the skull) preserved with several associated amphibians called *Lydekkerina huxleyi* (small arrows indicate their dorsoventrally flattened skulls). The specimen was found by James Kitching, in1964, near the small Free State town of Harrismith, South Africa. *Image courtesy of Saniye Guven.*

Plate L. An almost complete skeleton of the therocephalian *Ictidosuchoides* (SAM-PK-K 10704) found by Roger Smith in 2009 on the farm Krugerskraal in the Graaff-Reinet District, South Africa. The specimen is housed in Iziko South African Museum.

Plate M. (Top) A beautifully preserved Early Triassic cynodont, *Galesaurus planiceps* (SAM-PK-K10465), recovered from Fairydale (Donald 207) in the Bethulie District from the *Lystrosaurus* AZ. The skull length is 7 cm and the skeleton length (including the dislocated distal portion of the tail) is 33 cm. The specimen was collected by Roger Smith, in 2005, and prepared by Annelise Crean. It is housed in Iziko South African Museum., Cape Town. *Image courtesy of Fernando Abdala.* (Bottom) A pair of fully articulated juvenile *Thrinaxodon liorhinus* (SAM-PK-K10017) recovered at the top of Old Wapadsberg Pass in the Graaff-Reinet District. Skull length is 4 cm. The specimen was collected by Roger Smith, in 2000, and was prepared by Annelise Crean. It is housed in Iziko South African Museum, Cape Town. *Image courtesy of Fernando Abdala.*

Plate N. A reconstruction of the *Cynognathus* AZ by Gerhard Marx. In the middle distance, we can see a group of large herbivorous dicynodonts called *Kannemeyeria*. A large predatory archosauromorph *Erythrosuchus* feeds on a *Kannemeyeria* carcass, while two *Cynognathus* cynodonts approach cautiously. In the foreground is a small archosauromorph *Euparkeria*. Lurking in the shallows is a large capitosaurid amphibian, while two smaller amphibians crawl onto the river bank. There are very few fossils of the plants that grew at this time, but we know that ferns, sphenophytes, lycopods, conifers, ginkgos, and cycadophytes were present. Only one species of *Dicroidium* has been found in this biozone, although in the Late Triassic these plants are extremely diverse and are the dominant flora. *Image courtesy of Billy De Klerk and the Albany Museum, Grahamstown.*

Plate O. A reconstruction of the carnivorous cynodont *Cynognathus* that lived during the Early to Middle Triassic. This approximately 2-m long nonmammalian therapsid is well represented in the *Cynognathus* AZ of the Beaufort Group of South Africa. *Artist, Luis Rey.*

Plate P. (Top) An artistic rendition by Luis Rey of the prosauropod dinosaur, *Massospondylus carinatus,* being attacked by the theropods *Coelophysis rhodesiensis* (also known as *Syntarsus* and *Megapnosaurus*). Fossils of both these dinosaurs have been recovered from the Early Jurassic, Elliot Formation of the Karoo Basin of South Africa. (Bottom) The exquisite embryonic remains of *Massospondylus carinatus.* Several nests and eggs (some with embryos) of *Massospondylus* were collected by James Kitching in the 1970s from the Early Jurassic, Elliot Formation of the Stormberg Group of South Africa (in the Golden Gate National Park). These eggs and the associated embryos are the world's oldest. These delicate embryos were recently prepared by Diane Scott. The specimen is housed in the Bernard Price Institute for Palaeontological Research, Johannesburg, South Africa.

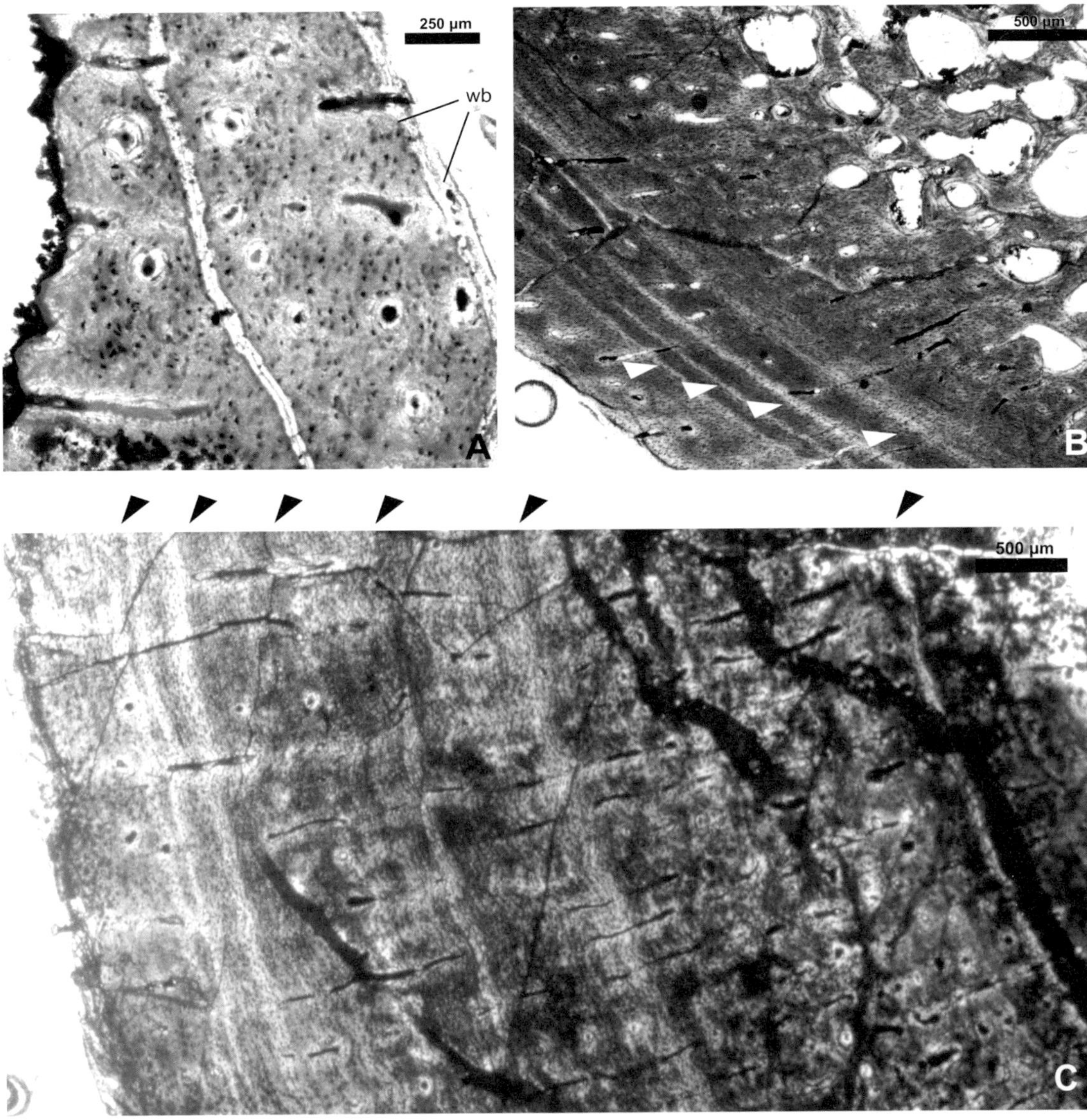

More recently, similar dorsal neural spine pathologies have been described in other specimens of *Dimetrodon* (Rega et al. 2005; Huttenlocker, Rega, and Sumida 2010) and *Sphenacodon ferocior* (Huttenlocker, Rega, and Sumida 2010). In *Dimetrodon giganhomogenes*—which manifests the markedly different quadrangular to figure-8 cross-sectional geometry from the base to more distal regions of the neural spine—a mature specimen (FMNH UC 1134) manifests seven apparent healing calluses in a consecutive series of dorsal spines, all at approximately the level of the shoulder-like constriction, or changing point. Serial sectioning of one at the proximal, mid-lesion and distal ends demonstrated a surprising finding. Instead of subperiosteal bone deposition, as expected in a normal healing fracture callus, the highly fibrolamellar and zonal structure demonstrated an unmistakable deformation of lamellae, likely due to torsional failure between lamellae, which shifted the position of the posterior groove markedly (Fig. 4.5A). In as few as one to two growth seasons, osteoclastic resorption of bone and additional lamellar deposition had re-established the proper anatomical orientation of this groove (Rega et al. 2005), underscoring the mechanical importance of the anterior and posterior grooves.

Callus-like structures in a dorsal neural spine of *Sphenacodon* cf. *S. ferocior* (CM 73367) and other specimens attributed to this species were recently described by Huttenlocker, Rega, and Sumida (2010). As in *Dimetrodon*, these abnormal surface lesions were generally located about 55–60 mm from the base of the spine, near the anatomical changing point. The authors suggested that the microstructural organization of proximal muscle scars and Sharpey's fibers indicated that approximately 50 percent or less of the neural spine served as anchoring sites for epaxial muscles (Figs. 4.6A, 4.6B), whereas the remaining distal portion was "free" (although may have been embedded in a collagenous webbing as in extant crested tetrapods, such as the corytophanid *Basiliscus* or the chamaeleonid *Trioceros*). Given the variability in the lengthening of the distal portion of the spine observed across sphenacodontids (e.g., *S. ferocior*, *Ctenospondylus*), this region clearly would have been exposed to large bending moments and, thus, greater probability of failure low in the neural spine. Likewise, the histomorphometric data suggest that the spines of *Sphenacodon* were well adapted for resisting fore and aft bending, but were not selected as strongly to resist lateral bending (Table 4.2). Given these requisites, it is not surprising to find occasional fractures in the spines of *Sphenacodon*, all of which appear to be located within the changing point of the spine as frequently found in *Dimetrodon*.

Evolution of Life History and Habitat Inferred from Normal Histology

Growth and Histovariability

The findings of Enlow and Brown (1956, 1957), Enlow (1969), de Ricqlès (1974b), and subsequent studies indicated that the major bone tissue types (lamellar, parallel-fibered, woven-fibered) and their vascular patterns were at least as diverse in Late Paleozoic stem amniotes as they are in recent tetrapods. The broad spectrum of histovariability known in large extant reptiles was evident in extinct large land-living tetrapods, including the temnospondyl *Parioxys*, the diadectomorph stem amniote *Limnoscelis*, and the captorhinid reptiles *Captorhinus* and *Labidosaurus*. The Permian reptiles *Captorhinus* and *Labidosaurus* even possessed localized fibrolamellar tissues, common to mammals and birds in the present day (Enlow 1969). The discovery of these various bony tissues in early synapsids and contemporary tetrapods led Enlow (1969) to suggest that the diversity of body sizes and ecological niches in extant reptiles were made possible by physiological attributes already in place during the Late Paleozoic. That is, the mechanical adaptations and prolonged, cyclic growth strategies with variable seasonal response, as inferred from bone microstructure and histomorphologic traits of extant reptiles, are plesiomorphic for amniotes because they were also present in basal synapsids and anamniotes. Decades later, however, the establishment of a workable cladistic framework might allow consideration of these features within their evolutionary and ecological contexts.

Inadequacy in the breadth of phylogenetic sampling may mislead one to arrive at an oversimplified view of the evolution of life history strategies in some lineages. The immature *Ennatosaurus* specimen studied by de Ricqlès (1974b) presents such a case. It is interesting to note that the cortical bone is composed of only a few longitudinal vascular canals within an entirely lamellar matrix in areas of the shaft devoid of muscle attachments. In fact, the bone is almost avascular. As was correctly pointed out by de Ricqlès (1974b) such a pattern does not corroborate a period of rapid juvenile growth found in some other large-bodied amniotes. This finding might at first suggest (as it did to de Ricqlès) that, in spite of dispersal into cool, temperate high latitude environments by the Middle Permian, caseasaurs such as *Ennatosaurus* retained a plesiomorphic growth pattern typical of earlier Permo-Carboniferous tetrapods. The low apparent growth rate, if typical of the genus, may represent retention of a plesiomorphic growth strategy typical of smaller-bodied tetrapods, or a secondarily evolved life history strategy specific to the ecology of the highly derived caseids. However, as the evolutionary history of caseasaurs spans the same 30–40 My that encompasses the radiation of eupelycosaurs and the origins of therapsids (compare branch durations in Fig. 4.1), there may be much hidden diversity in histomorphology and life habits of this clade not encompassed by *Ennatosaurus*. More complete ontogenetic and taxonomic sampling of caseasaurs is needed for an adequate comparative basis.

Histologic sampling of varanopids provides a number of opportunities for skeletochronology (given the distribution of LAGs in mycterosaurines and large sample sizes available), as well as studies concerning locomotion and biomechanics (given the unique cross-sectional properties of their long bones among basal synapsids). Variation in bone microstructure and compactness profiles between large-bodied varanodontines (e.g., *Varanops*) and the mycterosaurines may also have important ecological implications for the two subfamilies. The thinness of the bone wall in *Varanops* and the scarcity of LAGs are noteworthy. However, due to the limited sampling at present, caution should be practiced when interpreting the nature of the histological profiles. Several conditions may account for the thin cortex. For example, the thin cortex and coarse trabeculae within the medullary cavity may reflect resorption, remodeling, and growth at the metaphysis specifically, a region that is generally thinner-walled than the mid-diaphysis of long bones (Enlow 1963; Currey 2002). More complete serial sections would resolve this query. Likewise, midshaft sections in other sampled material attributable to the mycterosaurine varanopid are comparatively thicker, suggesting functional considerations. More sampling is needed to understand the complete range of histovariability in *Varanops* and how it compares to other varanopids.

Habits and Interplay between Growth and Mechanical Function

Other aspects of the histomorphology may be more appropriately viewed as biologically informative traits characteristic of varanopid bone. Notably, the presence of a well-defined medullary cavity, common to *Varanops* and the mycterosaurine from Richards Spur (one of few Lower Permian localities confidently interpreted as a completely terrestrial upland environment), is not currently known in other eupelycosaur limb bones and may represent an ecologically informative characteristic of at least some varanopids. In addition to being present in terrestrial mammals and birds, this trait is shared in many small extant reptiles and has been suggested to correlate with a terrestrial lifestyle (Currey 1987; Germain and Laurin 2005; Kriloff et al. 2008). Long bones having a more occluded lumen, on the other hand, may be more typical of predominately aquatic or amphibious vertebrates (the high degree of compactness in the lumen often being attributed to ballast [see, e.g., Germain and Laurin 2005]). Interestingly, the compactness profiles of sampled varanopid limb bones are nearly identical to that of the tubular humeral shaft described by Enlow (1969) for the early reptile Captorhinus, co-occurring in upland faunas of the late Early Permian (e.g., Richards Spur).

What is the primary contributor of internal histomorphology: growth or mechanical regimen? The studies of de Ricqlès (1974b) and Germain and Laurin (2005) suggested that the size of sampled *Ophiacodon* specimens were likely indicative of adults and emphasized the functional role of the cross-sectional distribution of bone compactness (rather than the ontogenetic history), giving particular attention to its inferred amphibious lifestyle. Size, however, is not always an accurate predictor of age in fossil assemblages; specimens in the upper ranges of a size distribution may just as likely represent early or late subadults, as fossil assemblages are generally skewed toward subadults (Varricchio et al. 2008). Enlow (1969) suggested earlier that the extensive distribution of primary osteons in the compacta of *Ophiacodon*, along with the highly occluded nature of the lumen (both traits that are in stark contrast to sampled varanopids), indicated rapid skeletal growth and, perhaps, immaturity of the studied specimens. At the time of these publications, this type of tissue organization had not been identified in any other species of pelycosaur. More recent examination of *Sphenacodon ferocior* long bones has indicated such extensive vascularization throughout early ontogeny, with only very large individuals approaching the maximum known sizes exhibiting reduced vasculature and other evidence of much slowed growth (indicative of an onset of asymptotic growth). More importantly, studies focused on bone compactness have not considered the bone tissue type itself, which in *Ophiacodon* was highly disorganized with a non-lamellar structure and lacking in LAGs or annuli. It is therefore likely that the organization of the compacta is at least as much a reflection of the animal's growth pattern as it is its lifestyle. Age at death and death assemblage constitution raise substantial concerns with definitive conclusions regarding *Ophiacodon* microstructure, as indeed will all conclusions derived from a data set that is not constructed from a more complete ontogenetic series.

The present review exposes further biases in the literature regarding histological sampling of basal synapsid bone; for example, more is known about the histology of the hyperelongated neural spines of edaphosaurids and sphenacodontids than any other skeletal element in either of those taxa. Emphasis on neural spine bone microstructure has likely been a combined result of the charismatic nature of the dorsal sail in these two families and wide availability of neural spine material for thin-sectioning. Biophysical analyses incorporating these structures have attempted to model a mathematical relationship between sail allometry and thermoregulatory capacity, but disagree in their allometric exponents and the sail's inferred efficacy as a heat exchanging organ (Tracy, Turner, and Huey 1986; Haack 1986; Tomkins et al. 2010). Histologic evidence, however, provides a more explicit test of hypotheses that incorporate aspects of growth and mechanics. Such data allows for variation in neural spine architecture to be assessed, exposing key differences in cross-sectional geometry and bone compactness between families and genera (Fig. 4.6F, Table 4.3) and further supporting the convergent origins of hyperelongated neural spines in synapsids, as similar gross architectural designs were achieved via different microstructural solutions (Huttenlocker 2008; Huttenlocker, Rega, and Sumida 2010; Huttenlocker, Mazierski, and Reisz 2011).

Rega, Wideman, and Brochu (2002) and Rega et al. (2005) were the first to consider the mechanical significance of the figure-8 cross-sectional profile of neural spines in *Dimetrodon*. Unlike those of *Edaphosaurus*, many large-bodied *Dimetrodon* species (e.g., *D. giganhomogenes*) exhibited mid-dorsal spines that were laterally expanded and deeply grooved longitudinally along the anterior and posterior face. This may at first seem to imply that the neural spine of *Dimetrodon* was not particularly resistant to bending (in contrast to that of edaphosaurids). Architecturally, however, the shape was likely highly resistant to bending, compensating by constructing a dense network of fine-cancellous and multilamellar compact bone tissue in the form of a figure-8 (Figs. 4.5A, 4.6F) that is functionally equivalent to an I-beam. Finite element analysis and estimates of second moment of area (Table 4.2) demonstrated the resistance to bending to be greatest laterally. This architectural design is observed in many biological systems (e.g., *Thunnus obesus* and billfish fin spines, feathers, leaf petioles) and may reduce bending (and thus breakage) imposed by tensile forces while still allowing some torsional flexibility (Etnier 2001; Vogel 1988, 2003). In addition, the highly lamellated bone microarchitecture with large numbers of resorption cavities (subcircular and oblate in section) lined by one to two deposited lamellae after resorption would have been highly resistant to crack propagation under bending loads, in much the same way that both plywood and portholes (or Haversian systems in mammals) function to blunt and contain cracks. In *Dimetrodon*, these features likely represent a necessary modification of the narrow crested condition observed in contemporaneous sphenacodontids (*Ctenorhachis*, *Ctenospondylus*, *Sphenacodon*) in order to resist mechanical insult imposed by lateral bending moments propagated through the proportionately larger sail, either during terrestrial locomotion or other activities.

Conclusions

At present, sampling limitations restrict definitive interpretations of large-scale evolutionary trends in the bone microstructure of pelycosaurian-grade synapsids. However, the varieties of craniodental (dietary) and postcranial (locomotory) specializations evident in basal synapsids are reflected in their diverse osteohistological profiles, supporting that pelycosaurian-grade synapsids were varied in their life habits, growth tactics, and other paleobiological attributes. Histological evidence provides a useful tool for functional analysis and comparative inferences of growth and lifestyle. The broad spectrum of histovariability observed in extant reptiles was already represented in fossil synapsids and other Permo-Carboniferous amniotes (Enlow and Brown 1956, 1957; Enlow 1969). Recent histological studies have consequently forfeited breadth in phylogenetic sampling for in-depth sampling of variation in individual limb bone elements (see, e.g., Girondot and Laurin 2003; Germain and Laurin 2005; Kriloff et al. 2008) or bizarre skeletal structures such as the hyperelongated neural spines of edaphosaurids and sphenacodontids (see, e.g., Rega, Wideman, and Brochu 2002, Rega et al. 2005; Huttenlocker, Rega, and Sumida 2010, Huttenlocker, Mazierski, and Reisz 2011). Such focused efforts have the potential to elucidate aspects of functional anatomy, growth, and variation *within* specific clades of basal synapsids (or among skeletal elements within single individuals or species). We suggest that longitudinal ontogenetic sampling of individual genera and species is necessary to understand intrataxonomic histovariation and should be favored over broad taxonomic sampling to elucidate growth dynamics within basal synapsid taxa. It is only through these more focused approaches that observed histovariation can be better understood, and a broader synthesis of evolutionary trends in the bone microstructure of the pelycosaurian ancestors of therapsids can be established.

Acknowledgments

For donation of specimens for sectioning, we thank William May and Rich Cifelli of OMNH, Norman; Kevin Padian and Patricia Holroyd of UCMP, Berkeley; Dave Berman and Amy Henrici of the Carnegie Museum, Pittsburgh; Robert Reisz and Dave Mazierski of University of Toronto; and Bryan Small and Kenneth Carpenter of the Denver Museum of Nature and Science (DMNH). For their correspondence and invaluable contributions to this and previous manuscripts, we also thank our collaborators: Kenneth Angielczyk, Jennifer Botha-Brink, Anusuya

Chinsamy-Turan, Michel Laurin, Andrew Lee, Dave Mazierski, Sean Modesto, Kevin Padian, Jason Pardo, Valerie Pelletier, Sanghamitra Ray, Robert Reisz, Christen Shelton, Christian Sidor, Bryan Small, and especially Stuart S. Sumida. Sanghamitra Ray and Stuart Sumida are acknowledged for having reviewed this manuscript.

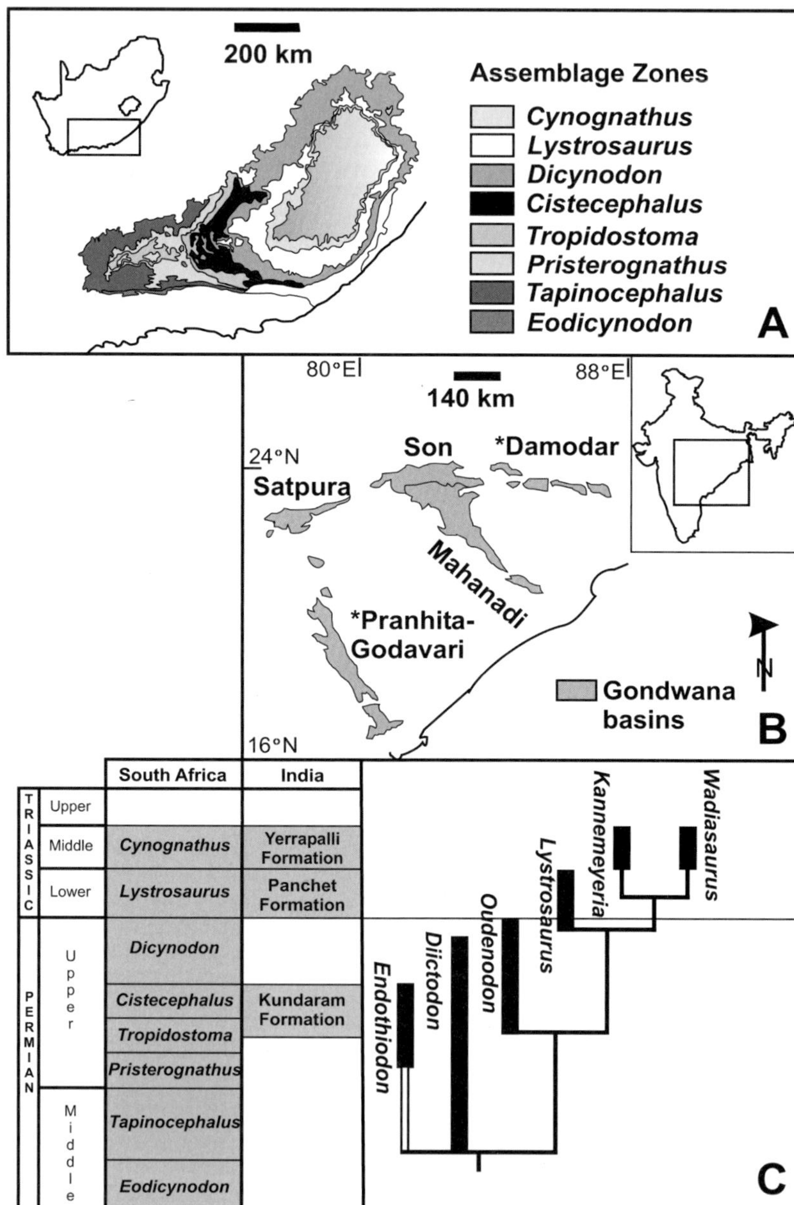

5.1. (A) Assemblage zones of the Beaufort Group, Karoo Supergroup, South Africa. (B) Major Gondwana basins of peninsular India. Dicynodonts discussed in the current study were collected from the basins marked by asterisks (*). (C) Biostratigraphic ranges of the dicynodont genera studied (after Angielczyk 2001; Ray 2006; Botha and Angielcyk 2007).

Dicynodont Growth Dynamics and Lifestyle Adaptations

5

Sanghamitra Ray, Jennifer Botha-Brink, and Anusuya Chinsamy-Turan

Introduction

Dicynodontia, a clade of non-mammalian therapsid, diversified into numerous genera and species during the Middle and Late Permian (Lucas 2002). These taxa are considered to be important biostratigraphic markers for global correlation between similar-aged strata (see Plate 1). Apart from Russia, Scotland, and China, these Middle-Late Permian forms are known from the Gondwanan countries such as South Africa, Tanzania, Zambia, Mozambique, Madagascar, Brazil, and India (King 1988; chapter 2 of this book). Although dicynodonts had an almost global distribution during the Triassic, their generic and specific diversity decreased markedly at the onset of this period. The genus *Lystrosaurus*, which appeared during the latest Permian, is the only dicynodont known to have crossed the Permo-Triassic boundary (Smith and Ward 2001; Smith and Botha 2005) (see Plates J and K), and it radiated widely during the Early Triassic. Except for fragmentary dicynodont remains from the Cretaceous sediments of Australia (Thulborn and Turner 2003), dicynodonts were relatively rare and almost extinct by the end of the Late Triassic.

Phylogenetic Context of the Dicynodontia

Following the initial work by Owen (1845, 1859) and Broom (1932), the taxonomy, osteology, and functional morphology of dicynodonts have been studied extensively (see, e.g., Camp and Welles 1956; Watson 1960; Crompton and Hotton 1967; Cluver 1971, 1978; Cluver and Hotton 1981; King 1981a, 1981b, 1988, 1993; Cluver and King 1983; Ray 2000, 2001, 2006; Angielczyk 2001; Surkov and Benton 2004; Vega-Dias, Maisch, and Schultz 2004; Surkov, Kalandadze, and Benton 2005). The complex phylogenetic interrelationships of dicynodonts are not well understood, although several attempts have been made to resolve this problem (see, e.g., Cluver and King 1983; King 1988; Angielczyk 2001). King (1988) recognized four subgroups based on differences in jaw structure, although the clade was weakly supported (Kemp 2005). A detailed phylogenetic analysis encompassing twenty dicynodont taxa and forty characters was carried out by Angielczyk (2001), but the clades were again weakly supported (Kemp 2005).

More recently, Kammerer and Angielczyk (2009) provided a new detailed taxonomic framework for anomodonts (dicynodonts and their relatives). Features that characterize this group are, among others, nasal bosses with a median swelling and continuous posterior margin, a caniniform process with long caniniform teeth, the absence of premaxillary and anterior maxillary teeth, a well-developed internal narial shelf formed primarily by the premaxilla and maxilla, a lateral palatal foramen at the level of the anterior expanded palatal exposure of the palatines, a vomer that contributes to the interpterygoid vacuity, a ventrally directed transverse flange of the pterygoid, which is relatively large and wing-like, and a relatively long interpterygoid vacuity, which does not reach the level of the palatal exposure of the palatines (Kammerer and Angielczyk 2009).

Past research has focused more on the Permian dicynodonts than the Triassic taxa, and thus some relationships remain unresolved: for instance, the relationships between *Lystrosaurus*, kannemeyeriiforms, and *Dicynodon*-like Permian taxa such as *Vivaxosaurus* and *Dinanomodon* (Kammerer and Angielczyk 2009). However, the monophyly of Lystrosauridae, which contains *Lystrosaurus* with at least four species (Grine et al. 2006) and the monospecific genus *Kwazulusaurus* (Maisch 2002) appears to have been accepted. Few rigorous phylogenetic analyses have focused on the Triassic kannemeyeriiforms and thus the validity of some genera and the monophyly of the clade are still the subject of intense debate (see, e.g., Maisch 2001; Surkov and Benton 2004; Vega-Dias, Maisch, and Schultz 2004; Surkov, Kalandadze, and Benton 2005; Ray 2006; Damiani et al. 2007; Fröbisch 2007).

Paleobiological Context of the Dicynodontia

Dicynodonts are considered to be the first successful terrestrial herbivores. They had various roles as browsers and foragers and consumed a variety of food such as leaves, stems, roots, fleshy parts of plants, and even small invertebrates (King 1993). The cranial morphology of the dicynodonts suggests that they had a propalinal movement of the lower jaw that resulted in efficient mastication (Crompton and Hotton 1967; King 1988; Angielczyk 2004), whereas their postcranial anatomy suggests that dicynodonts exhibited a postural dichotomy, which consisted of abducted forelimbs and nearly adducted hindlimbs (Watson 1960; Cox 1972; Cluver 1978; King 1981b; Ray and Chinsamy 2003). The size of the dicynodonts ranged from that of a mouse to a hippopotamus (King 1990). They had a barrel-shaped body and were strongly built, but slow-moving animals. A comprehensive examination of the Permian forms reveals a more sprawling stance with the body positioned near the ground compared with the Triassic forms (Ray 2006). The postcranial anatomy evolved toward a more upright stance, as is evident from their functional tendency to reduce the lateral component of the propulsive force of the forelimb while still in an abducted mode and upright hindlimb morphology with the body held well off the ground (Ray 2006).

Histological Context: Review of Previous Studies of Dicynodontia

Bone histology is a well-established tool for examining growth patterns and lifestyle habits of extinct animals (see, e.g., Chinsamy 1997; Botha 2003; Chinsamy-Turan 2005; Erickson 2005). It has been used extensively to understand various aspects of the biology of extinct animals such as dinosaurs (see, e.g., Chinsamy 1990, 1993a; Chinsamy, Rich, and Vickers-Rich 1998; Curry 1999; Horner, de Ricqlès, and Padian 1999; Horner et al. 2000; Chinsamy-Turan 2005), other archosaurs (see, e.g., de Ricqlès et al. 2008; Chinsamy, Codorniú, and Chiappe 2008), and non-mammalian therapsids (see, e.g., de Ricqlès 1969, 1972; Chinsamy and Rubdige 1993; Botha and Chinsamy 2000, 2004, 2005; Ray and Chinsamy 2004; Ray, Botha, and Chinsamy 2004; Ray, Chinsamy, and Bandyopadhyay 2005; Jasinoski 2008; Jasinoski, Rayfield, and Chinsamy 2010a).

Early studies on the bone microstructure of dicynodonts include those by Gross (1934), Enlow and Brown (1957), de Ricqlès (1972, 1975, 1976), and Chinsamy and Rubidge (1993). Although these studies, with the exception of Chinsamy and Rubidge (1993), involved the examination of isolated bones, they provided an important foundation for further research. These studies included a variety of dicynodonts ranging from basal Permian taxa such as *Diictodon* (see Plates F and G), *Endothiodon*, and *Cistecephalus* (Chinsamy and Rubidge 1993); the bidentalians *Oudenodon* and *Aulacephalodon* (de Ricqlès 1972; Chinsamy and Rubidge 1993); the more derived *Dicynodon*, Permo-Triassic *Lystrosaurus* (see Plate J) (de Ricqlès 1975; Chinsamy and Rubidge 1993); and highly derived Middle Triassic kannemeyeriids, *Kannemeyeria* (see Plate N) (Gross 1934; Enlow and Brown 1957; de Ricqlès 1975; Chinsamy and Rubidge 1993); *Dinodontosaurus* (Enlow and Brown 1957); and *Stahleckeria* (de Ricqlès 1972).

Subsequent recovery of additional material has led to more comprehensive studies on dicynodonts involving multiple individuals (see, e.g., Chinsamy and Rubidge 1993; Ray and Chinsamy 2004; Ray, Chinsamy, and Bandyopadhyay 2005; Green, Schweitzer, and Lamm 2010; Ray, Bandyopadhyay, and Appana 2010; Botha-Brink and Angielczyk 2010), multiple skeletal elements of the same individual (see, e.g., Ray and Chinsamy 2004; Botha-Brink and Angielczyk 2010; Ray, Bandyopadhyay, and Appana 2010; chapter 6 of this book), and biomechanical considerations (see, e.g., Jasinoski, Rayfield, and Chinsamy 2010a, 2010b). All researchers noted the presence of fibrolamellar bone, a variable incidence of growth rings (annuli and lines of arrested growth [LAGs]), and in the larger, more derived kannemeyeriids, a prevalence of secondary reconstruction. Green, Schweitzer, and Lamm (2010) examined the Upper Triassic genus *Placerias* to reveal that the cortices of the limb bones contain zonal fibrolamellar bone tissue in the inner cortex and peripheral layers of parallel-fibered bone. Peripheral rest lines were reported by Green, Schweitzer, and Lamm (2010), suggesting complete cessation of growth and attainment of maximum size. Histological sampling of several *Oudenodon* and *Lystrosaurus* skulls have shown that there are four types of suture morphology which may be correlated with finite-element–predicted strain (Jasinoski, Rayfield, and Chinsamy 2010a). Both of these dicynodonts had scarf sutures in the infraorbital bar, zygomatic arch, and postorbital bar that suggest small adjustment movement; however, in *Lystrosaurus*, small adjustment movement was also present in the anterior surface of the snout, at the premaxilla-nasal suture (Jasinoski, Rayfield, and Chinsamy 2010a, 2010b).

It is evident from the aforesaid sections that considerable advances have been made in recent years on dicynodont bone histology research, which has resulted in a better understanding of their growth regimes, histovariability, and other paleobiological aspects. Since the dicynodonts were the first successful group of herbivores to occupy varied ecological niches, deduction of their paleobiology constitutes an important component in understanding terrestrial Permo-Triassic ecosystems. Hence, the current chapter reviews and discusses the current state of knowledge regarding dicynodont bone histology and compares their growth patterns and lifestyle adaptations. We focus on five genera that are known from South Africa and India and on the histological sampling of multiple skeletal elements and from multiple individuals of different ontogenetic stages.

Institutional Abbreviations

This chapter uses abbreviations for the following institutions:, Indian Statistical Institute, Kolkata, India (ISIR); National Museum, Bloemfontein, South Africa (NMQR); and Iziko, South African Museum, Cape Town, South Africa (SAM-PK).

Geological Context, Morphology, and Paleobiology of Dicynodonts Assessed

The dicynodonts discussed in the current work were recovered from South Africa and India (Figs. 5.1A, 5.1B) and comprise examples of basal to more derived taxa in order to provide an overview of the changes in dicynodont growth patterns through their lineage. These taxa include the Permian *Endothiodon*, *Diictodon*, and *Oudenodon*; the Permo-Triassic *Lystrosaurus*; and the Middle Triassic genera *Kannemeyeria* and *Wadiasaurus*.

Specimens of *Endothiodon*, *Oudenodon*, and *Lystrosaurus* are known from both South Africa and India; whereas *Diictodon* and *Kannemeyeria* are South African forms, and *Wadiasaurus* is only known from India.

Geological Background

The Karoo Basin of South Africa is a large intracratonic basin and is characterized by sediments ranging from Upper Carboniferous to Lower Jurassic (Hiller and Stavrakis 1984). The most complete dicynodont record of the Permian Period and Mesozoic Era is found in the Karoo Basin, mainly from the Beaufort Group and the Elliot Formation of the Stormberg Group (chapter 2 of this book). The Beaufort Group contains a diverse vertebrate fauna which resulted in the biostratigraphic subdivision into several biozones from the base upward: *Eodicynodon*, *Tapinocephalus*, *Pristerognathus*, *Tropidostoma*, *Cistecephalus*, *Dicynodon*, *Lystrosaurus*, and *Cynognathus* assemblage zones (Rubidge 1995a; chapter 2 of this book) (Plate 1B; Fig. 5.1A). In contrast, the Elliot Formation of the Stormberg Group contains a diverse assemblage of reptiles including dinosaurs and is subdivided into the informal *Euskelosaurus* and *Massospondylus* range zones. Comparison of these vertebrate fauna show that the lower portion of the Elliot Formation is Late Triassic (Carnian-Norian), and the upper part of the Elliot Formation and overlying Clarens Formation is Early Jurassic in age.

In India, the Gondwana sediments are preserved in several discrete and isolated basins (Fig. 5.1B).The basal part of the Gondwana successions show uniformity in lithological features and in general begin with basal diamictite and glacial outwash deposits, followed successively by coal-bearing siliciclastic units characterized by *Glossopteris* flora. The upper part of the succession differs considerably and the formations are defined depending on distinctive physical attributes and fossil flora and fauna. These sediments are rich repositories of varied vertebrate fauna. Dicynodonts are known from different stratigraphic horizons of these basins, especially the Pranhita-Godavari, Damodar, and Satpura basins (Fig. 5.1B). *Endothiodon* is known from the Late Permian Kundaram Formation of the Pranhita-Godavari Basin (Ray 2001), whereas *Lystrosaurus* is known from the Lower Triassic Panchet Formation of the Damodar Basin (Tripathi and Satsangi 1963) and *Wadiasaurus* is from the Middle Triassic Yerrapalli Formation of the Pranhita-Godavari Basin (Roy Chowdhury 1970). All these formations contain fluvial deposits and are characterized by sandstone-mudstone units, which suggest channel-interchannel facies that were deposited during the semi-arid Triassic climate of India with a highly seasonal rainfall (Robinson 1970). Ray and Bandyopadhyay (2003) proposed a warm climate with seasonal rainfall for the Late Permian period during which the trophic composition of the Kundaram scenario was dominated by the herbivorous dicynodonts. An open habitat such as the savanna woodlands was proposed for the Kundaram sediments (Ray and Bandyopadhyay 2003). The vertebrate fauna of these Indian fossiliferous horizons bear marked similarity with those of the Karoo Supergroup in South Africa, except for a few endemic forms, and have resulted in the direct correlation between these two now widely separated regions.

In South Africa, the genus *Endothiodon* is known from the Late Permian *Tropidostoma* and *Cistecephalus* assemblage zones of the Lower Beaufort Group, Karoo Supergroup, South Africa (Rubidge 1995a) (Fig. 5.1C). *Endothiodon* (*E. mahalanobisi* and *E. uniseries*) is known from the Late Permian Kundaram Formation of the Pranhita-Godavari Basin of India (Ray 2000). The other Late Permian genus discussed here,

Diictodon, is the most prolific Late Permian form and constitutes approximately 40–50 percent of the total fossil finds from the *Tropidostoma* and *Cistecephalus* assemblage zones of South Africa (Rubidge 1995a). The genus has a particularly long stratigraphic range (Fig. 5.1C), ranging from the *Tapinocephalus* Assemblage Zone (AZ) to the upper levels of the *Dicynodon* AZ (Rubidge 1995a). However, *Diictodon* has not yet been reported from India.

The genus *Oudenodon* is known from the upper levels of the *Tropidostoma* AZ through to the upper levels of the *Dicynodon* AZ of the Karoo Supergroup, South Africa (Botha and Angielczyk 2007). It is also known from Zambia (King 1990) and India (Ray 2001).The genus *Lystrosaurus* appeared during the latest Permian (Smith and Botha 2005) and continued well into the Early Triassic. It is known from the Lower Triassic sediments of India, China, Mongolia, Russia, Antarctica, Australia, and Laos (Tripathi and Satsangi 1963; Kemp 1982; Gubin and Sinitza 1993; King 1993; Retallack and Krull 1999), but it is particularly abundant in the Karoo Basin of South Africa. The various species of this genus have been used as biostratigraphic markers for the upper levels of the Permian *Dicynodon* AZ (*L. maccaigi*),[1] Permo-Triassic boundary (*L. curvatus*), and Lower Triassic *Lystrosaurus* AZ (*L. murrayi* and *L. declivis*) of the Beaufort Group in South Africa (Botha, Abdala, and Smith 2007). In India, *L. murrayi* is known only from one stratigraphic horizon, namely the Lower Triassic Panchet Formation of the Damodar basin (Tripathi and Satsangi 1963; Ray 2005).

Numerous *Kannemeyeria* specimens have been recovered from the Middle Triassic *Cynognathus* AZ of the Karoo Supergroup (see Plate N), South Africa, and from the equivalent Lower Ermaying Formation of China (Kemp 2005). Although *Kannemeyeria* was reported from South America (Bonaparte 1966), recent work by Renaut and Hancox (2001) on the cranial osteology shows that this may not be the case. *Kannemeyeria* is not known from India, although a closely related endemic kannemeyeriid form *Wadiasaurus* was recovered from the Middle Triassic Yerrapalli Formation of the Pranhita-Godavari Basin of India (Roy Chowdhury 1970; Bandyopadhyay 1988).

Morphology and Paleobiology

ENDOTHIODON

Endothiodon was a small- to large-sized dicynodont depending on the species (skull length ranging between 160 to 570 mm) and is characterized by longitudinal ridge(s) on the snout, a pineal foramen situated on an elevated boss, posteroventrally directed boss formed by the maxilla and jugal (Fig. 5.2A), large palatines with concavities, a narrow intertemporal bar with respect to the interorbital region, a high parietal crest, a single row of maxillary teeth placed posteriorly and internally, and multiple rows of teeth on the dentary (Cox 1964; Ray 2000). The South African species,

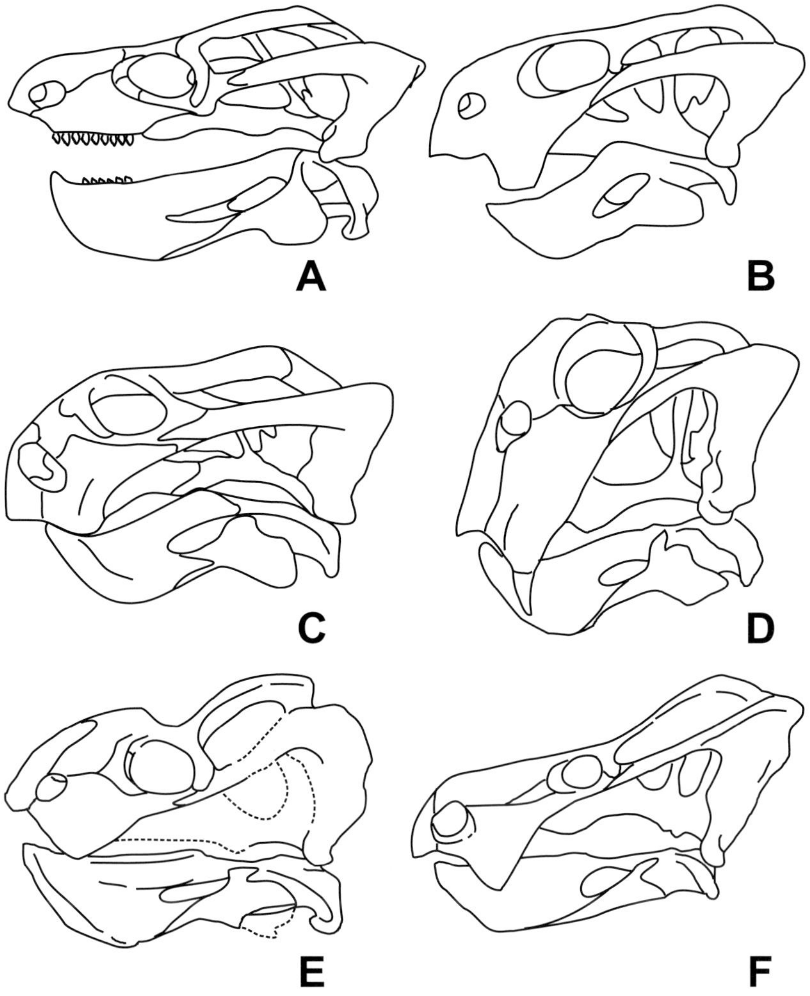

5.2. Restored skull and lower jaw of the dicynodont genera examined in lateral view. (A) *Endothiodon*. (B) *Diictodon*. (C) *Oudenodon*. (D) *Lystrosaurus*. (E) *Kannemeyeria*. (F) *Wadiasaurus*. Sources of information: King (1988), Bandyopadhyay (1988), Ray (2000, 2006). Not to scale; the line drawings have been brought to the same height for comparative purposes.

Endothiodon bathystoma is relatively larger in size than *E. mahalanobisi*, an endemic Indian form (Ray 2001). The presence of oblique wear facets on the upper and lower teeth and possibility of a masseter-like muscle based on the cranial anatomy suggest mediolateral movement of the lower jaw, which enhanced the slicing of plant matter (Ray 2000).

DIICTODON

The taxonomy, description, and functional morphology of *Diictodon* are well established (King 1988, 1993; Ray and Chinsamy 2003). *Diictodon* is a relatively small animal (skull length ~110 mm) and is characterized by the absence of postcanine teeth, and a prominent caniniform process (Fig. 5.2B; see Plates F and G), short vomer, and a large ectopterygoid (King 1988). The forelimb of *Diictodon* was articulated in a near horizontal plane, however, the humerus was caudolaterally orientated and resulted in a reduced transverse component of propulsive thrust during locomotion. In contrast, the hindlimb gait was similar to the high walk of crocodilians (Ray and Chinsamy 2003). The presence of skeletons inside large helical burrows (Smith 1987a) and a detailed functional analysis of its postcranial skeleton suggest a fossorial (burrowing) life habit for *Diictodon* (Ray and Chinsamy 2003) (see Plates F and G).

OUDENODON

This genus is a large dicynodont (maximum skull length 300 mm) (Fig. 5.2C) and is characterized by the absence of teeth, a wide intertemporal region, a recessed septomaxilla within the external naris, a weak caniniform process, narrow dentary tables, and a deep sulcus on the dorsal surface of the dentary (King 1988). Based on the cranial anatomy, it was suggested by King (1990), that the propalinal action of the lower jaw was much more enhanced in *Oudenodon* compared with that of other dicynodonts. Based on bone histology, Botha (2003) suggested a fossorial lifestyle for *Oudenodon*.

LYSTROSAURUS

The cranial morphology of *Lystrosaurus* is distinctive and easily distinguishable from that of other dicynodonts. Some of its characteristic features include a strong ventral elongation of the snout (Fig. 5.2D), a short temporal area compared with the snout, widely exposed parietals on the skull roof, the absence of a contact between the anterior portion of the palatine and premaxilla, the presence of a labial fossa, faceted stapes, and the absence of an ectopterygoid (Cluver 1971; Ray 2005; Grine et al. 2006; Angielczyk 2007; Fröbisch 2007; Fröbisch and Reisz 2008).

In South Africa, the largest *Lystrosaurus* species comprises the Permian *Lystrosaurus maccaigi* (maximum skull length 345 mm). It is characterized by relatively large orbits and prominent pre- and postorbital bosses, and the snout slopes forward and down from its junction with the frontal plane, similar to *L. declivis*. However, the premaxillary plane lies at a sharper angle and the orbits of *L. maccaigi* are more upward and forward facing compared with *L. declivis* (Brink 1951; Cluver 1971; Grine et al. 2006). *Lystrosaurus curvatus* is characterized by particularly large orbits, a gently curved snout, and the absence or weak development of a frontonasal ridge, longitudinal premaxillary ridge, and postorbital and prefrontal bosses (Brink 1951; Cluver 1971; Grine et al. 2006). It is an important species as it is restricted to the uppermost Permian and lowermost Triassic and can be regarded as a Permo-Triassic boundary marker (Botha and Smith 2007). However, it is also the rarest *Lystrosaurus* species, and consequently, its bone microstructure has yet to be studied due to a lack of available material. The Triassic *Lystrosaurus* species in South Africa comprise *L. declivis* and *L. murrayi*. The two species are similar in size and morphology, but *L. declivis* (maximum skull length 315 mm) differs from *L. murrayi* (maximum skull length 200 mm) in the presence of prefrontal bosses, a more prominent frontonasal ridge, and a longitudinal premaxillary ridge (Cluver 1971; Grine et al. 2006). Furthermore, the length of the snout in *L. murrayi* is approximately equal to the length of the skull roof (from the frontonasal ridge to the occiput), whereas the snout of *L. declivis* is notably longer than the length of the skull roof (Botha and Smith 2007).

In India, Tripathi and Satsangi (1963) reported the presence of four *Lystrosaurus* species from the Lower Triassic Panchet Formation of the Damodar Basin. These species included *L. murrayi*, *L. platyceps*, *L. maccaigi*, and *L. rajurkari*, the latter being an endemic form. Subsequently, Ray (2005) suggested that the Indian specimens pertain to only a single species, *L. murrayi*, and that the species *L. rajurkari* is a junior subjective synonym of *L. murrayi*. Bivariate scatter plots of different parameters of approximately fifty crania of *L. murrayi* from India and South Africa yielded high correlation coefficients and showed that all specimens constitute one morphologically homogeneous group in various stages of ontogenetic development. Allometric changes were mostly related to the strengthening of the trophic apparatus and probably suggest a shift in diet toward more resistant rhizomes and other plant matter as the individual matured (Ray 2005).

In a recent study using finite-element analysis, Jasinoski, Rayfield, and Chinsamy (2009) found that the specialized cranial morphology of *Lystrosaurus* under orthal bite simulations produced less stress than the more generalized *Oudenodon* cranium. It also appears that *Lystrosaurus* may have used a snapping bite to process tough fibrous vegetation, and that perhaps the added mobility in the anterior region of the snout of *Lystrosaurus* (i.e., the scarf suture between the premaxilla and nasal) may have acted to reduce strain in this region (Jasinoski, Rayfield, and Chinsamy 2010a, 2010b). These findings support the original hypothesis by King and Cluver (1991) that the unusual snout morphology of *Lystrosaurus* may have acted as a "shock absorber."

KANNEMEYERIA

One of the major groups of Middle Triassic dicynodonts were the large kannemeyeriids with body sizes ranging 1–3 m in length (Walter 1988), an average skull length of 400 mm, an elongated and pointed snout, anteriorly placed jaw articulation, and an oblique occiput and high parietal crest (Bandyopadhyay 1988; King 1990). Of these kannemeyeriids, the genus *Kannemeyeria* (Fig. 5.2E) is characterized by large size (skull length about 400 mm) (see Plate N), an almost parallel alignment of the zygomatic arches, a high and narrow parietal crest, large orbits, and a well developed ridge medial to the nasal bones. Pearson (1924a, 1924b) and King (1990) suggested that the mastication in *Kannemeyeria* mostly involved the straight up and down action of the lower jaw, where food processing was concentrated in the anterior region of the mouth.

The taxonomy, morphological description, and functional anatomy of *Wadiasaurus* are well established (Roy Chowdhury 1970; Bandyopadhyay 1988; Ray 2006). The genus is represented by at least seven hundred cranial and postcranial elements pertaining to approximately twenty-three individuals of varying ontogenetic stages and collected from a single bone bed (Bandyopadhyay 1988). *Wadiasaurus* (Fig. 5.2F) was a large dicynodont with a skull length of approximately 400 mm. Characteristic features include a low nasal ridge, vertically projected maxillary flanges, a wide interpterygoid vacuity (Roy Chowdhury 1970; Bandyopadhyay 1988), narrow intertemporal (IT) region compared with a wide interorbital (IO) region ($IO/IT = 4.7$), a tall, narrow scapular blade with a short acromion process and a prominent olecranon process. A functional anatomical study by Ray (2006) showed that *Wadiasaurus* had an almost sagittal articulation of the humerus, an efficient forearm movement (extension and flexion), a reduced/restricted lateral undulation of the vertebral column, and a semi-erect/adducted stance of the hindlimbs.

Methodology and Terminology

Ideally, a comprehensive analysis of the bone microstructure should involve multiple skeletal elements of several individuals at different ontogenetic stages in order to document the growth pattern and inter-elemental histovariability of the animal (Chinsamy-Turan 2005). Since the analysis is destructive in nature, complete skeletons are generally not available for study. Thus, efforts are made to study the long limb bones as they exhibit the least secondary remodeling compared with other skeletal elements and provide the best record of the primary bone tissue. Before preparing the thin sections, photographs were taken of all the specimens, morphological variations were noted, and standard measurements were recorded using digimatic calipers with a precision of 0.01 mm. Specimens were embedded under vacuum in a clear resin, sectioned in the required direction, ground and polished, and examined under ordinary and polarized light following the procedures outlined by Chinsamy and Raath (1992).

In the case of the genera *Diictodon*, *Lystrosaurus*, and *Wadiasaurus*, the largest individual of each taxon was standardized to compare the long bone dimensions of other individuals and to assign the bones to different size or age classes (Ray and Chinsamy 2004), which are expressed as a percentage of the adult sizes (Chinsamy 1993a; Horner, de Ricqlès, and Padian 2000). Histological terminology and definitions in this chapter follow Francillon-Vieillot et al. (1990), de Ricqlès et al. (1991), and Reid (1996) unless specifically mentioned otherwise.

Starck and Chinsamy (2002) have shown that the channels in a fossil bone would have included lymph tissue and nerves as well as vascular canals when the individual was alive. The number of channels in the bone as a percentage of total cortical area (Chinsamy 1993b) was quantified using either the image analysis program NIS Elements D 3.0

(Nikon Instruments, Inc., Melville, New York) or Jandel SigmaScan Pro (version 4.0, Systat Software, Inc., Chicago, Illinois) and two terms are used: "cortical porosity" (Ray and Chinsamy 2004) and "vascular density" (Botha-Brink and Angielczyk 2010).

The cortical or relative bone wall thickness (RBT) was calculated as the ratio of the cross-sectional bone wall thickness to the cross-sectional diameter measured across the transverse section following Bühler (1986), Botha (2003), and Ray and Chinsamy (2004) and was expressed as a percentage.

Results of the Assessment of the Bone Histology of Specified Dicynodont Taxa

Endothiodon

Chinsamy and Rubidge (1993) examined the humeri of the genus *Endothiodon* from South Africa, whereas several long bones of the Indian *Endothiodon* (*E. mahalanobisi*) were recently examined by Ray, Bandyopadhyay, and Bhawal (2009). Subsequently the long bones of the South African form *Endothiodon bathystoma* were examined by Botha-Brink and Angielczyk (2010). These studies reveal that the limb bones contain a small medullary cavity, which is surrounded by a thick cortex. The bone tissue in the smaller individuals consists of highly vascularized (13 percent vascular density) fibrolamellar bone ranging from a subreticular to plexiform pattern (Fig. 5.3A), similar to the initial findings of Chinsamy and Rubidge (1993). Botha-Brink and Angielczyk (2010) observed poorly vascularized (4 percent vascular density) parallel-fibered bone at the periosteal periphery of the largest individual studied (SAM-PK-6618a, estimated to be 100 percent adult size) (Fig. 5.3B). The smaller South African specimen (SAM-PK-K5605), estimated to be approximately 77 percent of the maximum adult size, exhibits thick cortices (RBT averaging 39 percent [Botha-Brink and Angielczyk 2010]), whereas the limb bones of the Indian form exhibit relatively thinner cortices with RBT values averaging 30 percent (Ray, Bandyopadhyay, and Bhawal 2009). The bones of SAM-PK-K6618 contain completely infilled medullary cavities, thus the cortical thickness was not measured in this individual. Growth rings are absent in all the elements belonging to SAM-PK-K5605, whereas multiple annuli are observed in the bones of SAM-PK-K6618 (Botha-Brink and Angielczyk 2010). ISIR 771/1 shows a distinct LAG in the midcortical region (Fig. 5.3A). In the Indian forms, the limb bones have large, irregular resorption cavities in the perimedullary region. Few secondary osteons are present in the perimedullary region of tibia ISIR 771/2.

Diictodon

A humerus of *Diictodon* was examined by Chinsamy and Rubidge (1993) to show that the cortex was composed of uninterrupted fibrolamellar bone tissue where growth rings (LAGs and annuli) were absent. Ray and Chinsamy (2004) carried out a comprehensive osteohistological examination of *Diictodon* based on multiple skeletal elements of several individuals,

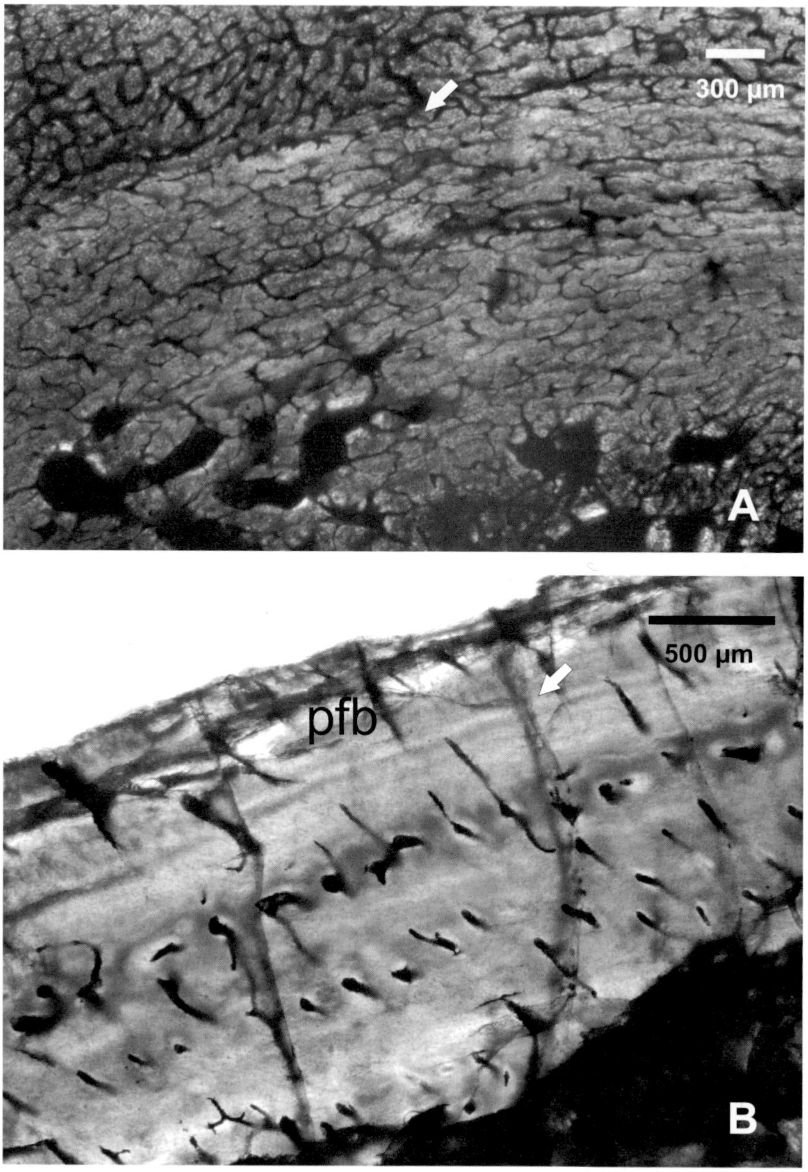

5.3. *Endothiodon.* (A) ISIR771/1, humerus showing vascularized fibrolamellar bone tissue. (B) Unidentified limb bone showing parallel-fibered bone (pfb). Arrows indicate LAGs.

which was later substantiated by Ray, Bandyopadhyay, and Bhawal (2009) and Botha-Brink and Angielczyk (2010). All these studies show that the skeletal elements of *Diictodon* are characterized by moderately vascularized fibrolamellar bone tissue (Figs. 5.4A, 5.4B). The average cortical porosity of the adult humeri (SAM-PK-K6716a and SAM-PK-K7015) was 7 percent (Ray and Chinsamy 2003). The longitudinally oriented primary osteons have radial and circumferential anastomoses and show differing patterns depending on the skeletal elements and age classes. *Diictodon* had uninterrupted fibrolamellar bone tissue in the juvenile individuals, whereas in the subadult and adult stages, there was a decrease in cortical porosity, primary osteons became more organized, and multiple LAGs and annuli appeared (Chinsamy and Rubidge 1993; Ray and Chinsamy

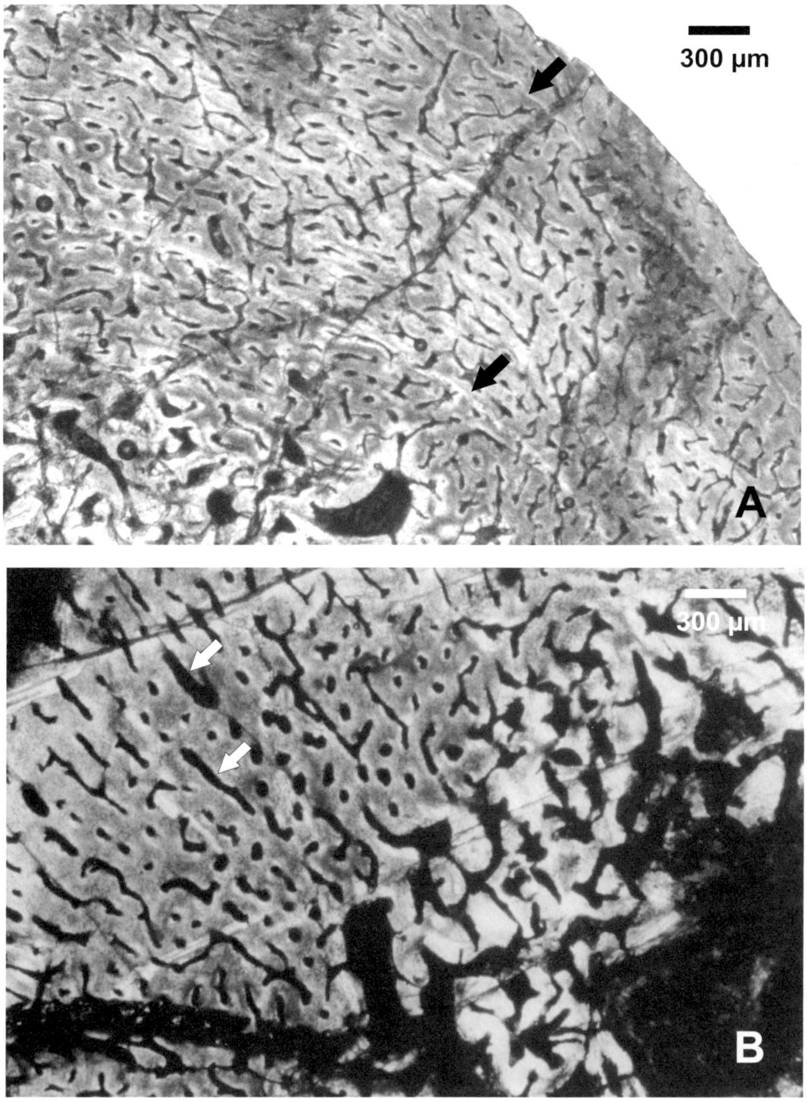

5.4. *Diictodon.* (A) SAM-PK-K6716a, femur showing zonal fibrolamellar bone tissue separated by annuli (arrows). (B) SAM-PK-K6717, humerus showing radially oriented vascular channels (arrows) and a small but distinct medullary cavity.

2004) (Plate 8). A small and distinct medullary cavity (Fig. 5.4B) is seen in most of the skeletal elements examined, which is surrounded by a thick cortex. The cortical thickness of the limb bones is high, particularly in the adults where RBT is well above 30 percent (Ray and Chinsamy 2004). The cortical thickness of the humeri (SAM-PK-K6716a and SAM-PK-K6717) is 41.79 and 58.55 percent of the diameter, whereas the RBT of the radius of the latter individual (SAM-PK-K6717) is 60.6 percent (Ray, Bandyopadhyay, and Bhawal 2009).

Oudenodon

De Ricqlès (1972) and Chinsamy and Rubidge (1993) showed that zonal fibrolamellar bone tissue was predominant in the limb bones of *Oudenodon* (see Plate 4). In addition, the onset of avascular lamellar

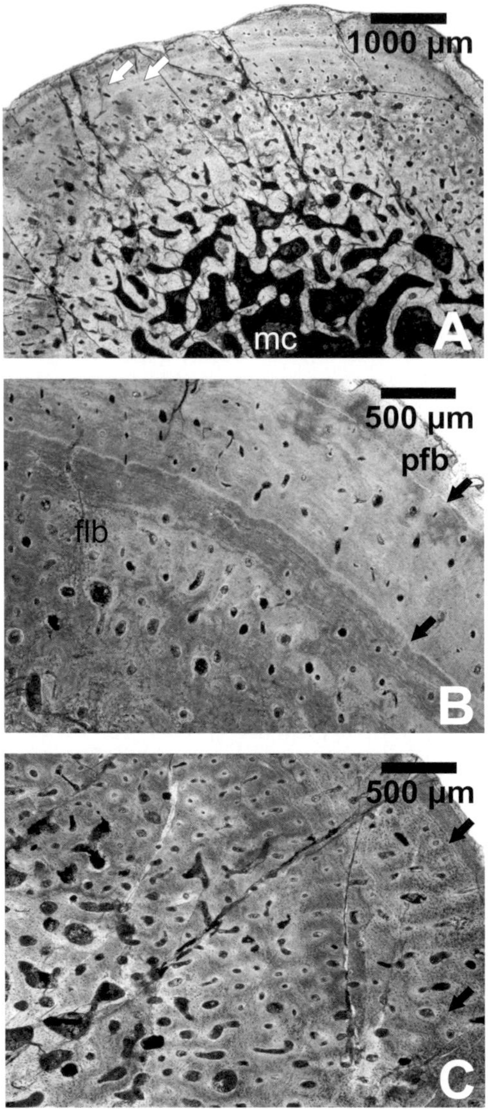

5.5. *Oudenodon.* (A) SAM-PK-K10107b, radius showing a thick cortex surrounding a medullary cavity (mc) partially filled with trabecular bone. (B) SAM-PK-4807c femur showing fibrolamellar bone (flb) in deeper cortex and thick zone of parallel-fibered bone (pfb) in the outer cortex. (C) SAM-PK-11141, rib showing erosionally enlarged resorption cavities in the deeper cortex. Arrows indicate LAGs in all the elements.

bone toward the periosteal margin was reported by de Ricqlès (1972). Osteohistological analysis of the genus *Oudenodon,* based on multiple limb bones, was conducted by Botha (2003) and later re-examined by Botha-Brink and Angielczyk (2010). These later studies, based on multiple limb bones, revealed moderately vascularized fibrolamellar bone (Figs. 5.5A, 5.5B) (vascular density varies from 12.6 percent in younger individuals to 5.7 percent in older individuals [Botha-Brink and Angielczyk 2010]), although in later ontogenetic stages, poorly vascularized parallel-fibered bone toward the periphery is observed (Fig. 5.5B; see Plate 5G, 5H). The primary osteons are longitudinally oriented (Figs. 5.5B, 5.5C) and in some areas form a laminar pattern (also observed by de Ricqlès [1972]). The limb bones of *Oudenodon* exhibit a relatively thick cortex (average femoral RBT of 41 percent [Botha-Brink and Angielczyk 2010]).

Extensive secondary remodeling with large resorption cavities is present in the fibulae (SAM-PK-10019c and SAM-PK-K4807d). Interruptions by annuli of lamellar bone and LAGs (Figs. 5.5A to 5.5C) were noted in the cortex of all the elements examined.

Lystrosaurus

Early studies show that the humeri of *Lystrosaurus* were composed of fibrolamellar bone tissue showing a plexiform arrangement (de Ricqlès 1972; Chinsamy and Rubidge 1993). A comprehensive examination of the thin sections of various skeletal elements was carried out by Ray, Chinsamy, and Bandyopadhyay (2005) and later substantiated by Botha-Brink and Angielczyk (2010), who examined the genus to species level. Jasinoski, Rayfield, and Chinsamy (2010a, 2010b) studied the cranial suture morphology of *Lystrosaurus* and showed that the premaxilla-nasal scarf suture provided sutural mobility in the anterior part of the snout, suggesting that *Lystrosaurus* employed a different biting regime than other dicynodonts. In chapter 6 of this book, Jasinoski and Chinsamy-Turan provide detailed descriptions of the bone histology of various cranial elements of *Lystrosaurus* (and *Oudenodon*).

All elements of *Lystrosaurus*, regardless of size or species, comprise rapidly formed fibrolamellar bone (Fig. 5.6). Subadult and adult individuals of the large Permian *L. maccaigi* reveal thick cortices (average cortical thickness of 38 percent) and either relatively free medullary cavities containing loose bony trabeculae or completely infilled medullary cavities (Figs. 5.6A, 5.6B). The perimedullary region contains numerous enlarged channels that gradually decrease in size and form a laminar network of fibrolamellar bone. Annuli and LAGs occur in the mid and outer cortex of all individuals studied (Fig. 5.6B). The Triassic *L. murrayi* and *L. declivis* exhibit a similar bone microstructure. The specimens studied by both Ray, Chinsamy, and Bandyopadhyay (2005) and Botha-Brink and Angielczyk (2010) reveal relatively thin cortices compared with other dicynodonts (ranging from 10 percent in a tibia to 33 percent in a humerus). Extensive secondarily enlarged vascular channels were noted in *L. murrayi* by Ray, Chinsamy, and Bandyopadhyay (2005), who also examined the bone microstructure of Indian specimens and found it to be similar to that of the South African specimens (Figs. 5.6C to 5.6E). The primary cortex becomes thinner in larger individuals and is replaced by secondary reconstruction, similar to *Kannemeyeria*. Thus, a free medullary cavity is completely absent in the largest specimens studied (Ray, Chinsamy, and Bandyopadhyay 2005; Botha-Brink and Angielczyk 2010). The bone tissue of *L. murrayi* and *L. declivis* comprises rapidly forming uninterrupted fibrolamellar bone in younger individuals, which is interrupted by annuli and LAGs in subadult and adult individuals (Ray, Chinsamy, and Bandyopadhyay 2005; Botha-Brink and Angielczyk 2010). Notably large, channels, which have been noted in other dicynodonts

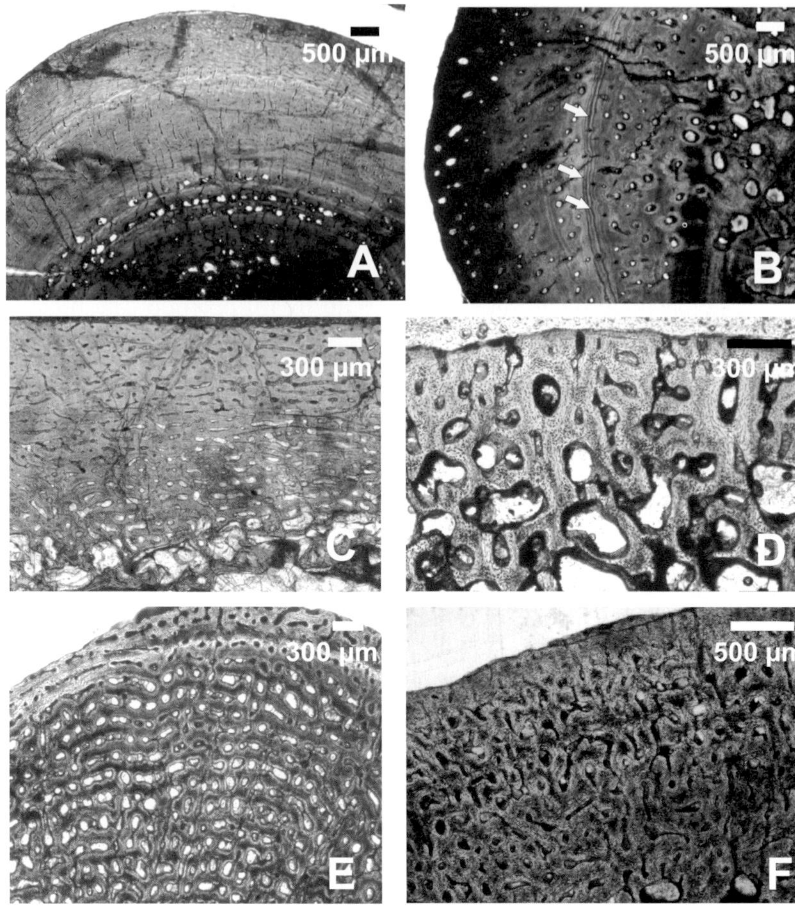

5.6. (A, B) *Lystrosaurus maccaigi.* (A) NMQR 3663b, ulna showing a thick cortex containing fibrolamellar bone tissue. Note the enlarged vascular channels are found in the deeper cortex. (B) NMQR 3663, rib showing enlarged vascular channels in the outer cortex. Multiple LAGs are indicated by the arrows. (C–F) *Lystrosaurus murrayi.* (C) SAM-PK-11184, femur showing fibrolamellar bone tissue and erosionally enlarged vascular channels in midcortical region. (D) ISIR759, tibia showing narrow cortex with erosionally enlarged vascular channels. (E) ISIR752, rib showing a thick cortex with highly vascularized fibrolamellar bone tissue and extensive occurrence of erosionally enlarged vascular channels. (F) NMQR 835b, tibia showing fibrolamellar bone tissue with the primary osteons arranged in plexiform-reticular pattern.

(e.g., *Eodicynodon, Rhachiocephalus, Aulacephalodon, Dicynodon,* and to some extent, *Kannemeyeria* [Botha-Brink and Angielczyk 2010]) are particularly prominent in *L. murrayi* and *L. declivis.* The vascular density ranges from 22 percent in younger individuals to 12 percent in older individuals (Botha-Brink and Angielczyk 2010).

Kannemeyeria

Gross (1934) demonstrated highly vascularized fibrolamellar bone in a laminar network in a large humerus. Similarly, Enlow and Brown (1957) reported fibrolamellar bone in a plexiform network. Their images reveal a humerus with highly vascularized bone and the medullary cavity appears to be infilled with bony trabeculae (Enlow and Brown 1957, Plate 23, Figs. 8 and 9). Highly vascularized fibrolamellar bone arranged in a plexiform network was also observed by de Ricqlès (1972). Several skeletal elements (Figs. 5.7A to 5.7D) were examined by Botha-Brink and Angielczyk (2010) who found that the cortices are composed of highly vascularized (13 percent vascular density) fibrolamellar bone (Fig. 5.7B) with longitudinally oriented primary osteons and short anastomoses (similar to

Chinsamy and Rubidge [1993]). Due to the absence of a free medullary cavity (Fig. 5.7E) in the specimens examined by Botha-Brink and Angielczyk (2010), the cortical thickness has not been quantified. Multiple LAGs and annuli (Figs. 5.7B to 5.7D) are present throughout the cortex of all the elements examined by Botha-Brink and Angielczyk (2010) and agree with observations made by Chinsamy and Rubidge (1993) in a humerus of *Kannemeyeria*. However, de Ricqlès (1972) did not report growth rings in the limb bones in his study and suggested that due to the continued rapid growth, his specimens represent immature individuals. Secondary reconstruction (also noted by de Ricqlès [1972]), in the form secondary osteons, is present in the inner and mid cortex. Sharpey's fibers were observed in several elements, particularly in the rib (NMQR3631), where secondary reconstruction also reached the outer cortex. Peripheral parallel-fibered bone is present in the ulna (NMQR2674a) (Fig. 5.7A), and multiple LAGs at the subperiosteal surface, which may represent peripheral rest lines, were observed on one side of an unidentified limb bone (NMQR789) (Fig. 5.7D).

Wadiasaurus

Various limb bones (humerus, femur, radius, ulna, fibula) as well as several dorsal ribs and scapulae were examined by Ray, Bandyopadhyay, and Bhawal 2009 and Ray, Bandyopadhyay, and Appana 2010). The diaphyseal section of the humerus (ISIR 175/81) shows peripheral parallel-fibered bone with fibrolamellar bone in the deeper part of the cortex (Fig. 5.8A). The primary osteons are longitudinally oriented and isolated, although in some places they are arranged in a laminar pattern, showing an almost parallel network. The cortices of the limb bones are moderately thick (>20 percent RBT) with the femur ISIR 175/68 having the maximum RBT of 31.7 percent. In contrast, the cortical thickness of the tibia is comparatively low (RBT, 8.05 percent) (Ray, Bandyopadhyay, and Appana 2010). As in other dicynodonts, all the skeletal elements examined show a predominance of fibrolamellar bone tissue (Fig. 5.8B). Ray, Bandyopadhyay, and Appana (2010) identified three size classes based on the overall size and morphological characteristics of the skeletal elements examined. Multiple annuli and LAGs are present in most of the elements (Fig. 5.8B), which have attained more than 30 percent of adult size. Other features include secondary reconstruction in the form of resorption cavities and secondary osteons in the perimedullary region.

Discussion

Dicynodont Bone Microstructure

All the bones examined contain either a small relatively clear or partially or completely infilled central medullary cavity surrounded by an outer relatively thick compact cortex. The infilling of the medullary cavity varies between genera, where smaller taxa such as *Eodicynodon*, *Diictodon*, *Cistecephalus*, *Dicynodontoides*, *Oudenodon*, and *Tropidostoma* often

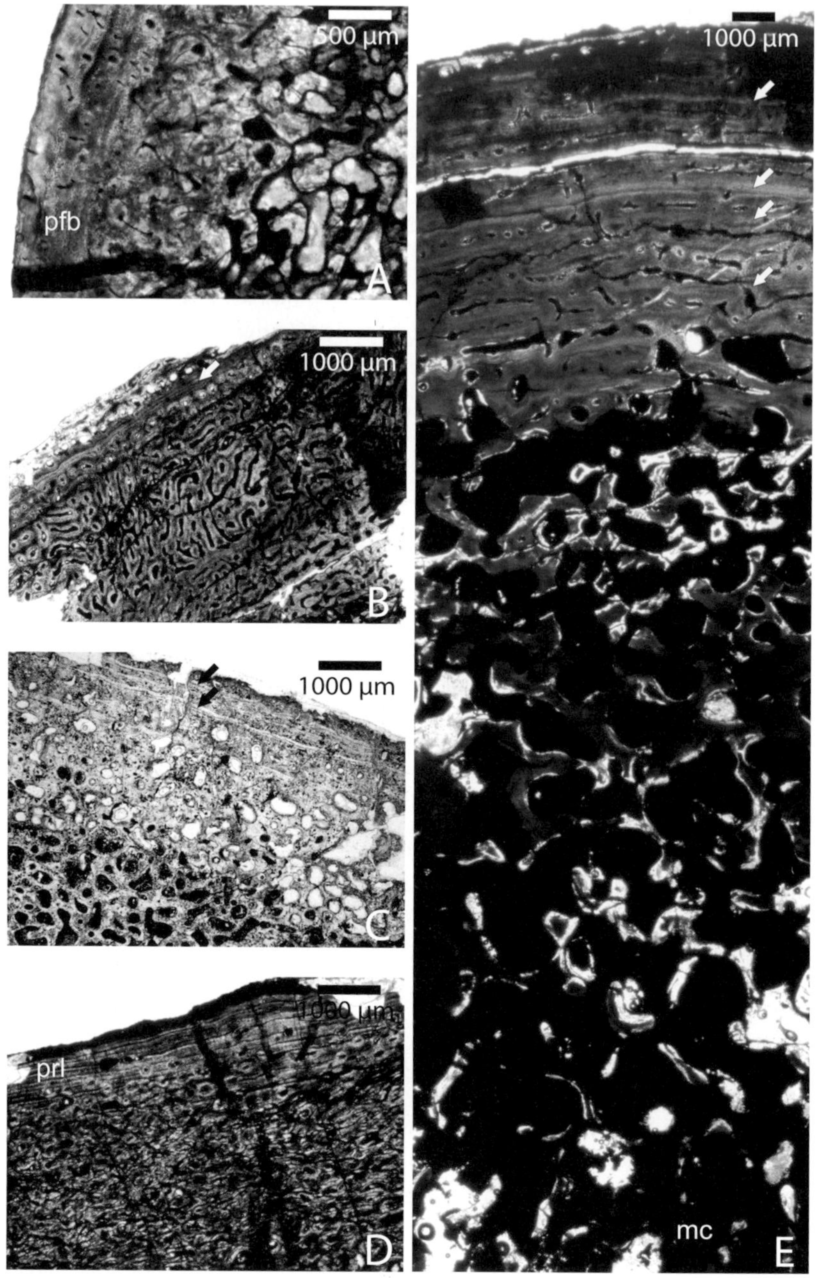

5.7. *Kannemeyeria.* (A) NMQR 2674a, ulna showing parallel-fibered bone (pfb) in the outer cortex, (B) NMQR 2674b, femur showing highly vascularized fibrolamellar bone in the cortex. A LAG is indicated by an arrow. (C) SAM-PK-1055d, femur showing extensive occurrence of enlarged vascular channels and infilling of the medullary cavity by trabecular bone. (D) NMQR 789, an unidentified limb bone showing peripheral rest lines (prl). (E) NMQR 3631, rib showing absence of free medullary cavity (mc). Arrows indicate multiple LAGs.

contain free or only partially infilled medullary cavities (Chinsamy and Rubidge 1993; Ray and Chinsamy 2004; Botha and Angielczyk 2007; Botha-Brink and Angielczyk 2010) and larger taxa such as *Rhachiocephalus, Aulacephalodon, Dicynodon, Lystrosaurus,* and kannemeyeriids contain completely infilled medullary cavities (de Ricqlès 1972; Chinsamy and Rubidge 1993; Green, Schweitzer, and Lamm 2010; Botha-Brink and Angielczyk 2010; chapter 7 of this book).

The dicynodont genera examined exhibit cortical bone of predominantly fibrolamellar bone tissue and support previous studies on

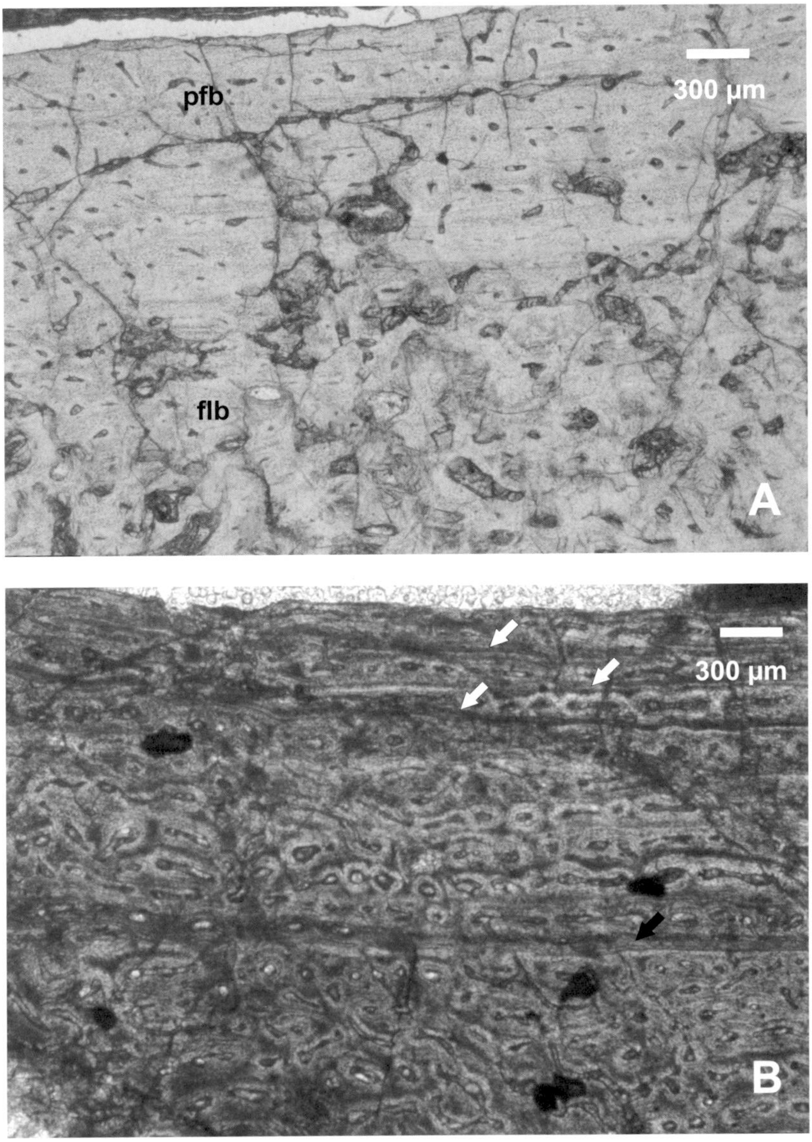

5.8. *Wadiasaurus.*
(A) ISIR175/68, femur showing discrete and isolated longitudinally oriented primary osteons in the deeper fibrolamellar (flb) cortical bone. The peripheral region of the bone wall consists of more parallel-fibered (pfb) bone. (B) ISIR175/75, ulna showing highly vascularized zonal fibrolamellar bone separated by a distinct annulus composed of lamellar bone (black arrow) and multiple LAGs (white arrows).

dicynodont bone microstructure (see, e.g., Gross 1934; Enlow and Brown 1957; de Ricqlès 1972; Chinsamy and Rubidge 1993) (Figs. 5.3 to 5.8). The channels in the cortex range from longitudinally oriented primary osteons with radial and circumferential anastomoses to plexiform or laminar arrangements. The cortical porosity or "vascular density" is relatively high in all young individuals (ranging from 13 percent in *Diictodon* of 56 percent adult size to 22.2 percent in *Lystrosaurus* of 39 percent adult size [Ray and Chinsamy 2004; Botha-Brink and Angielczyk 2010]), but decreases toward the periphery in older individuals as many genera exhibit poorly vascularized parallel-fibered bone in the outer cortex. A profusion of radial channels running from the medullary cavity to the peripheral margin is seen in all elements in the deltopectoral crest, trochanter major, and cnemial crest regions. In most of the elements, the perimedullary region contains numerous erosionally enlarged resorption cavities and a few secondary osteons. This feature is particularly prominent in large *Kannemeyeria* individuals and is also seen in other large kannemeyeriids such as *Stahleckeria* (de Ricqlès 1972), *Dinodontosaurus* (rib [Enlow and Brown 1957]), and *Placerias* (Green, Schweitzer, and Lamm 2010) to the point where dense Haversian bone has been reported for the latter three taxa. Proposed causes for extensive secondary remodeling resulting in dense Haversian bone include old age, large size, fracture or microcrack repair, and bone grain change due to the presence of muscle insertions (Currey 2003). Sharpey's fibers were observed in most of the limb bones, a feature which has been noted in other studies on dicynodonts as well (Chinsamy and Rubidge 1993; Botha 2003; Ray and Chinsamy 2004; Botha-Brink and Angielczyk 2010), and their distribution in the crania of *Lystrosaurus* and *Oudenodon* is documented in chapter 6 (see Plate 9). Annuli and LAGs are present in all the dicynodont genera, but usually only appear in larger individuals, suggesting that growth rings appeared late in ontogeny.

Growth Patterns Exhibited

All dicynodonts are characterized by a predominance of fibrolamellar bone tissue in the cortex. Fibrolamellar bone is indicative of rapid osteogenesis (Amprino 1947; Buffrénil 1980; Reid, 1990; Chinsamy 1997; de Margerie, Cubo, and Castanet 2002), which suggests an overall fast growth for dicynodonts. Interestingly, many genera exhibit uninterrupted fibrolamellar bone early in ontogeny with annuli and LAGs only appearing at approximately 50 percent of adult size or in some taxa only at 70 percent of adult size (Ray and Chinsamy 2004; Botha-Brink and Angielczyk 2010). Several taxa also exhibit peripheral parallel-fibered bone in larger individuals, suggesting that the overall growth rate decreased, possibly with the onset of reproductive maturity (Castanet and Smirina 1990; Castanet and Baez 1991; Sander 2000; chapter 3 of this book). Additionally, peripheral rest lines, suggesting asymptotic growth have also been observed in *Placerias* (Green, Schweitzer, and Lamm 2010) and *Kannemeyeria*.

Different growth stages have been determined for *Diictodon* (Ray and Chinsamy 2004), *Lystrosaurus* (Ray, Chinsamy, and Bandyopadhyay 2005), and *Wadiasaurus* (Ray, Bandyopadhyay, and Appana 2010). These growth stages are summarized in Table 5.1. The early rapid growth of *Diictodon* was first reported by Chinsamy and Rubidge (1993) and later supported by Ray and Chinsamy (2004) and Botha-Brink and Angielczyk (2010). Annuli occur in the compacta only once the individual attained 70–90 percent of adult size (Table 5.1, growth stage C), whereas LAGs only appeared at even later stages (>90 percent of adult size, growth stage D). As such, Ray and Chinsamy (2004) deduced that *Diictodon* experienced sustained fast growth as juveniles, which slowed down probably once sexual maturity was reached in the subadult stage. In the adult stage, growth continued yet was periodically interrupted, and so *Diictodon* had an indeterminate growth pattern (sensu Foote and Miller 2007).

In *Lystrosaurus*, four distinct growth stages are identified (Table 5.1) where growth stages C and D (subadult and adult stages) were characterized by extensive secondary reconstruction that resulted in an abundance of erosionally enlarged resorption cavities not only in the perimedullary region but also in the mid and outer cortex (Plate 8). In addition, the medullary region exhibits a dense network of bony trabeculae. Based on the bone microstructure, it was proposed by Ray, Chinsamy, and Bandyopadhyay (2005) that *Lystrosaurus* had a continuous fast growth during the juvenile stage (up to 30 percent of adult size), after which growth became periodic with intermittent slowing down or cessation of growth. However, growth continued and an indeterminate growth strategy was proposed for *Lystrosaurus*.

Three growth stages have been identified for *Wadiasaurus* where fibrolamellar bone tissue predominates in the juveniles and subadults. However, the adults show parallel-fibered and lamellar bone along the cortical periphery, suggesting that the overall fast growth decreased considerably when the adult stage was reached. The overall growth of *Wadiasaurus* was possibly much slower than that of other dicynodonts, as is evident from the relative predominance of parallel-fibered bone and laminar fibrolamellar bone tissue (Ray, Mukherjee, and Bandyopadhyay 2009). Although it was not possible to determine the different growth stages of either *Endothiodon* (Ray, Bandyopadhyay, and Bhawal 2009) or *Oudenodon*, Botha (2003) documented that *Oudenodon* grew fast initially, but that growth decreased markedly late in ontogeny as was evident from the presence of more slowly formed bone such as lamellar bone, parallel-fibered bone, and lamellated bone at the periosteal periphery of several skeletal elements (e.g., as illustrated in a jugal and fibula in Plate 5E to 5H). Botha (2003) suggested that the transition from fast to slow growth marked the onset of sexual maturity.

The presence of growth rings indicates alternating cycles of bone deposition and suggests that older individuals may have been susceptible to environmental fluctuations (Chinsamy 1997; Chinsamy-Turan 2005; chapter 3 of this book). Starck and Chinsamy (2002) attributed the variable incidence of annuli and LAGs among several dinosaurs, extant sauropsids, and non-mammalian synapsids as reflecting plasticity in their development in response to the environment, and thus the presence of such growth rings in dicynodonts suggests a flexible growth strategy and therefore a high degree of developmental plasticity (sensu Smith-Gill 1983; Starck and Chinsamy 2002; Ray, Botha, and Chinsamy 2004).

Another noteworthy feature is the difference in the type of tissues present in basal and more derived dicynodont genera. The more basal genera exhibit rapid uninterrupted fibrolamellar bone early in ontogeny, which becomes periodic during later development. However, many of the more derived genera (irrespective of body size) also reveal notably large "vascular" channels and vascular arrangements that imply rapid growth. These enlarged channels are absent in the more basal genera (although they have been observed in *Eodicynodon* [Botha-Brink and Angielczyk 2010]), but become increasingly prevalent in the Bidentalia. They have been observed in *Aulacephalodon, Dicynodon, Lystrosaurus,* and, to a lesser extent, *Kannemeyeria* (Botha-Brink and Angielczyk 2010). These researchers suggest that these enlarged channels allowed greater blood flow and would have facilitated rapid growth (for a counter perspective see chapter 6 of this book), and that their abundance in Triassic *Lystrosaurus* species may have played a role in allowing this genus to become the most abundant and successful tetrapod during the Early Triassic post-extinction recovery phase.

Mapping of osteohistological character states on a cladogram showing the interrelationship between available neotherapsids (dicynodonts and their relatives) shows a significant evolutionary trend toward an asymptotic growth strategy and reduced developmental plasticity within the neotherapsids (Ray, Mukherjee, and Bandyopadhyay 2009).

Lifestyle Adaptations

A functional relationship exists between bone wall thickness, habitat of the animal, and the forces that act on the bone (Chinsamy-Turan 2005). Wall (1983) considered the thickness of the cortex (RBT) of mammals to be correlated with a specific mode of life. He suggested that if the cortex exceeds 30 percent of the average bone diameter in most of the limb bones, then the animal was at least semi-aquatic (e.g., hippopotamus, manatee, and beaver). The cortical thickness of sirenian, cetacean, crocodilian, and certain aquatic bird bones have particularly high bone densities to counteract buoyancy (Buffrénil et al. 1990; Hua and Buffrénil 1996). The high RBT of aquatic/semi-aquatic animals suggests that the high bone density aids in overcoming buoyancy. High values of RBT have been observed in the crocodilian *Crocodylus* and *Alligator* (36.3 and 36.45 percent, respectively [Ray and Chinsamy 2004]) and that corroborated Wall's findings.

Magwene (1993) studied the cortical thickness of numerous extant crocodilian, lizard, mammalian, and non-mammalian synapsid femora and found that, on average, non-mammalian synapsids and mammals have thinner cortices than crocodilians and lizards do. He argued that non-mammalian synapsids and mammals have lighter bones as a weight-saving modification because they are subject to greater bending and torsional stresses due to their higher activity levels. In contrast, crocodilian and lizard limb bones are more robust and stiff as they have less active lifestyles (Magwene 1993). Magwene examined each group collectively, but when the results are examined according to lifestyle, 75 percent of the semi-aquatic crocodilians and lizards and all of the arboreal/fossorial lizards studied had a cortical thickness of more than 30 percent. Similarly, 38 percent of the fossorial/arboreal mammals had a cortical thickness that exceeded 30 percent and several of the other fossorial mammals had a cortical thickness that was at least 28 percent. Thus, Wall's "30-percent" threshold appears adequate for indicating the approximate minimum cortical thickness required for aquatic habits, as well as for fossorial or arboreal lifestyles.

More recent studies on the humerus, radius, and tibia of various extant vertebrates quantified bone compactness using the computer program Bone Profiler (Girondot and Laurin 2003; Germain and Laurin 2005; Kriloff et al. 2008; Canoville and Laurin 2010). The findings of these studies generally support Wall's (1983) observations that a thick, particularly compact cortex with a small medullary cavity is usually present in animals living in shallow water, and highly cancellous bones with completely infilled medullary cavities are usually present in active deep sea divers. However, it should be noted that body size and taxonomic position also play a role in the degree of bone compactness (Germain and Laurin 2005; Laurin, Canoville, and Quilhac 2009).

In general, dicynodonts are characterized by thick cortices irrespective of body size, as is evident from the ten dicynodont genera that Botha-Brink and Angielczyk (2010) examined (RBT exceeds 30 percent in subadult and adult individuals). Botha-Brink and Angielczyk (2010) suggested that as dicynodonts are generally regarded as fully terrestrial (based on their functional morphology) and not semi-aquatic, the high cortical thickness was indicative of a fossorial lifestyle where thick bone walls would have given extra strength and support for digging. Ichnological evidence for burrowing has been found for some dicynodonts such as *Diictodon* (Smith 1987a) and Triassic *Lystrosaurus* (Retallack, Smith, and Ward 2003; Modesto and Botha-Brink 2010), and cranial and postcranial characteristics of others such as the cistecephalid dicynodonts suggest a fossorial ability (Cox 1972; Cluver 1974, 1978). Other dicynodonts such as *Rhachiocephalus*, *Aulacephalodon*, *Dicynodon*, *Lystrosaurus maccaigi*, and *Kannemeyeria* were probably too large to be active burrowers, but as they exhibit features such as spatulate unguals (present in extant digging vertebrates), digging may still have played an important role in their ecology. Alternatively, Botha-Brink and Angielczyk (2010) suggest that the high cortical thickness of dicynodonts may reflect a phylogenetic constraint and that a fossorial behavior was plesiomorphic for the clade and the more derived or larger taxa retained these adaptations due to ancestry. That bone microstructure may be phylogenetically informative was also suggested by Green, Schweitzer, and Lamm (2010).

Ray and Chinsamy (2004) showed that cortical thickness measured at the diaphyseal sections of multiple limb bones of *Diictodon* vary even within the same individual. For example, the humeral RBT of SAM-PK-K6716a (41.79 percent) is more than the femoral RBT (36.28 percent) of the same individual. It was further observed that for an increase in the estimated body mass of 8 kg from juvenile to adult, the cortical thickness of the humerus more than doubled, which corresponds to the humeral lengthening of 33 mm. In contrast, the cortical thickness of the femur increases 1.5 times for the same increase in body mass and femoral lengthening of about 38 mm (Ray and Chinsamy 2004). Since the increase in cortical thickness with ontogeny was parabolic and does not maintain a linear relationship, it was suggested that the humeral and femoral appositional growth was variable. This may be attributed to different biomechanical constraints. The hindlimbs in general supported a slender pelvic girdle and provided the main locomotory thrust by extension-flexion arc (Ray and Chinsamy 2003). In contrast, the forelimbs had to support a massive thorax and large head resulting in an anterior shift of the center of gravity with the forelimbs supporting more weight than the hindlimbs. The forelimbs were also used for digging with humeral excursion playing a dominant role (Ray and Chinsamy 2003). *Diictodon* is considered to be fossorial based on Smith's (1987a) findings of deep helical burrow casts containing *Diictodon* skeletons (see Plate F).

In *Wadiasaurus*, the cortical thickness of all the adult limb bones is low to moderate (except for the femur [Ray, Bandyopadhyay, and Appana 2010]). During the subadult stage, the cortical thickness of different limb bones ranges between 20.8 percent (radius) and 26.6 percent (ulna), whereas in the adult stage, the cortical thickness of the femur is 31.7 percent and that of the ribs varied from 18.9 percent to 26.5 percent (Ray, Bandyopadhyay, and Appana 2010). In addition, there are no morphological adaptations (Ray 2006) to suggest that *Wadiasaurus* had any specialized life habit. However, RBT of the forelimb bones of *Wadiasaurus* is comparable with that of the extant megaherbivores such as *Ceratotherium* (white rhinoceros) (26 percent RBT for the radius and ulna), although the femoral thickness of *Wadiasaurus* is much higher compared with that of *Ceratotherium* (25 percent). Ray, Bandyopadhyay, and Appana (2010) suggested that *Wadiasaurus* was a generalized terrestrial animal, a megaherbivore (which is further corroborated by its skeletal design [Ray 2006]) that lived in herds (Bandyopadhyay 1988). However, the diaphyseal section of the femur shows a high cortical thickness, especially along the lateral and medial edges, which may be related to the dorsoventral flattening of the femoral midshaft (Ray 2006) for enhanced resistance to mediolateral bending during the nearly upright posture of the hindlimbs.

Lastly, the Permo-Triassic *Lystrosaurus* needs special mention, since its lifestyle adaptation has been the source of much debate. It is regarded as amphibious (Cluver 1971, Ray, Chinsamy, and Bandyopadhyay 2005; Germain and Laurin 2005), fully aquatic (Canoville and Laurin 2010), or fully terrestrial (King 1991; King and Cluver 1991; Botha and Smith, 2006, 2007; Botha-Brink and Angielczyk 2010). The bone microstructure of juvenile and subadult *Lystrosaurus* typically comprises highly vascularized fibrolamellar bone with notably large vascular canals throughout the cortex. Furthermore, the inner cortex and perimedullary region contains extensive resorption cavities and the medullary cavity is almost completely infilled with trabecular bone (Ray, Chinsamy, and Bandyopadhyay 2005; Botha-Brink and Angielzyk 2010). Although Triassic *Lystrosaurus* conforms to the usual dicynodont skeletal structure, there are certain distinctive features in its gross postcranial anatomy (Ray 2006). The distal end of the radius is extended lateromedially and strongly elliptical in cross section. The ratio between the long and short axes is about 4:1 (e.g., SAM-PK-K1415), which is unlike that seen in any other dicynodont. This feature resulted in a flattened distal end of the radius. Ray (2006) suggested that the flattened and lateromedially expanded radius and ulna of Triassic *Lystrosaurus* aligned with a broad and robust manus to produce a paddle-like distal forelimb that probably helped in swimming and/or digging. Ray, Chinsamy, and Bandyopadhyay (2005) suggested that the lack of a free medullary cavity, prevalence of extensive resorption cavities in the perimedullary region that often extended to the outer cortex, and high cortical thickness of the ribs in comparison to the limb bones throughout ontogeny were indicative of amphibious behavior in *Lystrosaurus*. Germain and Laurin (2005) also proposed an amphibious lifestyle for this genus by assessing the bone compactness (relative to extant taxa with known lifestyles) using the analytical program Bone Profiler, and Canoville and Laurin (2010) inferred an aquatic lifestyle using the same technique.

However, Botha-Brink and Angielczyk (2010) suggested that the above-mentioned histological features are also present in other dicynodonts such as *Dicynodon* and *Kannemeyeria*, both of which are considered to be fully terrestrial, indicating that an alternative reason for the presence of such structures must be considered. The limb bones of *Dicynodon* contain highly vascularized fibrolamellar bone (vascular density as much as 33 percent) with numerous enlarged vascular canals that reach the outer cortex in younger individuals. The size and abundance of these channels is similar to that seen in *Lystrosaurus* (Botha-Brink and Angielczyk 2010). Ribs of both *Dicynodon* and *Kannemeyeria* have medullary cavities that are completely infilled, and in *Kannemeyeria*, secondary reconstruction is extensive up to the outer cortex. Extensive secondary remodeling is also seen in the limb bones of *Placerias* (Green, Schweitzer, and Lamm 2010). Kriloff et al. (2008) initially proposed an amphibious lifestyle for *Dicynodon* using the Bone Profiler porosity program, but on the basis of functional morphology, they proposed a terrestrial lifestyle. Thus it may be hypothesized that these histological features could reflect phylogenetic constraints within the lineage containing Triassic dicynodonts. In addition, the presence of large channels may be attributed to lightening of the skeleton in the larger and heavier animals. For example, their robust bone morphology, prevalence of Sharpey's fibers in the bone microstructure, and frequent absence of a free medullary cavity may reflect biomechanical constraints in these animals.

A good test of this hypothesis would be to assess the bone microstructure of *Dicynodon*, *Lystrosaurus maccaigi*, and the Triassic *Lystrosaurus* species in an ontogenetic framework. The bone microstructure of large *L. maccaigi* limb bones is similar to that of *Dicynodon*, but the growth patterns of juvenile *L. maccaigi* is as yet unknown. A comparison of similar-sized individuals to rule out biomechanical constraints and juvenile versus adult specimens would help to shed light on this problem. For example, most Triassic *Lystrosaurus* material comprises juvenile and subadult individuals as these are the stages most frequently collected during field expeditions (J. Botha-Brink, pers. observation 2007). When more material becomes available, the comparison of large Triassic individuals with that of *L. maccaigi* and *Dicynodon* will hopefully provide more information regarding the lifestyle and growth of *Lystrosaurus*.

Concluding Remarks

It is apparent from the foregoing discussion that the examination of dicynodont bone microstructure enabled an assessment of their growth dynamics and other aspects of their paleobiology. The dicynodonts exhibit cortical bone of predominantly fibrolamellar bone tissue, which suggests overall fast rates of growth. Many genera exhibit uninterrupted fibrolamellar bone early in ontogeny with annuli and LAGs only appearing after attaining approximately 50 percent of adult size or in some taxa only after 70 percent of adult size. Several taxa exhibit peripheral parallel-fibered bone in larger individuals, suggesting that the overall growth rate decreased, possibly with the onset of reproductive maturity, whereas peripheral rest lines suggesting asymptotic growth have also been observed in a few genera. Using bone microstructural features, the different growth stages such as juvenile, subadult, and adult were identified for *Diictodon*, *Lystrosaurus*, and *Wadiasaurus*. Dicynodonts are generally characterized by thick cortices irrespective of body size, which may be indicative of a fossorial or at least digging lifestyle. *Wadiasaurus* is distinctive in having relatively thin cortices, whereas the bone architecture of *Lystrosaurus*, in terms of lifestyle adaptations, still remains perplexing.

Note

1. The correct spelling of the specific epithet of *Lystrosaurus* is "*maccaigi*." *Rhabdotocephalus maccaigi* (Seeley 1898) is the first use of the name "*maccaigi*." and Broom's (1903) subsequent use, with a change in the spelling to "*mccaigi*" is an objective synonym.

Acknowledgments

A. Chinsamy-Turan and J. Botha-Brink acknowledge the National Research Foundation (NRF), South Africa, and S. Ray acknowledges the Department of Science and Technology (DST), India, for research support. This manuscript has benefited from constructive comments made by reviewers, Jeremy Green and Sandra Jasinoski.

6.1. Cranium of *Oudenodon* (outlines of OUM.TSK 67) in lateral (A) and dorsal (B) views with section planes in transverse, longitudinal, or horizontal orientation (modified from Jasinoski, Rayfield, and Chinsamy [2010a]). Refer to Table 6.2 for a complete description of the location of the numbered thin sections. *Abbreviations:* E = ectopterygoid; EP = epipterygoid; F = frontal; J = jugal; L = lacrimal; M = maxilla; N = nasal; P = premaxilla; Pa = parietal; Pal = palatine; Par = parasphenoid; PO = postorbital; PoF = postfrontal; PP = preparietal; PrF = prefrontal; Pt = pterygoid; Pt ar = anterior palatine ramus of the pterygoid; Pt m = main body of the pterygoid; Pt qr = posterior quadrate ramus of the pterygoid; Q = quadrate; QJ = quadratojugal; S = squamosal; SM = septomaxilla; Sp = sphenethmoid complex; V = vomer.

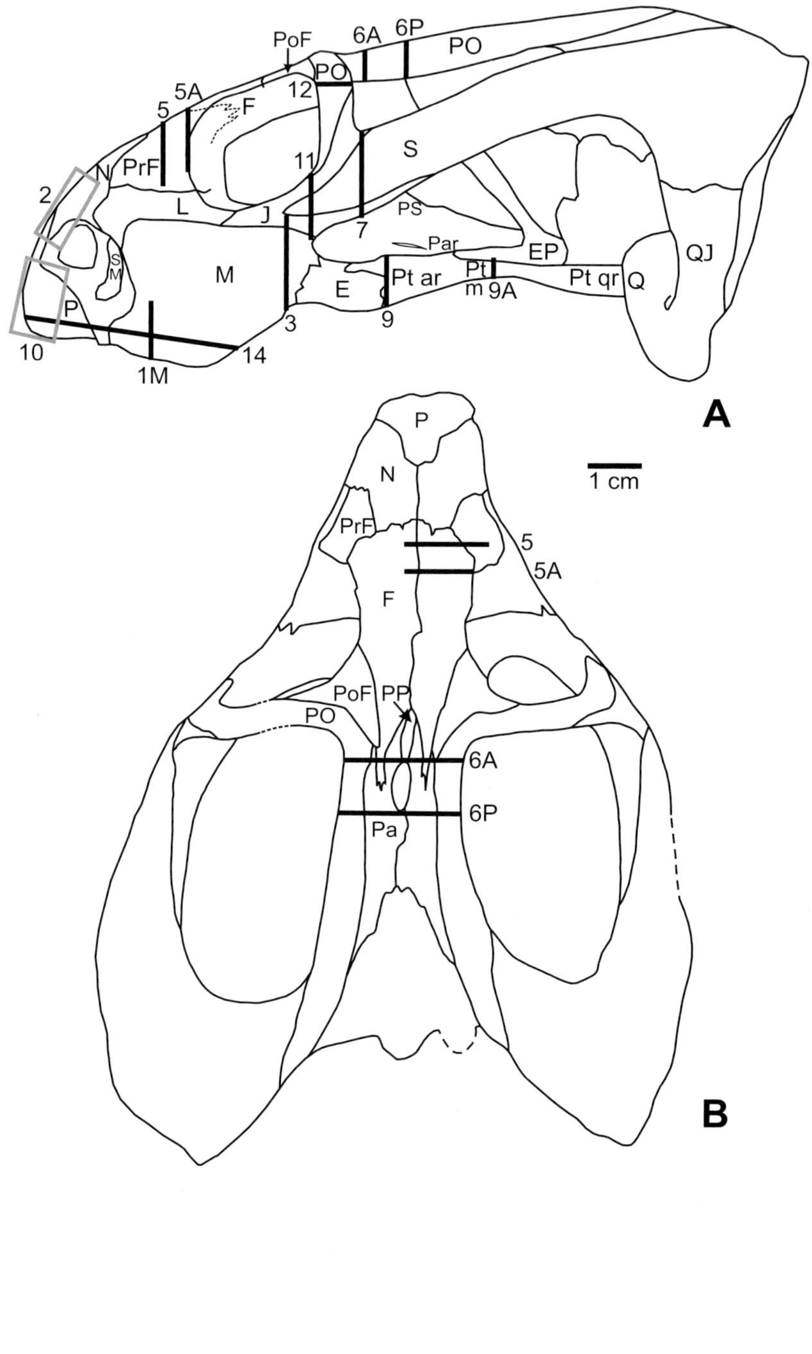

Biological Inferences of the Cranial Microstructure of the Dicynodonts *Oudenodon* and *Lystrosaurus*

6

Sandra C. Jasinoski and Anusuya Chinsamy-Turan

Introduction

Dicynodonts were nonmammalian therapsids and the most common herbivores during the Permo-Triassic. The remains of dicynodonts are well documented from therapsid bearing deposits worldwide, and because of their abundance, they have been used as biostratigraphic markers (e.g., Kitching 1977) (see Plate 1, Fig. 2.2). Their derived feeding apparatus, which allowed orthal and propalinal jaw movement, may have accounted for their success as herbivores (Crompton and Hotton 1967). Further details of their morphology and paleobiology are given in the following sections and chapter 5 of this book.

Studies of bone microstructure of various extant tetrapods have demonstrated its usefulness in deducing various aspects of the biology of the animal (e.g., Enlow 1969; Starck and Chinsamy 2002) and have also provided insight into biomechanical adaptations of the skeleton (e.g., Bouvier and Hylander 1981; de Margerie 2002). Here we undertake a comparative examination of the cranial histology of the dicynodonts *Oudenodon* and *Lystrosaurus* (see Plates J and K) in order to assess whether functional signatures were recorded in their bone microstructure. This study represents the first comprehensive cranial histological analysis of an extinct tetrapod and documents the microstructure as well as the histovariability that exists among various cranial elements. In conjunction with previous quantitative (e.g., finite element analysis [Jasinoski, Rayfield, and Chinsamy 2010a]) and qualitative functional analyses (e.g., King and Cluver 1991), we evaluate to what extent histology provides insight into dicynodont cranial function.

Phylogenetic Context

Within the Dicynodontia, *Oudenodon* and *Lystrosaurus* are both part of the higher taxonomic group Bidentalia (Kammerer and Angielczyk 2009). This group was split into the clade Cryptodontia, which contains the family Oudenodontidae (*Oudenodon*), and the more derived clade Dicynodontoidea, containing the family Lystrosauridae (*Lystrosaurus*) (Kammerer and Angielczyk 2009).

The genus *Lystrosaurus* occurred from the Late Permian to the Early Triassic and consists of four species: *L. curvatus, L. declivis, L. maccaigi,* and *L. murrayi* (Grine et al. 2006). Among these, *Lystrosaurus curvatus* is the only confirmed species to have crossed the end-Permian extinction boundary (Smith and Botha 2005). The genus *Oudenodon* occurred only during the Late Permian and is represented by three species, with *O. bainii* being the genotype and most common species (King 1988).

Paleobiological Context: Dicynodont Cranial Function

Except for early dicynodonts, a keratin layer covered the surfaces of the upper and lower jaws and replaced the function of teeth (King 1990). Small foramina on the surface of the bones, similar to the pitting on regions of horn attachment in extant chelonians, were used to infer the presence of a keratin beak (King 1990; Angielczyk 2001). Using this correlation, it was deduced that keratin covered the premaxilla and maxilla, forming an effective shearing blade against the lateral edge of the mandible, and the anteroventral (rugose) part of the palatine, which provided an effective grinding surface during retraction of the mandible (Cluver 1971; Jasinoski, Rayfield, and Chinsamy 2010b). A bony secondary palate was also present, allowing simultaneous food processing and breathing (King 1990), as well as strengthening of the rostrum (Thomason and Russell 1986).

Early studies of dicynodont cranial function (Watson 1948; Crompton and Hotton 1967) indicated that the specialized jaw joint of dicynodonts accommodated propalinal, or horizontal, movement of the mandible. The dicynodont masticatory cycle consisted of a beak bite, where shearing of food occurred near the front of the keratin-covered beak, followed by a retraction of the mandible, where food was ground against the rugose palatines (Crompton and Hotton 1967). Jaw movement was facilitated by two large jaw adductor muscles: the M. adductor mandibulae externus lateralis and medialis (Crompton and Hotton 1967). Food processing did not occur during jaw protraction due to the relatively small size of the posterior and internal adductor muscles (Crompton and Hotton 1967). Studies of tooth wear suggested that the tusks of some dicynodonts were used to rake or grub in the substrate (Cluver 1971; Hotton 1986), although not all researchers agree with this conclusion. In *Diictodon* (see Plate F), the tusks may have been a sexual dimorphic feature, serving a display or armament function (Sullivan, Reisz, and Smith 2003). However, the tusks of other dicynodonts, such as *Lystrosaurus* (see Ray 2005), may not have been used for sexual display.

Changes in cranial structure across Dicynodontia have been correlated to differences in muscle force and masticatory function, which in turn may reflect differences in dietary preference (Crompton and Hotton 1967; Cluver 1971; Hotton 1986; King, Oelofsen, and Rubidge 1989; King and Cluver 1991; Cox 1998). In many of these functional studies, *Lystrosaurus* was the focus because of its peculiar cranial morphology relative to other dicynodonts (Figs. 6.1, 6.2). Several cranial features of *Lystrosaurus*, such as the scarf premaxilla-nasal suture and the shortening and posterior deepening of the cranium, were hypothesized to have increased the efficiency of its masticatory system relative to purely Permian dicynodonts such as *Oudenodon* (Cluver 1971; King and Cluver 1991; Jasinoski, Rayfield, and Chinsamy 2009, 2010a). These specialized cranial features possibly allowed *Lystrosaurus* to exploit more resistant vegetation, which may have facilitated its survivorship across the Permo-Triassic extinction boundary (King and Cluver 1991; King and Jenkins 1997). Indeed, a comparative finite element (FE) analysis of *Oudenodon* and *Lystrosaurus* showed that lower stress occurred in the cranium of *Lystrosaurus* in response to masticatory loads (Jasinoski, Rayfield, and Chinsamy 2009). In addition, the mechanical advantage of adductor musculature, a scarf premaxilla-nasal suture, a broad mandibular symphysis, and other palatal features suggested that *Lystrosaurus* utilized a snapping bite to cope with fibrous vegetation (Jasinoski, Rayfield, and Chinsamy 2009, 2010a). Because of its downturned snout, *Lystrosaurus* was likely a grubber, but it may have also fed on vegetation in shallow water (Hotton 1986). In *Oudenodon*, the postcaniniform crest that is posterior to the caniniform process, may have allowed more effective shearing of leaves during propalinal jaw movement (Jasinoski, Rayfield, and Chinsamy 2009), which is in agreement with the hypothesis that *Oudenodon* was a browser (Hotton 1986).

Besides the obvious involvement in mastication and protection of the brain, the dicynodont cranium may have served other roles. For instance, the broad posterior skull roof and the reduced or absent interpterygoid vacuity of cistecephalids were interpreted as adaptations for strengthening the skull during burrowing (Cluver 1974). The differential occurrence of cranial protuberances within populations of *Aulacephalodon* (Tollman, Grine, and Hahn 1980) and *Diictodon* (Sullivan, Reisz, and Smith 2003) were considered as evidence of sexual dimorphism. In addition, sexually dimorphic features in *Lystrosaurus* were noted by Tripathi and Satsangi (1963) and Cluver (1971) and later corroborated by Ray (2005), who distinguished two morphotypes (robust and gracile forms) within *L. murrayi* based on bivariate determination. Based on stress distribution, bosses and ridges in *Lystrosaurus declivis* may have served a display function rather than strengthening the cranium; whereas the nasal, prefrontal, and posterior pineal bosses of *Oudenodon* decreased bone stress in those cranial regions, suggesting that they imparted some strength to its cranium (Jasinoski, Rayfield, and Chinsamy 2009).

Previous Studies of Dicynodont Cranial Histology

Although the dicynodont skull has been surveyed in terms of serial sectioning (Sollas and Sollas 1914, 1916; Cluver 1971; Keyser 1975), comprehensive histological studies of the cranial elements have not been documented before. Instead, several studies have documented the microstructure of the postcranial skeleton of various dicynodonts (see, e.g., de Ricqlès 1972; Chinsamy and Rubidge 1993; Botha 2003; Ray and Chinsamy 2004; Ray, Chinsamy, and Bandyopadhyay 2005; Botha and Angielczyk 2007). These studies revealed that dicynodont postcranial bones are generally composed of fibrolamellar bone, which may alternate with lamellar or parallel-fibered bone tissue. Comprehensive surveys of specific dicynodont taxa have also shown that the histology can vary between different postcranial elements, and this histovariability has been attributed to differences in function or growth rates of the bones (Ray and Chinsamy 2004; Ray, Chinsamy, and Bandyopadhyay 2005; Botha-Brink and Angielczyk 2010). Except for brief and isolated histological descriptions of the postorbital (Davies 1981) and tusks (Poole 1956; Thackeray 1991), dicynodont cranial histology remained undescribed until recently. Jasinoski (2008) and Jasinoski, Rayfield, and Chinsamy (2010a) resolved this lack of knowledge by histologically sampling several crania of *Oudenodon* and *Lystrosaurus*. The crania were sectioned in regions that had high strain as predicted by FE analysis (Jasinoski, Rayfield, and Chinsamy 2010a) in order to assess bone microstructure in relation to the type, orientation, and magnitude of masticatory strain. Functional correlations between suture morphology and masticatory strain are well established in miniature pigs (*Sus scrofa*) (Rafferty and Herring 1999; Herring and Teng 2000) and the fish *Polypterus* (Markey, Main, and Marshall 2006; Markey and Marshall 2007). These correlations were applied to dicynodonts to predict the strain environment that the cranial bones were subjected to during life. Several cranial sutures, ranging from butt-ended to interdigitated, correlated with the FE-predicted tensile or compressive strain in the dicynodont cranium (Jasinoski, Rayfield, and Chinsamy 2010a).

On the contrary, Jasinosky (2008) found it difficult to determine conclusively if variation of dicynodont cranial bone histology was related to phylogenetic differences, differential growth, metabolic function, and/or biomechanical function due to the lack of comparative cranial histological material of *any* tetrapod, fossil or extant. Thus it is uncertain whether cranial histology of the extinct dicynodonts were completely reflective of biomechanical function, as these features may have been overprinted by metabolic signatures (see Bouvier and Hylander 1996) or simply related to growth. These and other possible factors that affect bone histology will be discussed in the following sections.

Biomechanical Interpretation of Bone Microstructure

Experimental studies on a wide range of extant mammals and birds have documented a correlation between bone histology and biomechanical function (e.g., Bouvier and Hylander 1981, 1996; de Margerie 2002; de Margerie et al. 2005; Skedros and Hunt 2004). In fossils, such correlations are important for reconstructing the type of strain environment that the bones were subjected to during life. However, the majority of these functional studies only considered the postcranial skeleton, with the exception of Bouvier and Hylander's (1981, 1996) studies of the mandible, zygomatic arch, and supraorbital bar of the macaque. In the aforementioned studies, channel orientation and secondary osteons were the two main histological features focused on. These features were considered to reflect the type, magnitude, and/or orientation of strain experienced in each region of the bone.

Studies of primary cortical bones of birds showed that the orientation of primary vascular channels/canals was indicative of biomechanical function, independent of growth (de Margerie 2002; Skedros and Hunt 2004; de Margerie et al. 2005). A laminar organization of vascular canals, consisting of elongated channels oriented parallel to the periosteal surface, were found in areas that experience torsion and may also be adapted to resist pure compression; whereas longitudinal canals, which are oriented perpendicular to the periosteal surface, can resist tensile loads (de Margerie 2002; de Margerie et al. 2005). This suggests that the structure of primary bone may be adaptive and independent of secondary remodeling (Skedros and Hunt 2004). These microstructural correlations with function were statistically significant in the limb bones of skeletally mature birds (de Margerie 2002; Skedros and Hunt 2004; de Margerie et al. 2005). However, channel orientation in the crocodilian femur did not correlate with strain type, but instead reflected growth processes (Lee 2004).

The size and shape of secondary osteons in skeletally mature mammals may correspond to the type, magnitude, and/or orientation of strain experienced in each cross-sectional region of the limb bone (Skedros, Mason, and Bloebaum 1994; Skedros et al. 1996; Skedros, Su, and Bloebaum 1997; Skedros, Sorenson, and Jenson 2007). In macaques, a higher density of secondary osteons correlated with a higher level of feeding stress in the mandible (Bouvier and Hylander 1981). In a subsequent cranial study, it was found that there was a stronger correlation between biomechanical function and secondary osteonal remodeling in the juvenile and subadult macaque individuals; whereas bone remodeling did not correlate with either high or low strain in the adult individuals, but seemed instead to be partially related to metabolic factors (Bouvier and Hylander 1996).

One problem with applying these microstructural correlations to cranial bone is that strain can have a different orientation and/or magnitude in the cranium during different bite regimes or muscle activity (see, e.g., Herring and Teng 2000); whereas limb bones, on which many of the histological-biomechanical correlations have been based, experience a similar direction of bending during different loading orientations because of the curvature of the long bone (Herring and Ochareon 2005). In addition, the majority of these functional studies of bone histology focused on mammals and birds, which have a determinate growth strategy (see, e.g., Bouvier and Hylander 1981; de Margerie 2002). These correlative features, therefore, have not taken into account growth dynamics for an indeterminate (continuous) growth strategy, which has been proposed for most dicynodonts (Ray and Chinsamy 2004; Ray, Chinsamy, and Bandyopadhyay 2005; Ray, Bandyopadhyay, and Bhawal 2009). In addition, the question of how phylogeny and body size affect bone remodeling is not well understood among extant taxa (see Currey 2003).

However, given these limitations, several functional observations can be applied in this study of dicynodont cranial function: (1) primary bone without internal organization of vascular channels is biomechanically weaker (possibly having a lower Young's modulus and bending strength) than primary bone with preferential and well-oriented organization (Currey 1987, 2003); (2) primary bone with vascular channels that have a certain orientation may be adapted to certain modes of strain: tension, compression, and/or shear (de Margerie 2002; Skedros and Hunt 2004; de Margerie et al. 2005); (3) resorption spaces or erosion cavities are potential weaknesses or stress concentrators in the bone (Currey 2003), and thus are more predominant in regions that experience low strain (Bouvier and Hylander 1981). However, resorption spaces have been associated with removal of microcracks (see review by Currey [2003]), but it is expected that their presence would be relatively short-lived as these resorption spaces would be rapidly infilled with secondary osteonal deposits; (4) lamellar bone is mechanically superior to woven-fibered bone (Carter and Spengler 1978; Currey 2003); (5) remodeled bone, composed of extensive secondary osteons, is weaker and less fatigue-resistant than organized primary bone (Reilly and Burstein 1975; Carter, Hayes, and Schurman 1976; Currey 1987), but remodeled bone may decrease the propagation of microcracks (Currey 2003). Secondary remodeling, the replacement of primary bone with secondary bone, may not always have a causal link with biomechanical function, but instead may be linked to metabolic factors (Bouvier and Hylander 1996; Currey 2003) or the renewal/removal of older bone (Currey 2003); (6) when comparing remodeled bones, a region with higher density of secondary osteons may indicate that the region experienced higher loading (Bouvier and Hylander 1981, 1996); (7) cancellous bone is biomechanically weaker than cortical bone (Martin, Burr, and Sharkey 1998), and porous bone experiences a higher strain magnitude than more compact bone (Herring et al. 1996). Thus, it is expected that compact bone would occur in regions of higher FE-predicted strain.

Materials and Methods

Several cranial specimens (cranium [= skull without the mandible] or mandible) of both *Oudenodon* and *Lystrosaurus* were thin-sectioned for histological analysis (Table 6.1). Sections were taken at specific regions of the cranium (Table 6.2; Figs. 6.1, 6.2), according to the strain levels predicted by FE analysis (Jasinoski 2008; Jasinoski, Rayfield, and Chinsamy 2010a).

Specimens of *Oudenodon* were not identifiable to the species level, although they most likely represented *O. bainii*, the most common species of *Oudenodon* (King 1988). These specimens were at least 61 percent of the adult size, and they likely represented an adult or mature ontogenetic stage (Table 6.1). This classification was supported by the tight suturing of most of the cranial sutures in the *Oudenodon* specimens. However, the specimens were smaller in size than the individuals sampled by Botha

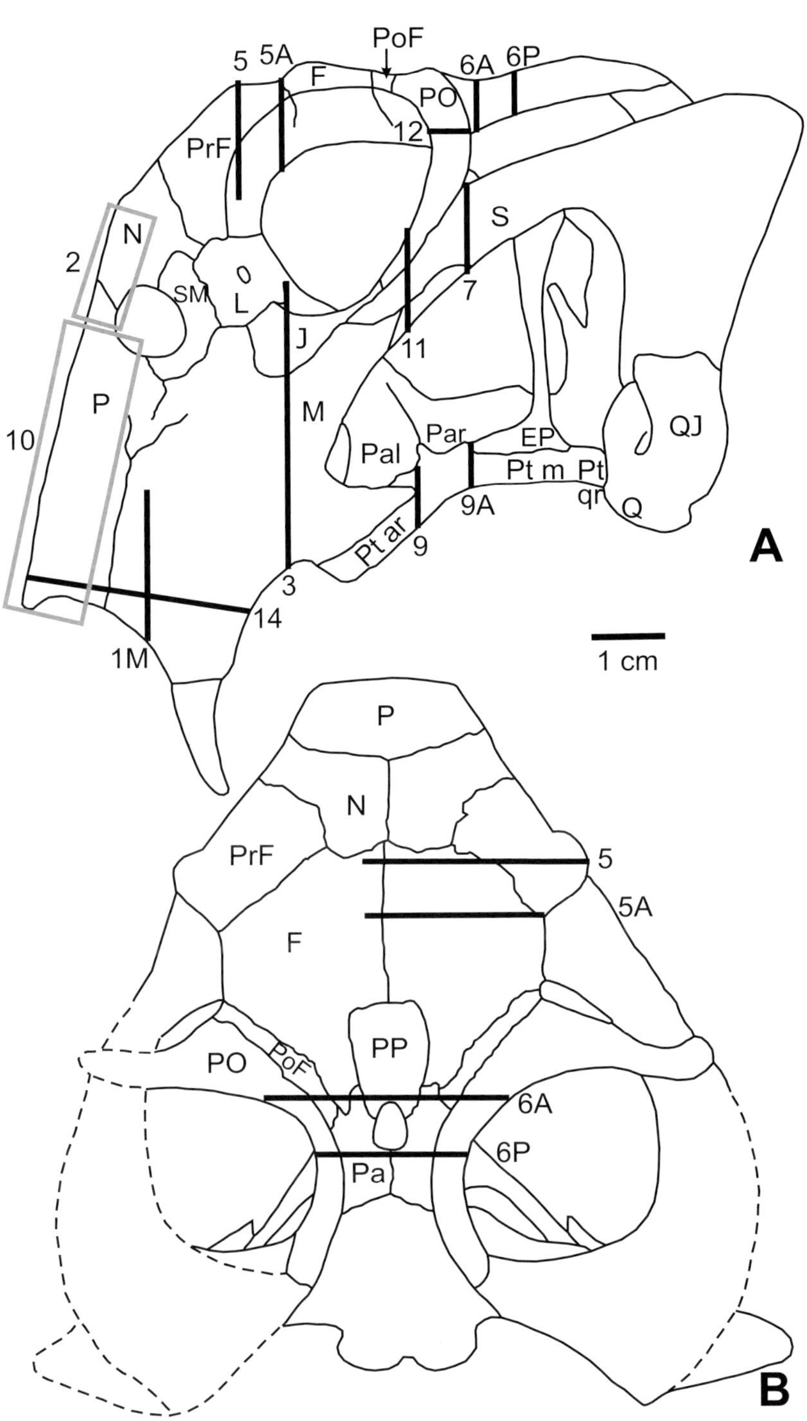

6.2. Cranium of *Lystrosaurus* (outlines based on Cluver's illustrations of *L. declivis* [1971, Figs. 22A and 23]) in lateral (A) and dorsal (B) views with section planes in transverse, longitudinal, or horizontal orientation (modified from Jasinoski, Rayfield, and Chinsamy [2010a]). Refer to Table 6.2 for a complete description of the location of the numbered thin sections. Abbreviations are the same as in Figure 6.1.

Table 6.1. *Oudenodon* and *Lystrosaurus* specimens sampled for cranial histology

Taxon	Specimen number SAM-PK-	Cranial elements sampled	Locality	Cranial length (mm)	Approx. % of adult size[a]	Possible ontogenetic stage[b]
Oudenodon sp.	K5216	cranium	Eastern Cape, ZA	183[c]	61	adult
Oudenodon sp.	K6043	cranium	Western Cape, ZA	N/A	N/A	adult
Oudenodon sp.	OUM. TSK 96	cranium	Luangwa Valley, ZM	187	62	adult
Oudenodon sp.	K5065	cranium	Western Cape, ZA	N/A	N/A	adult
Oudenodon sp.	K5343	mandible	Western Cape, ZA	N/A	N/A	adult
Oudenodon sp.	K5288	mandible	Western Cape, ZA	N/A	N/A	adult
Oudenodon sp.	K5396	mandible	Western Cape, ZA	N/A	N/A	adult
L. murrayi	K1402	cranium	Free State, ZA	106	67[d]	late juvenile
L. curvatus	K1422	cranium	Free State, ZA	N/A	N/A	subadult
L. murrayi	K1342	cranium	Eastern Cape, ZA	135	85[d]	subadult
L. curvatus	K1269	cranium	unknown	125[e]	54	subadult
L. murrayi	K8	cranium	Eastern Cape	150	94[d]	late subadult
Lystrosaurus sp.	K28 (extra)	cranium	Eastern Cape, ZA	N/A	N/A	adult
L. maccaigi	K28	skull	Eastern Cape, ZA	N/A	N/A	adult

Note: All specimens are from Iziko South African Museum (prefix SAM-PK-) except OUM.TSK 96. The basal cranial length was measured from digital photographs, and the relative order of the cranium specimens listed here reflects their increasing size.
[a] Calculated from the approximate maximum basal cranial length of 300 mm for *Oudenodon* (Cluver and Hotton 1981) and those established in Grine et al. (2006) for each species of *Lystrosaurus*.
[b] Based upon gross osteological features, Ray, Chinsamy, and Bandyopadhyay's (2005) ontogenetic categories, and cranial histological features (Jasinoski 2008). See text and Jasinoski, Rayfield, and Chinsamy (2010a) for further details.
[c] Cranium is compressed dorsoventrally.
[d] Calculated from a relatively low maximum basal cranial length of 159 mm (Grine et al. 2006).
[e] Cranium is compressed anteroposteriorly.

(2003). The two adult crania associated with the postcranial material of her study had basal cranial lengths of at least 230 mm (SAM-PK-10019 [two individuals] and SAM-PK-11141).

At least three species of *Lystrosaurus*, identified on specific cranial characteristics (Grine et al. 2006) and ranging in size from juvenile to adult (with the majority being subadult), were included in the current study (Table 6.1). The cranial histology of this study was compared with an earlier assessment of postcranial histology of four ontogenetic stages of *Lystrosaurus* (Ray, Chinsamy, and Bandyopadhyay 2005, Table 6). For the most part, the cranial histology of specimen SAM-PK-K8 was not well preserved and prevented comparison to other specimens.

A thorough histological survey of several cranial bones, including description of patterns of vascularization and the amount of secondary reconstruction, was undertaken in Jasinoski (2008). Here, we have taken a more functional approach and focused on particular cranial microstructural characteristics and their paleobiological implications.

Terminology

The term channel rather than vascular canal was used, as channels in living bone house vascular tissue as well as other soft tissue, and the shape of the vascular canal may not match the shape of the bony channel (Starck

Table 6.2. Location of histological sections in the *Oudenodon* and *Lystrosaurus* skulls

Section number	Location of thin section	*Oudenodon* SAM-PK-	*Lystrosaurus* SAM-PK-
1M	One-half distance between the P-M suture on palatal rim and the middle of the base of the tusk/caniniform process.	K5065-1M, K6043-1M	K1342-1M, K1422-1M, K8-1M
2	Longitudinal section through the midline of premaxilla-nasal suture.	K5065-2, K5216-2, K6043-2	K1269-2, K1342-2, K1422-2
3	Section through the midlength of the anterior (palatal) part of the palatine. Intersects Pal, M, J, V and possibly L, E.	K5065-3[a], K5065-3 II[ab]	K1342-3
4	Section through tusk/caniniform process at one-half of the width at its base.	K5065-4, K6043-4	K1342-4, K1422-4, K8-4
5	Section through center/apex of prefrontal boss. Intersects prefrontal and anterior part of frontal.	K5065-5, K5216-5, K6043-5	K1269-5, K1342-5, K1402-5[c], K28-5, K28-5(extra)
5A	Section through the prefrontal-frontal suture near its contact with the dorsolateral orbital rim.	K5065-5A, K5216-5A, K6043-5A	K1269-5A, K1342-5A, K1402-5A
6A	Section through the anterior extent of the pineal foramen. Intersects PO, Pa, PP, and possibly PoF, F.	K5216-6A, K6043-6A(1), K6043-6A(2)[b]	K1402-6A, K1342-6A, K28-6A[b]
6P	Section through the posterior extent of the pineal foramen. Intersects the postorbital and parietal.	K5216-6P	K1342-6P, K1402-6P, K28-6P[b], K8-6P
7	Section through the zygomatic arch at the deflection point posterior to postorbital bar. Intersects J, S.	K5065-7, K5216-7	K1269-7, K1342-7
9	Section at posteriormost projection of the maxilla/ ectopterygoid and its contact with the anterior palatine ramus of the pterygoid.	OUM.TSK 96-9	K1342-9, K1402-9[b], K1422-9, K8-9
9A	Section through main body of pterygoid. Intersects pterygoid and parasphenoid.	OUM.TSK 96-9A	K1402-9A, K1422-9A, K8-9A
10	Longitudinal section through anterior palatal ridge of premaxilla.	K5216-10, K6043-10	K1342-10, K1402-10, K1422-10
11	Section through the infraorbital bar at the deflection point anterior to postorbital bar. Intersects J, S, and possibly PO, M.	K5065-11, K5216-11	K1269-11, K1342-11
12	Horizontal section at the midlength of the postorbital bar. Intersects PO, possibly J.	K5065-12	K1269-12, K1342-12
14	Horizontal section through palatal rim. Intersects the premaxilla and maxilla, and the tusk in *Lystrosaurus*.	K5216-14	K1269-14[d], K1342-14, K1422-14
R	Section through mandibular body (ramus) at the anterior extent of longitudinal groove.	K5343-RR, K5288-LR, K5396-LR	K28-LR
SY	Longitudinal section through the mandibular symphysis.	K5343-SY, K5288-SY	N/A

Note: All sections have a transverse orientation unless indicated otherwise. See Figures 6.1 and 6.2 for locations of the sections and bone abbreviations.

[a] Sectioned obliquely: the right lateral side is at the correct position of sectioning.

[b] Section taken further posteriorly.

[c] Only sectioned through the prefrontal bone.

[d] Section taken ventral to other sections.

and Chinsamy 2002). Elongated channels that are oriented roughly parallel to the periosteal surface have a laminar arrangement (= circumferential or circular orientation of de Ricqlès et al. [1991] and de Margerie [2002]). Elongated channels oriented perpendicular to the periosteal surface have a radial orientation, and those without a preferential orientation have an oblique/reticular orientation (see de Margerie [2002]). Channels that are circular to oval in shape in transverse section have a longitudinal orientation relative to the sagittal plane of the cranium.

Results and Discussion

Overall Cranial Histology of *Oudenodon*

The cranial histology of *Oudenodon* was characterized by fibrolamellar bone tissue with extensive secondary remodeling (Fig. 6.3A). Alternating with the zones of fibrolamellar bone were thinner annuli composed of either lamellar or lamellated bone, which indicated cyclical growth (Fig. 6.3A). Lines of arrested growth (LAGs), indicating periodic cessations in growth, were common, and some of the cranial bones also had closely spaced rest lines in the peripheral layer of lamellar bone. Many of these histological features were also observed in the postcrania of *Oudenodon* (Botha 2003) and suggest that the specimens in the current study were either subadult or adult individuals.

Overall Cranial Histology of *Lystrosaurus*

The cranial histology of *Lystrosaurus* consisted of fibrolamellar bone with large, open channels. Toward the peripheral edges of the bone, the channels were small and primary, but within a few hundred microns, they were enlarged due to remodeling (Fig. 6.4D). Some of these enlarged channels had secondary infillings, but few fully formed primary and secondary osteons were present. In fact, most of the cranial elements sampled from the three ontogenetic stages of *Lystrosaurus* (Table 6.1) had distinctly porous bone histology overall. Channel orientation, relative porosity, and the level of remodeling (including resorption cavities and osteonal infillings) varied between cranial bones of a single individual, and in some cases, within the same bone. In order to discriminate between growth characteristics and functional signatures, the cranial histology was compared with Ray, Chinsamy, and Bandyopadhyay's (2005) survey of postcranial histology.

IMMATURE (LATE JUVENILE AND SUBADULT) CRANIAL HISTOLOGY

Gross osteological features of *Lystrosaurus* specimen SAM-PK-K1402, including suture patency in the skull roof, relatively large orbits, and short basal cranial length relative to the larger subadult crania of Ray, Chinsamy, and Bandyopadhyay's (2005) specimens, indicated that it was a late juvenile (Table 6.1). Cranial histological features observed in this

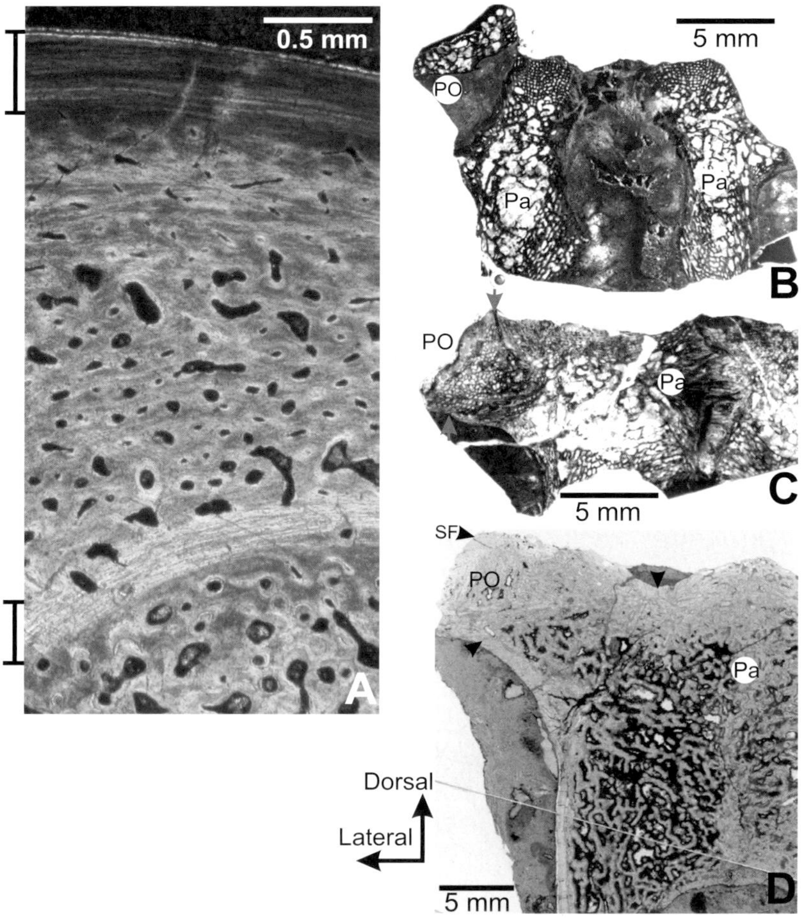

6.3. General cranial histology of *Oudenodon* (A) and *Lystrosaurus* (B–D). (A) SAM-PK-K5065-11, showing the detail of the annuli (indicated by brackets, composed of lamellated bone) and fibrolamellar zones of the jugal bone. (B–D) Transverse sections of the posterior skull roof of *Lystrosaurus* showing ontogenetic changes in the parietal and postorbital bones of (B), SAM-PK-K1402-6P, a late juvenile; (C) SAM-PK-K1342-6P, a subadult; and (D) SAM-PK-K28-6P, an adult individual. The level of porosity is high throughout all ontogenetic stages of *Lystrosaurus.* Arrows in (C) indicate the ectocranial and endocranial extent of the postorbital-parietal suture; arrowheads in (D) indicate location of Sharpey's fibers. *Abbreviations:* SF = Sharpey's fibers; see Figure 6.1 for bone abbreviations.

specimen supported this designation when compared with Ray, Chinsamy, and Bandyopadhyay's (2005, Table 6) ontogenetic table. These features included the overall porosity of the bone histology, the presence of mainly woven-fibered bone rather than completely formed fibrolamellar bone, the absence of LAGs, and the presence of radially oriented channels in the ventral part of the pterygoid (SAM-PK-K1402-9A) that indicated rapid growth, and the low amount of secondary remodeling (Figs. 6.3B, 6.6A).

Histological features of the crania of subadult specimens were more difficult to compare with the postcranial features used by Ray, Chinsamy, and Bandyopadhyay (2005) to discriminate between the subadult and adult categories. However, the absence of peripheral parallel-fibered or lamellar bone and the lack of secondary osteons/secondary infillings in the skull roof (sections 5 and 6, the only sections that can be compared across all ontogenetic stages) (Tables 6.1, 6.2; Fig. 6.3C) suggest that these specimens most likely represent a subadult stage. The absence of radially oriented channels from all sampled bones, a juvenile feature identified by Ray, Chinsamy, and Bandyopadhyay (2005), also supported this age assignment. In addition, LAGs were not observed in any of the subadult cranial specimens, which may indicate that these subadult individuals were younger than those sampled by Ray, Chinsamy, and Bandyopadhyay (2005) or that secondary reconstruction may have obliterated earlier formed growth lines.

Overall, the bone histology of immature specimens of *Lystrosaurus* was characterized by open channels and a low density of secondary infillings. Such features were also observed in several postcranial elements of *Lystrosaurus* (see Ray, Chinsamy, and Bandyopadhyay 2005).

ADULT CRANIAL HISTOLOGY

Specimens SAM-PK-K28 and SAM-PK-K28 (extra) were the largest *Lystrosaurus* individuals sampled and were presumed to represent an adult stage (Table 6.1). These specimens had more secondary osteons (mostly incompletely formed, with thin secondary deposits of bone) and extensive reconstruction than the immature specimens of *Lystrosaurus*. Evidence of parallel-fibered bone and a decrease in vascularity near the periphery of the bones was prevalent in these adults, and agrees with the postcranial survey (Ray, Chinsamy, and Bandyopadhyay 2005). Even with extensive remodeling, the bones remained porous, with secondary cancellous bone forming in the center of the elements (Fig. 6.3D). Comparison of the two adult specimens indicated that specimen SAM-PK-K28-5 was from an older adult individual due to its larger physical size, the presence of peripheral lamellar bone with low vascularization, and more extensively reconstructed bony struts in the prefrontal bone. Peripheral rest lines were absent from both adult specimens, which suggests that although growth had slowed down it did not stop (it may also indicate indeterminate growth [see Ray, Chinsamy, and Bandyopadhyay (2005)]).

Comparison of Cranial Histology of *Oudenodon* and *Lystrosaurus*

Ontogenetic differences of the sampled *Oudenodon* and *Lystrosaurus* specimens (Table 6.1) made it difficult to directly compare their cranial histology. The specimens of *Oudenodon* sampled were at least 61 percent of adult size and are probably ontogenetically older than the sampled immature *Lystrosaurus* specimens (Table 6.1). This supposition is further supported histologically by the absence of LAGs and lamellar bone peripherally from the immature specimens of *Lystrosaurus;* whereas these features were present in most *Oudenodon* specimens.

Cranial histology of adult *Lystrosaurus* was limited to two partial and fragmentary specimens (Table 6.2), and the comparable sections were not well preserved in the adult *Oudenodon* specimens. Thus, a brief and qualitative comparison of the adult cranial histology of both taxa follows, and subsequently, a comparison between specimens of different ontogenetic stages is undertaken.

RELATIVE POROSITY AND SECONDARY RECONSTRUCTION OF BONE

The overall cranial histology of *Lystrosaurus* was distinctively porous; whereas most of the cranial elements examined in *Oudenodon* consisted of relatively compact bone with extensive development of secondary osteons. Despite these overall differences, several bones of the palatal and snout region, including the premaxilla (thin-sections 1M, 10) (Table 6.2; Figs. 6.4A, 6.4B), maxilla (sections 1M, 4) (Figs. 6.4A to 6.4D), palatine (section 3) (Figs. 6.4E, 6.4F), and premaxilla-nasal (section 2) (Fig. 6.6B) had a porous texture in both taxa. The former three sections were from the palatal region and form the shearing or grinding surfaces for the processing of food items. The porous cranial histology did not reflect the high FE-predicted strain in these areas of *Oudenodon* and *Lystrosaurus* (Jasinoski, Rayfield, and Chinsamy 2010a). The presence of an overlying keratin layer, which has a much lower stiffness than bone (Seki, Schneider, and Meyers 2005), may have helped to decrease the high strain in these bones and could explain their porous nature in both taxa. Interestingly, in immature specimens of *Lystrosaurus*, the main body of the pterygoid (section 9A) (Table 6.2) was composed of slightly more compact bone relative to the rest of its cranium, which may reflect the higher FE-tensile strain predicted for that region (see Jasinoski, Rayfield, and Chinsamy 2010a).

The other cranial elements of *Oudenodon* had compact bone consisting of numerous secondary osteons. Bones of the skull roof and infraorbital, zygomatic, and postorbital bars experienced high FE-predicted strain (Jasinoski, Rayfield, and Chinsamy 2010a); however, it is uncertain whether the increased density of secondary osteons was a response to a high magnitude of strain (e.g., Bouvier and Hylander 1981) or related to the renewal of older bone (Currey 2003). Future studies are needed to quantify the density of secondary osteons and erosion cavities across regions of high and low FE-predicted strain in a single individual. The prefrontal boss of *Oudenodon* had relatively lower FE-predicted strain than the surrounding regions of the anterior skull roof (Jasinoski, Rayfield, and Chinsamy 2010a), but it did not appear to have a lower density of secondary osteons. On the other hand, strain magnitude may be reflected in the relative compactness of the bones of the infraorbital bar (section 11) and zygomatic arch (section 7). In both of these regions, the squamosal (not figured) appears to have slightly larger cavities with secondarily infillings than the adjacent jugal (Fig. 6.7E). The FE results indicated that higher compressive strain occurred on the medial side of the infraorbital bar and zygomatic arch, which corresponds to the jugal, than on the lateral side that corresponds to the squamosal (Jasinoski, Rayfield, and Chinsamy 2010a). The relative compactness of the jugal may indicate a response to the higher magnitude of compressive strain in these areas during orthal biting. Correlation between bone density and strain may become more apparent in other cranial regions once quantification of these features is undertaken in the future.

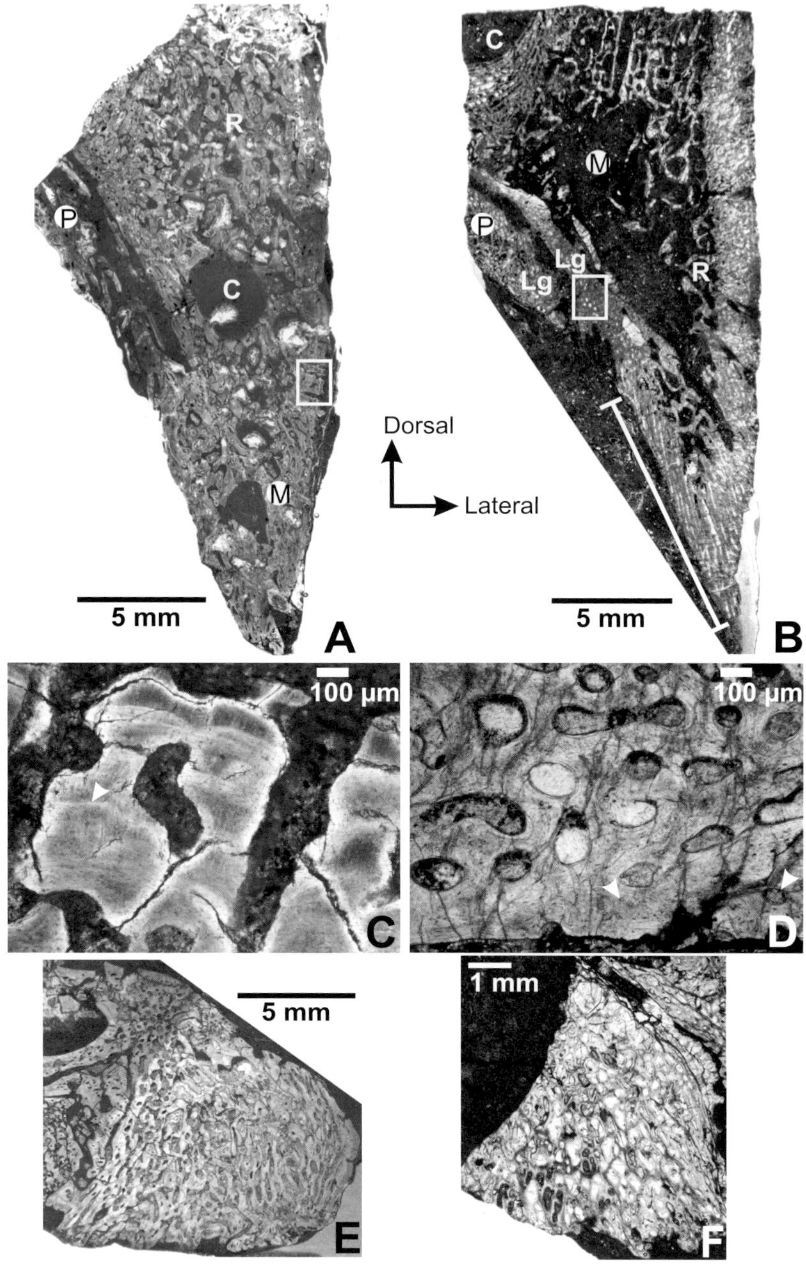

6.4. Porous bone histology in the palatal and snout region of *Oudenodon* (left column) and *Lystrosaurus* (right column). (A) Composite image of the right maxilla-premaxilla region of SAM-PK-K5065-1M. (B) Composite image of the right maxilla-premaxilla region of SAM-PK-K1422-1M. The ventral part of the maxilla has elongated channels in a laminar arrangement (bracket). (C) Enlargement of the boxed area in (A), lateral edge of the maxilla, showing porous histology with an unfinished edge, Sharpey's fibers, and at least one distinct LAG (arrowhead). (D) Enlargement of boxed area in (B), medial edge of the maxilla, showing primary channels (arrowheads) that become enlarged away from the edge of the bone. (E) Composite image of the palatal part of the right palatine of SAM-PK-K5065-3 II. (F) Composite image of SAM-PK-K1342-3, showing the porous nature of the palatal part of the left palatine. *Abbreviations:* C = cavity; Lg = channels with a longitudinal orientation relative to the sagittal plane of the cranium; R = resorption; see Figure 6.1 for bone abbreviations.

The level of bone compactness and density of secondary osteons in the anterior (section 5) and posterior (sections 6A, 6P) skull roof are the most striking histological differences between the adult specimens of *Lystrosaurus* and *Oudenodon*. Both taxa show evidence of remodeling and resorption of bone in these regions of the skull roof; however, the bone appears more compact in *Oudenodon* (e.g., Fig. 6.7C) in comparison to *Lystrosaurus*, which had extensive secondary cancellous bone with thin bony struts in the central part of the bones (e.g., Fig. 6.3D). Differences in the structural dimensions of the skull roof of both taxa may partially account for this discrepancy in bone histology. The skull roof of *Oudenodon* is proportionately thinner than the comparable region of *Lystrosaurus*. Thicker bone experiences less strain due to the increase in bone stiffness (Herring and Ochareon 2005). Thus, the biomechanical constraints placed upon the relatively thinner skull roof of *Oudenodon* may have been reduced in *Lystrosaurus*. This supposition is corroborated by the FE results, which show higher strain in the anterior and posterior skull roof regions of *Oudenodon* than in *Lystrosaurus* (Jasinoski, Rayfield, and Chinsamy 2010a).

The thick, compact cortical bone in the postcranial elements was interpreted as an adaptation for digging in *Oudenodon* and *Diictodon* (Botha 2003; Ray and Chinsamy 2004). However, compact bone in the cranium of *Oudenodon* would not have been necessary for this type of lifestyle, and it is unknown whether younger individuals had a porous cranial histology similar to immature *Lystrosaurus*.

The porous bone histology of *Lystrosaurus* may have served several functions. Organized primary bone is biomechanically stronger than remodeled bone with extensive secondary osteons (Carter and Spengler 1978; Currey 1987), and thus, several cranial elements in immature specimens of *Lystrosaurus* that have well-organized channels would retain stronger mechanical properties. The porous cranial histology, with secondarily enlarged channels, may indicate the need for lightening the skull. Ray, Chinsamy, and Bandyopadhyay (2005) suggested that the porous texture and narrow cortical bone characteristic of some *Lystrosaurus* postcranial bones, in combination with thickened dorsal ribs, may be an adaptation for a semi-aquatic lifestyle as it reduced the bony mass of the skeleton. In specimen SAM-PK-K8, the only individual sampled for both postcranial (Ray, Chinsamy, and Bandyopadhyay 2005) (Fig. 6.5B) and cranial histology (Table 6.2, Fig. 6.5A), the parietal (and possibly maxilla) were more porous along the periphery with secondarily enlarged cavities in comparison to the humerus (although the humeral cortex was particularly thick [Ray, Chinsamy, and Bandyopadhyay 2005]). In a recent study, Botha-Brink and Angielczyk (2010) attributed the high porosity in *Lystrosaurus* to rapid growth rates. However, it is worth noting that many of the channels in the *Lystrosaurus* bones are secondarily enlarged and, therefore, perhaps unrelated to rapid rates of growth. In addition, the enlarged channels formed by secondary reconstruction may not be related to increased assimilation of nutrients, but rather to various other factors such as calcium and phosphate metabolism and/or lifestyle adaptations (Ray, Chinsamy, and Bandyopadhyay 2005). The highly porous nature of the postcrania of *Lystrosaurus* could be interpreted as an adaptation for lightening the relatively thick-walled bones of the skeleton.

CHANNEL ORIENTATION

One of the most striking variants in the cranial histology of *Lystrosaurus* was the difference in orientation of vascular channels occurring within

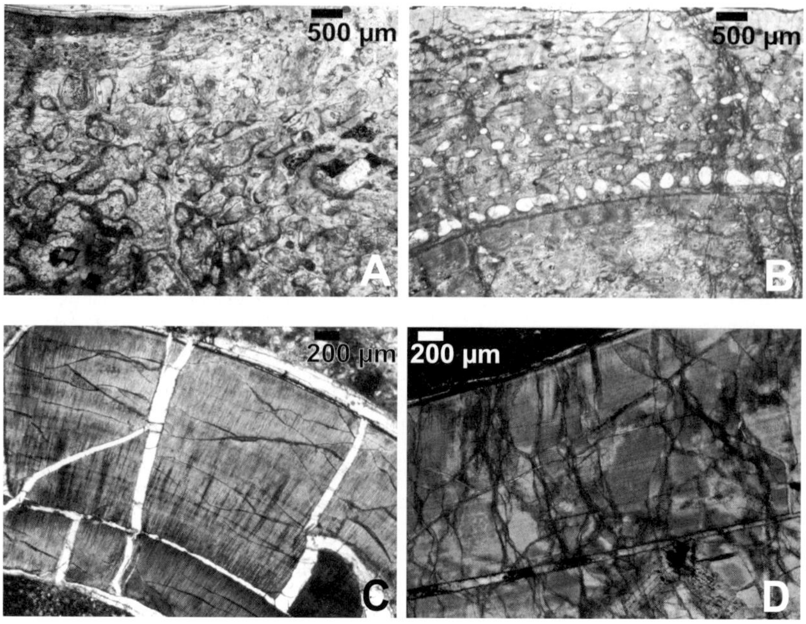

6.5. Comparison of the porosity of cranial (A) and postcranial (B) bones of specimen SAM-PK-K8 (*Lystrosaurus*). (A) SAM-PK-K8-6P, dorsal part of the parietal bone. (B) SAM-PK-K8 (H), humeral cortex. (C, D) Sections through the tusk of *Lystrosaurus.* (C) SAM-PK-K1422-14, horizontal section through the left tusk showing dentinal tubules, perpendicular to the edge of the tusk. Note that the thin, light-colored layer surrounding the outer edge of the dentine is not enamel (this layer is not even in thickness, and separated from the underlying dentine by a dark, extrinsic layer of amorphous material). (D) SAM-PK-K1422-4, longitudinal section through the right tusk showing closely spaced incremental lines of von Ebner.

the same bone or between different cranial elements. Although Ray, Chinsamy, and Bandyopadhyay (2005) assigned different orientations of primary osteons to the different ontogenetic stages of *Lystrosaurus* (see Ray, Chinsamy, and Bandyopadhyay 2005, Table 6), they later stated that the orientation of these structures can range from laminar to subplexiform to reticular across different postcranial elements or even change within the same element. This suggests that growth factors cannot adequately explain the variations of channel orientation. The variation observed was broadly attributed to several factors, including variable growth rate of the postcranial elements and/or functional constraints (Ray, Chinsamy, and Bandyopadhyay 2005). Here, we attempt to differentiate two possible factors that may influence channel orientation: biomechanical function and growth factors. However, it must be noted that many of the channels have been secondarily enlarged in *Lystrosaurus,* which may preclude the use of strain correlations employed by de Margerie (2002) that were based on primary bone tissues. Despite this remodeling, there appears to be a preferential direction to how the channels are enlarged. For instance, in some regions, the longitudinally oriented (i.e., circular-shaped) channels converge with each other to form elongated channels in a laminar orientation; whereas in other cases, the longitudinally oriented channels remain circular in shape even after they have been secondarily enlarged.

In several cranial bones of immature *Lystrosaurus* specimens, the long axis of the channels appeared to align to the overall three-dimensional shape of the bone, especially in restricted regions such as narrow arches (Fig. 6.6D). This may indicate the growth dynamics of the cranial bones, with the overall shape of the bone affecting channel orientation. This trend was also observed in the pterygoid (sections 9, 9A) (Fig. 6.6C),

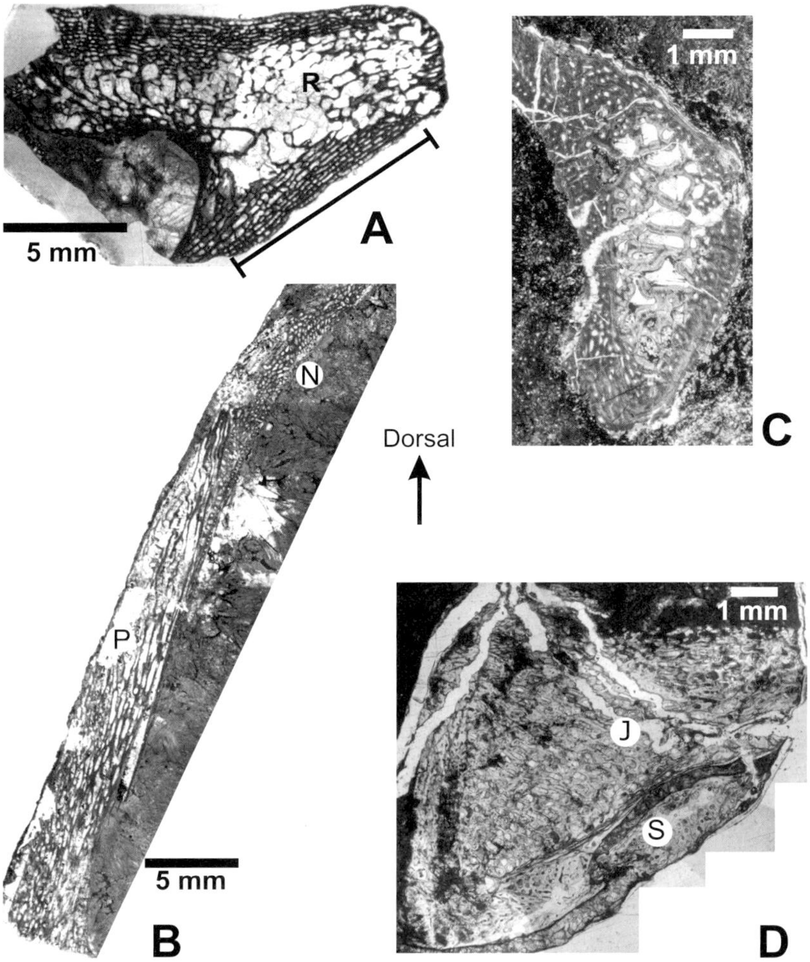

6.6. Channel orientation in the cranial bones of *Lystrosaurus.* (A) Composite image of the right prefrontal of *Lystrosaurus* (SAM-PK-K1402-5). Bracket indicates region of laminar-oriented channels. (B) SAM-PK-K1269-2, showing longitudinally oriented channels in the premaxilla and transversely oriented channels in the nasal bone. The open cavity within the center of the premaxilla is an artifact due to fracturing of the thin bony channels during histological preparation. (C) Anterior palatine ramus of the left pterygoid of SAM-PK-K1422-9 showing longitudinal channels. (D) SAM-PK-K1342-11, showing open channels in the left infraorbital bar region. In the squamosal, the channels are subcircular, and some have osteonal deposits. In the jugal, the large channels are oval to elongate in shape. Abbreviations are the same as in Figures 6.1 and 6.4.

but the orientation may also reflect the tensile strain oriented along the long axis of the pterygoid (Jasinoski, Rayfield, and Chinsamy 2010a). Further investigation of channel orientation in other irregularly shaped bones, such as the (i) premaxilla-nasal (section 2) (Fig. 6.6B), (ii) ventral maxilla (section 1M) (Fig. 6.4B), (iii) premaxilla (section 10), (iv) prefrontal (section 5) (Fig. 6.6A), and (v) parietal (section 6P) (Figs. 6.3B, 6.3C), are discussed in the following section.

There was a distinct difference in channel orientation across the premaxilla and nasal bones (section 2) (Fig. 6.6B): channels had a longitudinal orientation along the posterior (endocranial) surface of the premaxilla, but were mainly transversely oriented in the nasal. These orientations did not reflect the longitudinal orientation of the FE-predicted predominant compressive strain (Jasinoski, Rayfield, and Chinsamy 2010a), although the presence of a scarf suture may have changed the distribution and/or orientation of strain. Instead, the channel orientations may reflect a common growth feature and/or phylogenetic constraint of dicynodonts, as the orientations were similar in both *Lystrosaurus* and *Oudenodon* (Jasinoski 2008).

The preferential orientation of some of the open channels in the maxilla and premaxilla (section 1M) (Figs. 6.4B, 6.4C) may indicate that this region was better able to resist strain in particular orientations. The longitudinally oriented channels of the ventral region of the premaxilla and the adjacent maxilla may indicate a resistance to tensile strain oriented parallel to the sagittal plane of the cranium; whereas the laminar orientation of channels in the ventral tip of the maxilla may indicate an increased resistance to compressive strain in this area. The disorganized channel orientations and the large resorption spaces of the dorsal part of the maxilla could suggest that these areas experienced lower strain overall. These suppositions could not be investigated further because the compressive and tensile strain was artificially high since this region was constrained in the FE model (Jasinoski, Rayfield, and Chinsamy 2010a).

Sampling of different regions of the premaxilla in different orientations (sections 2, 10) (Table 6.2, Fig. 6.6B) showed that the channels have a similar longitudinal orientation in other parts of the premaxilla. In the ventral floor of the nasal aperture (section 10) and the near the dorsal edge of the nasal aperture (section 2), the premaxilla has elongated channels oriented parallel to the sagittal plane of the cranium, suggesting that this alignment of channels may be a strengthening feature around the nasal aperture. Cancellous bone occurred within the middle of the premaxilla, suggesting that this region experienced low strain, which is in contrast to the peripheral regions of the premaxilla that have more compact and organized bony channels and, therefore, may have experienced higher strain.

In the prefrontal, elongated channels were aligned with the dorsal and ventrolateral edges of the bone in immature specimens (Fig. 6.6A) (note that at the dorsal edge they are separate, small, primary circular-shaped channels that eventually merge to form enlarged laminar channels). A similar laminar orientation of channels occurred in the frontal bone of a juvenile *Metriorhynchus*, an extinct crocodilian (Hua and de Buffrénil 1996, Fig. 15A). Thus, the laminar channels may reflect the growth pattern of the prefrontal bone. In the adult *Lystrosaurus* specimens, the channels were instead mostly longitudinally oriented (i.e., appeared circular in transverse section).

In the parietal of immature specimens, longitudinally oriented channels (few with osteonal deposits) were organized roughly in dense rows along the dorsal part of the parietal (Figs. 6.3B, 6.3D); whereas the middle of the parietal consisted of erosionally enlarged irregularly shaped cavities and cancellous bone. The ventralmost part of the parietal became more compact with subcircular channels and semi-elongated channels that are oriented parallel to the medial and lateral edges (Fig. 6.3B).

Quantification of the channel orientation (as in de Margerie 2002) and further sampling of several regions from one individual may reveal a clearer pattern. A comparative, comprehensive survey of the cranial histology across different ontogenetic stages in any tetrapod is also still required.

LINES OF ARRESTED GROWTH AND ANNULI

Annuli and lines of arrested growth (LAGs) occurred in most of the sampled cranial elements of *Oudenodon* (Table 6.3; Figs. 6.7A, 6.7C, and 6.7E), indicating periods of slow growth and periodic cessations of growth, respectively. However, these growth features were not as prevalent in the snout and palatal regions (Table 6.3; Figs. 6.4A, 6.4C, and 6.4E). Variations in the occurrence of annuli and LAGs among bones may indicate differential growth rates (see Ray and Chinsamy 2004). However, the irregular shape, size, and growth strategy of cranial bones may preclude this utility, and further investigation is required. For instance, within a single individual, the jugal had multiple peripheral rest lines in the infraorbital bar (Fig. 6.7A); however, fewer rest lines were observed in the zygomatic arch where the jugal bone became narrower (Fig. 6.8E). Lines of arrested growth, or rest lines, were also not always continuous around the entire circumference of cranial bones and/or their appearance and number changed, possibly because the distinctive morphology of cranial bones requires significant remodeling that likely obliterates these histological features.

In nonperipheral annuli (representing an earlier stage of growth), LAG(s) usually occurred near the middle of the annulus (Fig. 6.7E), and this was also observed in the postcrania of *Oudenodon* (see Botha 2003). This indicates that during a period of slow growth, there was a complete cessation of bone deposition. Closely spaced LAGs in the jugal of the infraorbital bar (Fig. 6.7A) and posterior skull roof (Table 6.3) may represent an outer circumferential layer (OCL), in which appositional growth has nearly stopped.

On the contrary, LAGs in *Lystrosaurus* were only apparent in the dentary of an adult individual (Table 6.3; Fig. 6.7F) and did not occur within the cranium of any specimen. This was unexpected since LAGs were previously described in the humerus, radius, and ulna of two large subadult individuals of *Lystrosaurus* (SAM-PK-8991a, SAM-PK-K8) and the femur of one adult individual (SAM-PK-8796a) (Ray, Chinsamy, and Bandyopadhyay 2005). However, all of these postcranial elements were from either adults or individuals close to the adult stage, whereas in the current study these stages were not well represented. Also, the maximum size documented for *Lystrosaurus* can be quite large, depending on the species (Grine et al. 2006), suggesting that larger cranial specimens of *Lystrosaurus* need to be sampled. The porous, remodeled bone of *Lystrosaurus* also made it difficult to observe LAGs and bone formed during early ontogeny.

In summary, the presence of LAGs and annuli in many cranial bones of adult *Oudenodon* might indicate that the rate of growth in these cranial elements was slowing down relative to elements without LAGs/annuli or with fewer such features. However, comparison to modern taxa is required to substantiate this interpretation of differential growth in the cranium. Determinate growth may have occurred in *Oudenodon*,

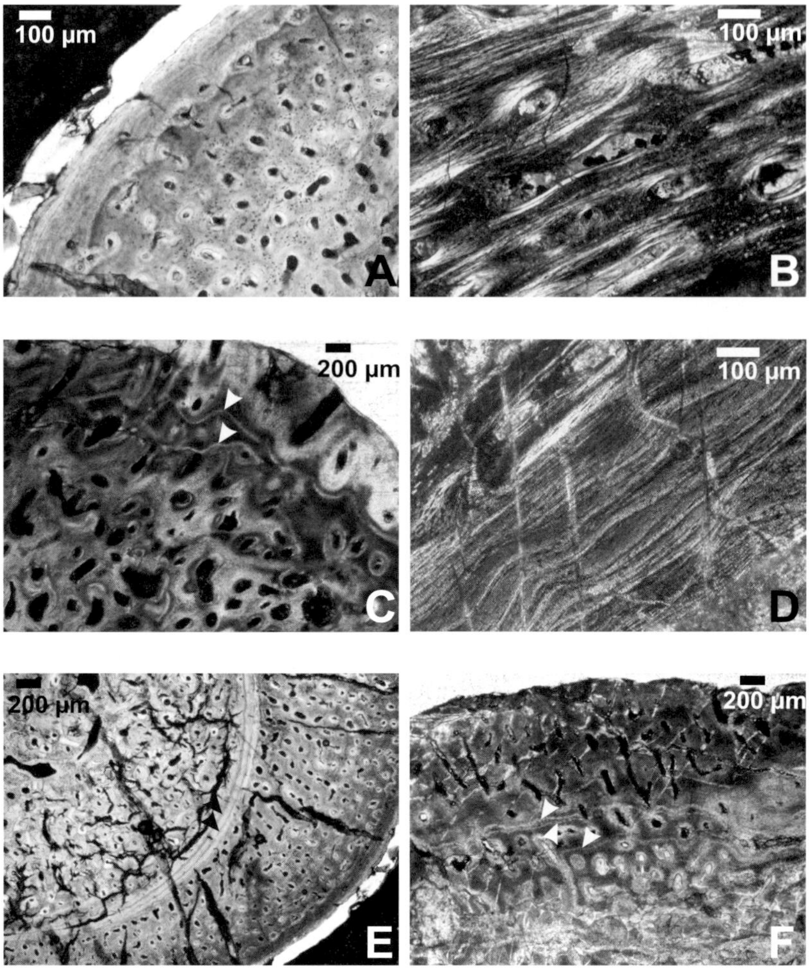

6.7. Examples of rest lines, LAGs, and slow growing bone tissues in *Oudenodon* (left column) and *Lystrosaurus* (right column). (A) SAM-PK-K5216-11, jugal with multiple rest lines at the periphery, which suggests that this represents an OCL. (B) SAM-PK-K28-5(extra) under polarized light (image is mirrored), showing parallel-fibered bone at the ventrolateral edge of the prefrontal. (C) SAM-PK-K5065-5 with possibly two LAGs (arrowheads) in the prefrontal. (D) SAM-PK-K28-5 under polarized light, showing parallel-fibered bone followed by lamellated bone near the ventrolateral edge of the prefrontal. (E) SAM-PK-K5216-11, jugal with two annuli of lamellar bone. The earlier annulus has two LAGs (arrowheads) within the central region of the annulus. (F) SAM-PK-K28-LR, dorsolateral edge of the dentary showing three LAGs (arrowheads).

as indicated by the presence of an OCL in some cranial bones (Table 6.3). This would corroborate the hypothesized growth strategy of another dicynodont, *Placerias*, which had an OCL (= external fundamental system) in the tibia of the largest sampled specimen (Green, Schweitzer, and Lamm 2010). On the contrary, an OCL was not observed in *Lystrosaurus*, which is in agreement with Ray, Chinsamy, and Bandyopadhyay's (2005) findings, although larger specimens may need to be sampled to ensure that this is not a sampling bias.

PARALLEL-FIBERED, LAMELLAR, AND LAMELLATED BONE TISSUES

The presence of a band of peripheral parallel-fibered bone following a zone of fibrolamellar bone in the postcranial bones of *Oudenodon* was suggested to indicate the onset of sexual maturity (Botha 2003; Botha and Angielczyk 2007), and its occurrence was used to discriminate between adult and subadult individuals in Botha's (2003) survey of four postcranial elements. However, the presence of lamellar bone, but not

Table 6.3. Summary of histological features in the skull bones of *Oudenodon* and *Lystrosaurus*

Specimen SAM-PK-	Genus	LAGs or growth marks	Tissue indicated by Sharpey's fibers
K1342-3	*Lystrosaurus*	absent	K, S
K1342-4		absent	K
K1342-6A		absent	S
K1342-6P		absent	K, S
K1342-7		absent	S
K28-5	*Lystrosaurus*	absent	K
K28-6A		absent	M
K28-6P		absent	K, M
K28-LR		few	K
K28-5(extra)	*Lystrosaurus*	absent	K
K5065-1M	*Oudenodon*	few	K
K5065-2		absent	K
K5065-3		few	N/A
K5065-3 II		few	K
K5065-4		few	K
K5065-5		few	K
K5065-7		few	S
K5065-11		few	S
K5065-12		few	absent
K5216-2	*Oudenodon*	absent	K, S
K5216-5		few	S
K5216-5A		few	S
K5216-6A		multiple	M, S
K5216-6P		multiple	M, S
K5216-7		few	S
K5216-10		absent	K
K5216-11		multiple	S
K5216-14		absent	K
K5288-SY	*Oudenodon*	absent	K
K5343-RR	*Oudenodon*	absent	S
K5396-LR	*Oudenodon*	few	K
K6043-1M	*Oudenodon*	few	K
K6043-2		absent	K
K6043-4		absent	K
K6043-5A		multiple	absent
K6043-6A(1)		multiple	M
K6043-6A(2)		multiple	M
K6043-10		N/A	K
OUM.TSK 96-9	*Oudenodon*	absent	S
OUM.TSK 96-9A		absent	S

Note: Thin-sections that did not show these features, or did not have sufficient preservation in order to observe these features, were listed as N/A or omitted. The number of LAGs or growth marks were listed as "few" (three or less) or "multiple." *Abbreviations:* K = keratin; LAGs = lines of arrested growth; M = muscle; S = sutural fibers.

parallel-fibered bone, was originally documented in cranial bones of *Oudenodon* in Jasinoski (2008). This discrepancy possibly results from the difficulty of discriminating between parallel-fibered and lamellar bone tissues (see Enlow 1969; chapter 3 of this book). The peripheral layer of lamellar bone observed in several cranial specimens of *Oudenodon* (e.g., Figs. 6.7E, 6.8G) in combination with the presence of multiple growth lines indicated that growth had slowed down considerably, and that these specimens were most likely adult individuals (Table 6.1).

Specimens SAM-PK-K28 and SAM-PK-K28 (extra), the largest *Lystrosaurus* individuals sampled for cranial histology (Table 6.1), had parallel-fibered bone near the periphery of the bone (Figs. 6.7B, 6.7D). Parallel-fibered bone was also observed in the outer cortex of the humerus of a late subadult and the rib of an adult *Lystrosaurus* (Ray, Chinsamy, and Bandyopadhyay 2005). The absence of peripheral parallel-fibered or lamellar bone in comparable cranial sections of the other *Lystrosaurus* specimens suggested that they had not yet reached sexual maturity. In addition to the parallel-fibered layer, specimen SAM-PK-K28-5 had a lamellated layer along the ventrolateral edge of the prefrontal (Fig. 6.7D), suggesting that it was from an older adult individual relative to specimen SAM-PK-K28-5(extra).

The term "lamellated bone" is used to describe lamellar bone that has a distinctive banding under ordinary light (Reid 1996; chapter 3 of this book). Here, it was observed in a few adult cranial specimens of *Oudenodon* (Fig. 6.3A) and *Lystrosaurus* (Fig. 6.7D), suggesting slow rates of bone accretion.

SHARPEY'S FIBERS

Sharpey's fibers are mineralized extrinsic bundles of collagen fibers that are associated with the attachment of soft tissues such as muscle, tendon, ligament, and dermal derivatives (de Ricqlès et al. 1991; Francillon-Vieillot et al. 1990; chapter 3 of this book). From the sampled dicynodont specimens, Sharpey's fibers were observed in the snout, palatal region, skull roof, and at the sutural margins. The distribution of Sharpey's fibers across cranial elements, the potential type of soft tissue attachment, and the differential occurrence between taxa, are discussed in the following sections. It should be noted that the identification of Sharpey's fibers was somewhat dependent on the quality of the preserved bone histology.

Keratin attachment

Sharpey's fibers in the palatal and snout regions may indicate the attachment of a keratin beak, as suggested by pitting of the bony surface in gross osteological specimens. On the snout of *Oudenodon*, the premaxilla and maxilla had extensive Sharpey's fibers, oriented roughly perpendicular to the ectocranial surface, or in a few cases, obliquely (Plate 9A, Fig. 6.8C). There was only a small patch of short Sharpey's fibers oriented roughly perpendicular to the ectocranial surface of the nasal (Table 6.3), but the other specimens were not well preserved in this region. In the palatal region, distinct fibers occurred along the ventral edge of the premaxilla and maxilla (Plate 7A, Fig. 6.8A), and a fibrous texture along the ventral edge of the palatine was observed. Sharpey's fibers were also present in the dentary of *Oudenodon* (Table 6.3).

In contrast, Sharpey's fibers were rarely observed in the snout region in the sampled specimens of *Lystrosaurus* (Table 6.3), although extensive small foramina on the surface of the snout and palatal bones in this region indicated the existence of a keratin beak (Jasinoski, Rayfield, and Chinsamy 2010b). The porous palatine bone (Fig. 6.4F) was extensively remodeled, which may have obliterated any trace of Sharpey's fibers. However, in the adult mandible, distinctive Sharpey's fibers were abundant along the dorsal edge of the longitudinal groove (Fig. 6.8B) and other edges of the dentary bone, which indicated the attachment of a mandibular keratin beak.

In adult *Oudenodon* and *Lystrosaurus*, the edges of the bone in many sampled regions of proposed keratin beak attachment did not have a finished surface, that is, there were many openings (Figs. 6.4C, 6.8B), which may indicate passage of numerous blood vessels and nerves to the overlying keratin layer.

Sharpey's fibers in the bones of the skull roof may also indicate the attachment of keratin. In adult *Lystrosaurus* (Table 6.3), there were a few bundles of Sharpey's fibers in the dorsolateral part of the prefrontal boss. Most of these fibers were elongated and had a perpendicular orientation to the bone surface. There were faint Sharpey's fibers in the prefrontal boss region of a single specimen of *Oudenodon* (Table 6.3), although this region was not well preserved in most specimens.

Along the dorsal edge of the parietals of *Lystrosaurus*, long Sharpey's fibers were present in one or two patches (Table 6.3; Figs. 6.3D, 6.8D; Plate 9B). Their presence may indicate the attachment of keratinized ornamentation, which would corroborate Cluver's (1971) supposition that keratin covered this region. Cluver (1971) also surmised that a thick horny pad covered the highly rugose frontal bones, but further histological sampling of the skull roof of other *Lystrosaurus* specimens is necessary to confirm this supposition.

Sutural Collagen Fiber Attachment

In both *Oudenodon* and *Lystrosaurus*, Sharpey's fibers were found embedded along the sutural margins of several cranial elements. These Sharpey's fibers were most probably derived from the intrinsic collagen fibers that spanned the sutural gap (see Jones and Boyde 1974).

In *Oudenodon*, Sharpey's fibers occurred along the edges of scarf sutures in the infraorbital bar and zygomatic arch, mainly within the jugal bone (Table 6.3, Fig. 6.8E). A few isolated patches of Sharpey's fibers were preserved near the interdigitated premaxilla-nasal suture, but more extensive Sharpey's fibers occurred along the sutures of the

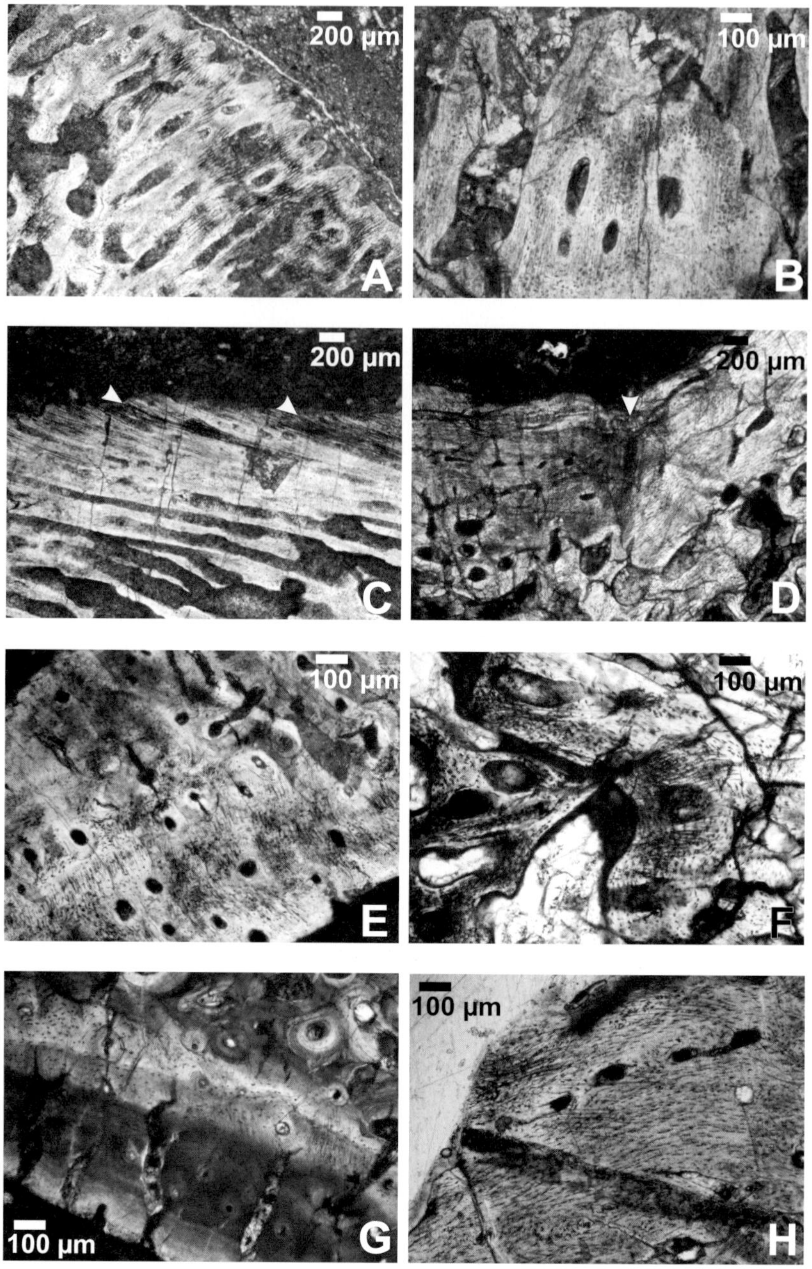

6.8. Examples of Sharpey's fibers distribution in cranial bones of *Oudenodon* (left column) and *Lystrosaurus* (right column). (A) Sharpey's fibers along the palatal edge of the premaxilla (SAM-PK-K5216-10), indicating the attachment of the keratin beak. (B) Sharpey's fibers on the dorsal surface of the dentary longitudinal groove (SAM-PK-K28-LR) indicate the attachment of the mandibular keratin beak. (C) Obliquely oriented Sharpey's fibers (arrowheads) along the anterior (outer) edge of the premaxilla (SAM-PK-K5216-2), indicating the attachment of keratin onto the snout. (D) Patch of elongated Sharpey's fibers (arrowhead) in the depression on either side of the median parietal ridge in the posterior skull roof (SAM-PK-K28-6P) (see also Figure 6.3D) may represent keratin horn attachment. (E) Sharpey's fibers along sutural margin of the jugal bone near its contact with the squamosal (located to the bottom of the image; SAM-PK-K5216-7). These extrinsic fibers would have associated with the intrinsic collagen fibers that filled the sutural gap. Note the absence of Sharpey's fibers along the medial edge of the jugal, away from the sutural contact. (F) Sharpey's fibers along margins of the postorbital-parietal suture (SAM-PK-K1342-6P) (see Figure 6.3C for entire suture morphology). (G, H) Sharpey's fibers on the ventrolateral (G: SAM-PK-K5216-6P) and lateral (H: SAM-PK-K28-6P) (see also Figure 6.3D) edge of the postorbital bone, which may indicate muscular attachment of the M. adductor mandibulae externus medialis complex.

skull roof (Table 6.3). In *Lystrosaurus*, Sharpey's fibers occurred near the sutural margin of several bones of one individual, SAM-PK-K1342 (Table 6.3, Fig. 6.8F).

Whether the orientation of Sharpey's fibers reflects that of the intrinsic sutural collagen fibers is debatable. Observations of rat and monkey cranial sutures suggested that orientations of these two types of fibers were not correlative (Jones and Boyde 1974). This requires further investigation across several different suture morphologies because more recent studies revealed that intrinsic fibers in compressed and tensed sutures can have a consistent orientation near the sutural margins (e.g., see Rafferty and Herring 1999), but the orientation of the corresponding Sharpey's fibers was not considered.

Muscle Attachment

Sharpey's fibers were also observed along the lateral edge of the posterior skull roof of adult *Oudenodon* and *Lystrosaurus* specimens, which may indicate the attachment site of the M. adductor mandibulae externus medialis complex. In some cases, this proposed region of attachment was marked by striations on the surface of the postorbital bones.

In the posterior skull roof of *Oudenodon*, extensive Sharpey's fibers were found along the ventrolateral edge of the postorbital bone (Table 6.3, Fig. 6.8G). In the adult *Lystrosaurus* specimen SAM-PK-K28-6P, some patches of Sharpey's fibers along the lateral edge of the postorbital (Table 6.3, Fig. 6.8H) had a similar oblique orientation to a patch observed ventrally along the lateral edge of the parietal (Fig. 6.3D). Interestingly, in regions of proposed muscle attachment in the postcrania of *Lystrosaurus*, such as the humeral deltopectoral crest and femoral trochanter major, primary osteons were radially oriented but Sharpey's fibers were not described (Ray, Chinsamy, and Bandyopadhyay 2005).

Overall Comparison of Sharpey's Fibers between *Oudenodon* and *Lystrosaurus*

There was a relatively low occurrence of Sharpey's fibers in cranial specimens of *Lystrosaurus* compared with those of *Oudenodon*. This observation agrees with previous surveys of postcranial bone histology, which documented Sharpey's fibers in several postcranial bones of *Oudenodon* (Chinsamy and Rubidge 1993; Botha 2003), but were not described in specimens of *Lystrosaurus* that spanned several ontogenetic stages (Ray, Chinsamy, and Bandyopadhyay 2005) nor in a recent study of three species of *Lystrosaurus* (Botha-Brink and Angielczyk 2010). The low occurrence of Sharpey's fibers in the current study may be partially related to ontogenetic age, as they were observed only in one subadult and two adult individuals of *Lystrosaurus* (Table 6.3), or that they were difficult to discern in woven-fibered bone. Previous observations of Sharpey's fibers in other tetrapods indicate that for the most part they are apparent in lamellar or parallel-fibered bone tissues, although it is possible for them to form in a woven bone matrix (Francillon-Vieillot et al. 1990). Sharpey's fibers are obliterated during extensive remodeling (Enlow 1990; Martin, Burr, and Sharkey 1998), but this does not explain their absence from the primary bone of some immature *Lystrosaurus* specimens.

Only in *Lystrosaurus* were there definitive Sharpey's fibers on the dorsal surface of the parietals, indicating keratin ornamentation on the posterior skull roof. Further sampling of additional specimens is required to confirm this possible anatomical difference between *Lystrosaurus* and *Oudenodon*.

In the regions of proposed keratin beak attachment in *Oudenodon*, Sharpey's fibers were numerous, occurring extensively along the edges of the snout and palatal bones. In *Lystrosaurus*, Sharpey's fibers representing keratin attachment occurred only in patches in the skull roof and were absent from the palate, but numerous fibers occurred along the lateral edge of the maxilla in the posterior snout (Table 6.3). At this time, it is not known what significance these differences in the distribution of Sharpey's fibers indicate for the overlying keratin attachment. In a transverse section of a turtle snout (*Chrysemys picta*), extrinsic fibers were present along the lateral edge of the maxilla only near the dorsal and posterior extent of the keratin beak, but they were absent from the palate that was covered by a thick layer of keratin (T. L. Hieronymus, pers. comm., 2010). This fiber distribution is similar to that observed in the snout of *Lystrosaurus*. In the beaks of modern birds, the extrinsic fibers may occur in patches or be continuous along the edge of the bone (Hieronymus et al. 2009). However, in the skull roof of pachycephalosaurs (Dinosauria), the Sharpey's fibers, which indicate keratin attachment, were very widespread (Goodwin and Horner 2004, Figs. 5B, 6C), unlike those on the skull roof of *Lystrosaurus*.

In addition, the relatively lower density and patchiness of Sharpey's fibers in the muscle attachment region of the posterior skull roof of *Lystrosaurus* may indicate a difference in muscular attachment (fleshy versus tendinous) and/or muscular stress at the attachment site (less versus higher) (see Hieronymus 2006). However, further sampling of well-preserved specimens of both taxa is necessary before these suppositions can be confirmed.

TUSK HISTOLOGY

The tusk of *Lystrosaurus* was thin-sectioned in the transverse and longitudinal plane (from a horizontal and transverse section, respectively, of the cranium) (Table 6.2). The tusk was composed of dentine. Dentinal tubules, perpendicular to the edge of the tusk, were clearly visible in most of the specimens (Figs. 6.5C, 6.5D). Incremental lines of growth, which are comparable to those found in *Diictodon* (Thackeray 1991) and a late Triassic dicynodont (chapter 7 of this book), were visible in longitudinal sections of SAM-PK-K1422-4 (Fig. 6.5D) and SAM-PK-K8-4. It was difficult to make accurate and precise measurements of these increments because they were not clearly visible at high magnification and were not continuous across the entire section (Fig. 6.5D), precluding a comprehensive analysis. The incremental thickness in both tusks was measured for ten well-defined lines, giving an average thickness of 17.4 μm. This thickness falls within the range for daily dentinal deposits in other amniotes (see Erickson 1996b), and thus they can be referred to as incremental lines of von Ebner.

The thicker concentric layers of dentine extinguished uniformly under polarized light at 90° intervals, thus there was no indication of an inner, secondary layer of dentine. This is unlike the condition described for

Diictodon, where an inward-deposited secondary dentine layer was surmised to have strengthened the elongated tusks for possible use in intraspecific conflict (Sullivan, Reisz, and Smith 2003). The surrounding maxilla had a predominant reticular pattern of channels, with some isolated patches of oriented elongated channels in specimen SAM-PK-K1342-4.

Concluding Remarks

This extensive histological survey of dicynodont cranial elements has revealed several important differences between *Oudenodon* and *Lystrosaurus*. Overall, the cranial bones were much more porous in *Lystrosaurus* throughout all ontogenetic stages, whereas *Oudenodon* had compact bone with extensive secondary osteons. Evidence of cyclical and periodic cessation of bone deposition in adult specimens of *Oudenodon* was common in most cranial elements. However, in the two sampled elements of adult *Lystrosaurus*, there was indication of slower deposition only near the periphery of the bone but not interspersed within the cortex of the bone, and there was no indication of LAGs within the cranium. Ray, Bandyopadhyay, and Bhawal (2009) stated that *Lystrosaurus* had a flexible growth strategy, that is, an ability to stop and continue growth depending on environmental conditions. The present survey may indicate that the cranium of *Lystrosaurus* had rapid cranial growth with no cyclical interruptions, or that the secondarily porous nature of the bone may have obliterated the growth lines. The presence of peripheral lamellated bone in specimen SAM-PK-K28 could, however, indicate a slowing down in bone formation due to attainment of adult size, or it could be an "annulus." Examination of larger *Lystrosaurus* specimens might help to resolve these observations.

In addition, secondary osteons were more abundant within cranial bones of the adult *Oudenodon* specimens than in specimens of *Lystrosaurus* of all ontogenetic stages. In fact, secondary osteons that were fully formed were somewhat rare even in adult cranial specimens of *Lystrosaurus*, which was also noted in many postcranial bones (Ray, Chinsamy, and Bandyopadhyay 2005). This difference could equally reflect a difference in mechanical or metabolic requirements between *Oudenodon* and *Lystrosaurus*, or perhaps a phylogenetic difference. The lack of infilling of erosional cavities or the erosion of existing secondary osteons may reflect the tendency toward an osteoporotic state, as was suggested for the postcrania (see Ray, Chinsamy, and Bandyopadhyay 2005).

Another difference is the distribution and amount of Sharpey's fibers in both taxa. For instance, Sharpey's fibers that indicate keratin beak attachment were absent from the palatal region and rarely preserved on the snout of *Lystrosaurus*, but they were abundant in both of these regions of *Oudenodon*. This discrepancy may reflect the porous nature of the *Lystrosaurus* cranial bones, in which the extensive bone remodeling may have obliterated the Sharpey's fibers or may have hampered observation of them. The noted differences could also reflect a difference in ontogenetic age and/or variation in the way the overlying soft tissue was anchored to the cranium.

The open channels or erosion cavities that were apparent in many of the cranial and postcranial bones of *Lystrosaurus* may have a biomechanical implication, indicating either a decreased loading of the region or an adaptation for a certain type of lifestyle. However, not all cranial regions characterized by high porosity had low strain, as predicted by FE analysis (e.g., the zygomatic arch and infraorbital bar regions, maxilla [Jasinoski, Rayfield, and Chinsamy 2010a]), although there was some organization of bony channels in these bones. Ray, Chinsamy, and Bandyopadhyay (2005) suggested that high cortical porosity in the proximal limb bones of *Lystrosaurus* compared with its pachyostotic ribs indicate a reduction of mass that may be related to a semi-aquatic lifestyle. Although this supposition regarding the lifestyle of *Lystrosaurus* could not be ruled out in the present study of cranial histology, it appears that the porous nature of the cranial bones may have been an adaptation for lightening the skull.

Our study has significantly contributed to understanding histovariability in the crania of two dicynodonts and, by extrapolation from studies on modern animals, has provided pertinent hypotheses to explain the histological differences evident in the bone microstructure. In addition, our study has highlighted the fact that several aspects of cranial histology require further research. First, histological sampling of more ontogenetic stages of both *Oudenodon* and *Lystrosaurus* is necessary. Second, quantitative analyses of particular histological features, such as channel orientation, the density and distribution of secondary osteons, and the intensity of bone resorption, is required to objectively ascertain whether these features correlate with biomechanical function. Third, it is imperative that comprehensive cranial histological analyses are conducted on extant animals to distinguish between growth effects and biomechanical influences.

Acknowledgments

Roger Smith (Iziko South African Museum, South Africa) and Tom Kemp (Oxford University Museum, United Kingdom) are thanked for granting permission to thin-section dicynodont specimens. Emily Rayfield (University of Bristol, United Kingdom) is acknowledged for having supervised the broader biomechanical aspects of S.C.J.'s doctoral research, and for the use of laboratory facilities for preparing some of the thin sections. We thank the technicians at the Department of Geology at University of Cape Town (South Africa) and University of Birmingham (United Kingdom), as well as the Earth Sciences Department at University of Bristol, for assistance with some of the histological sectioning. We would also like to thank the collections staff at the Iziko South African Museum and the Oxford University Museum. Jeremy Green and Sanghamitra Ray are acknowledged for having reviewed this manuscript. S.C.J. was supported by the Claude Leon Foundation.

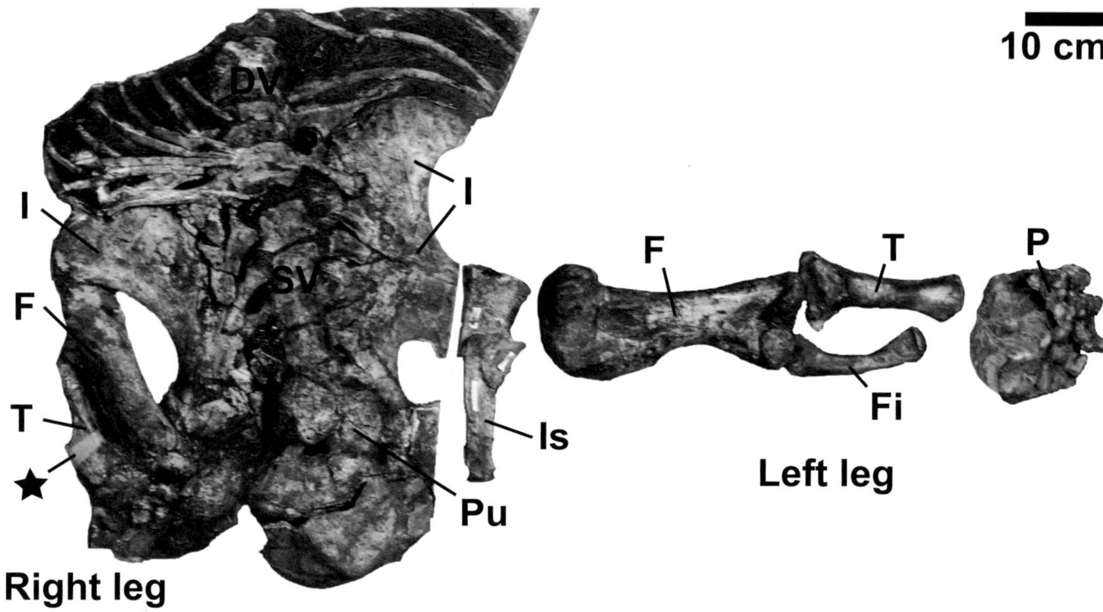

7.1. Ventral view of semi-articulated, semi-complete postcranium (NCSM 21719) of an unnamed species of dicynodont from the Pekin Formation (Upper Triassic) of North Carolina. Star indicates location of histological sampling of the right tibia for this study. *Abbreviations:* DV = dorsal vertebrae; F = femur; Fi = fibula; I = ilium; Is = ischium; SV = sacral vertebrae; P = pes; Pu = pubis; T = tibia.

Bone and Dental Histology of Late Triassic Dicynodonts from North America

7

Jeremy L. Green

Introduction

Dicynodonts were the dominant terrestrial herbivores during the Late Permian (Hotton 1986). Although diversity declined at the end of the Permian, at least four dicynodont lineages managed to survive the greatest mass extinction of the Phanerozoic (Fröbisch 2007; chapter 1 of this book), and this group continued to thrive until its eventual disappearance near the end of the Triassic (King 1990; although see also Thulborn and Turner 2003). Late Triassic dicynodonts were the last and largest (skull length [SL] > 400 mm) members of this clade (King 1990) and lived in a world of ever-changing climatic and environmental conditions associated with continental rifting in Pangaea (Simms and Ruffell 1990; Parrish 1993; Fraser 2006; Golonka 2007). Studying the life history (e.g., growth, lifestyle adaptations) of these animals may elucidate the strategies dicynodonts used to adapt to this dynamic setting and further expand our understanding of paleobiology in nonmammalian therapsids. Analyses of growth in dicynodonts have focused mainly on small- to medium-sized (SL ≤ 400 mm) taxa from Permian and Early to Middle Triassic deposits (Gross 1934; de Ricqlès 1972; Chinsamy and Rubidge 1993; Ray and Chinsamy 2004; Germain and Laurin 2005; Ray, Chinsamy, and Bandyopadhyay 2005, Ray, Bandyopadhyay, and Bhawal 2009; Botha and Angielczyk 2007; Kriloff et al. 2008; Botha-Brink and Angielczyk 2010), and our knowledge of growth in Late Triassic dicynodonts is poor relative to that of their geologically older, smaller ancestors. Additionally, previous studies have focused primarily on dicynodonts from Africa (Chinsamy and Rubidge 1993; Ray and Chinsamy 2004; Ray, Chinsamy, and Bandyopadhyay 2005; Botha and Angielczyk 2007; Botha-Brink and Angielczyk 2010) and India (Ray, Chinsamy, and Bandyopadhyay 2005, Ray, Bandyopadhyay, and Bhawal 2009), leaving North American taxa relatively unstudied. A recent study on growth in *Placerias* from Arizona (Green, Schweitzer, and Lamm 2010) has helped address these deficiencies. The current study expands on this previous work by conducting two separate analyses: (1) comparison of bone microstructure between *Placerias* (Green, Schweitzer, and Lamm 2010) and a previously unsampled, non-*Placerias* dicynodont (unnamed species) from North Carolina; (2) examination of the thickness and periodicity of growth increments in dicynodont tusks from North Carolina. Collectively, these studies will improve our knowledge of life history in Late Triassic dicynodonts from North America.

Institutional Abbreviations

The following institutions have been abbreviated in this chapter: North Carolina Museum of Natural Sciences, Raleigh, North Carolina (NCSM); University of California Museum of Paleontology, Berkeley, California (UCMP).

Phylogenetic Context

Our understanding of the diversity and evolution of Late Triassic dicynodonts from North America has changed drastically over the past hundred years. Before the phylogenetic context of this group can be discussed, it is necessary to review the taxonomy of dicynodonts from the Late Triassic of this continent. In the early part of the twentieth century, scientists recognized three distinct species from the Late Triassic of North America: *Brachybrachium brevipes* (Williston 1904); *Eubrachiosaurus browni* (Williston 1904); and *Placerias hesternus* (Lucas 1904). However, the first taxon is recognized as a nomen dubium because its holotype has been lost, and *E. browni* is now considered a junior synonym of *P. hesternus* (Lucas and Hunt 1993). Camp and Welles (1956) described another Late Triassic species, *Placerias gigas*, from the Placerias Quarry in Arizona, a locality so-named for its relative abundance of *Placerias* material (i.e., a minimum number of 41 individuals represented by over 1,600 skeletal elements [Fiorillo, Padian, and Musikasinthorn 2000]). However, this species has now also been synonymized with *P. hesternus* (Lucas and Hunt 1993).

Although *Placerias* is the most commonly recovered taxon from North America, remains of non-*Placerias* dicynodonts have also been identified from Upper Triassic deposits of this continent; Lucas and Hunt (1993) described skeletal material they identified as cf. *Ischigualastia* sp. However, subsequent researchers argue that this material differs from *Placerias* and refute its tentative assignment to *Ischigualastia*. These investigators suggest this material may instead belong to *Stahleckeria* or *Dinodontosaurus* (Long and Murry 1995; Langer 2005), whereas Fröbisch (2009) recommends it be classified as Kannemeyeriiformes gen. et sp. indet. Green et al. (2005) reported a partially complete, semi-articulated postcranium of an unnamed species of dicynodont (differentially diagnosed from *Placerias*) from the Late Triassic of North Carolina (NCSM 21719) (Fig. 7.1), but a full description and phylogenetic analysis of this specimen has yet to be completed. Similarly, fossils of another non-*Placerias* dicynodont have been recently reported from the Late Triassic of Texas (Mueller and Chatterjee 2007); a full description of this material is in progress (B. Mueller, pers. comm. 2010). These findings collectively paint a more complex picture of dicynodont diversity during this time frame in North America. However, we currently do not have a good understanding of the taxonomy of these animals or the phylogenetic relationships between these and other dicynodonts. *Placerias hesternus* is the only dicynodont species currently recognized from the Late Triassic of North America, with its remains being recovered from both western North America (Camp and Welles 1956, Lucas and Heckert 2002) and North Carolina (Baird and Patterson 1967; Huber, Lucas, and Hunt 1993; Lucas 1998). Therefore, our current understanding of evolutionary relationships among North American dicynodonts from the Late Triassic is limited to *Placerias*.

This taxon has been included in many different phylogenetic analyses, including those using cranial characters (Surkov and Benton 2004, 2008), postcranial characters (Ray 2006), and a suite of cranial and postcranial characters (Maisch 2001; Vega-Dias, Maisch, and Schultz 2004). In each of these independent studies, *Placerias* is recovered in a monophyletic clade, Kannemeyeriiformes, which includes all Triassic members of the Dicynodontia except for *Kombuisia*, *Myosaurus*, and *Lystrosaurus* (Fröbisch 2007). However, the exact position of *Placerias* within Kannemeyeriiformes varies, depending on the characters analyzed (Maisch 2001; Vega-Dias, Maisch, and Schultz 2004; Ray 2006; Surkov and Benton 2004, 2008). Here, I follow the most recent and comprehensive analysis of Triassic dicynodont relationships by Surkov and Benton (2008), where *Placerias* is derived within Kannemeyeriiformes and is the sister taxon to the group of *Ischigualastia* + *Jachaleria*, dicynodonts from the Middle to Late Triassic of South America. It is possible that the new, non-*Placerias* dicynodonts from North America occupy a similar position within Kannemeyeriiformes, but this can only be determined through future phylogenetic analyses. Understanding the phylogenetic relationships between *Placerias* and these other forms will provide more resolution concerning the evolution of Late Triassic dicynodonts from North America.

Paleobiological Context

Our understanding of the paleobiology of Late Triassic dicynodonts from North America is also currently limited to data from *Placerias*. Like other Late Triassic dicynodonts, *Placerias* was a large-bodied animal (SL = 680 mm [Camp and Welles 1956]) with a semi-sprawling posture (Ray 2006); the forelimbs were held in an abducted stance, whereas the hindlimbs were nearly upright (Ray 2006). Sexual dimorphism in *Placerias* was hypothesized by Camp and Welles (1956), who suggested, based on a large sample of material from the Placerias Quarry, that males had an enlarged, horn-like premaxillary process (unique relative to other dicynodonts), whereas females lacked this feature. They concluded that males used these premaxillary "horns" for intraspecific combat for females, although this hypothesis has yet to be properly tested.

An herbivorous diet in *Placerias*, as well as in most other dicynodonts, is supported by the replacement of teeth (except the maxillary tusks) with a keratinous beak and propalinal sliding of the lower jaw (Camp and Welles 1956; Crompton and Hotton 1967; Hotton 1986). Some studies have suggested that *Placerias* fed on vegetation at or below ground level, possibly using premaxillary horns to dig or root for food (Camp and Welles 1956;

Cruickshank 1978). However, new data from head kinematics suggest that this taxon, as well as other kannemeyeriiform dicynodonts from the Late Triassic, fed on high-level foliage (Surkov and Benton 2008). *Placerias* fossils are most common from the Placerias Quarry (Chinle Formation, Arizona), a locality that has been reconstructed as an upland floodplain environment with ephemeral standing water. This data, along with the relative rarity of *Placerias* fossils from fluvial-lacustrine deposits of the Chinle Formation (Fiorillo, Padian, and Musikasinthorn 2000), suggest that these animals most likely inhabited upland, forested habitats in North America during the Late Triassic (contra Camp and Welles 1956).

Histological Context

Green, Schweitzer, and Lamm (2010) recently published a detailed analysis of growth patterns from twenty-three limb bones of *Placerias hesternus* from the Placerias Quarry, thus improving our knowledge of growth in large-bodied, Late Triassic dicynodonts. Bone microstructure in this taxon is generally zonal fibrolamellar (supporting periodic rapid growth, similar to other dicynodonts [see, e.g., Ray, Bandyopadhyay, and Bhawal 2009; Botha-Brink and Angielczyk 2010]) with a peripheral layer of parallel-fibered (slow-growing) tissue (Green, Schweitzer, and Lamm 2010). An external fundamental system (EFS), which supports complete or near-complete cessation of bone growth, is present in the largest tibia recovered from the Placerias Quarry (Green, Schweitzer, and Lamm 2010); this is the first report of an EFS in dicynodont limb bones and is interpreted as the attainment of maximum body size for this species (Green, Schweitzer, and Lamm 2010). This histological pattern suggests that *Placerias* may have had determinate growth, in contrast to the indeterminate growth strategy seen in other sampled dicynodonts (see, e.g., Botha 2003; Ray and Chinsamy 2004; Ray, Chinsamy, and Bandyopadhyay 2005, Ray, Bandyopadhyay, and Bhawal 2009; Botha and Angielczyk 2007; Botha-Brink and Angielczyk 2010). In addition, peripheral slow-growing bone was found in elements of various sizes, which suggests a differential growth pattern between some individuals from the Placerias Quarry (Green, Schweitzer, and Lamm 2010). This variation may be attributed to sexual dimorphism and/or developmental plasticity within *Placerias*, or possibly the presence of two dicynodont taxa at this locality (Green, Schweitzer, and Lamm 2010). Further histological analysis and comparison with other similar dicynodont taxa is needed to test these contrasting, but equally viable hypotheses.

Mapping histological characters onto a cladogram may reveal the evolution of growth strategies within a lineage (e.g., Ray, Botha, and Chinsamy 2004, Ray, Bandyopadhyay, and Bhawal 2009). Here, histological characters for previously sampled dicynodonts (including *Placerias*) were mapped (Fig. 7.2) onto a cladogram (modified from Surkov and Benton [2008]) to hypothesize the evolution of growth patterns in the clade; skull length (SL) is indicated for each taxon to give an approximation for body size. The resulting cladogram (Fig. 7.2) suggests

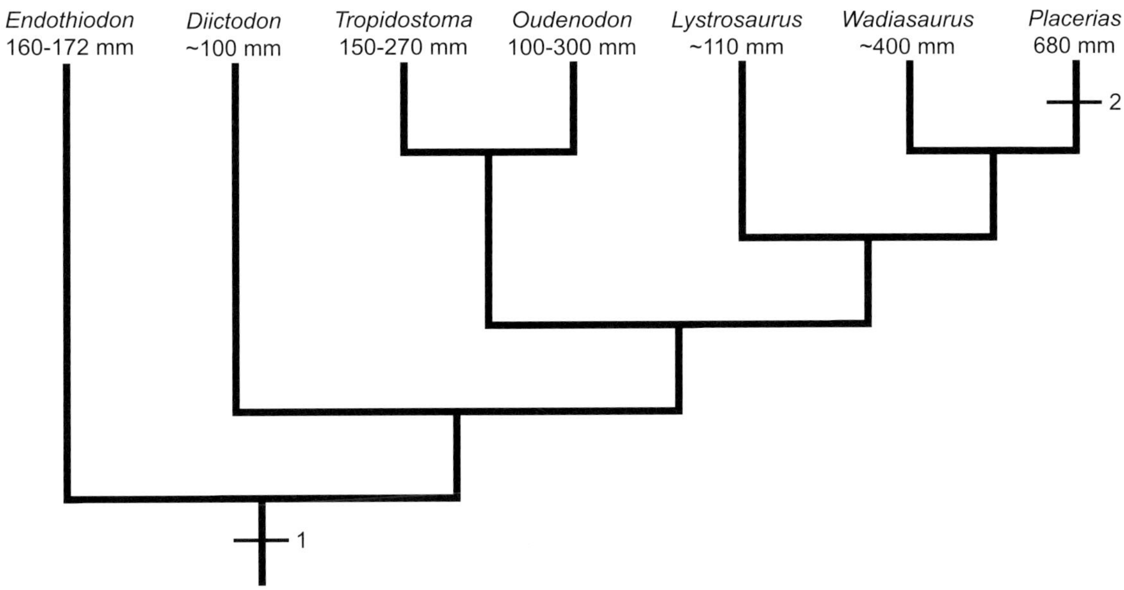

7.2. Cladogram (modified from Surkov and Benton [2008]) showing the evolution of growth strategies across Dicynodontia. Histological synapomorphies: (1) Zonal fibrolamellar bone with peripheral parallel-fibered bone in adult individuals, periodic rapid growth followed by slowed growth. (2) Zonal fibrolamellar bone with peripheral parallel-fibered bone, EFS present in largest individuals, periodic rapid growth with slowing in later ontogeny and eventual cessation or near-cessation of appositional bone growth. Histological data taken from Botha and Angielczyk (2007); Botha-Brink and Angielczyk (2010); Green, Schweitzer, and Lamm (2010); Ray and Chinsamy (2004); Ray, Chinsamy, and Bandyopadhyay (2005), Ray, Bandyopadhyay, and Bhawal (2009). Average SL taken from Camp and Welles (1956); Bandyopadhyay (1988); King (1988, 1990); Ray (2000); Ray and Chinsamy (2003); Botha and Angielczyk (2007).

that a complete or near-complete cessation of appositional growth may have evolved in larger, more derived dicynodonts, such as *Placerias*. The presence of an EFS in other derived, large-bodied taxa from the Late Triassic (e.g., *Ischigualastia, Jachaleria*) would support the evolution of determinate growth in the latest and largest dicynodonts. Alternatively, it is also possible there was a sampling bias in previous studies of Permian dicynodonts (i.e., adult specimens but not the largest individuals may have been sampled). Sampling of the largest individuals in more basal taxa would confirm whether an EFS was present or not and further resolve the evolutionary trends noted here.

Placerias also differs from other dicynodonts through other histological traits beyond the microstructure of primary bone. For example, extensive secondary remodeling was observed in the outer cortex in all sampled elements of *Placerias* (Green, Schweitzer, and Lamm 2010). With the exception of *Kannemeyeria* (Botha-Brink and Angielczyk 2010), other dicynodonts usually lack dense Haversian tissue in the outer cortex (Chinsamy and Rubidge 1993; Ray, Chinsamy, and Bandyopadhyay 2005, Ray, Bandyopadhyay, and Bhawal 2009; Botha and Angielczyk 2007). Extensive remodeling of the cortex in *Placerias* may be attributed to age, slowed appositional growth relative to rate of remodeling, and/or large body size, as well as possible biomechanical factors (see Green, Schweitzer, and Lamm 2010 for further discussion). The medullary cavity of all sampled limb bones is completely filled with bony trabeculae (Green, Schweitzer, and Lamm 2010), which possibly supports an amphibious or aquatic lifestyle for *Placerias* (similar to that hypothesized for *Lystrosaurus* based on histological data [Germain and Laurin 2005; Ray, Chinsamy, and Bandyopadhyay 2005]). However, the depositional environment of Placerias Quarry (Fiorillo, Padian, and Musikasinthorn 2000) is not consistent with this interpretation. Green, Schweitzer, and

Lamm (2010) suggested that infilling of the medullary cavity in *Placerias* may be more phylogenetically than ecologically informative, yet this hypothesis remains to be tested in an evolutionary context.

Current Study: Comparison of Bone Microstructure in Late Triassic Dicynodonts

The analysis of Green, Schweitzer, and Lamm (2010) illuminated new, perplexing questions about dicynodont growth: Is the differential pattern seen in *Placerias* from the Placerias Quarry related to taxonomic, gender, or environmental differences? Why is secondary remodeling so extensive in cortical bone of *Placerias* relative to most other dicynodonts? These questions will only be answered through continued study of bone histology in Late Triassic dicynodonts other than *Placerias*. To help address this need, this study describes growth (deduced from bone microstructure) in a previously unsampled non-*Placerias* dicynodont from the Late Triassic of North Carolina. Results are directly compared with bone microstructure in *Placerias* (Green, Schweitzer, and Lamm 2010) to further elucidate growth patterns in Triassic dicynodonts.

Specimens and Sampling

The distal mid-shaft of the right tibia (Fig. 7.1) of NCSM 21719 (unnamed dicynodont species) was sectioned for histological analysis. This specimen may represent a holotype, so a partial transverse section that includes only anterior, posterior, and medial aspects was removed; additional elements were not available for destructive sampling. NCSM 21719 was recovered from NCSM locality NCPALEO1902 (Deep River Basin, Pekin Formation, Upper Triassic [Reinemund 1955]) in the southeastern corner of Chatham County, North Carolina. For comparative purposes, a transverse diaphyseal section from a *Placerias* tibia of similar size (UCMP 24864) from the Placerias Quarry, Arizona (Chinle Formation, Upper Triassic; UCMP locality A269) was also analyzed. These tibiae have similar dimensions at the diaphysis (NCSM 21719: mediolateral diameter at the mid-diaphysis of long bones [MLD] = 34.21 mm; anteroposterior diameter at the mid-diaphysis of long bones [APD] = 39.84 mm; UCMP 24864: MLD = 34.84 mm; APD = 40.32 mm). Specimens were measured, molded, and cast prior to destructive sampling. Thin sections were prepared using standard histological techniques (chapter 3 of this book). Transverse sections were examined for fiber arrangement, primary osteonal canal orientation, and presence/absence of growth lines in anterior, posterior, and medial aspect. Histological terminology and definitions follow Francillon-Vieillot et al. (1990). Standard measurement of relative bone wall thickness and porosity was not possible in either section because extensive secondary remodeling obscured the boundary between the medullary cavity and cortex (Ray, Chinsamy, and Bandyopadhyay 2005; Green, Schweitzer, and Lamm 2010). The degree of secondary remodeling in each specimen was estimated using the methods of Green,

Schweitzer, and Lamm (2010). The thickness of primary bone was first measured (to nearest 0.01 mm) from the periosteal surface until dense Haversian bone appeared in four regions (A = anterior; B = posterior; C = medial; D = lateral). Total APD (E) and MLD (F) were also measured. Percent secondary remodeling was calculated for each transect, X and Y, and then total remodeling for the whole section using the following equations:

$X = (1-[(A+B)/E])\times 100$

$Y = (1-[(C+D)/F])\times 100.$

Total secondary remodeling = $(X)[E/(E+F)] + (Y)[F/(E+F)]$

The lateral region of the tibia in NCSM 21719 was not sectioned, thus precluding mediolateral measurements of primary bone thickness; calculations of total percent secondary remodeling for this specimen are therefore based only on anteroposterior data (see Green, Schweitzer, and Lamm 2010). The thin section from UCMP 24864 was deposited in the collections at UCMP, whereas the section from NCSM 21719 was deposited at NCSM.

Histological Description and Comparison

In transverse section, both the *Placerias* specimen, UCMP 24864 (Fig. 7.3), and NCSM 21719 (Fig. 7.4) have a compact cortex surrounding a large medullary region; the central cavity is entirely filled with trabeculae, which grades into dense Haversian bone with enlarged resorption cavities in the perimedullary region. The inner cortex is completely remodeled in both specimens (Figs. 7.3, 7.4). Secondary remodeling is confined to the deep cortex in anterior and posterior view in NCSM 21719 (Figs. 7.3A, 7.3C) and extends further in medial aspect (73.49 percent total remodeling) (Fig. 7.3B, Table 7.1). In the *Placerias* specimen, dense Haversian tissue extends radially to the outer cortex (98.52 percent total remodeling [Green, Schweitzer, and Lamm 2010]) (Table 7.1, Fig. 7.4). Primary bone in the outer cortex of NCSM 21719 consists of zonal fibrolamellar bone with primarily longitudinal and radial canals in both anterior and medial view (Figs. 7.3A, 7.3C); highly vascularized bone tissue arranged in a subplexiform network is present in posterior aspect (Fig. 7.3B). Where visible between patches of remodeling, the primary tissue in the outer cortex of the *Placerias* specimen reveals poorly vascularized zonal fibrolamellar bone (Fig. 7.4); a peripheral layer of avascular parallel-fibered bone is present in anterior and posterior view (Figs. 7.4A, 7.4B), yet is not visible in medial aspect. Bands of poorly vascularized/avascular bone are present in both specimens. Although it was not possible to trace these lines around the circumference of the bone, either because of heavy cortical remodeling (*Placerias*) or incompleteness of the section (NCSM 21719), they closely resemble growth lines that are common in other dicynodonts (see, e.g., Chinsamy and Rubidge 1993; Ray and Chinsamy 2004; Ray, Chinsamy, and Bandyopadhyay 2005, Ray, Bandyopadhyay, and Bhawal

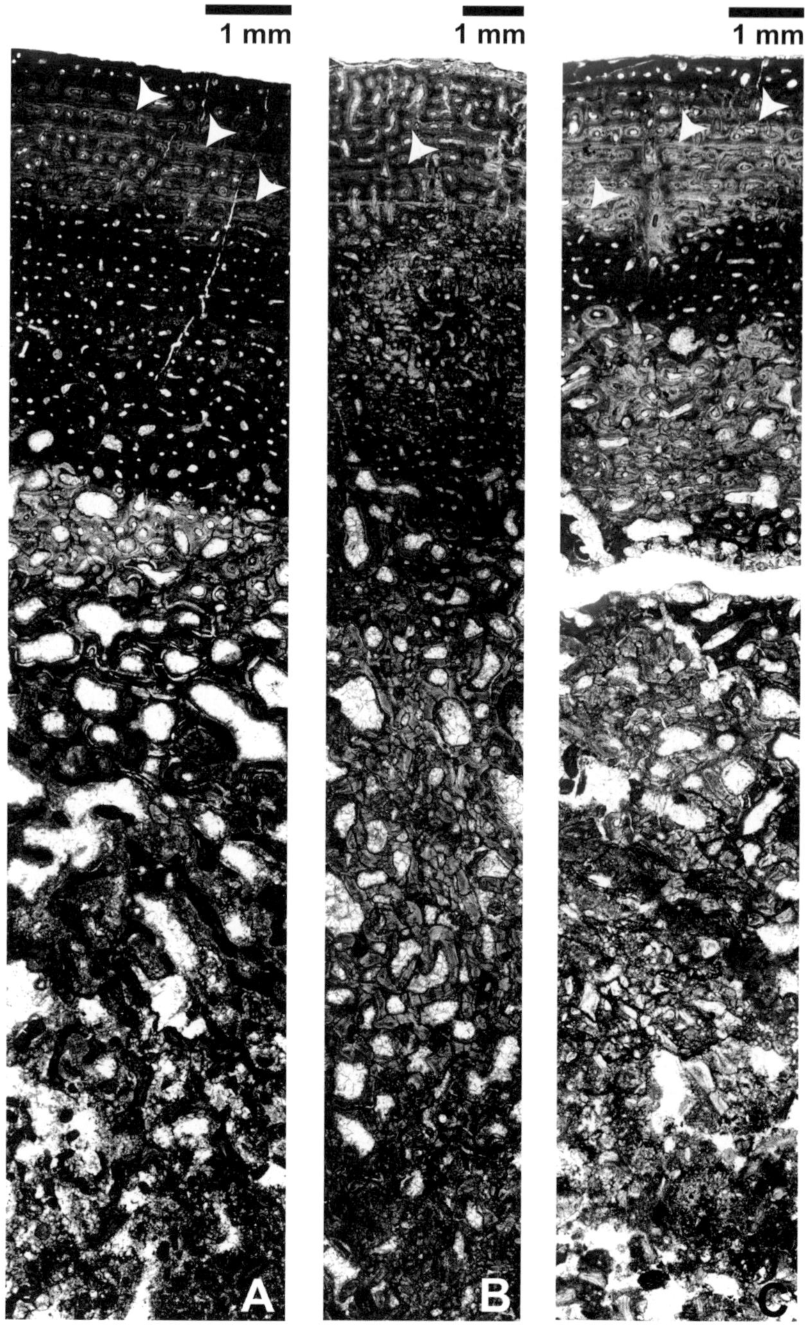

7.3. Low-magnification transverse sections showing microstructure of the right tibia of NCSM 21719. Primary tissue in the cortex is generally vascularized zonal fibrolamellar bone, grading to dense Haversian bone in the deep cortex, and eventually to an extensive network of bony trabeculae filling the medullary cavity. Dark areas in the mid-cortex likely represent diagenetic mineral precipitation. (A) Anterior aspect with three annuli (arrowheads). (B) Posterior region with sub-plexiform cortical tissue and one visible annuli (arrowhead). (C) Medial region with three annuli (arrowheads).

2009; Botha and Angielczyk 2007). As such, these features are here interpreted as annuli; no lines of arrested growth were visible. Two annuli (arrowheads) are present in the outer cortex of the *Placerias* specimen between areas of heavy remodeling (Fig. 7.4C), and three total annuli (arrowheads) are visible in NCSM 21719 (Figs. 7.3A, 7.3C).

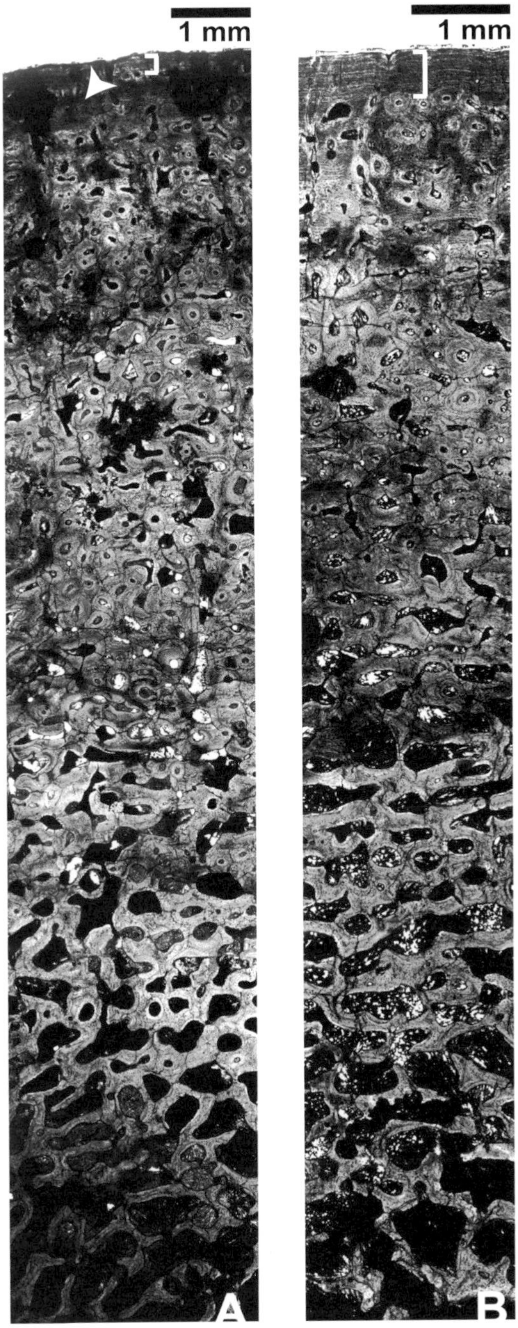

7.4. Low-magnification transverse sections showing microstructure of a *Placerias* tibia, UCMP 24864, from the Placerias Quarry (Upper Triassic), Arizona. Primary cortical bone is generally poorly vascularized parallel-fibered bone on the periphery grading to zonal fibrolamellar bone in the outer cortex. The inner-mid cortex is completely remodeled, transitioning into an extensive network of bony trabeculae filling the medullary region. (A) Anterior aspect showing peripheral parallel-fibered bone (bracket) and one visible annuli (arrowhead). (B) Posterior region showing a well-defined peripheral layer of parallel-fibered bone. (C) Medial aspect with zonal fibrolamellar bone bone and two annuli (arrowheads) visible amid regions of heavy remodeling.

GROWTH PATTERNS

The primary cortical bone in the tibia of NCSM 21719 is generally zonal fibrolamellar with annuli visible (Fig. 7.3), indicating rapid growth with periodic interruptions. This is similar to the primary tissue seen in the inner cortex of the *Placerias* specimen (Fig. 7.4), which is typical of dicynodonts (Gross 1934; de Ricqlès 1972; Chinsamy and Rubidge 1993;

Table 7.1. Measurements used to calculate total percentage of secondary remodeling (in bold) for sampled dicynodont tibia

Specimen#	MLD	MT	LT	%ML	APD	AT	PT	%AP	Total
NCSM 21719	—	—	—	—	36.78	3.61	6.14	73.49	**73.49**
UCMP 24864	40.32	0.48	0.12	98.51	34.84	0.30	0.21	98.54	**98.52**

Note: See Green, Schweitzer, and Lamm (2010) for equations; data for UCMP 24864 is taken from Green, Schweitzer, and Lamm (2010).
Abbreviations: MLD = mediolateral diameter of sectioned specimen; MT = thickness of primary bone in medial aspect; LT = thickness of primary bone in lateral aspect; %ML = percent of remodeling in mediolateral transect; APD = anteroposterior diameter of sectioned specimen; AT = thickness of primary bone in anterior aspect; PT = posterior thickness of primary bone in posterior aspect; %AP = percent of remodeling in anteroposterior transect; Total = total percent remodeling for the sectioned specimen.

Botha 2003; Ray and Chinsamy 2004; Germain and Laurin 2005; Ray, Chinsamy, and Bandyopadhyay 2005, Ray, Bandyopadhyay, and Bhawal 2009; Botha and Angielczyk 2007; Kriloff et al. 2008; Green, Schweitzer, and Lamm 2010; Botha-Brink and Angielczyk 2010) and fits within the overall framework of rapid growth in non-mammalian therapsids (de Ricqlès 1972, 1974a, 1976; Botha and Chinsamy 2000, 2004, 2005; Ray, Botha, and Chinsamy 2004; Ray, Bandyopadhyay, and Bhawal 2009). However, NCSM 21719 differs from the *Placerias* specimen in the absence of a peripheral, poorly vascularized layer of parallel-fibered bone (Figs. 7.3, 7.4A, 7.4B). This supports more rapid growth, prior to death, in NCSM 21719 relative to the *Placerias* specimen. There are several possible explanations for this difference in growth.

First, assuming both individuals have similar ontogenetic pathways, it is possible that NCSM 21719 is a younger individual than the *Placerias* specimen. However, there is not a complete ontogenetic series available for either *Placerias* (Green, Schweitzer, and Lamm 2010) or the unnamed species. The lack of peripheral slow-growing bone and the evolutionary trend toward slowed growth during later development in Triassic dicynodonts, first reported by Ray, Bandyopadhyay, and Bhawal (2009) and corroborated by Green, Schweitzer, and Lamm (2010), support the supposition that NCSM 21719 was a younger animal. Peripheral parallel-fibered bone has been proposed to correspond to the onset of sexual maturity in dicynodonts (e.g., Ray, Chinsamy, and Bandyopadhyay 2005; Ray, Bandyopadhyay, and Bhawal 2009). Under this potential explanation, NCSM 21719 may not have reached sexual maturity when death occurred. Unfortunately, there is no comparative material available to help assess the exact ontogenetic stage of NCSM 21719 to further test these hypotheses at this time.

Second, it is also possible that NCSM 21719 represents a faster growing taxon than *Placerias* (i.e., at a similar ontogenetic stage, this animal was maintaining more active growth relative to *Placerias*). Third, it is plausible that differences in the surrounding environment could account for the observed variation in growth (i.e., *Placerias* and NCSM 21719 have similar ontogenetic pathways, yet growth in *Placerias* was influenced by harsh environmental conditions at the Placerias Quarry [Fiorillo, Padian, and Musikasinthorn 2000]). It is further possible that a combination of these influences may account for the variation described here. These contrasting hypotheses need to be tested in an ontogenetic framework, which is not currently available for taxa sampled here.

The limited histological data currently available preclude additional explanations of the observations made here. Further analyses of growth in other Late Triassic dicynodonts, using a complete ontogenetic series, will help resolve these matters. Growth patterns deduced from bone microstructure should ideally be based on examination of multiple sections from different elements (Horner, de Ricqlès, and Padian 2000). However, only the right tibia of NCSM 21719 was available for sampling for this study. Future histological analysis of other limb elements (e.g., femur, fibula) from this specimen may also affect the conclusions drawn here.

SECONDARY REMODELING AND INFILLING OF THE MEDULLARY CAVITY

Secondary remodeling in NCSM 21719 is most similar to patterns seen in *Lystrosaurus* and *Wadiasaurus*, where dense Haversian tissue usually extends only to the inner-mid cortex (Ray, Chinsamy, and Bandyopadhyay 2005; Ray, Bandyopadhyay, and Bhawal 2009). In contrast, extensive remodeling of the outer cortex is visible in the *Placerias* specimen (Fig. 7.4) and is common in limb bones of *Placerias* in general (Green, Schweitzer, and Lamm 2010). *Kannemeyeria* also exhibits heavy cortical remodeling in some elements (Botha-Brink and Angielczyk 2010), although not with the same intensity as *Placerias*. Differences in remodeling between specimens sampled here further support the extreme nature of secondary reconstruction in *Placerias* relative to most other dicynodonts (Green, Schweitzer, and Lamm 2010). Extensive cortical remodeling in *Placerias* may be attributed to age, slowed appositional growth relative to rate of remodeling, large body size, as well as biomechanical factors, and/or a combination of these variables (see Green, Schweitzer, and Lamm [2010] for further discussion). Again, there is no growth series available to test these possible explanations and further histological analysis in similar taxa (in terms of body size and temporal distribution) is needed to understand this disparity.

Although there are differences in the extent of remodeling in the cortex, *Placerias* (Green, Schweitzer, and Lamm 2010) and the dicynodont represented by NCSM 21719 are similar in that they show a gradual transition from dense Haversian tissue in the inner compacta to an extensive network of spongiosa that fills the medullary cavity (Figs. 7.3, 7.4). This microstructure is usually found in animals with aquatic adaptations (Wall 1983; Ray, Chinsamy, and Bandyopadhyay 2005). However, Green, Schweitzer, and Lamm (2010) did not use this data to draw lifestyle adaptations for *Placerias* for two reasons: (1) the depositional environment of the Placerias Quarry is not consistent with an aquatic environment (Fiorillo, Padian, and Musikasinthorn 2000); and (2) the potential phylogenetic effects or signal for this pattern have not yet been established. Extensive spongiosa in the medullary cavity are present in most dicynodonts, particularly those of large size (Botha-Brink and Angielczyk 2010). The shared presence of this feature in related animals may represent more of a phylogenetic signal than an ecological one (e.g., Laurin, Girondot, and Loth 2004), although there is additional evidence for at least a semi-aquatic lifestyle in some dicynodonts (e.g., *Lystrosaurus* [Ray, Chinsamy, and Bandyopadhyay 2005]). A detailed analysis that statistically tests for a phylogenetic signal with regard to a filled medullary cavity is needed to better explain these patterns.

IMPLICATIONS FOR GROWTH PATTERNS IN LATE TRIASSIC DICYNODONTS

Microstructure of a tibia from an unnamed species of Late Triassic dicynodont from North Carolina (NCSM 21719) suggests that this animal was growing rapidly at the time of its death. The lack of peripheral slow-growing tissue in this specimen, common in other Triassic dicynodonts (i.e., *Lystrosaurus, Wadiasaurus, Placerias*), suggests NCSM 21719 represents a young individual. Alternatively, absence of this tissue may also reflect differences in growth rate or external environmental influences between this taxon and other Triassic dicynodonts. Secondary remodeling is more extensive in *Placerias* than NCSM 21719, which may also be explained by ontogeny and the lack of slowed appositional growth. Both taxa have a medullary cavity filled with bony trabeculae, a trait that may be more phylogenetically than ecologically informative in dicynodonts. The small amount of data currently available and the lack of a complete ontogenetic series preclude further discussion about growth patterns in Late Triassic dicynodonts from North America. Additional histological study of other large-bodied, Late Triassic dicynodonts (e.g., *Ischigualastia, Jachaleria*) may provide the comparative framework needed to address the questions posed here.

Current Study: Incremental Growth Lines in Dicynodont Tusks from North Carolina

Besides bone, growth is also recorded in other biomineralized elements, such as teeth. Dicynodonts are named for a prominent pair of enlarged, ever-growing upper caniniform teeth, or tusks (King 1988), that are composed primarily of dentine (Poole 1956). Tusks were not replaced during ontogeny (Sullivan, Reisz, and Smith 2003), and because dentine was deposited continuously throughout life via biomineralization but was not subject to remodeling (Poole 1956), dicynodont tusks may record a more complete growth history than their bones. Among vertebrates, incremental growth lines in dentine are visible in ground sections as alternating light and dark bands (Fox 2000). Lines of von Ebner are the finest incremental features (30 μm) corresponding to higher periodicities have been noted in the dentine of some mammals (e.g., Fisher 1996). In proboscidean tusks, a three-order hierarchy of incremental growth lines is well established for living (*Loxodonta, Elephas*) and fossil taxa (*Mammut, Mammuthus, Gomphotherium* [Fisher 1987, 1988, 1990;

Fisher et al. 1998; Koch, Fisher, and Dettman 1989; Fox 2000]). These consist of annual "first-order" increments, fortnight to lunar monthly "second-order" increments, and daily (von Ebner) "third-order" increments. Annual and bi-weekly increments are also noted in the dentine of other mammalian groups, including pinnipeds (Mansfield 1958; Laws 1962), odontocetes (Hohn 1980), and sirenians (Marsh 1980). Changes in thickness between incremental lines of the same periodicity are a function of the depositional rate of dentine (Fisher 1996), and fluctuations in this rate are influenced by changes in the external environment (Fisher 1996; Fox 2000). Plots of incremental thickness as a function of time (i.e., youngest to oldest increments) have been used in fossil proboscidean tusks to extrapolate seasonality in growth and can function as an indicator of climatic fluctuations in deep time (Fisher 1996; Fox 2000).

Dicynodont tusks (see Plate F) resemble those of proboscideans in their composition, their macroscopic "cone-in-cone" growth pattern, and in having microscopic growth features consisting of alternating light and dark bands in the dentine (Thackeray 1991; Sullivan, Reisz, and Smith 2003; Thulborn and Turner 2003). Thackeray (1991) observed a statistically significant decrease in average thickness of incremental lines in *Diictodon* tusks during the Late Permian, which may reflect changes in environmental conditions. However, our understanding of tusk growth lines in Triassic dicynodonts is poor, and no study to date has established the periodicity of tusk growth increments or analyzed the potential of incremental thickness profiles as a function of environmental fluctuations during this period. Here, this gap in knowledge will be partially bridged through description and analysis of the thickness and potential periodicity of growth features in Late Triassic dicynodont tusks from North Carolina.

Specimens and Sampling

For histological analysis, a partial longitudinal section was taken from two isolated dicynodont (sp. indet.) tusks (NCSM 19585, 21735) from NCSM locality NCPALEO1902 (Deep River Basin, Pekin Formation, Upper Triassic [Reinemund 1955]). Specimens were measured, molded, and cast prior to destructive sampling. For each specimen, an approximate 2.5-cm long portion was removed from the proximal end of each tusk. Longitudinal sections (with respect to the longitudinal plane of the skull) were prepared from this portion, using standard histological techniques (chapter 3 of this book). Growth increments were analyzed at 100× magnification under transmitted light. A single increment was defined as the distance between successive dark bands. However, increments were not continuously visible across the entire section, making it difficult to directly measure the individual thickness of all lines in the tusk. Instead, I calculated mean growth increments (MGIs [sensu Thackeray 1991]) for different regions of each tusk and used this data as an estimate of growth line thickness per specimen. First, a series of digital images were taken along a transect from the internal pulp cavity to the external surface of

Table 7.2. Data for average thickness of growth increments in dicynodont tusks (sp. indet.) sampled in this study

Specimen #	MGI #	Line count	MGI (μm)
NCSM 19585	1	6	16.67
	2	7	14.29
	3	8	12.50
	4	7	14.29
	5	7	14.29
	6	5	20.00
	7	3	33.33
	8	4	25.00
	9	8	12.50
	10	4	25.00
Mean (st. dev.)			**18.79 (6.95)**
NCSM 21735	1	3	33.33
	2	2	50.00
	3	4	25.00
	4	2	50.00
	5	3	33.33
	6	5	20.00
	7	5	20.00
	8	3	33.33
	9	4	25.00
	10	4	25.00
Mean (st. dev.)			**31.50 (10.98)**

the tusk. Images were captured using an AxioCam MRc5 camera (Carl Zeiss AG, Oberkochen, Germany) and viewed using AxioVision (version 4.6.3) software (Carl Zeiss). Ten MGIs were calculated along this transect in areas where a successive series of growth lines were clearly visible. For each MGI, the number of visible growth increments was counted in a 100-μm moving window (positioned perpendicular to the orientation of growth lines); this number was then divided into 100 to yield the MCI for that region. For example, if five growth lines were visible in a 100-μm increment, then the MGI for that region was 20 μm. Growth lines did not overlap between measured MGIs along the transect. Then MGIs were averaged together for each specimen to yield an estimate of total mean thickness of growth lines.

Thickness and Description of Tusk Growth Increments

Growth lines in the dentine of tusks sampled here are visible as alternating dark and light bands oriented perpendicular to odontoblastic tubules (Fig. 7.5). On average, growth lines in NCSM 19585 were thinner (18.79 μm) (Table 7.2) than those in NCSM 21735 (31.50 μm) (Table 7.2). Based on relative appearance, there are two broad types of growth increments visible in both specimens: bold lines (black bands in Fig. 7.5) and faint lines (gray bands in Figs. 7.5A, 7.5B, and 7.5D). Both bold and faint lines are clearly visible in NCSM 19585 (Fig. 7.5A). However, bold features are more prevalent (Figs. 7.5B to 7.5D) than thin lines in NCSM 21735,

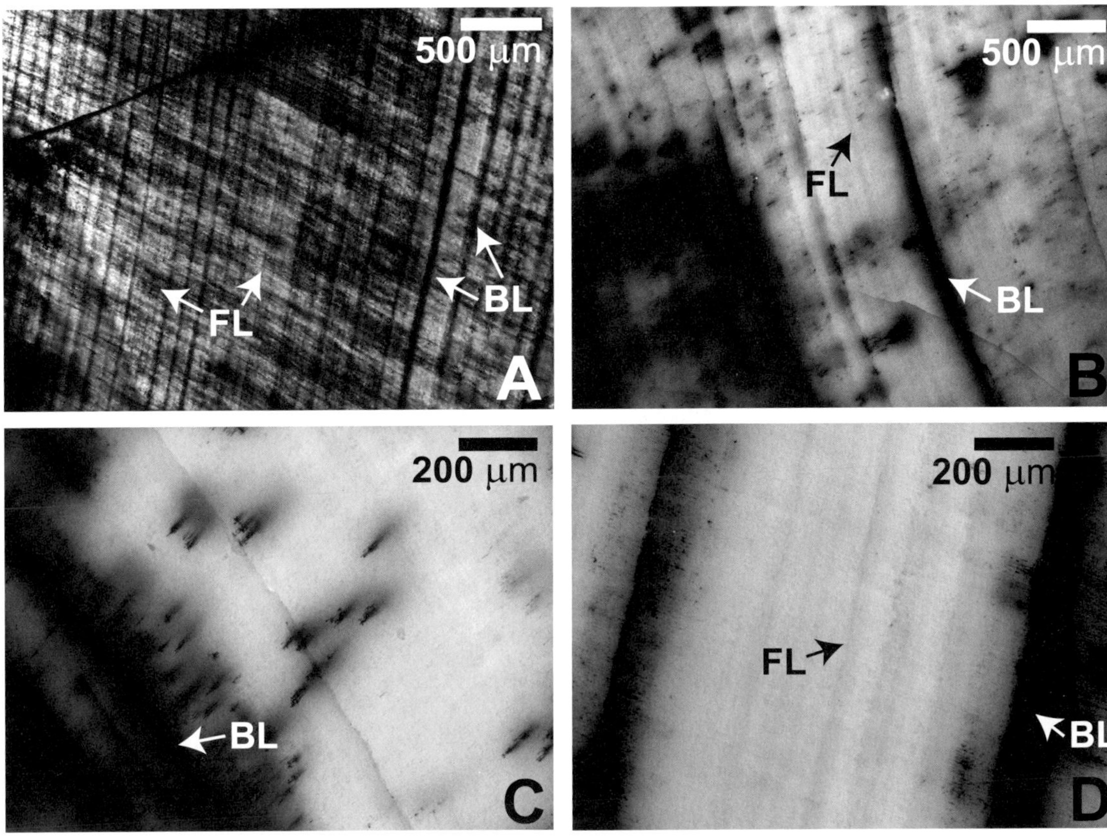

7.5. Longitudinal sections of dicynodont tusks (sp. indet.) from the Pekin Formation (Upper Triassic), North Carolina. Growth increments in tusk dentine are visible as alternating light-dark bands. Both bold lines and faint lines (defined in the text) are visible in both specimens. Regions of widespread dark material likely represent diagenetic mineral precipitation. (A) Low-magnification image, NCSM 19585, showing bold and faint growth lines visible across the field of view. (B) Low-magnification image, NCSM 21735, with bold and faint growth lines visible. (C) High-magnification detail of NCSM 21735, two bold lines are visible with no clear faint lines. (D) High-magnification detail of NCSM 21735, at least twelve faint lines are visible between two successive bold lines. *Abbreviations:* BL = bold lines; FL = faint lines.

as the latter structures are clear in some areas (Fig. 7.5D) and not visible in other regions (Fig. 7.5C). Variation in the appearance of growth increments and the relative absence of some lines in NCSM 21735 may be related to preservation.

Periodicity of Growth Increments

Overall thickness of MGIs in NCSM 19585 is consistent with the thickness of daily von Ebner lines found in the dentine of living vertebrates (Schour and Hoffman 1939a, 1939b; Erickson 1996a, 1996b). Therefore, these lines likely represent a diurnal periodicity. Growth increment thickness in *Diictodon* tusks (~20 µm [Thackeray 1991]) matches that of daily growth lines in living taxa, and von Ebner lines have also been detailed in a tusk of *Lystrosaurus* (chapter 6 of this book). However, MGIs in NCSM 21735 are thicker on average than the maximum rate of daily dentine deposition in vertebrates (30 µm [Erickson 1996a, 1996b]), and thickness in individual MGIs reached as high as 50 µm in this specimen (Table 7.2). These values do not support a daily periodicity in this tusk. Von Ebner lines are common in the dentine of living vertebrates and are also present in NCSM 19585, so they are expected to be present in this specimen. There are several potential explanations for their perceived absence.

First, it is possible that growth lines in NCSM 21735 were deposited under an overall higher periodicity (e.g., weekly, bi-weekly, monthly) than in NCSM 19585. In addition, it is possible that bold and faint lines observed here were deposited under different periodicities, which would correspond to a hierarchical pattern, similar to proboscidean tusks (e.g., Fisher 1996). In the latter animals, a clear and repeated pattern of fine-scale, daily features are consistently visible within coarser, weekly-monthly increments (Fisher 1996; Fox 2000). Indeed, there are regions of NCSM 21735 that portray a similar organized structure (e.g., Fig. 7.5D, where at least twelve faint lines are visible between bold lines). Thus, faint lines may represent daily increments, whereas the bold lines were deposited fortnightly. However, the pattern visible in Figure 7.5D is not consistently repeated in this specimen, plus a hierarchical relationship of growth increments is not visible in NCSM 19585; these observations refute this explanation. Alternatively, it is possible that not all growth lines were preserved in NCSM 21735. In this case, the bold, thicker lines may have been preferentially preserved and more of the finer lines were removed during fossilization (e.g., Fig. 7.5C), thus creating the illusion of thicker growth increments overall. The final possibility is the simplest explanation for the observations made here, and therefore, assignment of a specific periodicity to growth increments in this specimen was not undertaken.

Implications for Dicynodont Growth and Paleoclimate during the Late Triassic

Growth increments in NCSM 19585 were likely deposited under a diurnal periodicity, based on comparisons of increment thickness with living animals. However, growth increments were not uniformly preserved in the other tusk sampled here (NCSM 21735), which precludes an estimate of periodicity in this specimen and a comparison of relative growth using MGIs between tusks. Ideally, other Late Triassic dicynodont tusks should be analyzed to provide a more robust understanding of the periodicity of tusk growth features. In addition, growth lines in fibrolamellar bone represent an annual period of slowing or cessation of rapid osteogenesis in dicynodonts (e.g., Ray and Chinsamy 2004; Ray, Chinsamy, and Bandyopadhyay 2005; Ray, Bandyopadhyay, and Bhawal 2009). By comparing the numbers of growth lines preserved in the tusk and associated limb bone material, it may be possible to calibrate the periodicity of incremental lines in dicynodont tusks. This would also expand our understanding of differential growth rates and patterns between skeletal elements in dicynodonts and other non-mammalian therapsids.

Where growth lines of a consistent periodicity can be confirmed in a tusk (e.g., NCSM 19585), it should be possible to reconstruct a thickness profile of growth increments in the entire specimen to provide a measure of growth history. This in turn has the potential to reveal any cyclic patterns in growth that may correspond to environmental and/or seasonal influences on growth. The paleoclimate of Pangaea during the Late Triassic has been reconstructed as megamonsoonal, involving extreme seasonality of dry and wet cycles (Simms and Ruffell 1990; Parrish 1993; Fraser 2006). Because the rate of dentine deposition is influenced by changes in the external environment (see, e.g., Fisher 1996; Fox 2000), analysis of incremental thickness profiles in dicynodont tusks may provide support for extreme seasonality during the early Mesozoic and further our overall understanding of paleoclimate during the rifting phase of Pangaea.

Concluding Remarks

Late Triassic dicynodonts represent the last yet largest and most derived members of an extremely diverse and widespread clade of extinct terrestrial herbivores. In North America, *Placerias* is the most common dicynodont recovered from Late Triassic deposits, although new and forthcoming data suggest a much more diverse dicynodont assemblage during this time period than previously thought. Our knowledge of growth in Late Triassic dicynodonts is currently based on extensive histological sampling of limb elements in *Placerias* and a single section from the tibia of an unnamed, non-*Placerias* taxon (represented by NCSM 21719) from North Carolina. However, a complete ontogenetic series is lacking for both taxa, which limits the conclusions drawn from histology. Bone microstructure in Late Triassic dicynodonts sampled here supports periodic rapid growth in these animals, which is similar to other sampled dicynodonts from Permian and Early to Middle Triassic times. Peripheral parallel-fibered bone is present in *Placerias* and supports an evolutionary trend toward slowed growth during later development in Triassic dicynodonts. *Placerias* may also have had a determinate growth strategy, as suggested by the presence of an EFS in the largest known tibia of this taxon; this is in contrast to indeterminate growth seen in other dicynodonts, although there may be a sampling bias, as the largest adults may not have been sampled in previous studies. The lack of slow-growing peripheral tissue in the tibia of NCSM 21719 suggests that this animal may be a young individual, although it is also possible that differences in growth rate between taxa and/or environmental influences explain this result. Relative to other dicynodonts (excluding *Kannemeyeria*), secondary remodeling is abnormally high in limb bones of *Placerias*. The reason for more extensive remodeling in *Placerias* compared with most other dicynodonts is currently unresolved. Late Triassic dicynodonts sampled here also have a medullary region in limb elements that is completely filled with bony trabeculae, which is an adaptation seen in aquatic taxa. However, this feature is found in most dicynodonts, which suggests more of a phylogenetic signal than an ecological one.

Dicynodont tusks are biomineralized structures that were ever-growing without replacement and, therefore, may provide a more complete record of growth than bones. However, this potential record of growth first requires an understanding of the periodicity under which growth lines in dicynodont tusks were deposited. Based on comparison of dental growth line thickness in extant vertebrates, growth increments in the dentine of a tusk (NCSM 19585) from the Late Triassic of North America were deposited on a daily scale. This result is promising because it provides a measure of relative growth rate in dicynodonts. However, incomplete preservation of growth lines may be problematic in some specimens (e.g., NCSM 21735). Where growth line periodicity can be confirmed in a dicynodont tusk, a profile of increment thickness throughout the entire specimen may serve as a record of growth rate throughout an individual's life (similar to studies on growth in proboscidean tusks). This data can then be analyzed for fluctuations in tusk growth that may correspond to seasonal climatic conditions or other environmental influences, which may help us to better understand the paleoclimate and environment of Late Triassic North America.

The limited conclusions of this chapter highlight the incredible amount of work that remains to be completed in this field. Our understanding of life history in Late Triassic dicynodonts is growing, but the histology of bones and tusks in these animals needs to be further studied in an ontogenetic context (i.e., how growth changes in a complete growth series), an individual context (i.e., compare-contrast growth in associated skeletal elements), a broad evolutionary context (i.e., how osteogenesis in derived dicynodonts compares not only with more basal taxa, but also other non-mammalian therapsids), and a developmental plasticity context (i.e., how growth patterns relate to changes and fluctuations in the external environment). The methods and preliminary results presented here serve as a platform for this future research. Continued study in this area will provide a clearer picture of the evolution of growth patterns across Dicynodontia and further broaden our view of the biological adaptations and strategies of non-mammalian therapsids during the early Mesozoic.

Acknowledgments

I thank UCMP (K. Padian, P. Holroyd, M. Goodwin) and NCSM (V. Schneider, M. Schweitzer) for allowing destructive sampling of dicynodont material. E. Lamm assisted with preparation of thin sections, and K. Padian and J. Horner provided comments on histology. I also thank J. Botha-Brink, S. Jasinoski, A. Chinsamy-Turan, and E. Schroeter for providing constructive comments on an earlier draft of this manuscript. Partial funding was provided by National Science Foundation (EAR-0541744 and EAR-0548847, to M. Schweitzer), UCMP Doris O. and Samuel P. Welles Research Fund, and Sigma Xi Grant-in-Aid of Research.

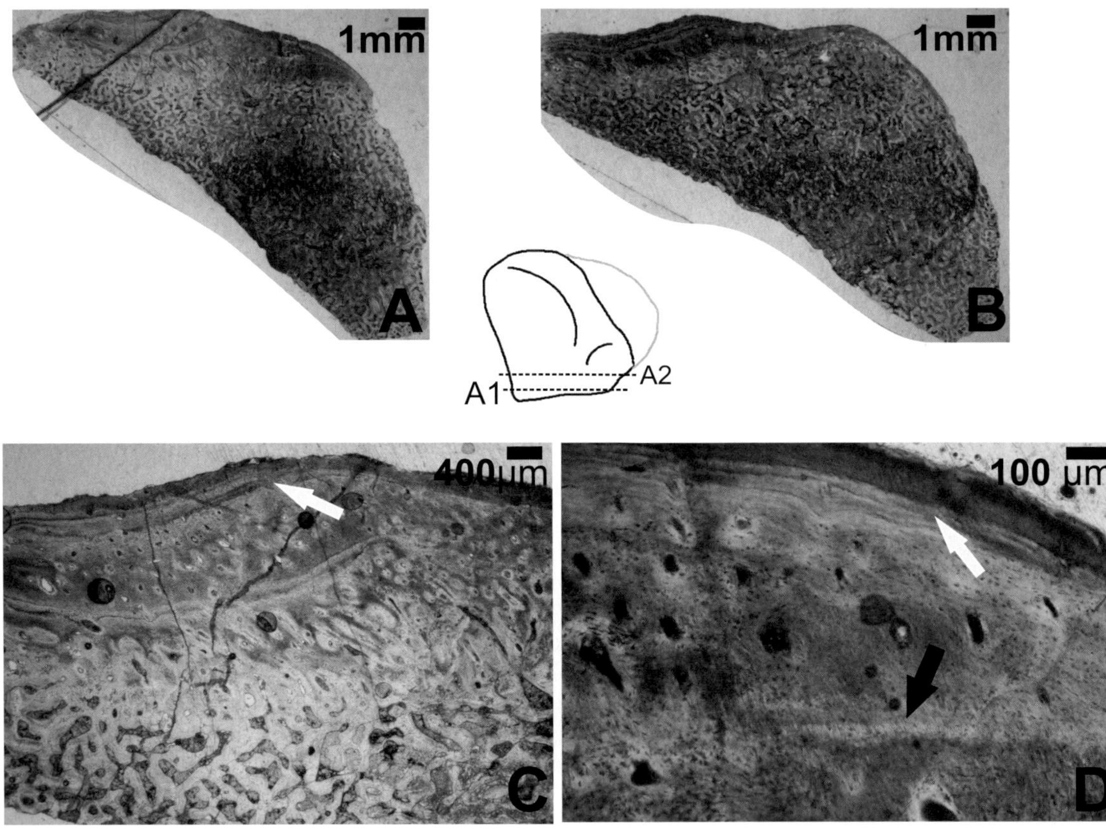

8.1. *Scylacops* humerus, SAM-PK-10188. (A,B) Shows the location and overall view of the two thin sections (A1, left, and A2, right). (C) Higher magnification of compact bone wall of A1. Arrow indicates peripheral rest lines. (D) Higher magnification of A2. The white arrow indicates peripheral rest lines, whereas the black arrows indicate Sharpey's fibers.

Bone Histology of Some Therocephalians and Gorgonopsians, and Evidence of Bone Degradation by Fungi

8

Anusuya Chinsamy-Turan and Sanghamitra Ray

Introduction

The Theriodontia are the carnivorous nonmammalian therapsids which include the Gorgonopsia, Therocephalia, and Cynodontia (the latter including mammals as a derived subgroup). With respect to crown Mammalia, the Gorgonopsia are considered to be the most plesiomorphic group of theriodonts, although they share several features with the early therocephalians (some of which may be considered convergences by some researchers [e.g., Hopson and Barghusen 1986]). Since at least the late 1930s, the Therocephalia were considered to be closely related to the cynodonts (see, e.g., Brink 1965), and over the years, there has been much debate about the specific nature of their relationship to the cynodonts. Whereas some workers have suggested a direct ancestral relationship between at least some therocephalians and cynodonts (Brink 1960, 1965; Kemp 1972; Botha, Abdala, and Smith 2007), others have advocated with increasingly robust support a sister-group relationship between the Therocephalia and Cynodontia (Romer 1969; Hopson and Barghusen 1986; Huttenlocker 2009; Huttenlocker, Sidor, and Smith 2011) in which the two form a monophyletic clade: the Eutheriodontia. Contention in their interrelationships has likely been a result of high levels of homoplasy and ecological convergence exhibited by theriodonts, among other explanations.

Although a considerable literature exists on the taxonomy and systematics of the Theriodontia (e.g., Boonstra 1964; van den Heever 1980; Hopson and Barghusen 1986; Sigogneau-Russell 1989), there is comparatively little research on their paleobiology, especially that of the gorgonopsians and therocephalians. Studies of bone microstructure of extinct animals provide a host of information about their biology, particularly in terms of their growth dynamics. Several studies have been conducted on a wide range of nonmammalian therapsids (de Ricqlès 1969, 1972, 1974a, 1974b), and in recent years, there has been much focus on the herbivorous dicynodonts (Chinsamy and Rubidge 1993; Botha 2003; Ray and Chinsamy 2004; Ray, Chinsamy, and Bandyopadhyay 2005; Ray, Bandyopadhyay, and Appana 2010; Jasinoski, Rayfield, and Chinsamy 2010a, 2010b; Botha-Brink and Angielczyk 2010), and on the ancestors of mammals, the cynodonts (e.g., Botha 2002; Botha and Chinsamy 2004,

2005). There have been far fewer studies on the therocephalian and gorgonopsian theriodonts (e.g., de Ricqlès, 1969, 1974a; Ray, Botha, and Chinsamy 2004) and there is still much that we do not understand about their biology and growth strategies.

The material assessed here was originally described by Ray, Botha, and Chinsamy (2004). However, in that paper, the material formed part of a larger study and the relatively brief histological descriptions did not discuss in detail all of the available thin sections. In this review, besides more extensive descriptions pertaining to the histological character of all the thin sections of the therocephalian (scylacosaurid) and gorgonopsian bones originally described by Ray, Botha, and Chinsamy (2004), we provide a reassessment of the gorgonopsian *Aelurognathus* and clarification of two previously illustrated images, and we document some intriguing evidence of fungal attack on the bones of at least one scylacosaurid, SAM-PK-5018.

Institutional Abbreviation

In this chapter, Iziko, South African Museum, Cape Town, South Africa, has been abbreviated as SAM-PK.

Phylogenetic Context

Gorgonopsia

Gorgonopsians were the dominant terrestrial carnivores during the Late Permian Period and are known from southern Africa and Russia (Kemp 2005) with fragmentary remains also known from India (Ray and Bandyopadhyay 2003). Apart from these regions, gorgonopsians are not known from elsewhere in the world (see chapter 1 of this book).

The gorgonopsians are characterized by particularly large canines, five upper and four lower well-developed incisor teeth and reduced postcanine teeth, a wide preorbital region, enlarged temporal fenestrae, a wide intertemporal region, fused vomers, and a deeply vaulted palate (Kemp 2005). Three subfamilies are recognized within the Gorgonopsidae, including Gorgonopsinae (the geologically oldest and most diverse group), Inostranceviinae (the largest gorgonopsians, with a long preorbital length, apparently endemic to Russia), and Rubidgeinae (heavily built skull, wide postorbital bar, and massive zygomatic arch) (Sigogneau-Russell 1989; Kemp 2005). Among these subfamilies, the Gorgonopsinae is likely paraphyletic (Gebauer 2007; see chapter 1 this book).

Therocephalia

The therocephalians are characterized by numerous synapomorphies, just a few of which include expanded temporal fenestrae that are widest posteriorly, anteriorly broad vomer, large suborbital vacuities in the palate, reduced laterally oriented lumbar ribs that are fused to the transverse processes, reduced number of caudal vertebrae (when compared to the number of vertebrae of gorgonopsians and cynodonts), specialized anterior notch and process of the iliac blade, and manual and pedal phalangeal formulae reduced to that of mammals (Sigogneau-Russell 1989; Kemp 2005). They share a number of characteristics with the cynodonts, suggesting a relationship between these two groups of therapsids (although the specific nature of this relationship is controversial due to high apparent levels of homoplasy and a poor record of Middle Permian therocephalians and cynodonts [Abdala, Rubidge, and van den Heever 2008; Huttenlocker 2009]). The earliest records of the Therocephalia are from the Middle Permian of South Africa, though younger records from the Upper Permian through Middle Triassic are known from other areas of southern Africa, east Africa, Russia, China, and Antarctica (Sun 1991; Abdala, Rubidge, and van den Heever 2008; chapter 1, this book).

The earliest forms (the scylacosaurids *Glanosuchus* and *Ictidosaurus*) are known from the Middle Permian *Eodicynodon* Assemblage Zone (AZ) of the Karoo Supergroup, South Africa, but the major radiations of eutherocephalians occurred later (e.g., Abdala, Rubidge, and van den Heever 2008; Huttenlocker 2009; Huttenlocker, Sidor, and Smith 2011). Although neglected for a long time, the systematics of the Therocephalia has received some attention in recent years (see, e.g., Hopson and Barghusen, 1986; van den Heever 1994; Botha, Abdala, and Smith 2007; Abdala 2007; Huttenlocker 2009), which has provided fresh insight into their relationship to the Cynodontia. Although there is still support for van den Heever's (1994) re-establishment of the two families of basal therocephalians (the broad-snouted, basal Lycosuchidae, and the long, slender-snouted Scylacosauridae), there is still much research needed to resolve the relationships within and between the scylacosaurids and lycosuchids.

The scylacosaurids were fairly large, though not as large as some lycosuchids, with adult skull lengths of about 20–30 cm, and dentition specialized for carnivory. The bones of two scylacosaurid individuals (previously identified as "*Pristerognathus*" [Ray, Botha, and Chinsamy 2004]) recovered from the Middle Permian *Tapinocephalus* AZ (see Plate 1B) of South Africa are examined here.

Paleobiological Context

During the Late Permian Period, the climate was semi-arid with seasonal rainfall (Keyser 1966; Smith 1993a) and the terrestrial ecosystem was dominated by the dicynodonts as primary consumers and the gorgonopsians as the dominant carnivores. The gorgonopsians were well adapted to a predatory lifestyle as suggested by their functional anatomy. The gorgonopsian *Leontocephalus*, for example, was studied by Kemp (1969), who suggested that the lower jaw was capable of two modes of action: the first mode consisted of opening of the jaws to almost 90° and adduction of the lower jaw (based on the presence of opposing wear facets on the canine teeth), whereas a second possible mode consisted of a propalinal shift of the lower jaw facilitated by a mobile articulation between the quadrate and squamosal bones (a condition known as "streptostyly"; but see Laurin [1998] for an alternative interpretation). Their postcranial skeletons reveal slender limbs and agility suggestive of a predatory lifestyle (Sigogneau-Russell 1989; Kemp 2005). Differences between gorgonopsian genera are relatively subtle, and they vary in terms of size, relative proportions of skull width, and number of postcanine teeth. Although contemporaneous with the gorgonopsians, the advanced therocephalians that lived during the Late Permian and Early Triassic were relatively smaller in size and are thought to have been small-prey carnivores and/or insectivores, although one Middle Triassic genus, *Bauria*, was likely a specialized herbivore (Kemp 2005). *Bauria* was the last surviving therocephalian and co-existed with herbivorous cynodonts in large burrow complexes (Groenewald, Welman, and MacEachern 2001). The therocephalian jaw mechanism was analyzed by Kemp (1972, 1982) to reveal that the earliest forms had a jaw mechanism similar to the gorgonopsians, but showing evidence of greater mobility of the quadrate-quadratojugal complex and no evidence of inter-digitation of the canines or incisors. Jaw mechanisms evident in therocephalians and their diverse dentitions reflect different dietary modes such as carnivory, insectivory, scavenging, and herbivory (Kemp 2005). Kemp (1986) showed that the ratio of the limb length with the body is comparable to that of extant cursorial mammals.

Histological Context

In general, bone microstructure of the nonmammalian therapsids is composed of various primary tissue types ranging from lamellar-zonal tissue to fibrolamellar bone with primary osteons arranged in longitudinal, laminar, and radial patterns (Gross 1934; Enlow 1969; de Ricqlès 1969, 1972, 1974b, 1975, 1976). Early work on the gorgonopsians and therocephalians was mainly done by Armand de Ricqlès (1969). He reported fibrolamellar bone in two therocephalians, a specimen attributed tentatively to the scylacosaurid *Pristerognathus* and the whaitsiid *Theriognathus* (see Plate H) (at the time referred to as "*Notosollasia*"), and in the long bones of five gorgonopsian taxa (*Aelurognathus, Scymnognathus, Dixeya, Lycaenops,* and in an indeterminate gorgonopsid). In a transverse section of a radius of *Theriognathus,* he described a fairly thick-walled compacta with rich vascularization. In the same section, he reported local differences in the histology: In one region of the bone, there was a distinctive radial orientation of the vascular channels, whereas in another area there was a patch of poorly vascularized circular "lamellations." However, in general, the tissue comprised a number of longitudinally oriented vascular channels in a predominantly fibrolamellar matrix, and the bone wall became spongy in texture nearer the poorly defined medulla, which may have been due to the somewhat distal location of the section. In a more distal femoral section of an indeterminate scylacosaurid, de Ricqlès (1969) found that the bone consisted of a narrow compact bone region with a number of primary vascular channels. More internally, secondary reconstruction occurred extensively. In the gorgonopsian *Aelurognathus,* he reported that the transverse section of the diaphysis of a femur showed a thick compact bone wall consisting of mainly primary vascular channels with a reticular orientation, and more internally, spongy bone occurred (de Ricqlès 1969). Although de Ricqlès (1969, Plate 1, Fig. 2) does not mention it, the higher magnification figure provided for *Aelurognathus* suggests that there is a narrow band of relatively avascular lamellar bone around the compacta in the outer cortex. The images provided for the sections of *Scymnognathus* suggest that the bone showed substantial diagenetic alteration. In *Dixeya*, de Ricqlès (1969, Plate 1, Figs. 5, 6) reported that the compacta mainly constituted Haversian bone. The cancellous bone is extensively developed and appears to occlude the medullary cavity. In the transverse section of *Lycaenops*, he describes a richly vascularized laminar type of bone with a large number of primary channels, with some perimedullary secondary reconstruction. In de Ricqlès's images (1969, Plate 2), the compacta appears to be stratified into a number of growth rings, although we cannot ascertain from these the actual nature of the growth rings (i.e., whether they comprise annuli and lines of arrested growth [LAGs] or just the latter). Finally in an indeterminate gorgonopsian, he found in the femur and tibia well-vascularized bone which showed a large amount of secondary reconstruction.

De Ricqlès (1969, 1976) noted a marked distinction between gorgonopsian and therocephalian bone microstructure based on the pattern of vascular channels. In the gorgonopsians, the vascular channels were mostly subplexiform to reticular, whereas he describes the therocephalian bones as "primaire lamellaire" (de Ricqlès 1969, 29). The dicynodonts, on the other hand, had primary osteons arranged in laminar pattern (de Ricqlès 1972). Furthermore, de Ricqlès (1969) showed that Haversian bone was more extensive in the therocephalians in comparison to that of the gorgonopsians where it was limited to the inner cortex. Recent findings of extensive Haversian reconstruction in the bones of adult *Moschorhinus* (A. K. Huttenlocker, pers. comm. 2010) corroborate de Ricqlès (1969) observations in therocephalians.

Ray, Botha, and Chinsamy (2004) examined several skeletal elements of an adult individual of the gorgonopsian *Scylacops* and found that the cortices contained wide zones composed of fibrolamellar bone that were separated by LAGs and annuli comprising avascular lamellar bone. These histological features suggested that *Scylacops* experienced rapid growth that was interrupted periodically by annuli, which reflected periods of slow growth, or complete cessation of growth as indicated by LAGs. In contrast to *Scylacops*, the gorgonopsian *Aelurognathus* was reported as having an absence of cortical stratification which suggested sustained rapid growth. However, the presence of calcified cartilage in the epiphyseo-metaphyseal sections of the limb bones suggested that the animal had not yet attained adult size (Ray, Botha, and Chinsamy 2004). Based on the bone microstructure, Ray, Botha, and Chinsamy (2004) proposed an indeterminate growth strategy for both gorgonopsian taxa.

The bone microstructure of two indeterminate scylacosaurids showed that, as in case of the gorgonopsians, fibrolamellar bone predominated in the cortex suggesting rapid growth (Ray, Botha, and Chinsamy 2004), although periodic interruptions in their growth were not documented. These observations concurred with those of de Ricqlès (1969) on a scylacosaurid femur. Ray, Botha, and Chinsamy (2004) suggested a fast uninterrupted growth early in ontogeny which decreased considerably later in ontogeny. They also proposed that the therapsid lineage showed an increasing trend toward fast growth, and an independent decrease or eventual loss of developmental plasticity as suggested by the absence of LAGs (Starck and Chinsamy 2002).

Current Analysis

Materials and Methods

The gorgonopsian taxa *Scylacops* and *Aelurognathus*, previously reported on by Ray, Botha, and Chinsamy (2004) were available for more detailed histological analysis. The genus *Scylacops* is known from the Late Permian *Cistecephalus* AZ (see Plate 1B) of South Africa. One individual (SAM-PK-10188 from the Dunedin locality) was available for the histological study from which six bones (humerus, radius, ulna, femur, fibula, and a rib) were sampled for thin-sectioning (Table 8.1) (see chapter 3 for methodology). The genus *Aelurognathus* is known from the middle Late Permian *Tropidostoma* AZ of South Africa (see Plate 1B). A humerus, femur, and tibia of specimen SAM-PK-K10000 (from the Leeukloof locality) were sampled for histology (Table 8.1).

The bone histology of two therocephalians from the *Tapinocephalus* AZ (see Plate 1B) was assessed: SAM-PK-5018 (from the Abrahamskraal locality) and SAM-PK-11557 (from Die Cypher locality). Originally, this therocephalian material was studied by Ray, Botha, and Chinsamy (2004) and was considered to be *Pristerognathus*, but since no diagnostic skulls were associated with the postcrania sectioned for histological analysis, it is more accurate to refer to them as Scylacosauridae indet. SAM-PK-5018 consists of several postcranial elements, and a humerus, femur, tibia, and fibula of this individual were sampled (Table 8.1). From the other scylacosaurid specimen, SAM-PK 11557, three ribs and a fibula were sampled (Table 8.1).

Scanning Electron Microscope Analyses

Carbon-coated thin sections of the femur (A2) and fibula (A2) (Table 8.1) of the therocephalian SAM-PK-5018 were chemically analyzed using a scanning electron microscope (FEI NOVA Nanosem 230, Hillsboro, Oregon) at the Electron Microscope Unit of the University of Cape Town.

Table 8.1. List of taxa studied and thin sections prepared

Taxon	Specimen number	Late Permian Karoo AZ	Element	Thin sections							
				Transverse (cross)				Longitudinal			
Scylacops	**SAM-PK-10188**	*Cistecephalus* **AZ**	**Proximal humerus**	**A1**	**A2**						
			Distal radius	**A1**	**A2**	**B1**	**B2**				
			Proximal ulna	**A1**	**A2**	**A3**					
			Partial rib	**A1**	**A2**	**B1**	**B2**				
			Mid-shaft femur	**A1**	**A2**	**B1**	**B2**				
			Proximal fibula	**A1**	**A2**						
			Distal fibula	**A1**	**A2**						
Aelurognathus	**SAM-PK-K10000**	*Tropidostoma* **AZ**	**Proximal humerus**	**A1**	**AII**			**B1a**	**B1b**	**B2a**	**B2b**
			Distal femur	**AI**	**AII**			**B1a**	**B1b**	**B2a**	**B2b**
			Proximal tibia	**A1**	**AII**			**B1a**	**B1b**	**B2a**	**B2b**
Scylacosaurids	**SAM-PK-5018**	*Pristerognathus* **AZ**	**Humerus**	**A1**	**A2**						
			Radius	**Ba1**	**Ba2**	**Bb1**	**Bb2**				
			Ulna (A)	**Aa1**	**Aa2**	**Ab1**					
			Femur	**A1**	**A2**						
			Tibia	**A1**	**A2**						
			Fibula	**A1**	**A2**	**B1**					
	SAM-PK-11557	*Pristerognathus* **AZ**	**Fibula**	**A1**	**A2**	**B1**	**B2**				
			Proximal rib 1	**A1**	**A2**	**B1**	**B2**				
			Partial rib 2	**A1**	**A2**	**B1**					
			Proximal rib 3	**A1**	**A2**	**B1**					

Gorgonopsia (*Scylacops*)

Humerus

Two thin sections from the proximal end of a humerus were prepared (Table 8.1; Figs. 8.1A, 8.1B). Both sections show a narrow region of richly vascularized, compact bone wall consisting of predominantly fibrolamellar bone (Figs. 8.1C, 8.1D). The outermost part of the bone wall comprises a lamellar type of bone tissue that is interrupted by a series of closely spaced rest lines (Fig. 8.1C). Internal to this layer is fibrolamellar bone with mainly longitudinally oriented channels, which is followed by a narrow annulus with a single LAG, below which is a layer of fibrolamellar bone with primary and secondary osteons, and several large erosion cavities (Fig. 8.1C). Following this, the erosion cavities reach cancellous dimensions, and trabeculae extend into the medullary cavity (Figs. 8.1A, 8.1B). The two thin sections have different amounts of periosteal bone around the periphery, and also differ in terms of the thickness of the zones in the compact bone, and number of LAGs or rest lines near the edge of the bone (Fig. 8.1). Sharpey's fibers are commonly observed in the sections (Fig. 8.1D).

Femur

Four thin sections were prepared from the femur (Table 8.1; Fig. 8.2). The histology in all the bones was similar with zonal bone tissue dominating the overall structure of the compacta (Figs. 8.2A, 8.2B). The bone consists of distinct cycles of bone deposition; in other words, fibrolamellar bone alternates with narrow bands of lamellar bone with or without LAGs (Fig. 8.2B). In the four thin sections studied, a variable number of cycles were observed. In the most distal section, B2 (i.e., closest to the mid-shaft), the maximum number of growth rings (seven) was recorded. The more proximal sections have fewer growth rings. Abundant Sharpey's fibers were observed in all the sections and were especially well-developed in section A2 (Fig. 8.2A), which is the most proximal section and nearest the trochanter major (Fig. 8.2C). In section B1 (Fig. 8.2A), near the medullary cavity, a narrow inner circumferential layer of lamellar bone is present and it appears to have followed a period of resorption. Near the medullary cavity, secondary reconstruction resulted in a large number of erosion cavities (Figs. 8.2A, 8.2B). Some of these cavities were infilled to form secondary osteons (Fig. 8.2B), whereas in some cases, the resorption resulted in the formation of cancellous bone in the medullary cavity. Figure 8.2A shows the variable extent of cancellous bone formation in the four serially taken femoral sections, which affected the preservation of the number of growth rings. Although there appears to be a narrowing of the zonal regions peripherally, no rest lines are observed in the outermost part of the compacta (Fig. 8.2B).

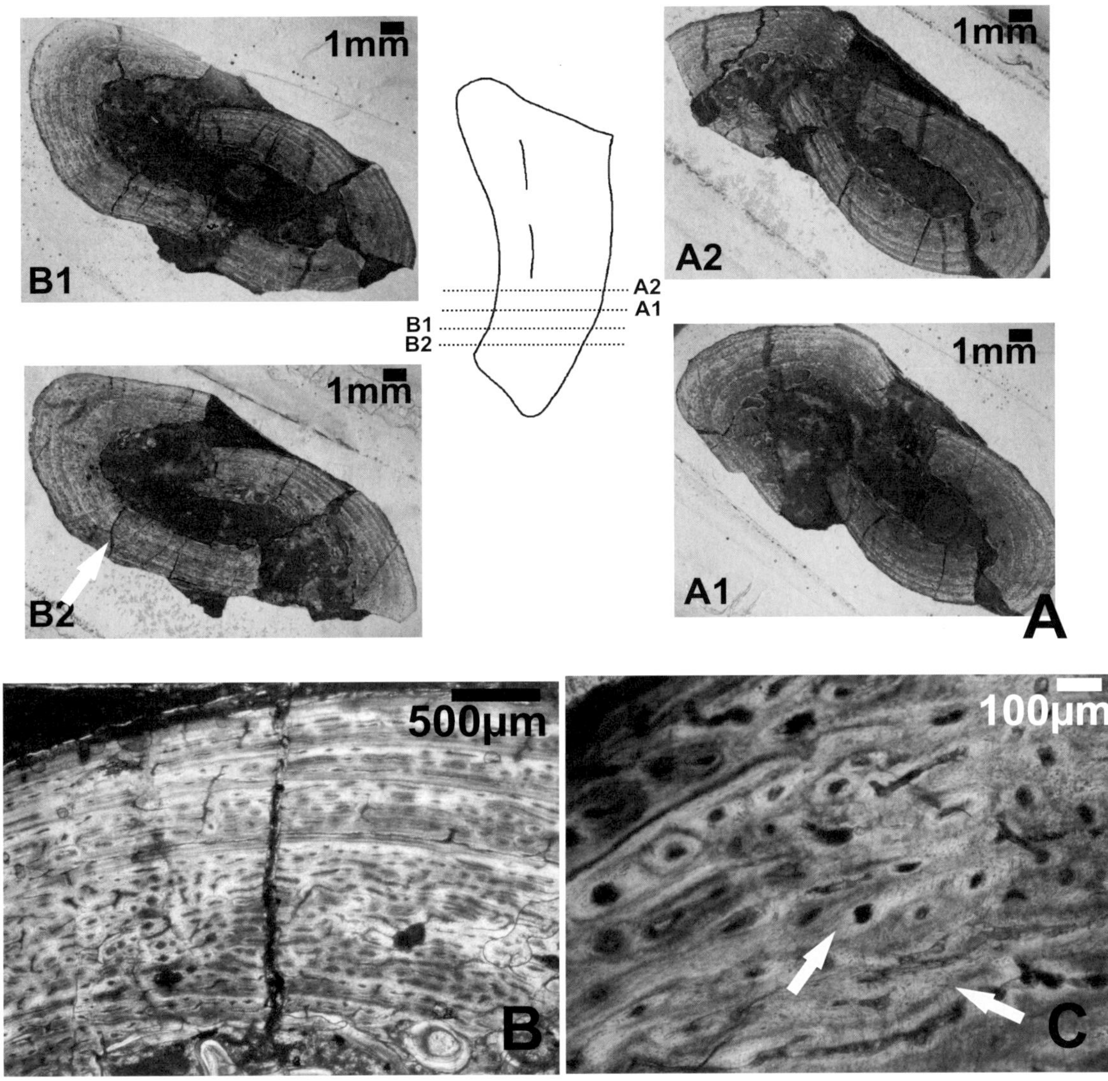

8.2. *Scylacops* femur, SAM-PK-10188. (A) The location and overall view of four femoral thin sections (A1, A2, B1, and B2). Arrow indicates growth rings in the compacta. (B) Higher magnification of B2 showing the zonal nature of the compacta. (C) Higher magnification of A2 showing Sharpey's fibers (arrows).

Fibula, Radius, Ulna, and Rib

These bones all showed a similar cortical bone tissue to that described for the humerus and femur. Additionally multiple sections of these bones were made (Table 8.1), which permitted subtle variations in the histology to be observed, particularly in terms of number of growth rings and amount of cortical and cancellous bone present. For example, the four serially sampled thin sections of the rib (Fig. 8.3) show striking variations in the overall architecture, extent of resorption of the primary compact bone wall, as well as in the development of cancellous bone in the medullary cavity. Predictably, the different rib sections expressed different numbers of growth rings. The maximum number of growth cycles observed in the fibula was seven, whereas the radius preserved eight growth cycles (seven of these are distinct, but the last one near the peripheral margin

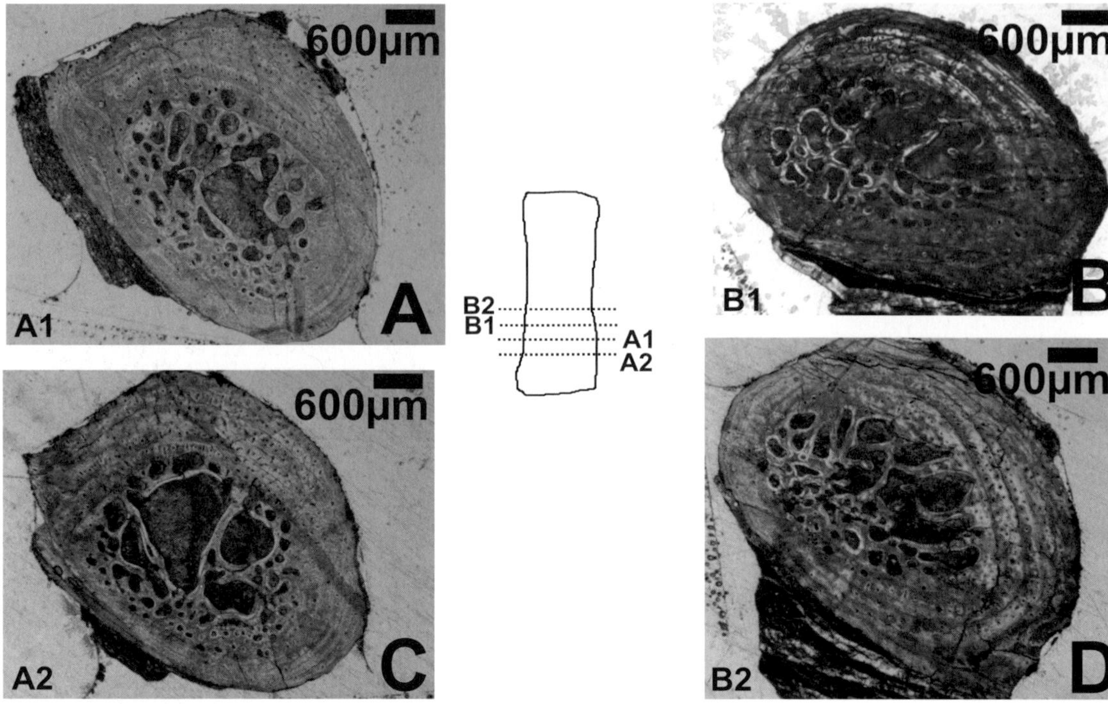

8.3. *Scylacops* rib, SAM-PK-10188. Overall histology and location of the four thin sections showing variation in the extent of cortical and cancellous bone present along the shaft of the rib.

of the bone wall is only visible in certain parts of the cross section) (Fig. 8.4A). Thin sections A1 and B1 from the radius (Table 8.1) also show a series of rest lines peripherally. Figure 8.4A presents a general view of the radius thin section A3, and quite distinctly depicts how extensive secondary reconstruction of the earlier formed periosteal bone, leads to an obliteration of the growth rings formed during early ontogeny. All the bones sampled show secondary osteons in the compacta, but in the radius this was developed to dense Haversian proportions (Fig. 8.4B). Figure 8.4C of the ulna (section B1) shows how the bone tissue can vary across a single cross section: that is, on the medial side the bone wall is almost entirely composed of secondarily formed bone tissues with only a narrow periosteally deposited layer of primary bone evident, whereas on the opposite side, more of the primary ontogenetic history of bone deposition is preserved.

SUMMARY OF THE HISTOLOGICAL FINDINGS OF *SCYLACOPS*

All the bones sampled from the *Scylacops* specimen exhibit a similar pattern of periodically interrupted fibrolamellar bone tissue. The bone was richly inundated with channels in which blood vessels and other connective tissue would have been located in the living animal (Starck and Chinsamy 2002). An interesting aspect of this study was the serial sectioning of the skeletal elements. These sections provided insight into how the bone microstructure changes within a single bone, including almost neighboring regions of a single bone, as well as across a single

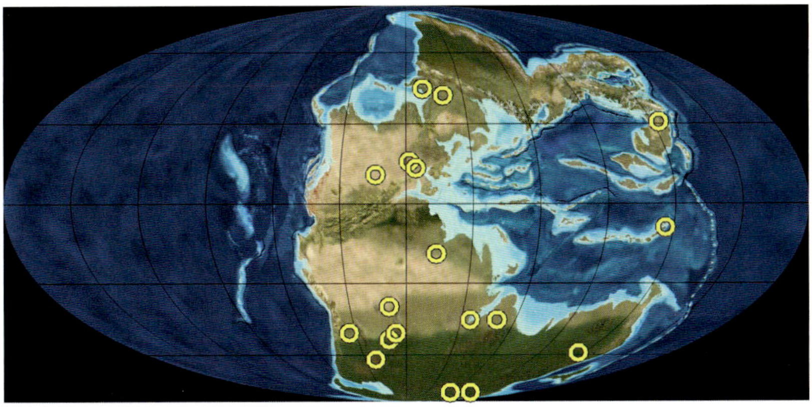

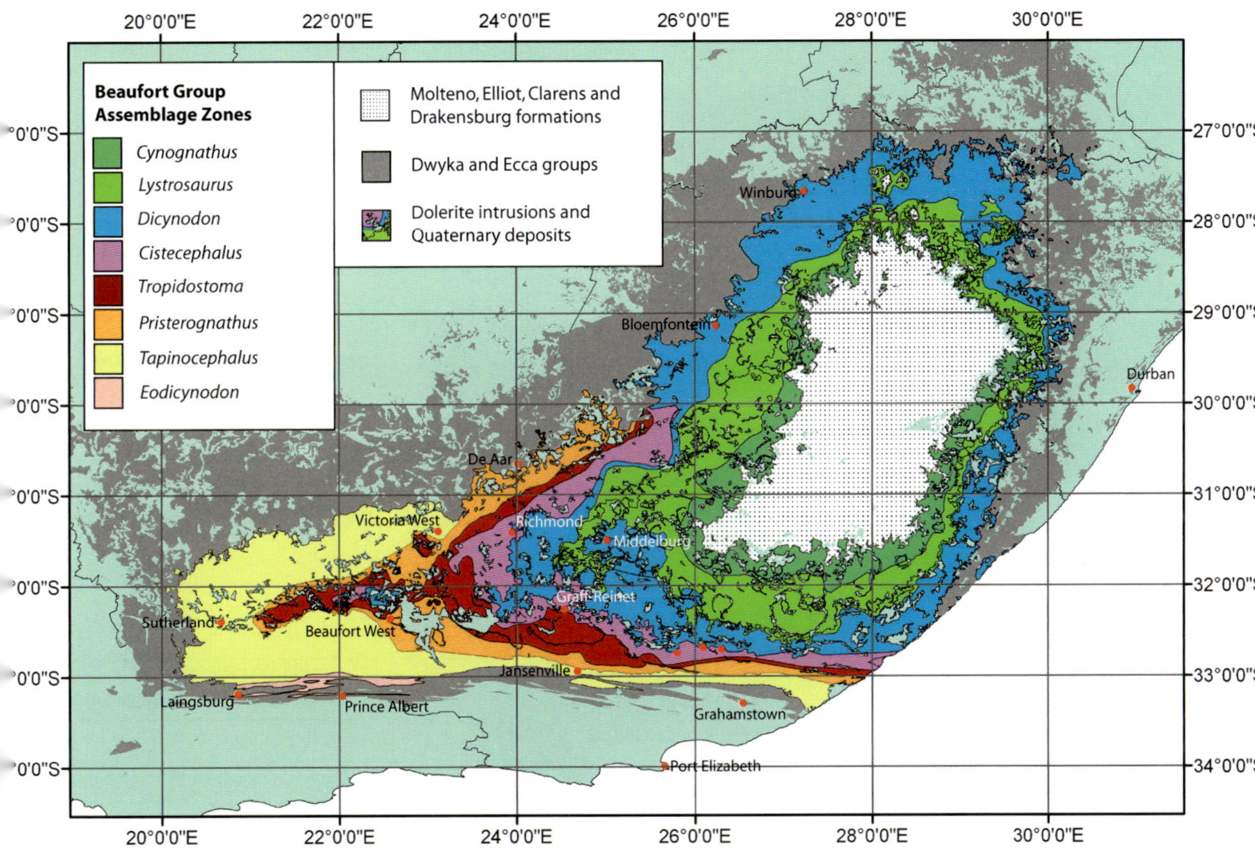

Plate 1. (Top) Distribution of therapsid-bearing rocks in Pangaea plotted on Ron Blakey's continental reconstruction of the Early Triassic. (Reproduced with permission.) (Bottom) Biozone map of the Main Karoo Basin, South Africa (after van der Walt et al. 2010).

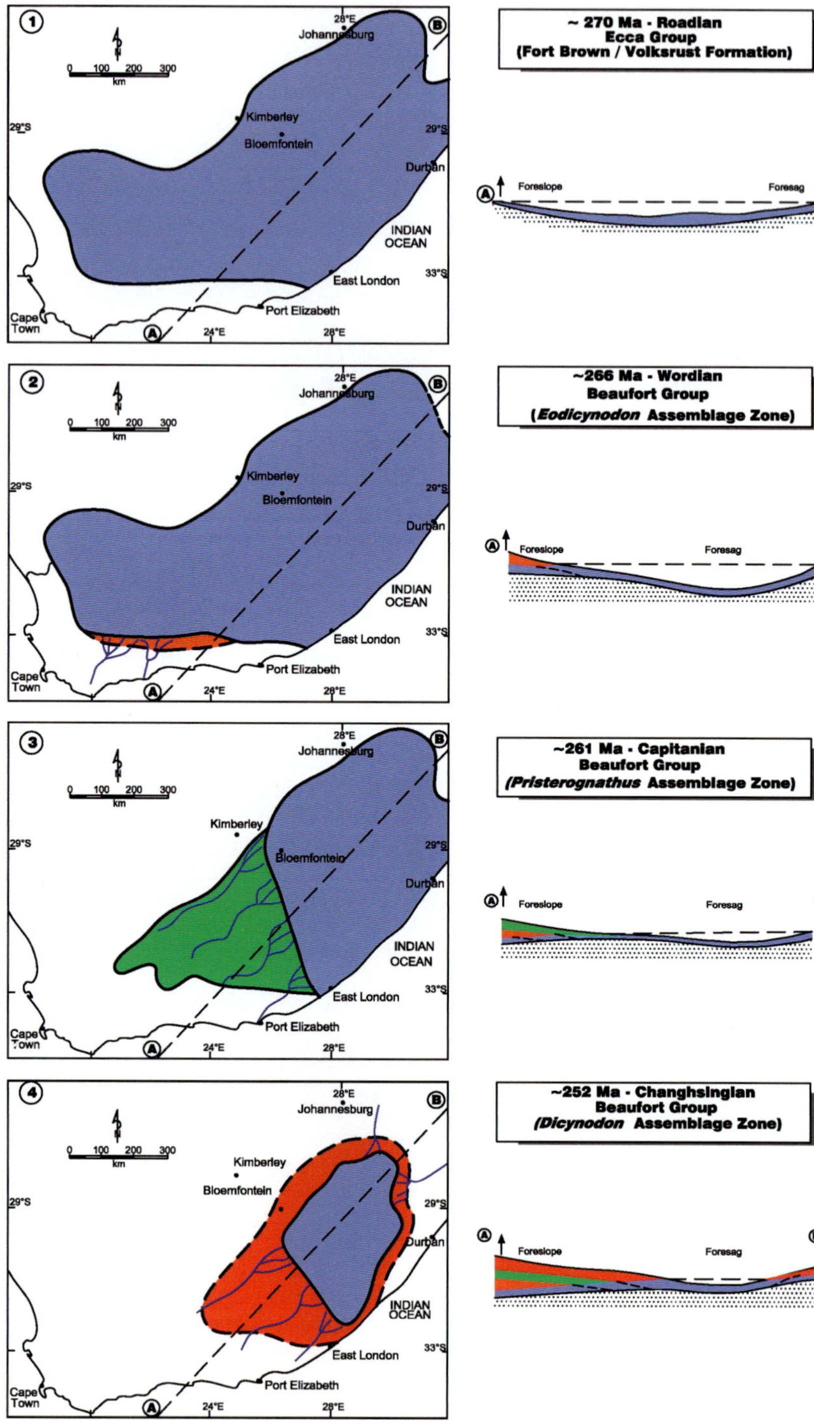

Plate 2. Tectonostratigraphic evolution of the Lower Beaufort Group in the main Karoo Basin during the Middle to Late Permian from 270–252 Mya (after Rubidge 2005).

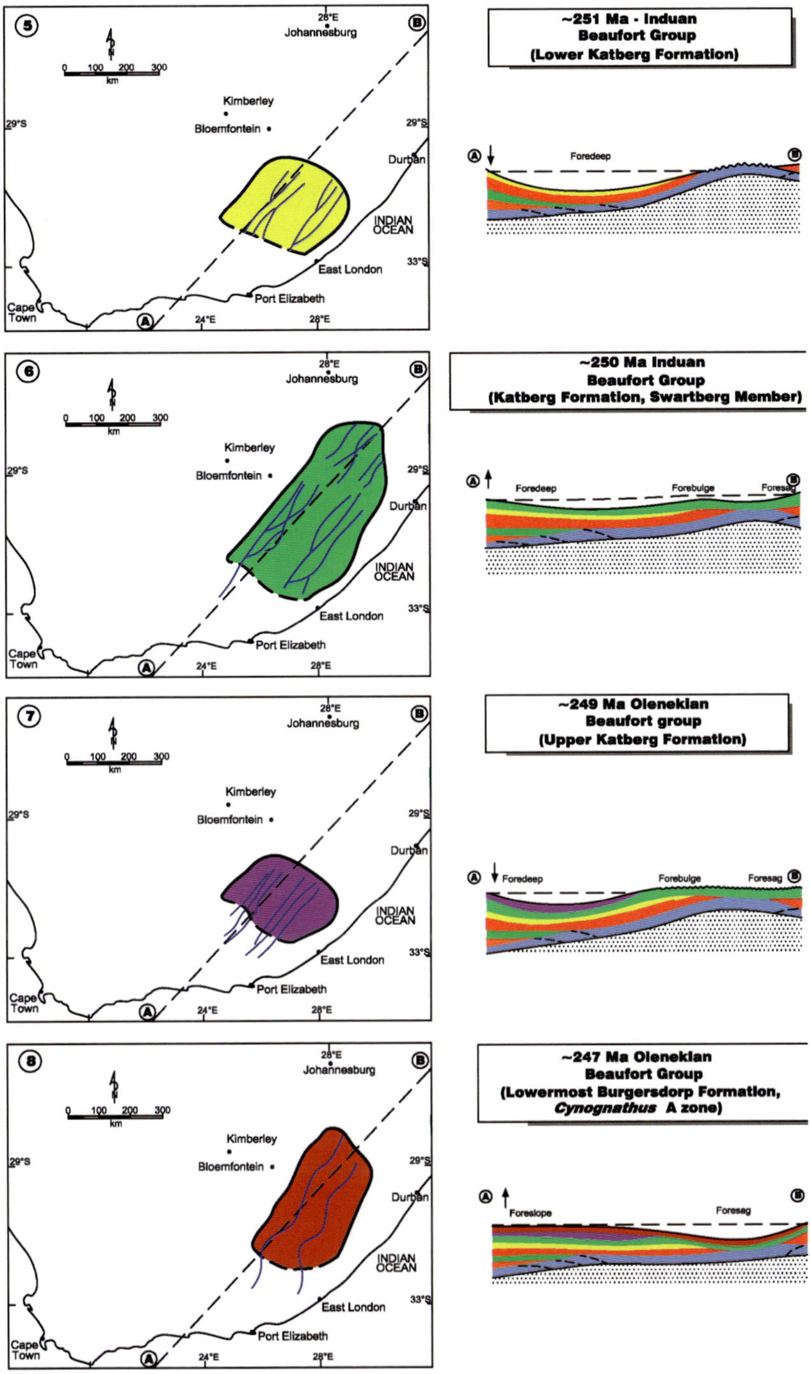

Plate 3. Tectonostratigraphic evolution of the Middle and Upper Beaufort Group in the main Karoo Basin during the Early Triassic from 251–247 Mya (after Rubidge 2005).

Plate 4. *Oudenodon* (SAM-PK-K4807) showing fibrolamellar bone. Note the globular-shaped osteocytes. There is a predominance of longitudinally oriented channels in the bone, which are surrounded by a distinct layer of lamellar bone that forms the primary osteon. The different colors indicate the varying orientation of the apatite, which reflects the arrangement of the collagen that was once present in the bone.

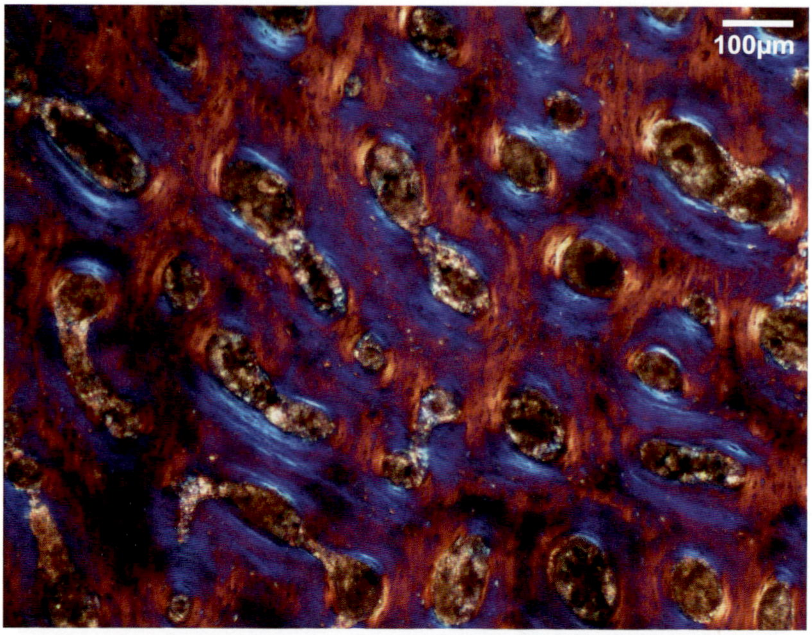

Plate 5. *Tropidostoma* (SAM-PK-K9960, tibia) illustrating lamellar bone tissue (arrow) in the peripheral regions of the bone wall under ordinary light (A) and under polarized light (B) showing the light and dark bands of the lamellae. *Oudenodon* tibia. Lamellar bone tissue (arrow) of the osteon (the ring of bone around the channel) under ordinary light (C) and polarized light (D). Lamellated bone tissue (arrow) under ordinary light (E) (note the lamellae are visible) and under polarized light (F) in *Oudenodon* (SAM-PK-K5065-11, jugal). Parallel-fibered bone tissue (arrow) in the peripheral region of a fibula of *Oudenodon* (SAM-PK-K 10019C1) under ordinary light (G) and under polarized light (H).

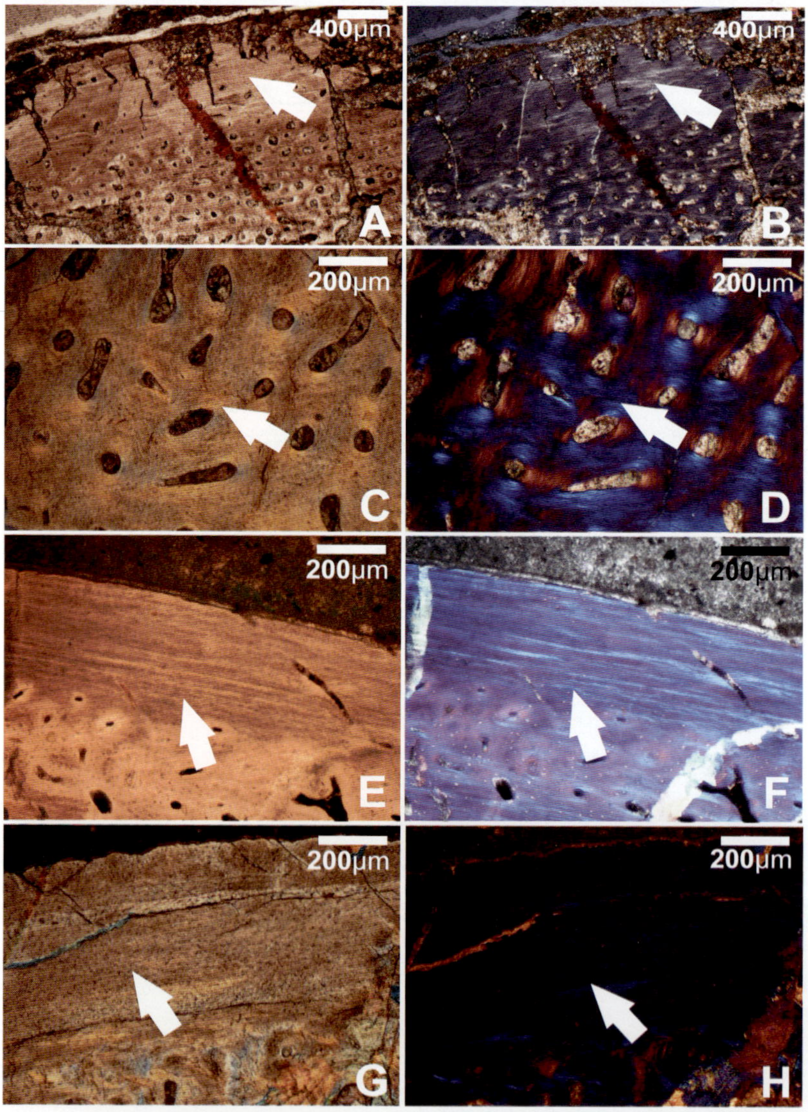

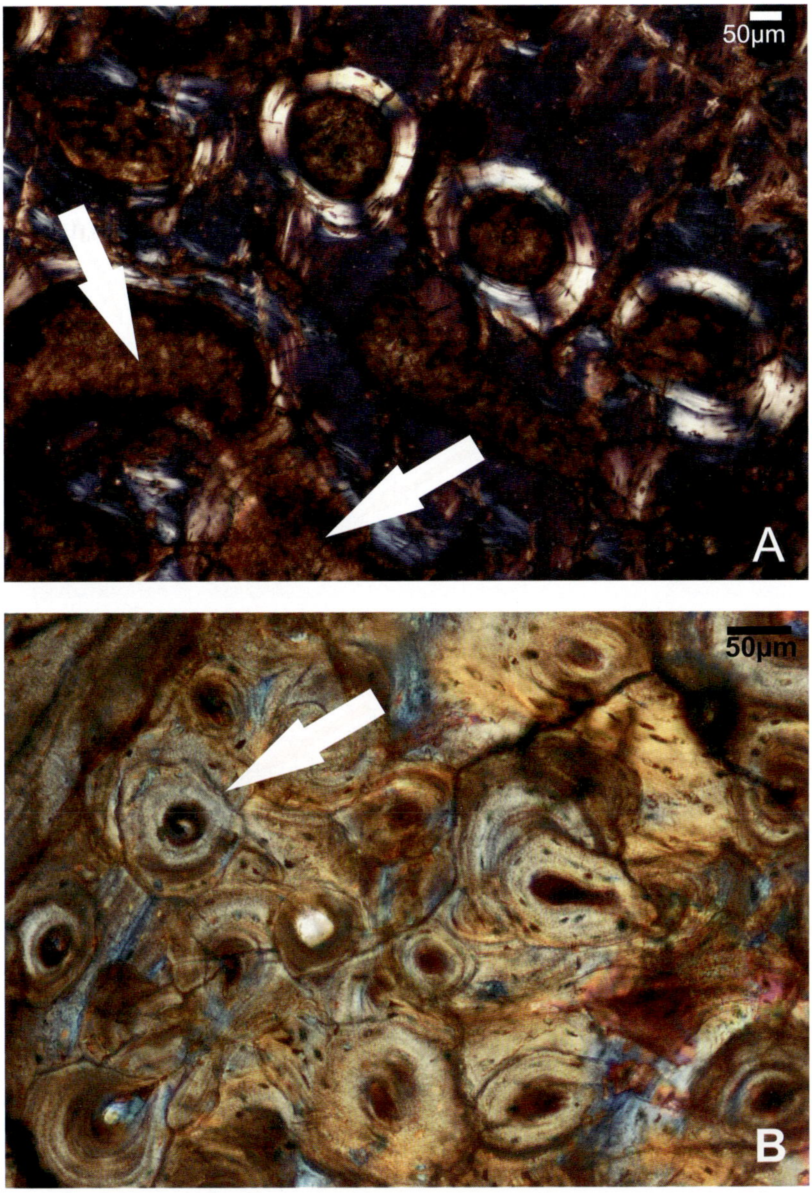

Plate 6. (A) Fibula of a therocephalian (SAM-PK-11557) showing enlarged erosion cavities (arrows) and three secondary osteons which display distinct axial crosses. (B) Dense Haversian bone in a *Scylacops* radius. Note that the bones between neighboring secondary osteons are remnants of older secondary osteons. Arrow indicates a cement line.

Plate 7. (A) *Oudenodon* (SAM-PK-K5216-10) premaxilla, showing a dense distribution of Sharpey's fibers (arrows) along the palatal edge which may be indicative of attachment of the keratin beak. (B) *Scylacops* femur showing lamellar bone (arrows) interrupting the deposition of fibrolamellar bone.

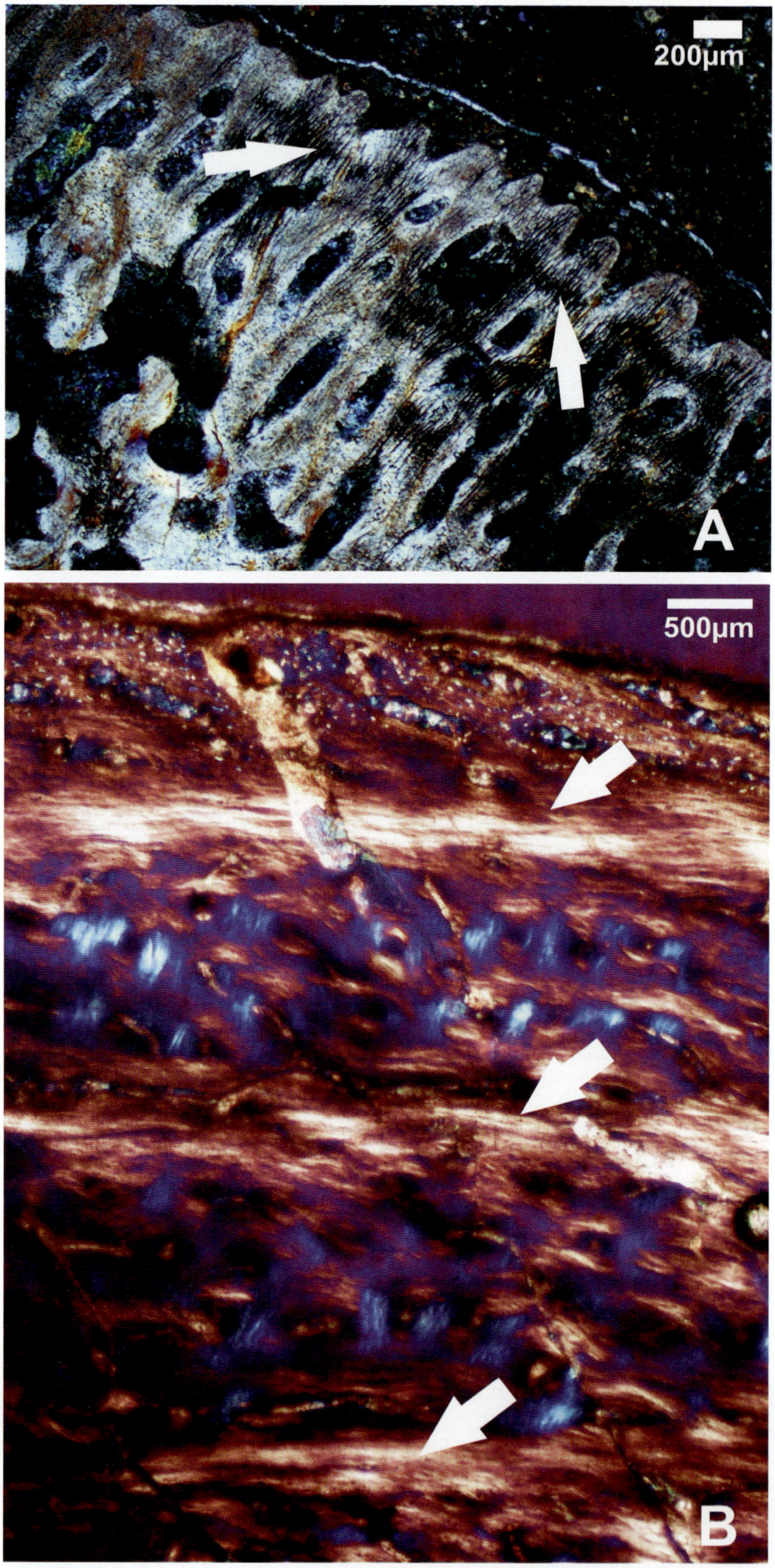

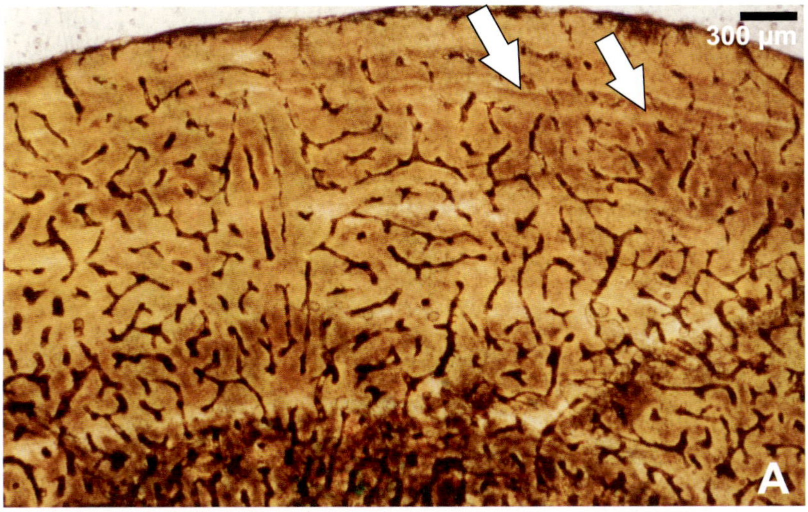

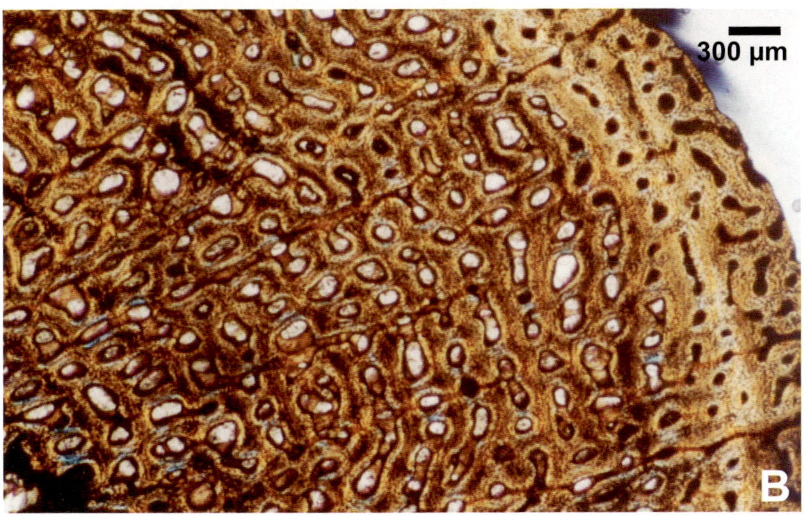

Plate 8. (A) *Diictodon feliceps* (SAM-PK-K6716a) femur, showing fibrolamellar bone and annuli (arrows) in the cortex. (B) Secondarily enlarged resorption cavities in the deep cortex of a clavicle of *Lystrosaurus murrayi* (ISIR 754). *Photomicrographs by Sanghamitra Ray.*

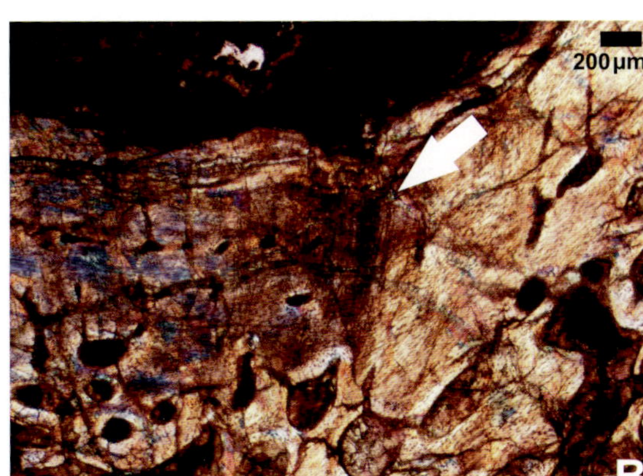

Plate 9. (A) Obliquely oriented Sharpey's fibers (arrows) along the anterior (outer) edge of the premaxilla of *Oudenodon* (SAM-PK-K5216-2), indicating the attachment of keratin onto the snout. (B) Elongated Sharpey's fibers (arrow) in the depression on either side of the median parietal ridge in the posterior skull roof of *Lystrosaurus* (SAM-PK-K28-6P), which may represent keratin horn attachment. *Photomicrographs by Sandra Jasinoski.*

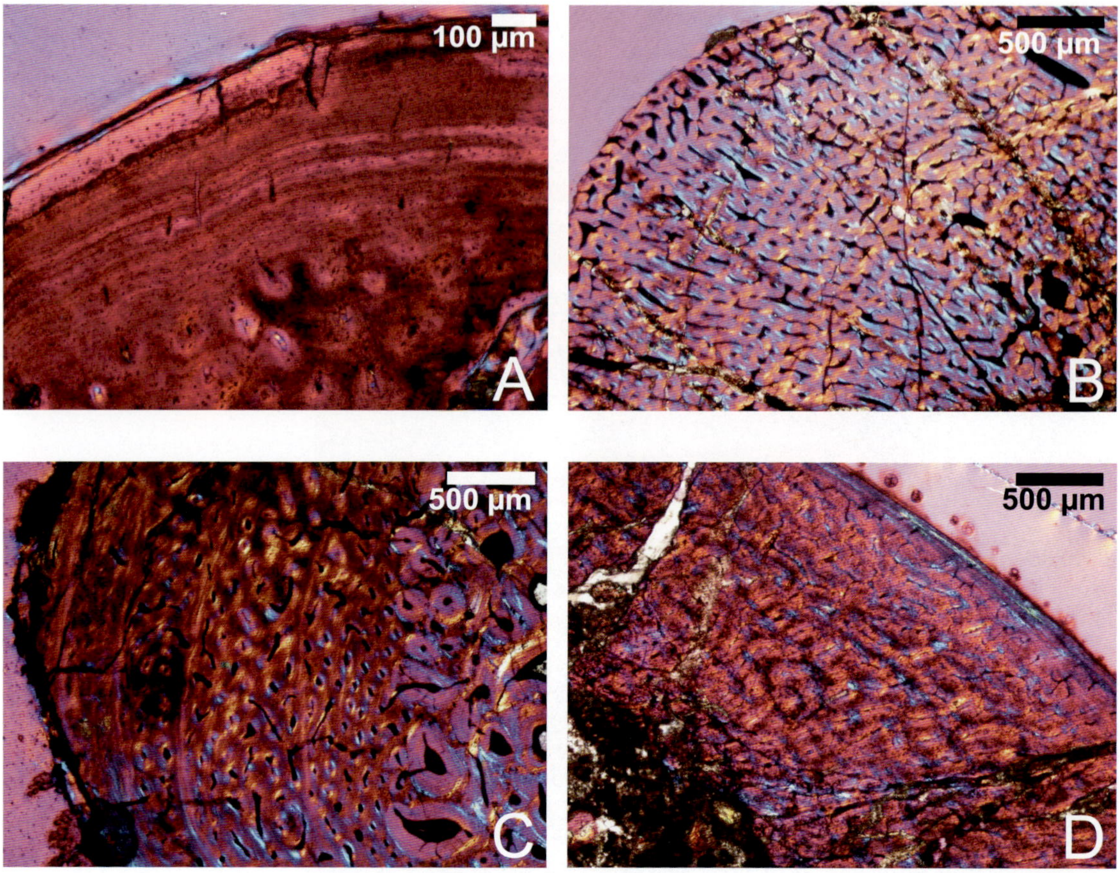

Plate 10. (A) The basal cynodont *Procynosuchus* (BP/1/3747, radius), showing lamellar-zonal bone toward the bone periphery. (B) The cynognathid *Cynognathus* (NMQR1208a, humerus), showing fibrolamellar bone in a plexiform network. (C) The gomphodont *Diademodon* (BP/13772, humerus), showing fibrolamellar bone interrupted by annuli of lamellar bone. (D) The tritylodontid *Tritylodon,* showing fibrolamellar bone and peripheral rest lines (blue area) at the subperiosteal surface. *Photomicrographs by Jennifer Botha-Brink.*

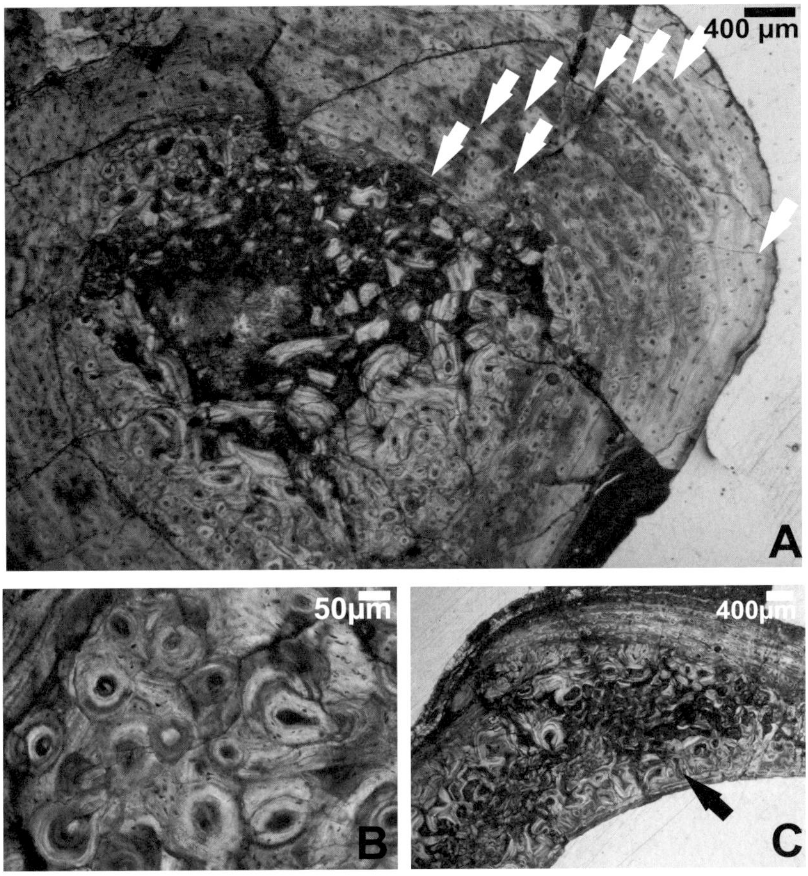

8.4. *Scylacops,* SAM-PK-10188. (A) Radius section A3 showing overall histology and eight growth rings (arrows) in the compacta. (B) Higher magnification showing dense Haversian bone in the radius. (C) Shows the overall view of the histology of thin section B1 of the ulna. Note the presence of secondarily formed bone on the medial side (arrow).

transverse section (Figs. 8.1, 8.2A, 8.3, 8.4A, 8.4B). These results emphasize the importance of examining multiple sections of individual bones, since variable numbers of growth rings were observed, and the structure of the compact bone wall can vary in terms of the amount of periosteal bone preserved, as well as the extent of secondary reconstruction and the development of cancellous bone within the medullary cavity. The variation in the number of growth rings observed between the different elements is also worth noting. From the femoral serial sections, it appears that the thin sections sampled nearest the mid-shaft of the bone kept the best record of the depositional history of the bone. However, we found that the radius recorded the highest number of growth rings. Other bones, such as the ulna and fibula, preserved far fewer growth rings in the compacta since the bones were highly remodeled.

Other ontogenetic features evident in the *Scylacops* bones studied are the reduction in the number of channels nearer the bone periphery as opposed to more internally located, richly "vascularized" plexiform fibrolamellar bone that was formed in earlier stages of ontogeny. However, it is apparent that throughout the life of the individual, fibrolamellar bone is deposited in the comparatively wider zonal regions of the compacta (e.g., Figs. 8.2B, 8.2D), whereas lamellar bone is deposited in the annuli. Only in the outermost regions of some bones (e.g., ulna, radius, and humerus) does the bone tissue change to a more poorly vascularized lamellar bone, and in the radial sections (A1 and B1) and the humerus, several closely spaced rest lines are observed (Figs. 8.1C, 8.1D). These peripheral rest lines in the humerus and radius suggest that this individual had attained adult body size and had virtually stopped growing, but since they are localized in particular regions of just two bones (of six bones studied), it may simply mean that that those particular elements have stopped growing (i.e., differential growth rates within the skeleton).

Mostly these results agree with Ray, Botha, and Chinsamy's (2004) previous descriptions of *Scylacops*, although here the analysis of multiple sections of the same skeletal elements provides insight into how the location of a section within the same element affects the histology, and we have shown how the bone histology can vary across a single thin section. Although we cannot be certain that the observations of lamellar deposits and rest lines in localized regions of the humerus (Figs. 8.1C, 8.1D) and radius indicate determinate growth for the taxon, the reduction in "vascularization" and narrowing of the zones seems suggestive that growth is slowing down in the individual. We are therefore not confident of the indeterminate growth previously ascribed to *Scylacops* (Ray, Botha, and Chinsamy 2004). More specimens, particularly of older individuals need to be examined histologically to confirm whether this taxon experienced determinate or indeterminate growth.

Gorgonopsia (*Aelurognathus*)

Humerus

Two thin sections were prepared from the proximal metaphysis of the humerus. In the transverse section A1 (Table 8.1; Figs. 8.5A, 8.5B), variation across the section is evident. In one part of the section, the cortical bone comprised mainly primary bone tissue with a large number of longitudinally oriented channels, although a large number of radial connections occurred (Fig. 8.5A). Two LAGs are visible near the medullary cavity on one side of the section (Fig. 8.5A). Since the endosteal region is highly resorptive, it is likely that LAGs have been obliterated in the other parts of the bone. This supposition is supported by the fact that one part of the bone wall consists almost entirely of compacted coarse cancellous bone. Figure 8.5B shows a region of the bone wall comprising a narrow region of primary periosteal bone and endosteally formed compacted coarse cancellous bone. Around the medullary cavity some enlarged resorption cavities occur and some isolated secondary osteons are present. The longitudinal sections of the humerus show a large number of bony struts that extend almost to end of the bone (Fig. 8.5C). These endochondrally formed struts are capped by an approximately 300-mm-thick layer of calcified cartilage. However, it is worth noting that the "articular" end

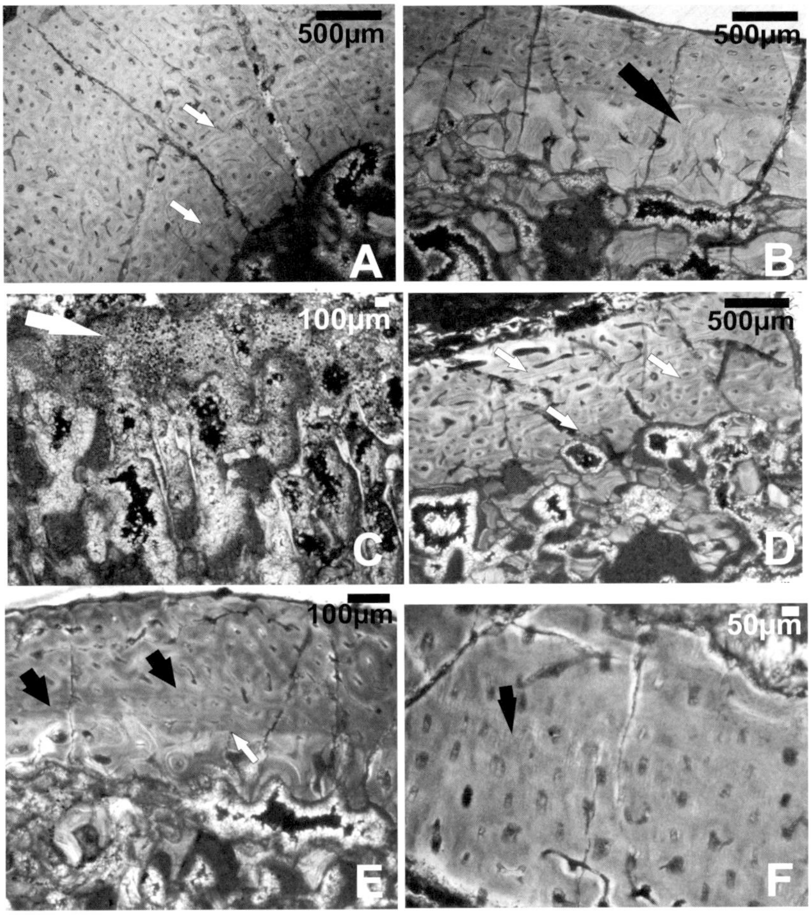

8.5. *Aelurognathus,* SAM-PK-K10000. (A) Humerus (section A1) showing two LAGs (arrows). (B) Humerus (section A1) showing periosteally formed primary bone, and endosteally formed compacted coarse cancellous bone (arrow). (C) Longitudinal section of the humerus showing endochondral bone in the humerus. Arrow indicates the layer of calcified cartilage. (D) Femur (section A1) showing three LAGs (arrows). (E) Tibial section showing two LAGs (black arrows) and a cement line (white arrow), below which is compacted coarse cancellous bone. (F) Higher magnification of a different part of the tibial section showing predominantly longitudinally oriented channels and a large amount of Sharpey's fibers (arrow).

preserved in the fossilized humerus may have been overlain by cartilage that would have decomposed early during fossilization (Chinsamy-Turan 2005).

Femur

Two thin sections were prepared from the distal metaphyseal region of the femur (Table 8.1). The cross sections showed only a thin rim of primary compact bone comprising fibrolamellar bone with abundant longitudinal and circumferential primary osteons with some radial connections between them (Fig. 8.5D); note in Ray, Botha, and Chinsamy (2004, Fig. 2B) a section of SAM-PK-K5018 was inadvertently figured as a femur of SAM-PK-K10000. Three growth rings interrupt this bone tissue. One of the growth rings can be followed around the compacta, but the innermost two are restricted in their occurrence around the circumference of the bone since in some places secondary remodeling impinges further into the primary bone. Some compacted coarse cancellous bone, as well as cancellous bone, is present in variable amounts in the different cross sections. The histology of the longitudinal section of the femur was not as well preserved as the humerus, although it showed similar histologic organization.

Tibia

Two transverse sections were prepared from the proximal metaphyseal region of the tibia. As such, the bone wall consists of an outer narrow rim of primary bone and internal to this are varying quantities of secondarily formed bone tissues such as secondary osteons, enlarged erosion cavities, and compacted coarse cancellous bone (Fig. 8.5E). The primary bone is inundated by longitudinally oriented channels and one to two growth rings are visible. The innermost growth ring is only visible for a limited extent since it is resorbed elsewhere in the cross section. In a localized region, the bone wall is composed principally of longitudinally oriented channels, and there is an abundance of Sharpey's fibers visible in the compacta (Fig. 8.5F). In both longitudinal sections of the tibia, thin endochondral struts are visible, and as in the humerus and femur, they extend almost toward the end of the bone and are capped by a layer of calcified cartilage (although the latter are not well preserved in these sections).

SUMMARY OF *AELUROGNATHUS* BONE HISTOLOGY

The multiple thin sections from the humerus, femur, and tibia of *Aelurognathus* specimens examined in this study provided the opportunity to thoroughly assess the bone microstructure of this individual. The most distinctive finding is that we document the presence of growth rings in the bones of all three long bones sampled (Figs. 8.5A, 8.5C, 8.5E). This is contrary to our earlier observations (Ray, Botha, and Chinsamy 2004) in which we only recorded two growth rings in the tibia of *Aelurognathus*. Here, in addition to those previously reported in the tibia, we document two LAGs in the humerus and three LAGs in the femur. The cross sections of the bones revealed that the periosteal bone consisted of a richly vascularized fibrolamellar bone. A distinctive feature of the three long bones sectioned is that the vascularization tends to be reticular with well-defined primary osteons; although in localized regions, even in a single section the channel orientation can change (Figs. 8.5E, 8.5F). The periodic growth marks that occur in the compacta are LAGs with or without an accompanying narrow band of lamellar bone (Fig. 8.5). All three bones sampled were from the metaphyseal regions of the skeletal elements, and since these regions are extensively involved in remodeling during normal growth processes (Enlow 1962, 1969), the bone microstructure that is visible does not provide a good track record of ontogenetic growth. In both the tibia and the femur, although just a narrow rim of periosteal bone is preserved, LAGs are evident. The humerus and the femur both show resorptive margins which directly suggest that the periosteal formed bone was actively resorbed at the time of death. Parts of the tibial sections show a region of compacted coarse cancellous bone under a reversal line (Fig. 8.5E).

Experimental studies using fluorochrome stains in extant animals, such as lepidosaurs and crocodilians, that deposit bone cyclically have shown that the bone tissues formed are influenced by seasonal conditions (e.g., Castanet et al. 1993) with the faster forming bone being deposited in the favorable season. Although more recent experimental studies on extant mammals suggest that cyclic periodicity may also be associated with endogenous circadian rhythms influenced by seasonal fluctuations in the photoperiod that affect foraging behavior, and not just environmental conditions alone (Castanet et al. 2004). Studies on living animals have also shown conclusively that fibrolamellar bone tissue is formed at rapid rates of osteogenesis (e.g., de Margerie, Cubo, and Castanet 2002; Starck and Chinsamy 2002; Montes, Castanet, and Cubo 2010). Thus, given that the two gorgonopsians studied here lived during the Late Permian of the Karoo Basin of South Africa, they both would have experienced similar environmental conditions, which paleoenvironmental analyses suggest were strongly seasonal (Smith 1993a). The growth patterns as deduced from our histological findings suggest that both individuals were influenced by the prevailing seasonal conditions. We suggest that during periods of favorable activity these animals grew relatively quickly and formed fibrolamellar bone, whereas the LAGs and annuli observed probably represent the effects of the unfavorable season or perhaps are related to abiotic seasonal effects on foraging activity (e.g., photoperiodicity).

Therocephalia (Scylacosauridae), SAM-PK-5018

Several skeletal elements of the two scylacosaurids (SAM-PK-5018 and SAM-PK-11557) were available for thin-sectioning and histological examination (Table 8.1). Please note that in the previous section pertaining to the gorgonopsians, we illustrated exactly where the different sections in the *Scylacops* humerus, femur, and rib were sampled, but here in the interests of brevity, we have simply tabulated the number of sections from each skeletal element in Table 8.1, and the differences in the histology will be described with selected images.

In SAM-PK-5018, all the bones sampled show extensive degradation of the bone microstructure. In some areas of the thin sections, the damage was severe, whereas in others parts of the same section the bone still preserved its histological integrity.

CHARACTER OF BONE DAMAGE

Distinctive clumps of dark matter were evident in all the thin sections of the bones studied. This material was predominantly associated with the channels in the bone, but also extended into the surrounding bone matrix and was more highly concentrated in the perimedullary regions (Figs. 8.6A, 8.6B). Under high magnification, individual "clear" structures could be observed (Figs. 8.6C, 8.6D). These appear to be short, branching structures/tunnels that measured about 4–10 μm in width (Fig.

8.6D). In some cases, the tunnels appear to connect to neighboring channels. Under polarized light, the clear tunnels showed a birefringence. These tunnel structures closely resemble those described by Jans et al. (2004) and Jans (2005) in archaeological bone, and this was confirmed (M. M. E. Jans, pers. comm. 2010). It appears that these structures are Wedl tunnels which are caused by saprophytic (saprobe) fungi (Jans 2005): fungi that feed on decomposing organic matter. These deductions also agree with that made by Marchiafava, Bonucci, and Ascenzi (1974), who conducted experimental studies on fungi of the family Mucoraceae that decompose bone. They found that in buried and unburied bone fungi of the genus *Mucor* bored through the bone tissue, which suggested that the erosive activity itself may have been associated with feeding. Given the similarity with the structures described by Jans et al. (2004), Jans (2005), and Marchiafava, Bonucci, and Ascenzi (1974), we deduce that the dense dark matter in the bones of this scylacosaurid was formed by saprophytic fungi and that the Wedl tunnels observed are the result of the degradation by the fungi. The distinctive birefringence observed in the tunnel structures suggests that they contain mineral deposits, which would agree with Marchiafava, Bonucci, and Ascenzi's (1974) finding that the channels in their experimental bones were well calcified. In an attempt to identify the nature of the mineral deposits in these Wedl tunnels, we conducted scanning electron microscope probe analysis.

Scanning Electron Microscope Probe Analysis of the Wedl tunnels

The scanning electron microscope probe analysis showed that the Wedl tunnels comprised calcium carbonate (Figs. 8.7A, 8.7C) while the undegraded bone nearby clearly consisted of fluorapatite (Figs. 8.7B, 8.7D). These results were verified by repeating the energy dispersive spectrocopy analysis of the walls of six different Wedl tunnels and three areas of bone. This result is consistent with the observation made by Marchiafava, Bonucci, and Ascenzi (1974) which showed that in the modern buried and unburied bones, fungi of the genus *Mucor* bored into the bone and resulted in pits and channels that were well calcified. They further postulated that perhaps *Mucor* utilized phosphates in the bone for metabolism, allowing it to spread through the bone. It is uncertain whether the fungus present in our scylacosaurid bones is indeed *Mucor*, especially since Machiafava, Bonucci, and Ascenzi (1974) acknowledge that there are possibly other fungal genera that are able to bore into bone, as well as attack rocks or solubilize phosphates.

Tibia

Sections of the tibia (Table 8.1) revealed that although there was a lot of dark matter and tunneling evident in the bone, some histological details were evident. The bone wall appears to be have been inundated by a large number of channels (Fig. 8.6A). Two LAGs are evident near the outer edge of the bone, and more internally, a narrow annulus of lamellar bone

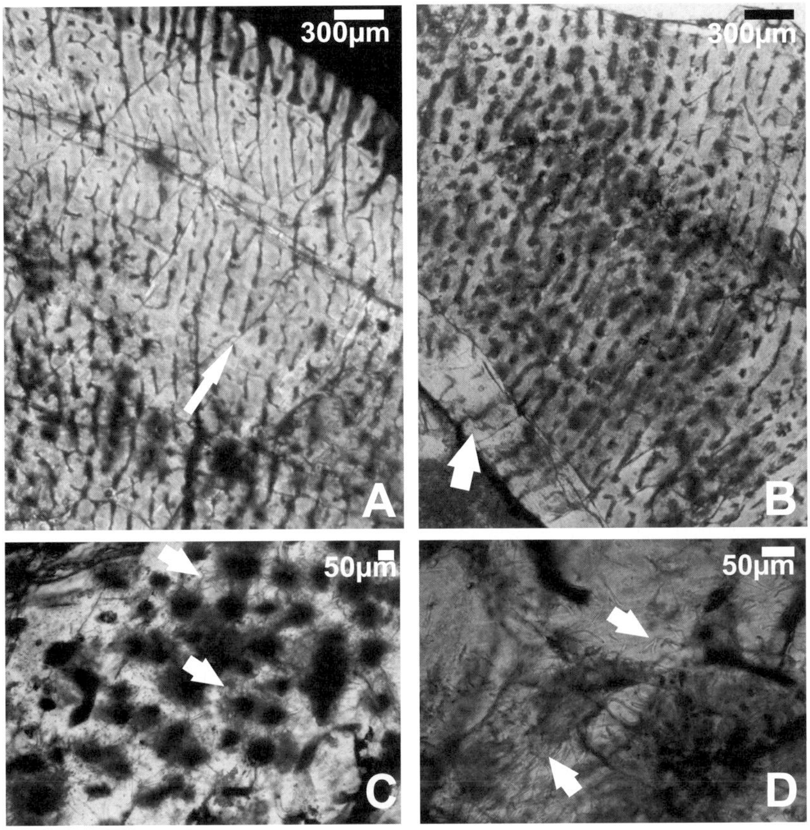

8.6. Scylacosaurid therocephalian, SAM-PK-5018. (A) Tibia (section A1) showing the radial orientation of the channels and the association of clusters of dark matter. Arrow indicates an annulus. (B) Tibial section showing the inner circumferential layer (arrow) of bone around the medullary cavity. (C) Higher magnification of the fungal clusters and the Wedl tunnels (arrows). (D) Higher magnification showing the Wedl tunnels (arrows).

is visible (Fig. 8.6A). The medullary cavity has a large amount of cancellous trabeculae, but on one side of the same section, there is a distinctive inner circumferential layer) comprising lamellar bone tissue (Fig. 8.6B).

Radius

Transverse sections of the radius (Table 8.1) revealed a compact thick bone wall that surrounded a small, vacant medullary cavity. The bone appears to have been richly vascularized. Since the tunneling seems to preferentially occupy the channels and extend from there into the bone, it is uncertain if the "reticular" arrangement is a normal feature or whether it is a consequence of the fungal tunneling. Interestingly, no osteocyte lacunae are preserved. The perimedullary region has a large amount of secondary reconstruction. Although there are many erosion cavities visible, secondary osteons were not evident.

Femur

Two thin sections of the bone were prepared (Table 8.1) and, as observed in radius sections, the fine histological details were not preserved, although the overall thickness and general orientation of the channels could be assessed. In Ray, Botha, and Chinsamy (2004), this section was misidentified in Figure 3A as a femur of *Pristerognathus* (SAM-PK-11557).

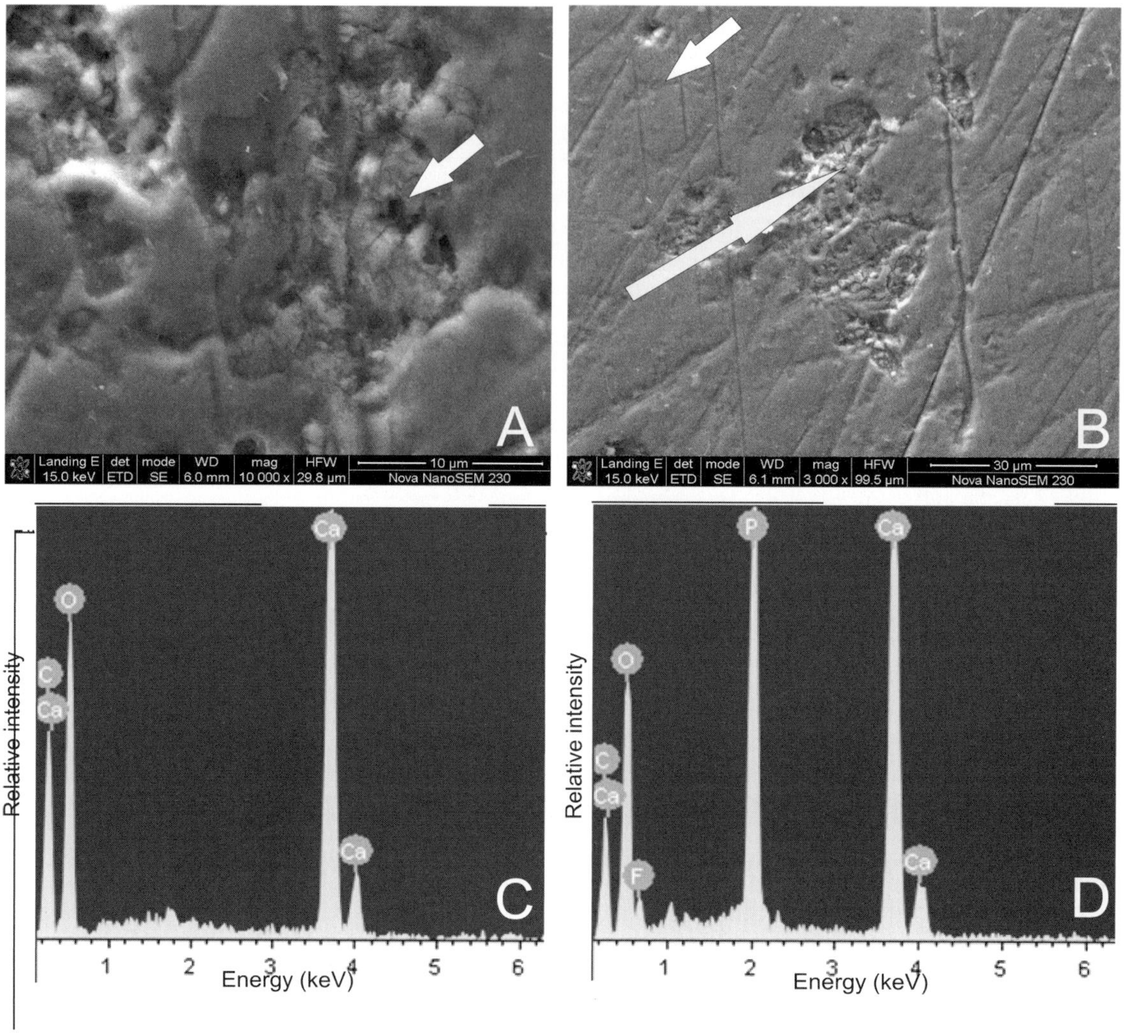

Except for one patch of the bone wall, osteocyte lacunae were not preserved. Around the medullary cavity, secondary reconstruction was clearly underway, but the fungal attack obliterated histological details, and we could not see any secondary osteons. A possible growth ring is evident toward the periphery, and there may be another one present near the peripheral margin.

Humerus

Two sections of the humerus were studied (Table 8.1). Even though the outermost (periosteal) surface of the bone is weathered, the remaining bone wall is fairly thick. The vascularization appears to be predominantly circumferential toward the periosteal surface, although more internally, the connections between neighboring vessels appear to be greater, resulting in a more plexiform arrangement. However, here again, it may not reflect the actual vascular channel connections, but rather the exploitation of the fungal growth in the bone. Near the peripheral edge of the bone, a growth ring is evident.

Fibula

This section was from the metaphyseal part of a fibula (Table 8.1). The histology of this bone is not well preserved because of the fungal growth.

Scylacosaurid, SAM-PK11557

Ribs

Multiple sections were taken of each of the three ribs (Table 8.1) and generally they show similar features. They all appeared elliptical in cross section and had variable thicknesses of compact bone wall that surrounded a central medulla that was occluded by cancellous bone (Figs. 8.8A, 8.8B). All the sections of the ribs are extensively secondarily reconstructed, and a large number of erosion cavities and secondary osteons are visible in the compacta (Figs. 8.8A to 8.8D). It is evident that besides the amount of periosteal bone present, the different sections varied also in terms of the extent of secondary reconstruction. In the perimedullary region, the erosion cavities are enlarged, and some show secondary deposits (Figs. 8.8A, 8.8C, 8.8D), whereas others reach cancellous dimensions (e.g. Fig. 8.8B). Figure 8.8A depicts a thin section of a rib (specimen 3, section A) showing a thick bone wall comprising fibrolamellar bone with predominantly longitudinal vascular canals with a few radial connections. The compacta is interrupted by three or four growth rings, whereas in another rib section (specimen 2A1), only two growth rings are visible. Interestingly the ribs tend to show groups of multiple LAGs (Fig. 8.8C) that interrupt the fibrolamellar bone tissue, which suggests that during the unfavorable season, the bones grew in fits and starts (i.e., a little bone was deposited then bone deposition stopped completely and thereafter it resumed). Most of the ribs show some weathering on the periosteal edge, so an accurate estimate of bone wall thickness could not be assessed. Abundant Sharpey's fibers are visible in one of the ribs sectioned (specimen 3, section A2; Fig. 8.8D) (Table 8.1), which may be related to intercostal muscle attachment.

Fibula

The fibula was serially sampled four times. Besides the expected differences in terms of periosteal bone, LAGs, and such, it was interesting to note that the histology across even single cross sections of the fibula was quite variable. For example, section A2 shows a large number of longitudinally oriented channels located in a fibrolamellar bone matrix; however, in a localized region, the channels tend to be radially oriented. In the same thin section, one part of the primary bone has three or four growth rings visible (Fig. 8.8F), whereas directly opposite this side of the bone

8.7. Scanning electron microscope energy dispersive spectroscopy analysis of Wedl channels and surrounding bone. (A) The arrow indicates the wall of the Wedl channel that was probed. (B) The short arrow indicates a region of the bone wall without any fungal growth that was probed. The long arrow points to a region with fungal growth which has a distinctive "rough" texture with tiny pores that are the Wedl tunnels. The elemental spectra for the walls of the Wedl tunnels and the undegraded bone are depicted in (C) and (D), respectively. Note that the unaltered bone (D) contains phosphorous and fluorine (indicative of fluorapatite), while the walls of the Wedl channel have a distinct lack of phosphorous and fluorine, which is suggestive of the presence of calcium carbonate.

8.8. Scylacosaurid therocephalian, SAM-PK-11557, (A)–(D), ribs. (A) Specimen 3A1 shows a general view of the compacta including growth rings (arrows) that interrupt the fibrolamellar bone, and the extensive development of secondary reconstruction that leads to cancellous bone around the medullary cavity. (B) Specimen 1A shows a narrow region of primary compact bone followed by a cancellous bone tissue that extends into the medullary cavity. (C) Specimen 2A1, here it is evident that the growth rings consist of groups of multiple LAGs (arrows). (D) Specimen 3A2, arrows indicate bundles of Sharpey's fibers (arrow). (E) Fibula, at high magnification, shows the secondary deposits of lamellar bone around the secondary osteons. (F) Fibula section showing an overview of the histology of the fibula with four LAGs (arrows) that interrupted the deposition of fibrolamellar bone. Note the development of erosion cavities more internally toward the medullary cavity. (G) This image is taken directly opposite the previous image (F), and it shows a thick layer of lamellar bone with several LAGs (arrow).

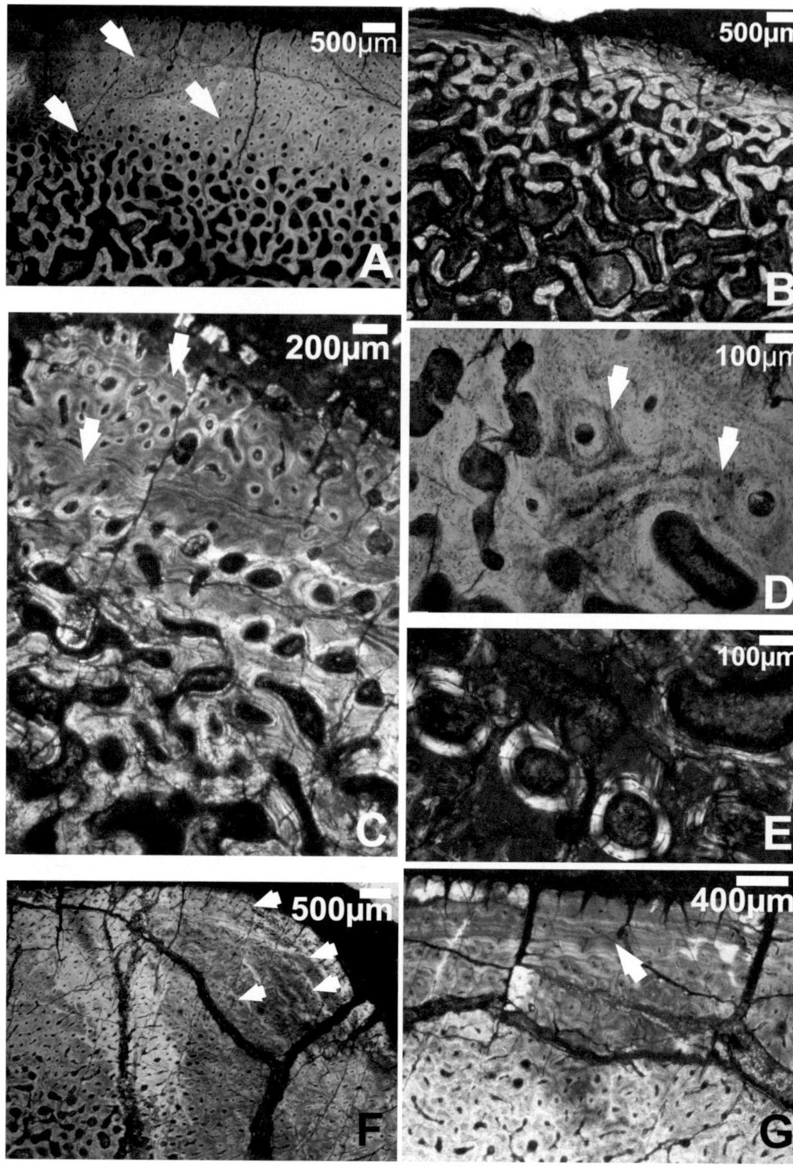

wall, a group of rest lines are evident (Fig. 8.8G). Toward the medullary cavity, extensive secondary reconstruction occurred, and there are a large number of enlarged channels and a few secondary osteons present (Figs. 8.8E, 8.8F). In some parts of the cross section, a region of compacted coarse cancellous bone is also visible. Section A1, which is more distal to section A2, shows a similar histology, although here only three growth rings are present.

SUMMARY OF THE SCYLACOSAURID HISTOLOGY

The presence of fibrolamellar bone tissue interrupted by LAGs and/or annuli in both individuals sampled suggests that they experienced alternating fast and slowed rates of growth. A micrograph of the rib of *Pristerognathus* in de Ricqlès (1969, Plate 3, Fig. 2) also appears to have several growth cycles. In none of the images of the *Theriognathus* radius illustrated by de Ricqlès (1969, Plate 4) are any growth rings visible. The compacta does show quite extensive secondary reconstruction, and it is possible that even if they were present, they would have been obliterated during remodeling. Unfortunately, there was no indication of the age of the individual, so the lack of rest lines may be as much a reflection of its immaturity. Thus, it appears that our two scylacosaurid individuals and the scylacosaurid specimens sampled by de Ricqlès (1969), except for the *Theriognathus* radius mentioned previously, all experienced cyclical patterns of growth. As in extant animals that show similar alternating rates of growth (e.g., lepidosaurs, crocodilians [Castanet el al. 1993; Chinsamy-Turan 2005]), it is possible that environmental conditions influenced the growth strategies of early therocephalians, such as the scylacosaurids. Except for the fibula that showed rest lines in one region but not in the opposite of the same section, none of the other bones sampled exhibit evidence of growth having virtually ceased. We therefore cannot assess whether the rest lines present in the fibula reflect growth ceasing in that particular element or whether it reflects determinate growth. Nevertheless, other characteristics of the bone tissues of the other elements (e.g., decrease in vascularization toward the periphery and extensive secondary reconstruction) suggest that the individual was well past its most rapid stage of growth and was perhaps reaching the plateau stage of its growth cycle. We also found significant histovariability in terms of the location of the section in a particular skeletal element, as well as within the cross section of a bone. Variable accounts of LAGs were evident since extensive secondary reconstruction was found to have obliterated much of the bone formed during the early stages of ontogeny. Interestingly, except for the fibula sampled (and only a localized region of that cross section), none of the other bones exhibited closely spaced rest lines. Between the two individuals sampled, there also appears to be differences in terms of the extent of secondary reconstruction: specimen SAM-PK-11557 showed more extensive secondary remodeling. Except for the fibula, the skeletal remains of the individuals comprised different parts of the skeleton, and we could not confidently assess the size difference of the two individuals. However, considering the fibula mid-shaft diameter of 18.59 mm for SAM-PK-5018 and 24.29 mm for SAM-PK-11557, the latter appears to be a larger/older individual. Perhaps this age difference is related to the more extensive development of secondary remodeling in the compacta of SAM-PK-11557.

This study also conclusively documented the presence of saprophytic fungi which acted on the bones of the scylacosaurid SAM-PK-5018 and caused severe degradation of the microstructure of the fossilized bones. This is the first evidence of the action of fungi on Permian-aged fossil bones.

Comparison of the Bone Histology of the Therocephalians and Gorgonopsians with that of Cynodonts

Like the cynodonts (see, e.g., Botha and Chinsamy 2004, 2005; chapter 9 of this book), it is evident that both the therocephalians and the gorgonopsians have a prominence of fibrolamellar tissues in the microstructure of their bones. Indeed, fibrolamellar bone tissue appears to be plesiomorphic for the nonmammalian therapsid lineage (Ray, Botha, and Chinsamy 2004; Botha-Brink and Angielczyk 2010). Furthermore, the cyclically formed bone tissues in the therocephalians and gorgonopsians, as well as in the majority of the nonmammaliaform cynodonts (chapter 9 this book), and among the nonmammalian synapsids in general are considered to reflect developmental plasticity in the clade (Starck and Chinsamy 2002; Ray, Botha, and Chinsamy 2004). The flexible growth patterns documented for therocephalians, gorgonopsians, and cynodonts show striking similarities in terms of the actual type of bone tissue that forms during the alternating favorable and unfavorable seasons, which suggests that their physiological responses to their environment were also similar. It is evident, however, that more intensive surveys of both basal and more derived therocephalians and gorgonopsians are needed to investigate whether there is a parallel trend toward increasingly rapid rates of growth in the lineages, as has been documented among the nonmammaliaform cynodonts (chapter 9 of this book). Interestingly, among all the nonmammaliaform cynodonts studied, only four taxa—*Cynognathus, Scalenodontoides, Traversodon,* and *Protuberum*—show rapid uninterrupted rates of growth (chapter 9 of this book). However, whether the rapid rates of growth are sustained throughout ontogeny still has to be verified in the latter taxa. As for the pattern of growth, determinate growth strategies are also evident among several of the nonmammalian cynodonts studied (e.g., *Diademodon* [Botha and Chinsamy 2000]; *Tritylodon* [chapter 9 of this book]). However, among the gorgonopsians and therocephalians, it is still uncertain whether they experienced determinate or indeterminate growth, although the *Scylacops* individual seems to suggest a determinate growth pattern. Thus, it is evident that more research is also needed to explore the histological changes during ontogeny for both gorgonopsians and therocephalians.

General Comments

Variations in histology have been documented before, especially between different skeletal elements (e.g., Horner, de Ricqlès, and Padian 1999). However, here we document not only histovariability in different elements of the skeleton, but also how the histology changes in serial sections across short distances of particular elements, as well as, within single cross sections (e.g., Figs. 8.1, 8.2, 8.3). This study has shown that the histology of a bone depends on a number of factors: such as, the particular skeletal element (i.e., which bone is sectioned), the location of the section within the element (i.e., proximal, distal metaphyseal, or epiphyseal), and also the actual location of the tissue within a cross section of a bone. Since the histological study of bones is necessarily destructive, collection managers are often reluctant to give permission for thin-sectioning (Chinsamy-Turan 2005). As such, studies of the histology of fossil bones often rely on opportunistic sampling of whatever material is available, and in many cases, only fragments of bones are available for thin-sectioning. The variation across a single thin section is also worth noting since often cores or only fragments of bones are available for histological analyses. Thus, histological assessments are often biased by these factors, and researchers need to be aware of this. This study highlights the importance of preparing multiple thin sections from single bones, as well as multiple bones from the skeletons of individuals (Horner, de Ricqlès, and Padian 1999) (as opposed to single sections or cores from fossil bones) to enable reliable and meaningful deductions about growth patterns and biology. Thus, as far as possible, sampling procedures need to be rigorous, and osteohistologists need to be aware of the limitations and inherent biases of their samples as well as the interpretations derived from them.

Acknowledgments

Miranda Jans is thanked for her comments on the occurrence of Wedl tunnels in the bones of the scylacosaurid SAM-PK-5018. Adam Huttenlocker and Jennifer Botha-Brink are acknowledged for having reviewed this manuscript. Roger Smith and Sheena Kaal of Iziko South African Museum are thanked for facilitating access to the gorgonopsian and therocephalian material. Daniel Thomas and Miranda Waldron are acknowledged for assistance with the scanning electron microscope analysis of the therocephalian bones. The National Research Foundation, South Africa, and Department of Science and Technology, India, are acknowledged for research support for A. Chinsamy-Turan and Sanghamitra Ray, respectively.

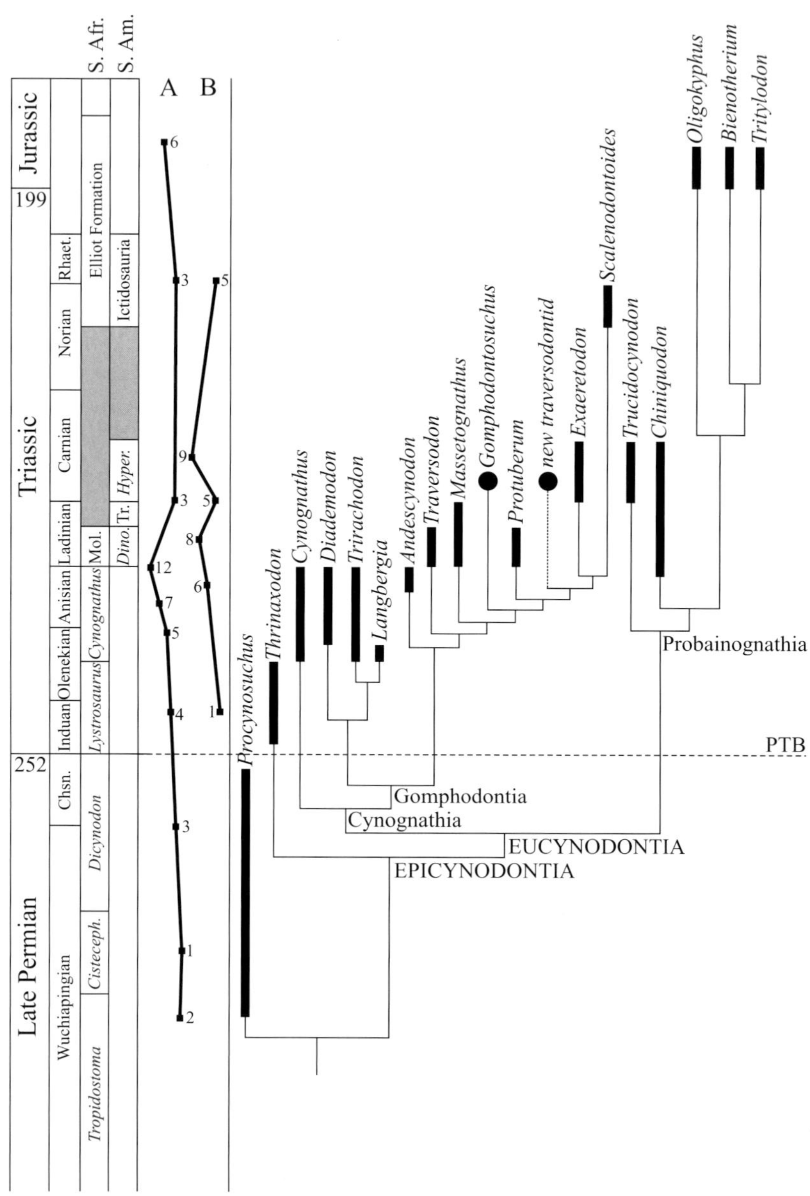

9.1. Phylogenetic relationships, stratigraphic ranges, and generic diversity of nonmammaliaform cynodonts. The phylogeny only includes cynodonts that have been examined histologically. Generic diversity is shown in (A) Africa and (B) South America. *Abbreviations:* Chsn. = Changhsingian; *Cisteceph.* = *Cistecephalus* AZ; *Dino.* = *Dinodontosaurus* AZ; *Hyper.* = *Hyperodapedon* AZ; Mol = Molteno Formation; PTB = Permo-Triassic boundary; Rhaet. = Rhaetian; S. Afr. = South African AZs; S. Am. = South American AZs. Numbers indicate millions of years ago. Shaded areas represent a hiatus. Stratigraphic chart follows Botha-Brink and Abdala (2008) and Abdala and Ribeiro (2010).

The Radiation and Osteohistology of Nonmammaliaform Cynodonts

9

Jennifer Botha-Brink, Fernando Abdala, and Anusuya Chinsamy-Turan

Introduction

Cynodontia is the last major therapsid lineage to appear in the fossil record (Rubidge and Sidor 2001) and includes mammals as living representatives. The extensive documentation of the acquisition of mammalian features in Paleozoic-Mesozoic nonmammaliaform cynodonts has prompted intense research into the origin and evolution of the Mammalia and the development of mammalian biology. The earliest nonmammaliaform cynodonts (herein referred to as cynodonts) are known from Upper Permian deposits in the Beaufort Group, Karoo Supergroup of South Africa (Botha, Abdala, and Smith 2007; Botha-Brink and Abdala 2008) (Fig. 9.1) and their record extends into the Lower Cretaceous of Russia (Tatarinov and Matchenko 1999). The oldest cynodonts are represented by only two small specimens (SAM-PK-K10369, the holotype of *Charassognathus gracilis*, and SAM-PK-K10138, a referred specimen of *Procynosuchus delaharpeae*) recovered from the middle Late Permian *Tropidostoma* Assemblage Zone (AZ) of South Africa (Botha, Abdala, and Smith 2007; Botha-Brink and Abdala 2008) (see Plate 1B). However, they became more common toward the close of the Permian, where they are best represented by *Procynosuchus* (Kemp 1979, 1980a; Abdala and Allinson 2005; Botha-Brink and Abdala 2008; Weide et al. in press). This taxon is the only cynodont known to have crossed two assemblage zones in the Karoo Basin (from the middle Late Permian to the end of the Permian) and is consequently one of the longest-ranging cynodonts known (Botha-Brink and Abdala 2008) (Fig. 9.1).

Cynodonts more derived than *Charassognathus*, *Procynosuchus*, and the Late Permian Russian *Dvinia* comprise the epicynodonts (including *Cynosaurus*, *Nanictosaurus*, *Progalesaurus*, *Galesaurus* (see Plate M, top), *Thrinaxodon* (see Plate M, bottom), *Platycraniellus*, and possibly *Bolotridon* [placement debatable] as basal members), as well as the more derived eucynodonts. *Cynosaurus* and *Nanictosaurus* are restricted to the uppermost Permian strata of South Africa, whereas *Progalesaurus*, *Galesaurus*, *Thrinaxodon*, and *Platycraniellus* are all from the Lower Triassic of South Africa, and in the case of *Thrinaxodon*, from equivalent strata in Antarctica and India as well (Brink 1954; Colbert and Kitching 1977; Hammer 1990; Rubidge 1995a; Abdala and Ribeiro 2010). As these basal epicynodonts (apart from the Late Permian *Cynosaurus* and *Nanictosaurus*) first appear above the Permo-Triassic boundary, they play a

Table 9.1. Localities and measurements (mm) of the new material examined in this study

Taxon	Accession number	Locality	Assemblage zone	Total length (mm)	Midshaft width (mm)	Proximal width (mm)	Distal width (mm)
Protuberum							
Rib	PV-1009T	Rincão do Pinhal, Rio Grande do Sul	*Dinodontosaurus*	—	15.9	68.7	—
Traversodontid							
Humerus	—	Arroio Cancela, Santa Maria, Rio Grande do Sul	*Hyperodapedon*	—	16.6	—	47.5
Femur	—	Arroio Cancela, Santa Maria, Rio Grande do Sul	*Hyperodapedon*	107.9	14.8	40.5	39
Scalenodontoides							
Limb bone	NMQR3053	Patriotsklip, Jamestown	*Euskelosaurus*	—	19.5	—	—
Truciodocynodon							
Radius	PV-1053T	Santa Maria, Rio Grande do Sul	*Hyperodapedon*	123.2	13.1	23.6	15.5
Tritylodon							
Femur	BP/1/5516	Mequatling 278, Clocolan	*Massospondylus*	—	8.7	—	22.3
Tibia	BP/1/5167a	Bramleyshoek 52, Bethlehem	*Massospondylus*	—	—	—	19.5
Rib	BP/1/5167b	Bramleyshoek 52, Bethlehem	*Massospondylus*	—	—	—	—
Fibula	BP/1/5167e	Bramleyshoek 52, Bethlehem	*Massospondylus*	—	8	—	11.1
Humerus	BP/1/4785d	Fouriesburg	*Massospondylus*	—	—	—	28.7
Ulna	BP/1/4785e	Fouriesburg	*Massospondylus*	—	—	—	15.4
Fibula	BP/1/4785f	Fouriesburg	*Massospondylus*	—	6.1	7.2	—

Note: The traversodontids *Protuberum*, the new unnamed traversodontid and *Trucidocynodon* are from Brazil, whereas the traversodontid *Scalenodontoides* and the tritylodontid *Tritylodon* are from South Africa. The *Euskelosaurus* and *Massospondylus* zones of the Elliot Formation are informal range zones and not formal assemblage zones.

fundamental role in the Early Triassic recovery after the end-Permian extinction event and consequently, are important for understanding the dynamics of the recovery phase (Botha and Smith 2006) and the effect of the extinction on cynodont biology.

All the remaining epicynodonts form a monophyletic group known as the Eucynodontia. The earliest representatives of this clade are known from the late Early Triassic, but they exhibit an important radiation during the Middle Triassic. Several modifications toward the mammalian condition appeared in members of this group at this time (Kemp 2005; chapter 1 of this book). The eucynodonts comprise two major divisions—the Cynognathia and the Probainognathia (Fig. 9.1). The Cynognathia include the large carnivorous *Cynognathus* (see Plates N and O) and the Gomphodontia (sensu Abdala and Ribeiro 2003). This latter group comprises the families Diademodontidae, Trirachodontidae, and Traversodontidae and is characterized by specialized bucco-lingually expanded postcanines known as gomphodont teeth, indicative of an herbivorous/omnivorous diet (Kemp 2005; Abdala, Neveling, and Welman 2006). The second major division comprises the primarily carnivorous Probainognathia, which includes among others, the diverse and highly specialized herbivorous Tritylodontidae (Sues and Jenkins 2006), the small carnivorous Tritheledontidae (Sidor and Hancox 2006), and Mammaliaformes (Kemp 2005). Although the monophyly of Mammaliaformes (= Mammalia of Hopson 1994; Luo 1994) is well established, there is still a lack of consensus as to which eucynodont group is its sister taxon (Kemp 2005; Liu and Olsen 2010; chapter 1 of this book).

In this chapter, we provide an overview of the diversity and radiation of the nonmammaliaform cynodonts, and we provide information pertaining to aspects of their paleobiology. We then focus on the paleohistology of cynodonts considering previously published research (e.g., de Ricqlès 1969; Botha and Chinsamy 2000, 2004, 2005; Ray, Botha, and Chinsamy 2004; Chinsamy and Abdala 2008), as well as, novel material representative of five nonmammaliaform cynodont taxa (Table 9.1).

Institutional Abbreviations

This chapter abbreviates the names of the following institutions: Bernard Price Institute for Palaeontological Research, University of the Witwatersrand, Johannesburg (BP/1/); National Museum, Bloemfontein (NMQR); Iziko South African Museum, Cape Town (SAM); Tom S. Kemp collection, Oxford University (TSK); Universidade Federal do Rio Grande do Sul, Brazil (UFRGS PV).

Phylogenetic Context: Diversity and Radiation of the Nonmammaliaform Cynodonts

All therapsid lineages (i.e., Dinocephalia, Anomodontia, Gorgonopsia, Therocephalia) first appear in Middle Permian deposits apart from the Cynodontia, which is first recorded at the base of the Upper Permian (Botha-Brink and Abdala 2008). Two cynodont taxa, *Charassognathus* and *Procynosuchus*, are represented at this time in the *Tropidostoma* AZ of the Karoo Basin (Botha-Brink and Abdala 2008) (Fig. 9.1; see Plate 1B), implying an even older origin of the lineage. The Cynodontia are the least represented therapsid group in the Permian and their most significant radiations only occurred during the Triassic and Jurassic periods (Abdala and Ribeiro 2010), with an important contribution of stem mammaliaforms during the Jurassic. Apart from the first records of *Charassognathus* and *Procynosuchus* from the *Tropidostoma* AZ, only *Procynosuchus* is known from the overlying *Cistecephalus* AZ (represented by two individuals [Botha, Abdala, and Smith 2007]). The first cynodont radiation only becomes apparent in the overlying Upper Permian *Dicynodon* AZ (see Plate 1B) and is represented by three taxa, with *Procynosuchus* again being well represented by several specimens. *Procynosuchus* is also known from Tanzania, Zambia, Germany, and Russia, and thus represents the first cosmopolitan cynodont in the fossil record (Abdala and Ribeiro 2010). In Russia, Permian cynodonts are represented by *Procynosuchus* and *Dvinia*, the latter having several autapomorphies that make it a most peculiar basal cynodont (Tatarinov 1968). Other Russian taxa, *Nanocynodon* and *Uralocynodon* (see Battail and Surkov 2000), were examined by one of us (F. A.) and are here considered to be juvenile specimens of *Procynosuchus*. We also consider *Cyrbasiodon vladimiriensis* described by Tatarinov (1994) to be a synonym of *Procynosuchus* (see Kammerer and Abdala 2009).

The end-Permian extinction had a dramatic effect on most tetrapod lineages in South Africa. All gorgonopsians disappeared, only one dicynodont genus, namely *Lystrosaurus* crossed the Permo-Triassic boundary (although at least three ghost lineages of Triassic dicynodonts can be traced to the Permian [Fröbisch 2007]), and there was a complete faunal turnover of therocephalians and cynodonts (Smith and Ward 2001; Abdala and Ribeiro 2010; Botha-Brink and Modesto 2011). In terms of diversity, three cynodont taxa are recorded from the uppermost Permian and four from the lowermost Triassic in South Africa. When their global diversity is considered, the number of taxa (four) remains the same before and after the extinction event. From a taxonomic perspective, however, none of the Permian cynodonts survived into the Triassic and there was a complete taxic replacement. Some of the taxa such as the Permian *Cynosaurus* and the Triassic *Galesaurus* (see Plate M, top) and *Progalesaurus* are closely related (Sidor and Smith 2004) and when phylogenetic hypotheses and ghost lineage extensions are considered, at least two cynodont lineages crossed the Permo-Triassic boundary (Sidor and Smith 2004; Abdala 2007). *Thrinaxodon* (see Plate M, bottom) is the most common cynodont from the Lower Triassic of the Karoo Basin and, similar to *Procynosuchus* during the Permian, is the only Early Triassic cynodont represented in extra-African deposits (Colbert and Kitching 1977; Satsangi 1987).

The Late Olenekian and Anisian is represented in South Africa by the *Cynognathus* AZ (see Plate 1B), in which there is a remarkable increase in cynodont diversity (Fig. 9.1) (chapter 2 of this book). This faunal assemblage is informally subdivided into three portions (Hancox et al. 1995): Subzone A is interpreted as Late Olenekian, Subzone B (the classical *Cynognathus* AZ previous to this subdivision) as Early Anisian, and the geographically restricted Subzone C is interpreted as Late Anisian. Cynodont diversity remains the same between the lowermost Triassic *Lystrosaurus* AZ and the overlying *Cynognathus* Subzone A. However, an important addition is observed in the latter assemblage zone, namely the first record of gomphodont cynodonts, represented by the trirachodontids. This group is also recorded in an Olenekian fauna from China (Gao et al. 2010) and is the only Laurasian cynodont record of this age. Cynodont diversity in the Karoo Basin increased slightly during the Lower Anisian *Cynognathus* Subzone B, but thereafter decreased again during the Upper Anisian *Cynognathus* Subzone C. It is during the Late Anisian, however, that cynodonts reached an important peak in their diversity, as at least seven putative faunas are recorded in Gondwana at this time, faunas that were boosted mainly by the explosive record of the traversodontids in Africa and South America (Abdala and Ribeiro 2010). In contrast, cynodonts of this age are represented only by one trirachodontid skull from China in Laurasia (Sun 1988), as previous records of gomphodont cynodonts from the Middle Triassic of Russia have been re-identified as bauriid therocephalians (F. Abdala, pers. observation 2009).

Ladinian cynodonts are well represented in South America (Schultz 2000; Abdala and Ribeiro 2010) and from the Late Ladinian/Early Carnian, in Madagascar as well (Flynn et al. 2000; Kammerer, Flynn, and Ranivoharimanana 2010). An isolated traversodontid postcanine is also known from the Ladinian of Germany (Hopson and Sues 2006). Several Gondwanan Carnian cynodonts, both probainognathians and traversodontids, are known from South America, and they have also been recovered from India (Chatterjee 1982). Carnian cynodonts also became more consistent in the fossil record in Laurasia, from basins in the eastern United States, but they are represented by the small traversodontid *Boreogomphodon* (Sues and Olsen 1990; Sues and Hopson 2010; Liu and Sues 2010) and the tiny sectorially toothed dromatherids (Sues 2001).

The Norian (probably Upper Norian-Rhaetian, see Abdala and Ribeiro 2010) is represented in South Africa by the lower Elliot Formation, which also records an important dinosaur fauna (Yates and Kitching 2003) and contains three cynodont taxa. This formation contains the last record of traversodontid cynodonts from Gondwana and the first tritheledontid from Africa (Battail 2005; Sidor and Hancox 2006). Faunas of similar age are also represented in South America and include several small cynodonts from southern Brazil (Bonaparte, Ferigolo, and Ribeiro. 2001, Bonaparte et al. 2003) and a restricted record of Tritheledontidae from Argentina (Bonaparte 1980; Martinelli and Rougier 2007). Norian and Rhaetian faunas are well represented in Laurasia (particularly Europe) where several taxa are represented by isolated teeth (Godefroit and Battail 1997). This European record signifies the highest diversity peak of nonmammaliaform cynodonts (more than twenty taxa), although the phylogenetic affinities of some of these taxa are currently under debate (Hopson and Sues 2006).

Early Jurassic Gondwanan cynodonts are represented by the Upper Elliot and Clarens formations in the Karoo Basin and the Kota Formation in India. At this time, the lineage was also common in Laurasia (North America, Europe, and Asia) and there was yet another important diversity peak, which was influenced almost equally by both the tritylodontids and mammaliaforms. Tritylodontids from the Early Cretaceous of Russia and Japan (Tatarinov and Matchenko 1999, Setoguchi, Matsuoka, and Matsuda 1999), represent the youngest record of nonmammaliaform cynodonts (when excluding the dubious record from the Paleocene of Canada [Fox, Youzwyshyn, and Krause 1992; Sues 1992]).

Paleobiological Context

The Cynodontia have an extensive fossil record, which has provided the opportunity for numerous morphological studies to be conducted (e.g., Brink 1955; Barghusen 1968; Jenkins 1971; Kemp 1982; Bennett and Ruben 1986; Hillenius 1992; Hopson 1994; Sidor 2003, Abdala 2007). These studies have shown that the cynodonts acquired increasingly mammal-like morphological characteristics during their evolution, including an enlarged dentary, reduced postdentary bones, a bony secondary palate, complex occluding multicuspid postcanines, a supplementary contact between the surangular and squamosal lateral to the quadrate-articular craniomandibular joint, an acromion process on the scapula, a mammalian phalangeal formula, and an increasingly parasagittal gait (Brink 1955; Jenkins 1971; Kemp 1982; Hopson 1994; Rubidge and Sidor 2001). This research has increased our knowledge on the phylogeny, neurosensory system, food acquisition and processing, locomotion, biomechanical adaptations, and various other aspects of cynodont biology.

Based on studies of three-dimensional reconstructions of the cranial endocast (Macrini, Rowe, and van de Berg 2007) and investigations of the nasal cavity of cynodonts, researchers have been able to deduce information pertaining to the neurosensory system of these animals (Kemp 2005, 2009a). Although reptile brains do not completely fill the cranial cavity and thus do not reflect the actual size or shape of the brain (Edinger 1951), the brains of mammals largely fill the endocranial space (80 to 100 percent [Hopson 1980]; 75 to 100 percent [Quiroga 1980]) and may even leave impressions of gross structures on the internal surface of the skull (de Beer 1937). Earlier studies on paleoneurology (i.e., the study of the nervous system of extinct animals [Jerison 1973]) relied on the preservation of natural endocasts of the cranial cavity or fragmentary braincases that exposed the internal region to investigate cynodont brains (see, e.g., Hopson 1979; Kemp 1979; Quiroga 1979, 1980, 1984; Kielan-Jaworowska 1986). However, recent advances in computed tomography have facilitated the study of cynodont cranial endocasts of relatively complete, intact individuals (e.g., Kemp 2009a), and has provided considerably more detailed information about the evolution of the cynodont brain. Although it is not possible to determine the absolute shape and size of cynodont brains, the various regions, that is, the olfactory bulbs, cerebral hemispheres, and hindbrain, can be distinguished to varying degrees (Kemp 2005). This information reveals that cynodonts had an enlarged cerebellum (hindbrain), whereas the cerebral hemispheres remained small, forming a long, narrow tubular structure in dorsal view (Kemp 2009a). This development of the cerebellum in cynodonts (hypothetic stage 1 in mammalian brain evolution of Kemp [2009a]) has been tentatively interpreted as an indication of enhanced neuromuscular control of the movement of the lower jaw, head, and body (Kemp 2009a). The enlargement of the olfactory bulbs also indicates a well-developed olfactory capability (Jerison 1973; Quiroga 1980; Kemp 2009a). The combined information from cranial endocasts, nasal cavities, and features associated with a process of miniaturization such as a reduced density of the lower jaw bone, which increased sensitivity to high frequency sound (Kemp 2009a), suggests that olfaction and hearing became more important in the derived nonmammaliaform cynodonts and mammaliaforms. Consequently, the development of these features has been used to infer nocturnality in the small derived cynodonts and earliest mammaliaforms (Kielan-Jaworowska, Cifelli, and Luo 2004; Kemp 2005).

The postcranial skeleton of cynodonts has been studied in less detail than the skull. Studies such as those of Kühne (1956), Jenkins (1971), Kemp (1980a, 1980b, 1982), and more recently, Oliveira (2010) and Oliveira, Soares, and Schultz (2010) suggest a progressive increase in locomotor agility during the evolution of the cynodonts. The pelvic girdle and hind limb developed the mammalian parasagittal form of locomotion earlier than the pectoral girdle and forelimb. This is clearly seen in the more basal eucynodonts, where although the musculature had increased in size relative to more basal therapsids, the forelimb retained a fairly sprawling orientation. However, changes in bone morphology in the later cynodonts such as the tritylodontids resulted in an almost entirely mammalian appendicular skeleton, facilitating a virtually mammalian-grade of agility (Jenkins and Parrington 1976; Kemp 2005).

Whereas the paleobiology of cynodonts has undergone intense scrutiny, their paleoecology is less well studied. As Permian cynodonts are relatively rare and are primarily known from skull material, their lifestyle habits remain poorly understood. Kemp (1980a) studied the postcrania of a *Procynosuchus* specimen (TSK 34), and on the basis of horizontally oriented zygapophyses on the lumbar vertebrae, particularly flat limb bones and feet and elongated hemal arches on the caudal vertebrae, he suggested that *Procynosuchus* was semi-aquatic (Kemp 1980a). Nothing is known regarding the lifestyle habits of other Permian cynodonts. Triassic and Early Jurassic cynodonts, however, are known from numerous postcranial skeletons and at times their exceptional taphonomic preservation has facilitated deductions regarding their lifestyle preferences. For example, fossilized burrows containing cynodont remains have been recorded for the basal epicynodont *Thrinaxodon* and the trirachodontid *Langbergia* (Groenewald, Welman, and MacEachern 2001; Damiani et al. 2003). More recently, a complete skeleton of the basal epicynodont *Galesaurus* was recovered associated with an owenettid procolophonoid (Abdala, Cisneros, and Smith 2006). The degree of articulation and preservation of delicate structures in both skeletons led to the suggestion that these animals were most likely entombed within a burrow. Several fully articulated, well-preserved *Galesaurus* skeletons, preserved as aggregations, have also recently been recovered, supporting the possibility of burrowing behavior (J. Botha-Brink, pers. observation 2008). Others, such as the tritylodontids *Tritylodon* and *Kayentatherium* exhibit morphological features typical of active burrowers (e.g., long, robust olecranon process on the ulna [Sues 1985; Sues and Jenkins 2006]), although a recent study of *Kayentatherium* postcrania suggests that these morphological adaptations reflect a semi-aquatic lifestyle (Egberts et al. 2009).

Stable light isotope analysis was recently applied to assess the paleoecology of the Triassic cynodonts *Diademodon* and *Cynognathus* (see Plate O) (Botha, Lee-Thorp, and Chinsamy 2005). The study showed differences in their microhabitat preferences and water-related behavior with *Diademodon* possibly preferring more shady or closed habitats than *Cynognathus.*

Various morphological adaptations in cynodonts have also been used to assess their thermal physiology (see, e.g., Brink 1955; Heath 1968; Hopson 1980; Kemp 1982; Carrier 1987; Hillenius 1992; Hillenius and Ruben 2004). Typically, mammals and birds are homeothermic endotherms (with some exceptions) that maintain relatively high and constant body temperatures, despite ambient temperature variation, and generate their own internal heat resulting in elevated resting metabolic rates (Hillenius and Ruben 2004; Kemp 2006b; see chapter 11 this book). In contrast, other vertebrates (with a few exceptions) are generally poikilothermic ectotherms. They obtain heat from an external source and have relatively low resting metabolic rates compared with those of mammals and birds. Poikilothermic ectotherms are more reliant on ambient temperature variation, although many engage in thermoregulatory behavior (e.g., heliothermy: heat gain from solar radiation, or thigmothermy: heat gain from contact with a warm object such as a rock) in an attempt to maintain constant core body temperatures (Pough, Heiser, and McFarland 1996). As all extant mammals are endothermic and nonmammaliaform cynodonts are closely related to mammals, the evolutionary origin of endothermy in this lineage has been and continues to be the subject of intense research (see chapters 10 and 11 of this book). Although individual morphological features cannot be used to unequivocally demonstrate endothermy, when considered together, they suggest that the later cynodonts had relatively higher ventilation and metabolic rates than reptiles.

Osteohistology of Nonmammaliaform Cynodonts

Armand de Ricqlès (1969) was the first researcher to investigate the bone microstructure of nonmammaliaform cynodonts: *Thrinaxodon;* the traversodontids *Traversodon* and *Gomphodontosuchus;* the chiniquodontid *Belesodon* (now considered to be a junior synonym of *Chiniquodon* [Abdala and Giannini 2002]); and the tritylodontids *Tritylodon, Bienotherium,* and *Oligokyphus.* This early work formed a foundation for future research. More recent studies have considered the ontogenetic changes in bone microstructure, as well as histological variation due to local differences in the skeleton (Botha and Chinsamy 2000, 2004, 2005; Chinsamy and Abdala 2008; Ray, Botha, and Chinsamy 2004). Here we provide a comprehensive overview of previous studies on cynodont bone histology, as well as new data from several taxa, to better understand cynodont growth dynamics and biology.

Current Study: New Materials Assessed and Methods

The new nonmammaliaform cynodont material examined in the current analysis comprises: (1) six limb bones and a rib of *Tritylodon*, the most common tetrapod from Lower Jurassic faunas in South Africa and Lesotho; (2) the midshaft of a limb bone fragment (possible fibula) of the Late Triassic South African traversodontid *Scalenodontoides*, the youngest known traversodontid from Gondwana (Abdala and Ribeiro 2010); (3) a radius of the recently described Late Triassic basal probainognathian *Trucidocynodon* (Oliveira, Soares, and Schultz 2010); (4) a rib of the Brazilian traversodontid *Protuberum* (Reichel, Schultz, and Soares 2009); (5) a humerus and femur of an as yet undescribed Late Triassic Brazilian traversodontid, closely related to *Exaeretodon*.

All limb bones were measured and photographed using a Canon EOS 500D digital camera (Canon, Inc., Lake Success, New York) prior to thin-sectioning. The thin-sectioning method follows that described in chapter 3 of this book. Thin sections were viewed and photographed using a Nikon Eclipse 50i Polarizing microscope (Nikon Instruments, Inc., Melville, New York) and DS-Fi1 digital camera (Nikon), and then analyzed using the image analysis program NIS Elements D 3.0 (Nikon). As the organization and density of the vascular canals in cortical bone is closely related to bone apposition rate (Amprino 1947), vascularization can be compared between different taxa (Chinsamy 1993b). The relative area of the channels within the cortex was quantified using NIS Elements D 3.0. Measurements were standardized using images from every third field of view in the mid-cortical region of the midshaft of each element. As the channels in fossil bone included lymph and nerves as well as vascular canals when the individual was alive (Starck and Chinsamy 2002), quantification reflects the maximum possible vascularization for each element.

Studies have shown that a direct relationship exists between an animal's lifestyle and the structural design of its bones (Bou et al. 1990; Fish 1993; Stein 1989; Wall 1983; Germain and Laurin 2005; Kriloff et al. 2008; Canoville and Laurin 2010), and that, compared to other animals, amphibious and fossorial vertebrates often contain thicker bone walls (cortex ≥30 percent of the average diameter) (Wall 1983; Magwene 1993; Botha 2002; for a comprehensive explanation see Botha 2003). Aquatic animals have also been shown to contain particularly thick, compact cortices (e.g., the manatee, *Dugong dugon*) or have completely infilled medullary cavities with broad transition zones between the compacta and medullary cavity (e.g., the dolphin, *Delphinus delphis*) depending on the extent of the aquatic lifestyle (Canoville and Laurin 2010). The relative bone wall thickness or cortical thickness was quantified in this study using the midshaft region, as it is the least secondarily remodeled and thus best preserves the compact bone wall.

Histological Results and Interpretations

Bone Tissues in Nonmammaliaform Cynodonts

The Permian *Procynosuchus* is the most basal cynodont studied histologically. Ray, Botha, and Chinsamy (2004) found that the bone microstructure of this genus consists of a zonal type of bone tissue (Plate 10A; Fig. 9.2). The radius of *Procynosuchus* consists of rapidly growing fibrolamellar bone with closely spaced lines of arrested growth (LAGs), but peripherally the bone tissue changes to a more slowly deposited lamellar-zonal bone (1.4 percent channel density) (Plate 10A; Figs. 9.2A, 9.2B). The rest of the elements studied consist exclusively of lamellar-zonal bone (0.9 percent channel density) (Figs. 9.2C, 9.2D). The prevalence of lamellar-zonal bone suggests a relatively slow rate of bone formation and therefore slow overall growth (Amprino 1947; de Ricqlès et al. 1991; Chinsamy-Turan 2005). As indicated by the periodic presence of annuli and LAGs, *Procynosuchus* experienced a cyclical growth strategy. Initial growth was rapid (indicated by the presence of fibrolamellar bone), but slowed or ceased during the unfavorable season. The dramatic transition to solely lamellar-zonal bone indicates an overall decrease in growth rate and probably the attainment of reproductive maturity. Many extant species exhibit such a change in growth, either as a decrease in the width between successive growth rings or a change in bone tissue type, and although it does not necessarily represent attainment of maximum size, the condition may occur when sexual maturity is reached (Castanet and Smirina 1990; Castanet and Baez 1991; Sander 2000; chapter 3 of this book). The presence of such prominent slowly forming lamellar-zonal bone tissue in *Procynosuchus* indicates that it grew more slowly than the later cynodonts did. The climate during the Late Permian is thought to have been semi-arid with seasonal rainfall (Smith, Eriksson, and Botha 1993; Smith 1995) and thus, the cyclical growth observed in *Procynosuchus* may reflect sensitivity to environmental fluctuations (Ray, Botha, and Chinsamy 2004).

De Ricqlès (1969) briefly reported that the Early Triassic basal epicynodont *Thrinaxodon* had fibrolamellar bone, which changed to a more parallel-fibered bone subperiosteally. Later, a more ontogenetically constrained study of *Thrinaxodon* (Botha and Chinsamy 2005) confirmed de Ricqlès's (1969) initial findings (Figs. 9.3A to 9.3C) of uninterrupted fibrolamellar bone and peripheral parallel-fibered bone in the older individuals. The channel density ranged between 9 percent in the younger individuals and 2.6 percent in the older individuals. Botha and Chinsamy (2005) also suggested that the shift from fast to slowly forming bone (Fig. 9.3C) was indicative of the attainment of sexual maturity. These researchers also noted that growth rings are absent in *Thrinaxodon* (apart from an annulus in a single femur) (Fig. 9.3A), which indicates that this genus experienced sustained growth and was possibly less susceptible to environmental fluctuations than *Procynosuchus*. The annulus in a single element probably represents an individual response to a particularly stressful, unfavorable environmental stimulus during which the growth rate decreased. The general absence of growth rings in *Thrinaxodon*,

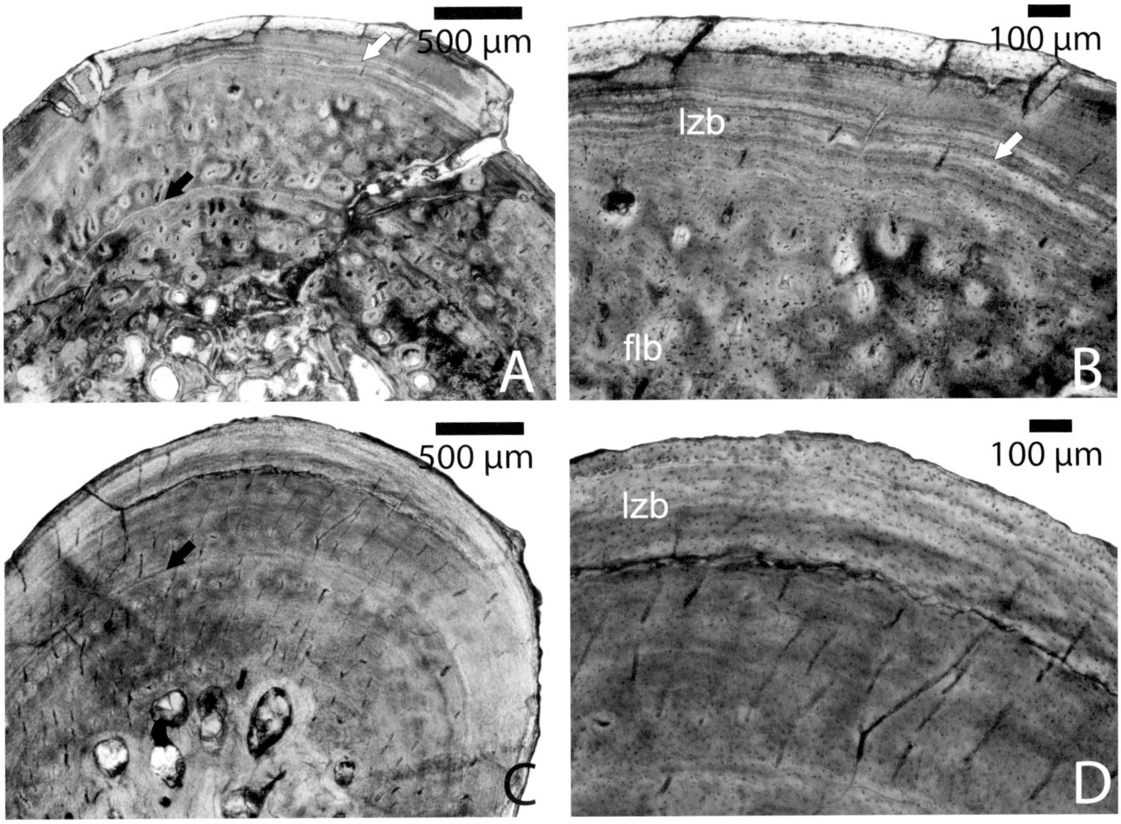

9.2. The bone histology of the Late Permian basal cynodont, *Procynosuchus.* (A) Adult radius BP/1/3747 showing fibrolamellar bone in the inner and midcortex. (B) High magnification of the bone periphery of radius BP/1/3747 showing poorly vascularized lamellar-zonal bone with multiple LAGs. (C) Adult clavicle SAM-PK-K8511 showing poorly vascularized lamellar-zonal bone. (D) High magnification of the bone periphery of clavicle SAM-PK-K8511 showing sparse simple vascular canals. Black arrows indicate single annuli or LAGs and white arrows indicate multiple LAGs. *Abbreviations:* flb = fibrolamellar bone; lzb = lamellar-zonal bone.

however, is noteworthy because the environment in which it lived was particularly unstable (Roopnarine et al. 2007). A dramatic change in climate occurred during the end-Permian extinction and, consequently, the Early Triassic environment became increasingly hot and arid with highly seasonal, unreliable rainfall (Retallack, Smith, and Ward 2003). *Thrinaxodon* forms part of the Early Triassic recovery phase following the end-Permian extinction and is the most abundant cynodont known from the *Lystrosaurus* AZ. Its rapid, sustained early growth may have been advantageous during this time of environmental instability, allowing it to reach sexual maturity relatively quickly. This strategy has been observed in extant animals that often reach sexual maturity at smaller body sizes in warmer environments or environments that retard growth (Berrigan and Charnov 1994; Angilletta, Steury, and Sears 2004) and has also been attributed to the success of the multituberculates during the Mesozoic Era (Chinsamy and Hurum 2006).

The limb bones of *Cynognathus,* which was a large (400 mm skull length), carnivorous Middle Triassic eucynodont (see Plate O), revealed highly vascularized (13.5 percent average channel density) uninterrupted fibrolamellar bone in a plexiform arrangement (Plate 10B, Fig. 9.4A). This bone tissue structure indicates a rapid, sustained growth strategy and suggests that *Cynognathus* grew independently of environmental

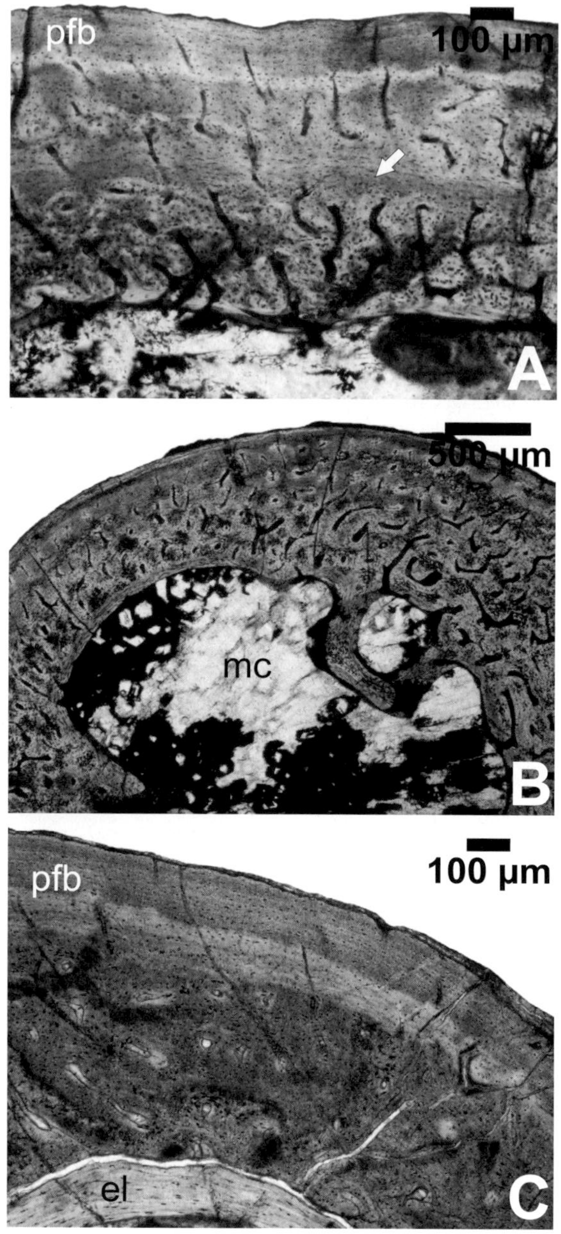

9.3. Bone histology of the Early Triassic basal epicynodont, *Thrinaxodon*. (A) Subadult femur SAM-PK-K1395, white arrow indicates single annulus in midcortex. (B) Subadult humerus BP/1/5208. (C) Adult radius BP/1/5208. All elements reveal moderately vascularized fibrolamellar bone that becomes poorly vascularized parallel-fibered bone toward the subperiosteal surface in ontogenetically older individuals. *Abbreviations:* el = endosteal lamellar tissue; mc = medullary cavity; pfb = parallel-fibered bone.

conditions (Botha and Chinsamy 2000). It is noteworthy that the phalanx of an adult individual (NMQR 3019) reveals possible peripheral rest lines (Fig. 9.4B) on one side of the bone and suggests that the growth in this element had begun to slow down.

Several gomphodont cynodonts, contemporaneous with *Cynognathus*, have also been examined in recent years. These genera include the large (300 mm skull length) herbivorous/omnivorous *Diademodon* (Botha and Chinsamy 2000) and smaller (100 mm skull length), herbivorous trirachodontids, *Trirachodon* and *Langbergia* (Botha and Chinsamy 2004). Although *Diademodon* and *Cynognathus* occupied different niches and

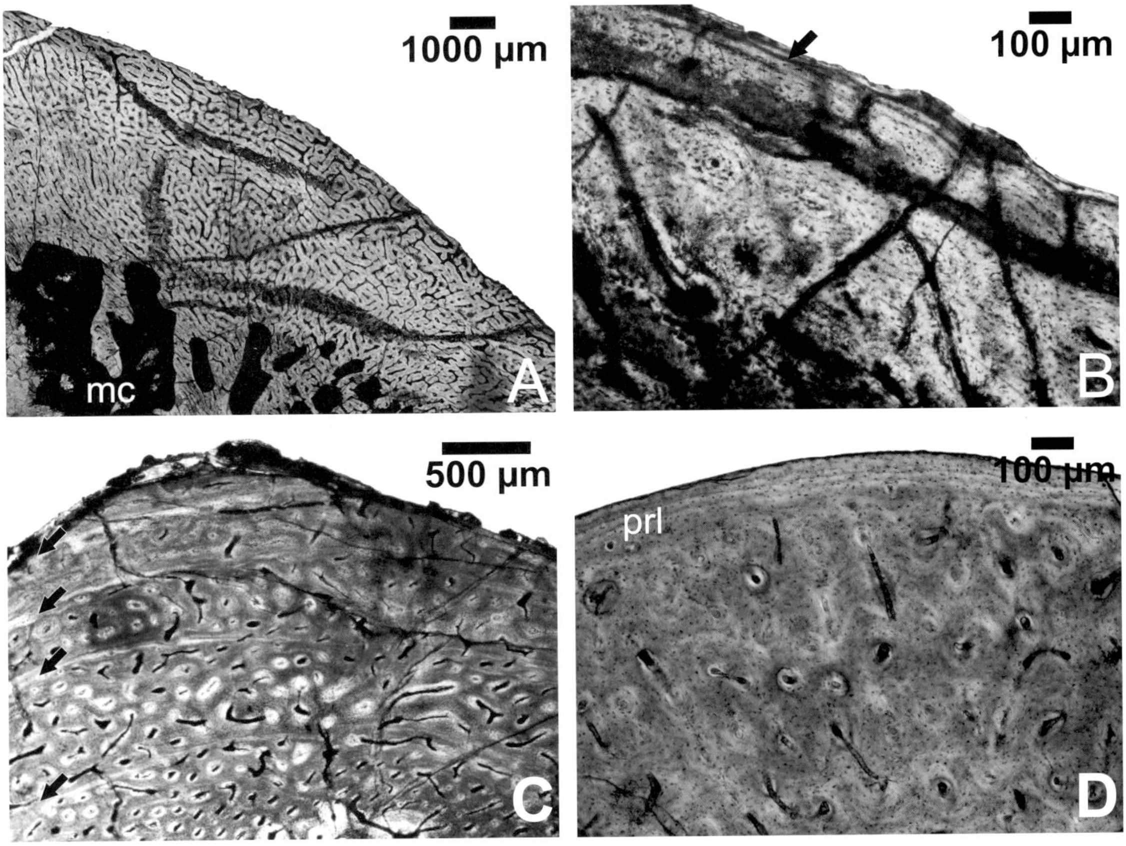

exhibit distinctive cranial characteristics, these genera have remarkably similar postcranial skeletons. As they are of similar body size and are often found together in the same strata, it is difficult to positively identify their postcrania in the absence of associated cranial material. However, a potential solution to this problem is in the analysis of their bone microstructure as the bone tissue patterns of these two taxa differ, allowing the positive identification of fossil material when associated skulls are absent (Botha and Chinsamy 2000). The tissue organization of *Diademodon* is zonal, consisting of alternating fibrolamellar bone and annuli of lamellar bone and/or LAGs (4 percent average channel density) (Plate 10C, Fig. 9.4C). Thus, the growth of this genus is cyclical and may have been influenced by the seasonal rainfall proposed for the Burgersdorp Formation (Smith, Eriksson, and Botha 1993; Smith 1995; Rubidge 1995). Rapid growth would have taken place during the favorable growing season and slow growth, or perhaps a pause in growth would have occurred during the unfavorable season (Botha and Chinsamy 2000). Peripheral rest lines have also been observed in several tibiae (Fig. 9.4D) and fibulae, indicating that these individuals had virtually stopped growing.

Botha and Chinsamy (2004) studied the bone histology of the gomphodont trirachodontid *Trirachodon* and reported a moderately vascularized, periodically interrupted, fibrolamellar bone tissue (Figs. 9.5A to

9.4. Bone histology of the contemporaneous cynognathid *Cynognathus* and gomphodont *Diademodon*. (A) *Cynognathus* subadult humerus NMQR1208a, showing highly vascularized uninterrupted fibrolamellar bone in a plexiform network. (B) *Cynognathus* phalanx NMQR3019, from a large adult individual, showing peripheral rest lines at the bone periphery (arrow). (C) *Diademodon* subadult humerus BP/1/3772, showing fibrolamellar bone interspersed with annuli and/or LAGs. (D) High magnification of the bone periphery of tibia UCMZ T447 showing peripheral rest lines. Black arrows indicate annuli or LAGs. *Abbreviations:* mc = medullary cavity; prl = peripheral rest lines.

9.5E), similar to that of *Diademodon*. Two limb bones in that study, originally identified as *Trirachodon* (a femur and tibia from a single individual NMQR3282 from Kestell District, *Cynognathus* Subzone A) may more likely belong to the recently described closely related trirachodontid *Langbergia* (based on its recovery from the same locality, Eerste Geluk 131 [Abdala, Neveling, and Welman 2006]). The bone histology of *Langbergia* also consists of fibrolamellar bone alternating with annuli of lamellar bone. However, Botha and Chinsamy (2004) did note that the bones now identified as *Langbergia* (11 percent channel density) were considerably more vascularized than the similar sized *Trirachodon* (5 percent channel density). Originally, the difference in vascularization was attributed to individual, gender, or taxonomic variation (Botha and Chinsamy 2004), but now it appears to be more likely because of generic differences. The bone tissue patterns of both taxa indicate rapid growth during the favorable season, which decreased or ceased during the unfavorable season. Skeletons of trirachodontids originally identified as *Trirachodon* were found in terminal chambers associated with burrow complexes from the Early Triassic Driekoppen Formation of South Africa (Groenewald, Welman, and MacEachern 2001). As these specimens were collected from Subzone A of the *Cynognathus* AZ, they are most likely representatives of *Langbergia* (Abdala, Neveling, and Welman 2006) and may have been obligate burrowers.

The bone histology of the South American traversodontids *Traversodon* and *Gomphodontosuchus* was examined by de Ricqlès (1969) and *Andescynodon*, *Massetognathus*, and *Exaeretodon* were studied by Chinsamy and Abdala (2008). The bone microstructure of *Andescynodon*, characterized by simple, radiating vascular canals in a lamellar matrix, suggests a relatively slow growth rate. A LAG was also observed at the bone periphery indicating interrupted growth (Chinsamy and Abdala 2008) (Fig. 9.6A). *Traversodon* reveals uninterrupted, highly vascularized fibrolamellar bone, with abundant, longitudinally oriented primary osteons, all features that indicate rapid, sustained growth (de Ricqlès, 1969, Ricqlès, Castanet, Francillon-Vieillot 2004; Chinsamy-Turan 2005). In contrast, Chinsamy and Abdala (2008) found that the bone tissue of *Massetognathus* (Fig. 9.6B) comprised parallel-fibered bone with longitudinally oriented simple vascular canals, indicating a slower growth rate compared with that of *Traversodon*, but faster growth than *Andescynodon*. *Gomphodontosuchus*, which is known only from the holotype housed in Tübingen (Hopson 1985; Abdala and Ribeiro 2003), contains a relatively thick cortex of poorly vascularized fibrolamellar bone. The medullary cavity contains many convoluted trabeculae, and secondary remodeling is extensive in the perimedullary region. The vascular canals are generally longitudinally oriented and growth rings, including both annuli and LAGs, are present; the latter becoming multiple at the periphery in some cases (de Ricqlès 1969, Plate 7, Fig. 6). The prominence of annuli and LAGs indicates that growth was intermittently interrupted and that *Gomphodontosuchus* may have been sensitive to the arid and highly

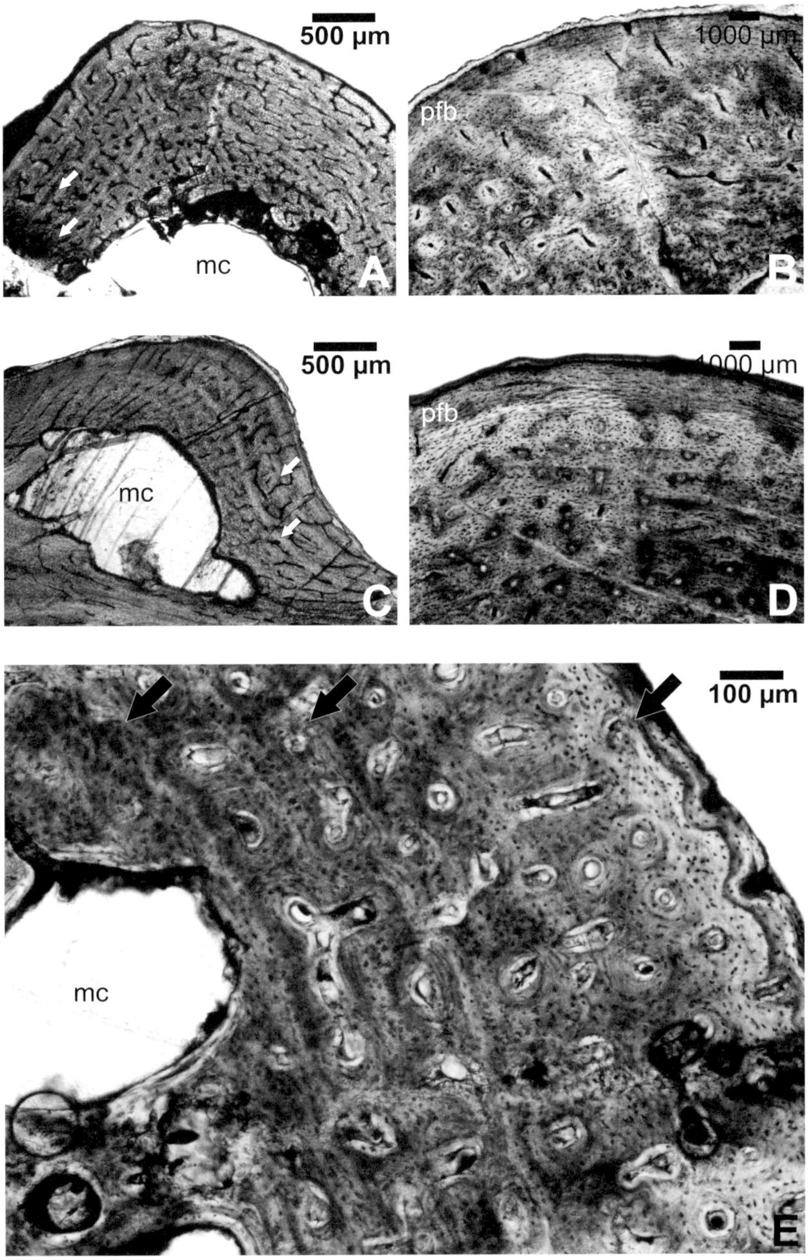

9.5. Bone histology of the Middle Triassic gomphodont trirachodontids, *Trirachodon* and *Langbergia*. (A) *Langbergia* subadult femur NMQR3282a. (B) *Trirachodon* subadult radius CGP1/79a. (C) *Trirachodon* subadult ulna CGP1/79b. (D) *Trirachodon* subadult femur SAM-PK-K5881a. (E) *Trirachodon* subadult tibia SAM-PK-K5881c. Both taxa exhibit fibrolamellar bone interrupted by annuli and/or LAGs, similar to the gomphodont *Diademodon*. Arrows indicate annuli and/or LAGs. Peripheral parallel-fibered bone is observed in some of the limb bones of *Trirachodon*. *Abbreviations:* mc = medullary cavity; pfb = parallel-fibered bone.

seasonal Middle Triassic climate (Tucker and Benton 1982; Abdala and Ribeiro 2010). *Exaeretodon*, a particularly large traversodontid with a skull length of 400 mm, contains rapidly forming fibrolamellar bone, followed by slowly forming parallel-fibered bone tissue with further periodic decreases in growth in the form of one to two faint LAGs at the subperiosteal surface. However, these growth rings cannot be traced around the entire circumference of the bone, thus the presence of LAGs requires confirmation with more material (Chinsamy and Abdala 2008) (Fig. 9.6C).

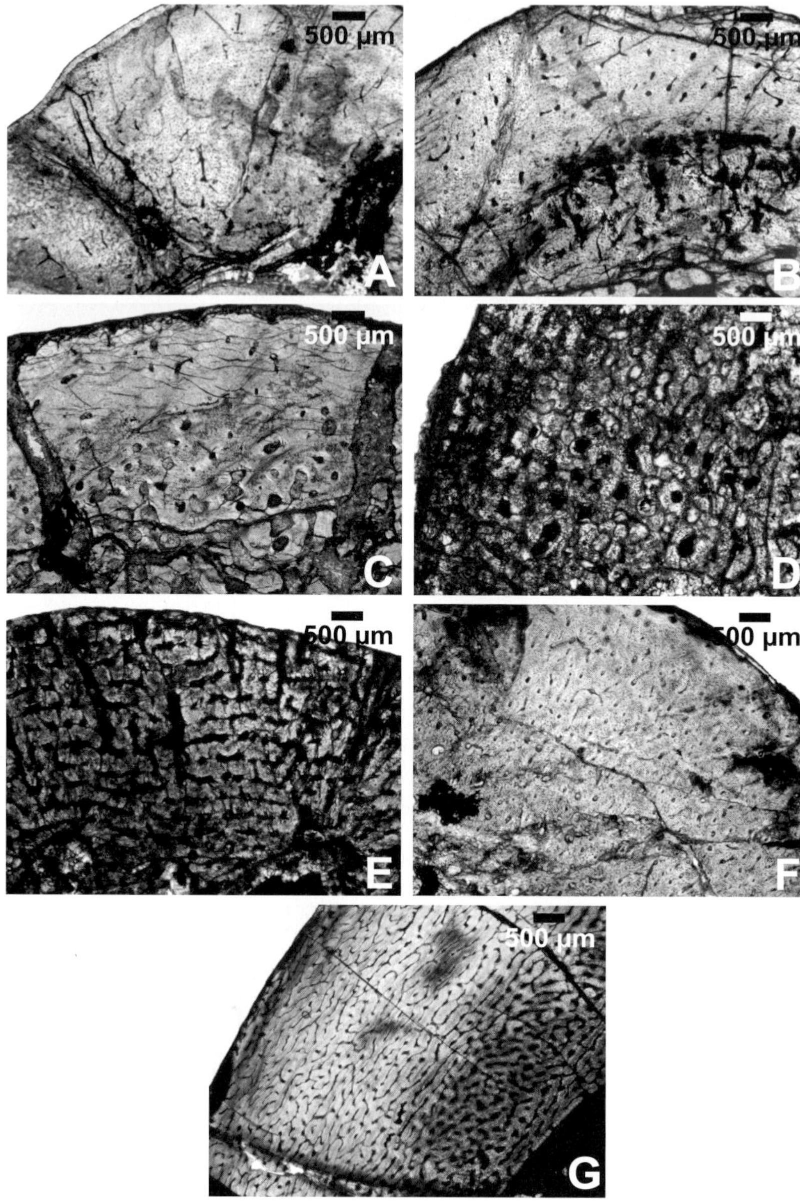

9.6. Bone histology of traversodontids. (A) *Andescynodon* femur PVL 3836 showing a lamellar matrix. (B) *Massetognathus* femur PVL 4613 showing parallel-fibered bone tissue. (C) *Exaeretodon* fibula PVL 2162 showing fibrolamellar bone. (D) Unnamed traversodontid humerus showing fibrolamellar bone with large longitudinally oriented vascular canals. (E) Unnamed traversodontid femur showing vascular canals in a plexiform arrangement. (F) *Protuberum* rib PV-1009T showing uninterrupted fibrolamellar bone with small, longitudinally oriented primary osteons. (G) *Scalenodontoides* limb bone NMQR3053 showing highly vascularized uninterrupted fibrolamellar bone in a subplexiform arrangement.

The analysis of the humerus and femur of the as yet unnamed traversodontid from the Late Triassic of Brazil reveals relatively thin cortices (femoral cortical thickness 18 percent; humerus too poorly preserved) of moderately vascularized (femur 8.7 percent channel density) uninterrupted fibrolamellar bone tissue (Figs. 9.6D, 9.6E). Both elements contain abundant globular osteocyte lacunae. The medullary cavities in both elements are infilled with bony trabeculae. The vascular canals form a plexiform network in the femur. The humerus is poorly preserved, but the vascular canals can be clearly seen in some regions. They either form a plexiform arrangement or occur as large, longitudinally oriented primary osteons. Secondary osteons are not observed in either element, but because of poor preservation in the perimedullary area, their absence cannot be confirmed. There is no peripheral decrease in vascularization indicating that this individual was actively growing at the time of death. The femoral length of this individual is 108 mm (Table 9.1). The size of the femur is remarkably smaller than that of other contemporaneous, closely related traversodontids (i.e., *Exaeretodon*; skull length ~350 mm). This evidence in conjunction with the observation that all-known Late Triassic Gondwanan traversodontids are relatively large animals suggests that the unnamed taxon may also be a large animal at maximum size. The absence of growth rings and relatively high vascular density suggests sustained growth and that this animal was not particularly susceptible to environmental fluctuations.

A rib of the Brazilian Middle Triassic traversodontid *Protuberum* (UFRGS PV-1009T) was sectioned for the purposes of this study. *Protuberum* is a large, robust traversodontid that is characterized by peculiar processes on the dorsal surface of the presacral ribs that progressively decrease in size distally and the presence of rugosities on the dorsal border of the iliac blade (Reichel, Schultz, and Soares 2009). The rib comprises a particularly thick cortex (cortical thickness 55 percent) and a small free medullary cavity. The cortex exhibits fibrolamellar bone with mostly longitudinally oriented primary osteons (Fig. 9.6F). The osteocyte lacunae are abundant and globular. Growth rings are absent. Secondary remodeling, with large resorption cavities, is extensive in the perimedullary region and secondary osteons extend into the midcortex. The osteocyte lacunae are large, globular, and abundant. There is a very slight decrease in vascularization at the subperiosteal surface, and a narrow region of parallel-fibered bone is observed in one region at the bone periphery, indicating an overall decrease in growth rate. The absence of growth rings indicates sustained growth, but the overall growth rate decreased later during ontogeny, as seen by the presence of peripheral slowly forming parallel-fibered bone. Reichel, Schultz, and Soares (2009) noted that *Protuberum* had particularly strong robust limbs that may have been used for burrowing or digging, and this suggestion is supported by the particularly high rib cortical thickness observed in this study. Studies on extant vertebrates have noted a correlation between thick bone walls (cortices) and an aquatic or fossorial lifestyle (Wall 1983; Magwene 1993; Germain and Laurin 2005; Kriloff et al. 2008; Canoville and Laurin 2010). As there are no morphological characteristics indicative of an aquatic lifestyle, we agree with Reichel, Schultz, and Soares (2009) in inferring a possible fossorial lifestyle for *Protuberum*.

The histological analysis of the undetermined limb bone of the large (skull length 400 mm) traversodontid, *Scalenodontoides* (NMQR 3503), from the Late Triassic of South Africa reveals a large free medullary cavity and a relatively thin cortex (cortical thickness 15 percent) containing highly vascularized (15 percent channel density, similar to that of *Cynognathus*) uninterrupted fibrolamellar bone (Fig. 9.6G). Large longitudinally oriented vascular canals with radial anastomoses surround the medullary cavity, but they gradually decrease in size and finally form a subplexiform network in the outer cortex. The osteocyte lacunae are large, globular, and abundant. A thin layer of circumferential endosteal bone surrounds most of the medullary cavity and secondary remodeling is absent in this region. However, the circumferential endosteal layer is destroyed by some secondary remodeling on one side of the bone and a reversal line separates this tissue from the primary bone. The skull of this individual is fragmentary; the largest preserved fragment consists of the snout. Thus, by comparing the distance between the upper canines of this specimen, which is ~135 mm, with that of a more complete skull (BP/1/5395a), which is ~70 mm, we interpret the specimen NMQR 3503 to be at least a late subadult as it is double the size of specimen BP/1/5395a. Yet, growth rings are absent, suggesting that this animal grew in a rapid, sustained manner and was not susceptible to environmental variation. If any overall decrease in growth did occur in this taxon, it occurred particularly late in ontogeny.

The radius of the Brazilian *Trucidocynodon*, which is a Middle Triassic carnivorous basal probainognathian (Oliveira, Soares, and Schultz 2010), comprises a thick cortex (cortical thickness 32 percent) and an essentially free medullary cavity. The bone tissue is poorly preserved and it is difficult to see the vascular canal orientation (Fig. 9.7A). However, there are a few well-preserved regions that reveal abundant longitudinally oriented primary osteons in a woven matrix. Many radial anastomoses are also present. Areas of the preserved osteocyte lacunae reveal them to be abundant and globular. Growth rings were not observed, but their absence cannot be confirmed because of poor preservation. Secondary remodeling is apparent in the perimedullary area and a few small secondary osteons are observed in this region.

The Brazilian *Chiniquodon* (previously referred to as *Belesodon* [Abdala and Giannini 2002b]) is the only chiniquodontid to have been examined histologically (de Ricqlès 1969). The bone tissue of this genus consists of moderate to highly vascularized fibrolamellar bone. Although de Ricqlès (1969) did not comment on the presence or absence of growth rings, from the published images they appear to be absent (see de Ricqlès 1969, Plate 5, Fig. 9) and, combined with the presence of fibrolamellar bone and numerous vascular canals, these characteristics suggest that *Chiniquodon* grew in a rapid, sustained manner.

De Ricqlès (1969) reported that the bone microstructure of the tritylodontid *Tritylodon* consists of uninterrupted fibrolamellar bone tissue with abundant longitudinally oriented primary osteons. Ray, Botha, and Chinsamy (2004) later confirmed these observations (Plate 10D). Although most of the bones in the latter study contained azonal bone, a radius revealed a distinct, wide annulus of lamellar bone, terminating in a LAG in the outer cortex (Fig. 9.7B). After this annulus, growth resumed at a relatively rapid pace on the anterolateral side of the bone. As growth rings are absent from all the other elements studied, this single LAG probably represents an unusually stressful period or local environmental stress

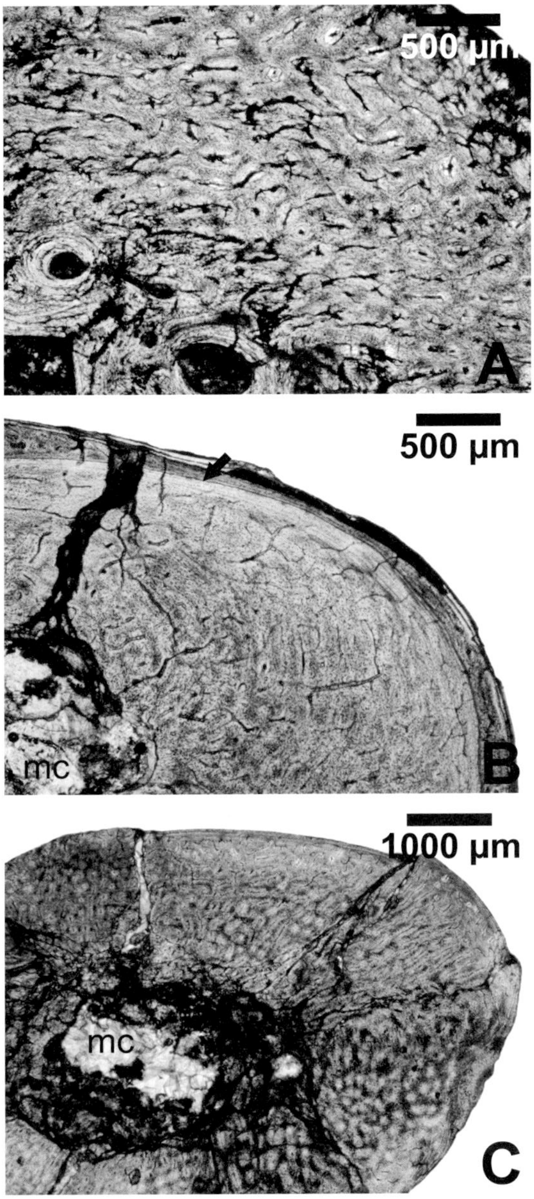

9.7. Bone histology of the Middle Triassic probainognathian *Trucidocynodon* and Early Jurassic tritylodontid, *Tritylodon*. (A) *Trucidocynodon* radius PV-1053T, showing poorly preserved fibrolamellar bone with longitudinally oriented primary osteons with short anastomoses. (B) *Tritylodon* subadult radius BP/1/5167. (C) *Tritylodon* subadult fibula BP/1/5167b. All *Tritylodon* elements exhibit uninterrupted fibrolamellar bone apart from radius BP/5167d, which exhibits a single annulus (arrow) near the subperiosteal surface and may represent a particularly stressful period for this individual. *Abbreviations:* mc = medullary cavity.

for this individual, and thus, interrupted growth is probably not representative of the genus. This proposal is supported by the analysis of the new material (six limb bones and a rib) in this study that confirms the presence of rapidly forming bone tissue. All these elements have a notably thick cortex, the average being 34 percent (including elements from Ray, Botha, and Chinsamy 2004). The bones also all contain fibrolamellar bone with a channel density varying from 3.2 percent in a radius (BP/1/4785e [Botha 2002]) to 6.5 percent in the humerus (BP/1/4785d). Vascular canals either extend radially from the medullary cavity or are arranged as longitudinally oriented primary osteons or in some places, simple canals. Growth rings are absent in all the elements studied. However, a fibula

(BP/1/5167e) reveals several closely spaced LAGs at the subperiosteal surface that may represent peripheral rest lines (Fig. 9.7C; Plate 10D), which is similar to the gomphodont *Diademodon* and suggests asymptotic growth in *Tritylodon*. All fourteen elements studied to date (Ray, Botha, and Chinsamy 2004; this study) exhibit sustained growth apart from a single radius, which displays an interruption in growth. Thus, we deduce that *Tritylodon* had a sustained, uninterrupted growth strategy. *Bienotherium*, a tritylodontid from China, contains a large, completely free medullary cavity and a relatively thick cortex. The bone tissue consists of fibrolamellar bone and the vascular density appears to be similar to that of *Tritylodon*. Vascular canals include radial simple canals and longitudinally oriented primary osteons (de Ricqlès 1969). De Ricqlès (1969) also examined the bone histology of the tritylodontid *Oligokyphus* from the United Kingdom and reported a large, free medullary cavity and relatively thin cortex. The cortical bone tissue comprises fibrolamellar bone with small longitudinally oriented simple vascular canals and primary osteons. Secondary remodeling appears to be negligible. Annuli and LAGs are absent (de Ricqlès 1969), indicating sustained growth; however, the relatively lower vascular density and prevalence of simple vascular canals suggest a slower growth rate compared with that for *Tritylodon*.

Concluding Remarks: Implications of the Bone Microstructure

To date, the bone microstructure of nineteen cynodont genera have been studied. These genera represent a diverse range of basal to more derived taxa comprising approximately 45 My of cynodont evolution, facilitating an assessment of changes in bone tissue patterns through the lineage. Fibrolamellar bone, indicative of rapid growth, is present in varying degrees in all cynodonts studied apart from the traversodontids *Andescynodon* and *Massetognathus*. Considerable variation in vascular density and orientation, as well as the presence/absence of cyclical growth among the cynodont taxa is evident, the cause of which is not immediately obvious as several factors, including phylogeny, biomechanics, ontogeny, body size, lifestyle preferences, and environment influence the microstructure of bones (see, e.g., Cubo et al. 2005; Kriloff et al. 2008).

When considering the observed bone tissue patterns within a phylogenetic context, there appears to be an overall increase in bone deposition rate, as indicated by the increasing prevalence of highly vascularized rapidly forming fibrolamellar bone tissue in the later cynodonts. Although fibrolamellar bone is present in the Permian basal cynodont *Procynosuchus*, slowly forming lamellar-zonal bone dominates once adulthood is reached. Fibrolamellar bone is more prevalent in the Early Triassic epicynodont, *Thrinaxodon*, but as in *Procynosuchus*, upon maturity, slowly forming parallel-fibered bone predominates. Fibrolamellar bone becomes increasingly prevalent in the eucynodont lineage in that it is the dominant bone tissue type until fairly late in ontogeny. The basal cynognathian, *Cynognathus*, exhibits highly vascularized uninterrupted fibrolamellar bone tissue in a plexiform arrangement, a condition only represented in other cynodonts by *Scalenodontoides* and the unnamed Brazilian traversodontid. The similarities in bone tissue structure (uninterrupted fibrolamellar bone with a plexiform arrangement) in this case may be related to similar body size. Both *Scalenodontoides* and *Cynognathus* were large and may have grown rapidly in an attempt to attain reproductive maturity as quickly as possible. This condition has been observed in extant vertebrates (Singh, Tonna, and Gandel 1974) as well as dinosaurs and pterosaurs (Padian, Horner, and de Ricqlès 2004). Based on the bone microstructure, the elements of the Brazilian traversodontid described here appear to be from a juvenile individual (femoral length 108 mm) and may thus also represent a large animal (although confirmation of maximum size is still required for this genus).

In contrast, although the Gomphodontia, which comprise the diademodontids, trirachodontids, and traversodontids, reveal a diverse array of bone tissue patterns, many of them are characterized by cyclical growth to some degree. *Diademodon*, *Trirachodon*, *Langbergia*, and the traversodontid *Gomphodontosuchus* exhibit a similar growth pattern of alternating rapidly forming fibrolamellar bone and annuli of slowly forming lamellar-zonal bone and/or LAGs. Although *Exaeretodon* experienced sustained growth early in ontogeny, its growth rate also decreased and may have become cyclical later in ontogeny. *Andescynodon* and *Massetognathus* grew more slowly than their traversodontid relatives as fibrolamellar bone is absent in both these genera. Instead, *Andescynodon* exhibits slowly forming lamellar-zonal bone, and *Massetognathus* exhibits parallel-fibered bone tissue, which is still formed more slowly than fibrolamellar bone. Several traversodontid cynodonts, including *Traversodon*, *Protuberum*, the new Brazilian taxon, and *Scalenodontoides*, do not appear to have experienced cyclical growth at any time during their ontogeny. *Traversodon* and *Protuberum* exhibit similar growth patterns in the form of uninterrupted fibrolamellar bone with small, longitudinally oriented primary osteons. In contrast, the new traversodontid and *Scalenodontoides* exhibit similar bone tissue patterns of relatively highly vascularized uninterrupted fibrolamellar bone in a plexiform arrangement and, in the case of the new traversodontid, enlarged longitudinally oriented vascular canals as well. This is worth noting as these taxa appear to be closely related to one another (Fig. 9.1). However, it should also be noted that the consistently high vascular density throughout the cortices of both individuals indicates that these animals were still actively growing at the time of death and an overall decrease or change in growth rate may have occurred later in ontogeny. More material is required to obtain a better grasp of the growth patterns of this newly discovered traversodontid and *Scalenodontoides* and to confirm the differences observed in *Exaeretodon*.

Trucidocynodon, *Chiniquodon*, and the tritylodontids all exhibit uninterrupted fibrolamellar bone and thus experienced sustained rapid growth to adult size. Additionally, *Tritylodon* exhibits peripheral rest lines, indicating a dramatic decrease in overall growth rate. Although the Tritylodontidae exhibit highly derived mammal-like morphological features (Kielan-Jaworowska, Cifelli, and Luo 2004; Kemp 2005), there is still disagreement regarding their relationship with mammaliaforms and traversodontids (chapter 1 of this book). The bone tissue patterns of the tritylodontids are more similar to the uninterrupted woven-fibered bone with longitudinally oriented primary osteons of the mammaliaform *Morganucodon* (Chinsamy and Hurum 2006) than to those of the Gomphodontia, which exhibit either strongly cyclical growth, parallel-fibered bone (*Massetognathus*) or fibrolamellar bone in a plexiform arrangement (unnamed Brazilian traversodontid, *Scalenodontoides*). However, it is important to note that *Tritylodon* (and *Morganucodon*) bone microstructure differs quite considerably from the Late Cretaceous multituberculate mammals *Nemegtbaatar* and *Kryptobaatar*, as well as the Late Cretaceous eutherian mammals *Barunlestes* and *Zalambdalestes*, many of which exhibit slowly forming parallel-fibered bone and, in general, grew more slowly than the tritylodontids (Chinsamy and Hurum 2006).

Many of the cynodonts studied exhibit an overall decrease in growth rate upon adulthood/attainment of sexual maturity. Interestingly, the basal epicynodont *Thrinaxodon*, the basal cynognathian *Cynognathus*, and probainognathians *Trucidocynodon*, *Chiniquodon*, and tritylodontids exhibit rapidly forming uninterrupted fibrolamellar bone early in ontogeny and appear to have been under strong selection pressure to attain reproductive maturity as quickly as possible. In contrast, although many gomphodonts also exhibit fibrolamellar bone early in ontogeny, their growth was also cyclical from an early stage of development. This condition is particularly apparent in the non-traversodontid gomphodonts *Diademodon*, *Trirachodon*, and *Langbergia*, indicating a higher level of developmental plasticity in this group, but also a slower growth pattern overall, as growth decreased or ceased periodically from an early age. Many of the traversodontids also exhibit a relatively slower growth strategy (e.g., *Andescynodon*, *Massetognathus*) or cyclical growth strategy, either from an early stage of development (*Gomphodontosuchus*) or once adulthood was reached (*Exaeretodon*, *Protuberum*). The unnamed Brazilian traversodontid and *Scalenodontoides* differ from this pattern and instead exhibit rapid sustained growth. *Scalenodontoides* was a large animal, and it is probable that the new traversodontid was as well. This factor may have played a role in their rapid growth as large animals frequently grow more quickly than their smaller relatives (Case 1978).

Other, non-phylogenetic causes such as biomechanical factors should also be considered when attempting to deduce the possible causes for observed bone tissue patterns. Although several cynodonts are considered to be fossorial either on the basis of gross morphology or the recovery of body fossils in burrows (e.g., *Galesaurus*, *Thrinaxodon*, *Langbergia*, *Tritylodon*), this type of lifestyle is not reflected in the bone microstructure. Instead, it appears to be reflected in the distribution of bone around the medullary cavity; burrowing animals tend to have thicker cortices (usually more than 30 percent of the total diameter [Magwene 1993; Botha 2002; Chinsamy and Hurum 2006]) and this has been observed in *Galesaurus* (Butler 2010) and *Tritylodon* (Botha 2002; this study). However, body size may play a more significant role in affecting bone tissue growth. For example, the basal traversodontid *Andescynodon* is a relatively small gomphodont with a maximum skull length of 137 mm (Liu and Powell 2009). Within a given clade, small taxa often take longer to reach adult size than larger taxa (Singh, Tonna, and Gandel 1974; Case 1978) and simple avascular or poorly vascularized lamellar or parallel-fibered bone has been observed in small mammals where size appears to be a limiting factor (e.g., rodents, bats [Enlow and Brown 1958]). It is possible that *Andescynodon* grew slowly throughout ontogeny because it was not under selective pressure to reach reproductive maturity particularly quickly. Interestingly, the Early Triassic epicynodont *Thrinaxodon* is of similar size to *Andescynodon*, but it experienced rapid growth early in ontogeny. Thus, its early rapid growth may have been a factor in allowing it to survive the unstable, harsh conditions of the post-extinction Early Triassic environment.

It has previously been suggested that cynodonts experienced a higher level of developmental plasticity (flexible growth strategy) than extant mammals, which allowed them to experience cyclical growth whereby their growth rate decreased or ceased periodically during unfavorable growing conditions (Ray, Botha, and Chinsamy 2004; Chinsamy and Hurum 2006; Chinsamy and Abdala 2008). The therocephalians were more diverse during the earliest Triassic after the end-Permian extinction event, but the cynodonts had essentially replaced them by the Middle Triassic (Abdala and Ribeiro 2010). Although several factors may have played a role in allowing the cynodonts to become one of the most successful therapsid clades during the Triassic, their high level of developmental plasticity may have been a factor in facilitating their success. Pörtner (2004) recently proposed that the development of eurythermy (the ability to withstand temperature extremes) was fundamental in allowing homeothermic endothermy to evolve. He suggested that as large animals went extinct during mass extinction events (such as the end-Permian extinction [Smith and Ward 2001]), maxillary turbinates evolved in the smaller survivors (e.g., *Thrinaxodon* [Hillenius 1992]) in response to eurythermy and not as a response to endothermy itself. The Early Triassic is considered to be a particularly harsh, arid environment, with desert-like conditions (i.e., extremely hot days and extremely cold nights); therefore, being able to withstand temperature extremes (eurythermy) and to vary growth rates in response to these harsh conditions (developmental plasticity), along with a complex masticatory apparatus, may have allowed cynodonts to become so successful during the Middle Triassic and finally evolve into mammaliaforms by the end of this period.

Acknowledgments

The Bernard Price Institute for Palaeontological Research, University of the Witwatersrand, Johannesburg, and Universidade Federal do Rio Grande do Sul, Brazil, are thanked for the loan of the material. We thank Sergio Kaminski for collecting the Brazilian material. Michel Laurin and Christian Kammerer are acknowledged for having reviewed this manuscript. J. Botha-Brink, F. Abdala, and A. Chinsamy-Turan are supported independently by the National Research Foundation of South Africa. F. Abdala is also supported by the Palaeontological Scientific Trust of South Africa.

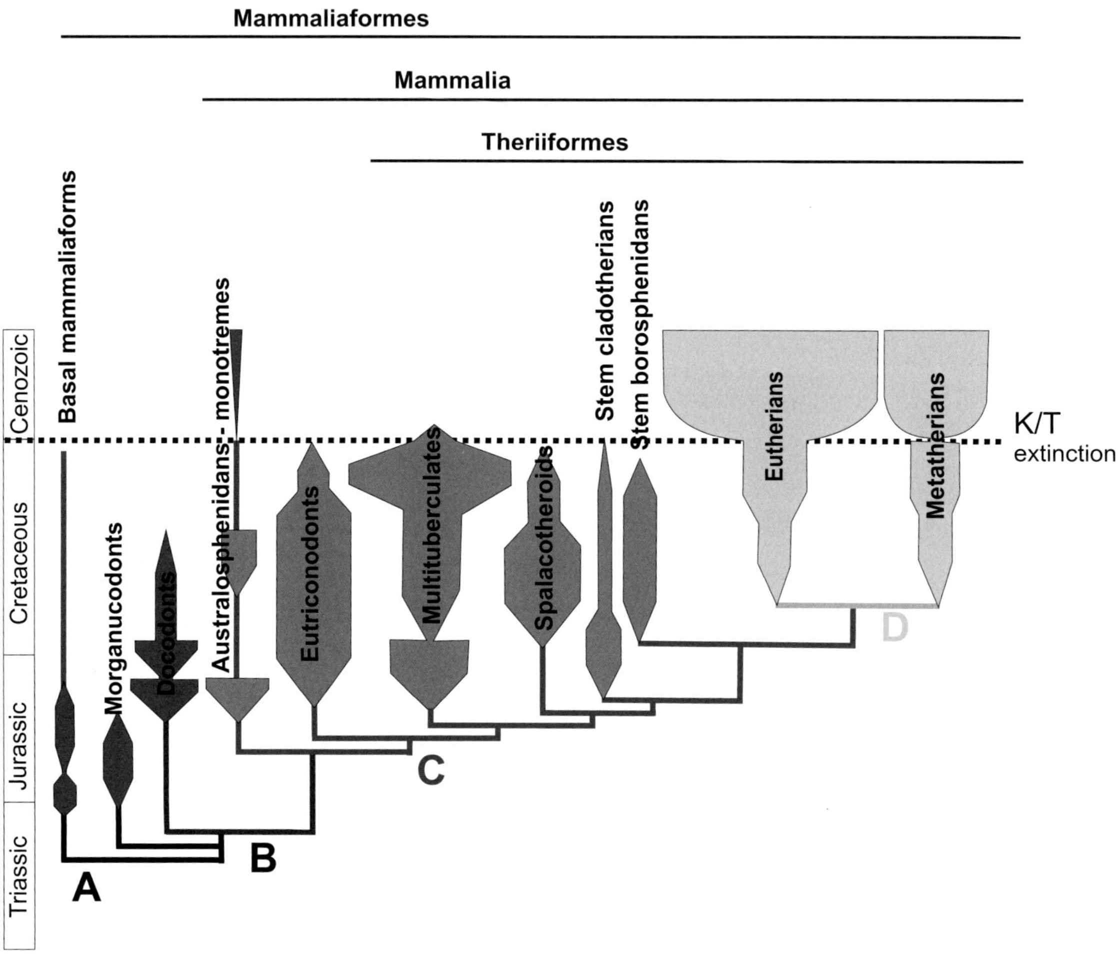

10.1. Simplified cladogram showing the main phylogeny and diversification of Mesozoic and major extant groups of mammals (simplified from Luo [2007] and Kielan-Jaworowska, Cifelli, and Luo [2004]): (A) the Late Triassic to Middle Jurassic diversification; (B) the diversification of docodonts; (C) the Late Jurassic diversification; and (D) the Early Cretaceous diversification.

The Radiation, Bone Histology, and Biology of Early Mammals

10

Jørn H. Hurum and Anusuya Chinsamy-Turan

Introduction

Modern mammals are represented by the monotremes, marsupials, and placentals. However, the fossil record provides a fascinating account of the radiation of mammals from their nonmammalian therapsid ancestors and shows evidence of other lineages of mammals. The evolutionary transition of mammals from the nonmammaliaform cynodonts involved complex morphological and physiological modifications. Many of the anatomical changes associated with this transformation are well documented; however, the physiological ones are much more difficult to assess from the fossil record (chapter 11 of this book). In this chapter, we present an overview of the radiation of mammals, and we briefly consider some of the morphological changes associated with their evolution from the nonmammaliaform cynodonts. Considering the bone microstructure of *Morganucodon* and Mesozoic multituberculates and eutherians, and through comparisons with extant monotremes, marsupials, and placentals, we provide information pertaining to the evolution of mammalian growth patterns and overall biology.

Institutional Abbreviations

This chapter uses abbreviations for the following institutions: Institute of Paleobiology, Institute of Paleontology, Warsaw, Poland (ZPAL MgM); University College of London, Department of Anatomy and Developmental Biology (UCL).

Phylogenetic Context

The evolutionary history of mammals extends to the Late Triassic when over a period of about 40–50 Mya, some nonmammaliaform cynodonts increasingly acquired a suite of "mammalian" characteristics such as a diaphragm, parasagittal gait, secondary palate, tooth differentiation, and an expansion of the dentary. Exactly when and where among the nonmammaliaform cynodonts these mammalian characteristics evolved has been the subject of much research (e.g., Kielan-Jaworowska, Cifelli, and Luo 2004; Kemp 2005). (Note that most researchers now use Mammalia for the crown group, and Mammaliaformes for the stem plus crown group). It is evident that by the Late Triassic, *Adelobasileus* from Texas (Lucas and Lou 1993) and *Sinoconodon* from the Early Jurassic of China (Crompton and Luo 1993) are recognizable as basal mammaliaforms

(Fig. 10.1 A). The earliest mammals are poorly represented in the fossil record, except for the morganucodontids, which are known from South Africa, Europe, and China (Kielan-Jaworowska, Cifelli, and Luo 2004). Around the Middle Jurassic, the docodonts, the first theriiform mammals (the group including multituberculates, spalacotheroids, through to the crown therians) and the australosphenidan mammals that are more basal to monotremes, represent the next major wave of radiation of mammals (Fig. 10.1B). This is followed by the Late Jurassic diversification of the extinct theriiform groups (Luo 2007) (Fig. 10.1C). Thus, by the Late Jurassic about eight mammalian lineages are recognized, including docodonts, eutriconodonts, spalacotheroids, multituberculates, and cladotherians. The multituberculates first appeared in the Middle Jurassic and became the dominant group of Mesozoic mammals in the Northern Hemisphere. They survived the Cretaceous-Tertiary (K/T) extinction event and are found right into the Late Eocene, whereupon they became extinct. The divergence of the metatherian and eutherian lineages occurred in the Early Cretaceous (Fig. 10.1D), and in the Cenozoic, they experience another diversification (see, e.g., Luo 2007).

Modern mammals are represented by the monotremes, which lay eggs; the marsupials, which give birth to tiny young that complete development in a pouch; and the placentals, which retain their young in the uterus to a more advanced stage of development. Modern placental mammals and their stem lineages together constitute the Eutheria, while modern marsupials and their stem lineages constitute Metatheria. The fossil record of the monotremes (Monotremata) extends to the Late Cretaceous (Rowe et al. 2008; Flannery et al. 1995) of Australia, and the fact that they are also known from the Paleocene of Argentina has led to the suggestion that they evolved in Australia and dispersed to Argentina (Rich et al. 2001). Today they are represented by two taxa, the platypus (*Ornithorhynchus anatinus*) of Australia and the echidnas (*Tachyglossus* and *Zaglossus*) of Australia and New Guinea. Genome analysis of the platypus revealed a branching point between Theriiformes and monotremes about 166 million years ago (Warren et al. 2008).

The earliest representative of Metatheria is *Sinodelphys szalayi* (Luo et al. 2003) from the Early Cretaceous Jehol biota in China. Marsupialia are thought to have originated in the Mid Cretaceous of North America and thereafter dispersed to South America and later (from the Eocene onward) to Australia (Rose 2006). Today the eutherians are by far the most diverse mammals and are represented by eighteen orders (Wible et al. 2007). The earliest representative of the eutherians is the Early Cretaceous *Eomaia* also from the Jehol biota in China (Ji et al. 2002). *Murtoilestes* from eastern Russia (Averianov and Skutschas 2000), *Prokennalestes* (Kielan-Jaworowska and Dashzeveg 1989) from Mongolia, and *Montanalestes* from North America (Cifelli 1999) are other Early Cretaceous genera. Approximately forty-two genera are known from the Cretaceous (Kielan-Jaworowska, Cifelli, and Luo 2004). Thus, by the Late Cretaceous about 65.5 Mya, just before the catastrophic mass extinction event, mammals were well represented in the terrestrial fauna of Laurasia, but very poorly documented in the southern continents of Gondwana. There is a major disagreement between molecular biologists and paleontologists in the dating of when the major lineages of mammals arose. For example, molecular dating invariably suggests a much earlier divergence of placentals from the marsupials and of the orders of placentals from one another than do the dates of first appearance of these taxa in the fossil record (see Archibald and Deutschman 2001; Foote et al. 1999). Four hypotheses have been proposed to explain the diversification of the mammalian clades: the "short fuse," "long fuse," "explosive," and "Garden of Eden" hypotheses (for a review see, e.g., Prasad 2009, Kitazoe et al. 2007).

The short fuse hypothesis is based on molecular data that suggests the origination of major superorders and orders of placental mammals in the Late Cretaceous, approximately 93–84 Mya (e.g., Bininda-Emonds et al. 2007; Kitazoe et al. 2007). The long fuse hypothesis claims early divergence in the Cretaceous of superorders and is based on a selected combination of molecular and fossil evidence (e.g., Archibald 1996; Archibald, Averianov, and Ekdale 2001; Springer et al. 2003). The explosive hypothesis claims that the divergence of most superorders and orders happened close to or just after the K/T. This is supported by Wible et al. (2007) who compiled all fossil evidence in a morphological study and concluded that the diversification occurred only after the extinction of the nonavian dinosaurs about 65.5 Mya. The "Garden of Eden" hypothesis tries to explain the findings of Early Cretaceous mammals argued by some to be eutherians in Australia (Rich et al. 1997; Woodburne, Rich, and Springer 2003). The original hypothesis suggested that eutherian mammals originated in an as yet unknown area, possibly in Gondwana, and that they subsequently dispersed to other continents without any descendants at their point of origination (e.g., Rich et al. 2001 contra Luo, Cifelli, and Kielan-Jaworowska 2001; Hunter and Janis 2006).

These alternative hypotheses directly influence our interpretations regarding the physiology of early mammals. Is it possible that all Cretaceous eutherian mammals were simply generalized eutherians (or side branches thereof), whereas the oldest known "true" placental is Paleocene in age (Wible et al. 2007)?

The mass extinction of nonavian dinosaurs at the end of the Cretaceous was a major impetus for the diversification of mammals in the Paleocene and Eocene, and this is evidenced by the fossil record (Cifelli and Gordon 2007; Rose 2006). Currently, however, there is no fossil evidence to support the molecular data for the deep origin of major superorders and orders of placental mammals. Indeed, when the first dinosaurs and mammals evolved approximately 230 Mya, what was the selective advantage that kept mammals mostly small and nocturnal until the dinosaurs became extinct? If placental mammals were physiologically like their extant relatives already in the Early Cretaceous, why did mammals not diversify before the extinction of the dinosaurs? Is it possible that besides the extinction of nonavian dinosaurs, the biology of Mesozoic mammals could be part of the explanation?

Paleobiological Context

Establishing the biology of extinct animals is wrought with controversy, especially since all that remains of these animals is generally their hard parts—their bones and teeth. Understanding early mammalian biology is compounded by the fact that the early radiation of mammals is not well represented in the living biota: Only three of the four major groups that were present in the Cretaceous survive today. It is also uncertain whether modern mammalian species are reliable models for their extinct ancestors and relatives. Therefore, the biology of Mesozoic mammals is poorly understood, and there are few studies that have attempted to address questions pertaining to the biology of our early mammalian ancestors. Questions such as When did endothermy or homeothermy evolve in the mammalian lineage?, or When did mammalian patterns of growth evolve? still remain. Were there inherent physiological traits in the multituberculates, as compared to other contemporary mammalian clades, that could explain their success in the Cretaceous? Is the explosive radiation of the eutherian mammals in the Early Cretaceous (Alroy 1999; Rose and Archibald 2005) linked to superior growth strategies or did this evolve later?

Falkowski et al. (2005) studied atmospheric oxygen levels during the last 205 My and deduced that raised oxygen levels can be correlated to major events in the radiation of vertebrates: in the Late Triassic, it was related to the evolution of endothermy; in the Late Cretaceous, it was associated with the origins and diversification of placental mammals; and in the Paleocene-Eocene, it was linked to the evolution of large body sizes of mammals. But is there any evidence that endothermy had already occurred in the Triassic? And if oxygen levels are the reason for the increase in body size of Eocene mammals, why did mammals not grow large in the Late Cretaceous?

Deciphering whether extinct nonmammalian therapsids were endotherms is extremely complex, especially since evidence of endothermy (such as high metabolic rates, increased aerobic respiration, higher concentrations of mitochondria) do not manifest as osteological traits (see, e.g., Chinsamy and Hillenius 2004). However, many morphological characteristics, such as the evolution of a diaphragm, parasagittal gait, diphyodonty, development of the secondary palate, respiratory turbinals, fur, determinate growth, enlarged brain and miniaturization, have been used to infer endothermy (see, e.g., Jenkins 1990; Kemp 2005; chapter 11 of this book).

The use of such morphological characteristics for assessing physiology (Hillenius and Ruben 2004; Chinsamy and Hillenius 2004) is further complicated by the complexity of applying a temporal scale to the evolution of a whole suite of "mammalian characters" from the nonmammalian cynodonts. Until about 10 years ago, the nature of the Mesozoic mammalian fossil record consisted predominantly of isolated bones and teeth, and only a few associated skeletons of morganucodontids (Jenkins and Parrington 1976) and Late Cretaceous mammals from the Gobi Desert (see summary in Kielan-Jaworowska, Cifelli, and Luo 2004). This material provided limited information about the transition from non-mammaliaform cynodonts to mammals, as well as the actual diversification of the mammals. New discoveries of complete skeletons of several different taxa, particularly from Mongolia and China in the last 10 years, have shown that the evolutionary history of mammals is far more complicated than previously envisioned. Many of these new findings suggest that features of mammalian biology are the end products of mosaic evolution of several traits and not the result of a straightforward acquisition of mammalian traits in their distant relatives. In the following, we provide a brief discussion of some of these features and how our understanding of them has changed in recent years (see also Kemp 2006b).

Differentiation of the Vertebral Column

The differentiation of the vertebral column into thoracic and lumbar regions with shortened lumbar ribs evolved in cynodonts like *Thrinaxodon* (Jenkins 1971) (see Plate M) and is regarded as evidence for the occurrence of a mammalian diaphragm. The development of the diaphragm is attributed to the need for more effective gas exchange and ventilation of the lungs due to an increased metabolic rate. The evolution of the mammalian diaphragm was considered to be a straightforward evolutionary development from reduced lumbar ribs in cynodonts to a fully developed diaphragm without any lumbar ribs in placental mammals. Recent findings of mobile lumbar ribs in the eutriconodonts *Yanoconodon* (Luo et al. 2007) and *Repenomamus* (Hu et al. 2005), the spalacotheriid *Akidolestes* (Li and Luo 2006), and the docodont *Castorocauda* (Ji et al. 2006) show many different Mesozoic mammals with primitive diaphragms.

Limb Posture

The limb posture in early mammals is still a matter of controversy: Did they have a sprawling "reptilian" posture, or did they have an erect (parasagittal) posture as is seen in modern mammals (for alternative views see Sereno [2006]; Kielan-Jaworowska and Hurum [2006])? Carrier (1987) suggested that the advanced gait in therapsids is linked to an elevated metabolic rate. Contrarily, the postcranial skeleton of the triconodont *Jeholodens* from the Early Cretaceous of China shows a primitive pelvic girdle and hind limb but a derived pectoral girdle that is closely comparable to those of derived therians (Ji, Luo, and Ji 1999). Supporting this evidence is the study of Kielan-Jaworowska and Hurum (2006) which demonstrated that early mammals had a sprawling posture, but with forelimbs more parasagittal than the hind limbs. Based on anatomical studies of complete skeletons. they concluded that parasagittalism developed in mammalian evolution only in the therians (see also Jin et al. 2006). The presence of an extra-tarsal spur in most nontherian Mesozoic mammals and extant monotremes (Hurum, Luo, and Kielan-Jaworowska 2006) also supports this hypothesis of sprawling hind limbs in early mammals. However, tracks from the Late Jurassic *Ameghinichnus patagonicus* of Patagonia (Casamiquela 1964) depict a more parasagittal gait.

Ossified Respiratory Turbinals

Ossified respiratory turbinals have not been reported for any of the nonmammalian therapsids, although bony ridges suggestive of attachment sites for turbinals within the nasal cavity have been observed in the therocephalian *Glanosuchus* and in several cynodonts (Hillenius 1994; see chapter 11 of this book). Similar ridges are also interpreted as evidence for maxilloturbinals in *Morganucodon* (Ruben 1995). Among other Mesozoic mammals, fragmented turbinals have been documented in serial sections of two Late Cretaceous multituberculates, *Nemegtbaatar* and *Chulsanbaatar* (Hurum 1994) (Fig. 10.2), and in the Late Cretaceous eutherian *Barunlestes* (Kielan-Jaworowska and Trofimov 1980), and also the Jurassic docodont *Haldanodon* (Lillegraven and Krusat 1991).

Fur

Fur was also proposed as suggestive of endothermy since the presence of insulation is considered important for maintaining homeothermy (e.g., Watson 1931). The pits on the premaxilla and maxilla of *Thrinaxodon* were considered as indicative of some form of vibrissae (Watson 1931). Recently, several Jurassic mammals from China have been described with impressions of fur (e.g., *Volaticotherium* [Meng et al. 2006]; *Castorocauda* [Ji et al. 2006]). Vullo et al. (2010) newly described the oldest mammalian hair preserved in 100 million-year-old amber. The documentation of fur in pterosaurs suggests that it is more basal among amniotes than previously thought (Sharov 1971; Wang et al. 2002; Unwin and Bakhurina 1994). More recently, Eckhart et al. (2008) identified cysteine-rich α-keratin genes in reptiles and birds and argued against hair keratins as an evolutionary innovation of mammals, but just a co-option of pre-existing structural proteins.

The Secondary Palate

The development of the secondary palate among the nonmammalian therapsids was long considered an important step toward the attainment of endothermy. However, more recent assessments have shown that it is not only linked to breathing while simultaneously processing of food, but that it is also important for suckling and, more generally, for strengthening the skull (Kemp 2005; Thomason and Russell 1986).

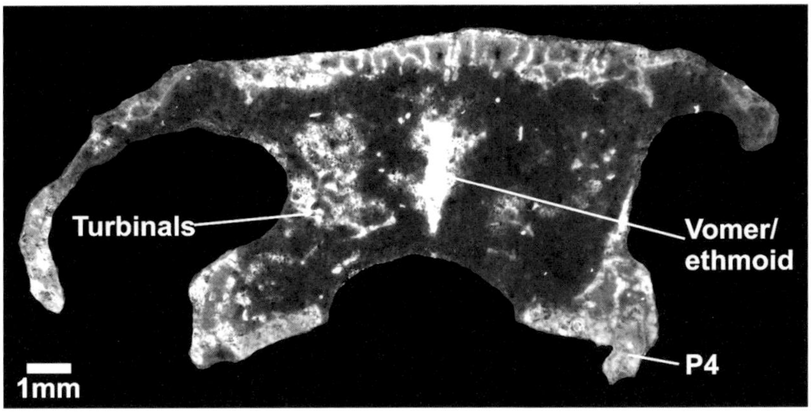

10.2. Microtome slice (section 840) of the skull of *Nemegtbaatar gobiensis* ZPAL MgM-I/76, showing well-developed nasal turbinates. *Abbreviations:* P4 = premolar four.

Diphyodonty

The evolutionary changes of the morphology and replacement patterns of teeth from more basal cynodonts to early mammals was considered as a step-wise addition of mammal-like dental characteristics. The evolution of diphyodonty in mammals from a pattern of multiple alternating replacement of teeth in most cynodonts to a single sequential replacement pattern of the postcanines in mammals is possibly related to specializations in feeding and growth (Luo, Kielan-Jaworowska, and Cifelli 2004).

Enlarged Brain and Body Miniaturization

The enlarged brain and body miniaturization of early mammals have both been used as characteristics indicating endothermy (Hopson [1973, 1979]; for reviews see Hayes and Garland [1995]; Rowe [1996]; Kielan-Jaworowska, Cifelli, and Luo 2004). Kemp (2009a) described the transition from cynodont to mammalian brain and proposed a correlation with miniaturization. The large expansion of the brain and the separation of the middle ear bones in the smallest Mesozoic mammal known, the early Jurassic Mammaliaform *Hadrocodium* (Luo, Crompton, and Sun 2001) support this contention. However, the finding of badger-sized carnivorous mammals with remains of small juvenile dinosaurs preserved in the region of the gut (Hu et al. 2005), and the large distribution range of relative brain size in other groups of Mesozoic mammals indicate that this is a more complex issue than was expected earlier (for a review see Hayes and Garland [1995]).

Bone Histology

Much of the earlier literature on fossil bone microstructure implied a direct link between bone histology and thermal physiology. Fibrolamellar bone tissue, with its woven-bone matrix and rich vascularization, was associated with high levels of metabolic activity and endothermy, while lamellar-zonal bone was deemed a characteristic of ectothermic vertebrates (see chapter 3 for a description of these bone tissue types). However, it has increasingly become evident that this dichotomy is not as clear-cut: There are documented examples of fibrolamellar bone in extant ectotherms, such as alligators and turtles (Reid 1993). This bone has also been widely documented in nonmammalian therapsids, for example, Dicynodontia (e.g., Chinsamy and Rubidge 1993; Ray and Chinsamy 2004; chapter 6 this book), that are not closely related to mammals, as well as, among the mammalian ancestors, the nonmammaliaform cynodonts (e.g., Botha and Chinsamy 2000, 2004; chapter 9 of this book). A host of studies on a variety of nonavian dinosaurs and other archosaurs have, however, shown that studies of bone microstructure of extinct animals can provide direct insight into the growth dynamics experienced by the animal (see Chinsamy-Turan 2005, references therein).

The preceding examples demonstrate the complexity and diversity of early mammalian evolution and biology not foreseen just 10 years ago (Martin and Luo 2005; Luo 2007). In an attempt to unravel the biology of early mammals, Chinsamy and Hurum (2006) undertook an investigation of the bone depositional patterns of five Mesozoic mammalian taxa to gain insight into their growth dynamics and compared this with extant mammals. Here, we expand on the comparisons made with extant monotremes, marsupials, and placentals, and we further expound on the earlier hypotheses proposed.

Histological Assessment of Mesozoic and Extant Mammal Bone

Chinsamy and Hurum (2006) examined the bone microstructure of the Early Jurassic *Morganucodon* (UCL); the Late Cretaceous multituberculates, *Kryptobaatar* (ZPALMgM-I/18) and *Nemegtbaatar* (ZPALMgM-I/81); and the Late Cretaceous eutherians, *Zalambdalestes* (ZPALMgM-I/43) and *Barunlestes* (ZPALMgM-I/77). In this review, we present an abbreviated version of the bone microstructure of these early mammals, and we provide more details pertaining to the bone microstructure of modern monotremes, marsupials, and placentals. We examine bones of:

1. Living monotremes: juvenile and adult long bones (femur and humerus) of the platypus (*Ornithorhynchus anatinus*), and a humerus of an adult echidna (*Tachyglossus aculeatus*).
2. Marsupials: femora of the kangaroo (*Macropus sp.*) and the North American opossum, *Didelphis virginiana.*
3. Placentals: femur of the groundhog or woodchuck, *Marmota monax.*

The monotreme bones were demineralized and stained with hematoxylin, while all the other bones thin-sectioned were undecalcified and unstained (Chinsamy and Raath 1992; chapter 3 of this book). All the sections were examined under a compound microscope using polarized and ordinary light. Whenever possible the relative bone wall thickness

([RBT], i.e., the percentage of the average thickness of the bone wall divided by the average cross-sectional diameter of the bone), was calculated. In instances where the cross-sectional diameter could not be measured (e.g., owing to incompleteness of the cross section) an average thickness of the bone wall is provided.

Results and Discussion

Descriptions of Extinct Mammalian Bone Microstructure

MORGANUCODONTA

The Morganucodonta were relatively small, shrew-like animals, having adult skull lengths of about 20–30 mm and total body lengths of about 150 mm (Kermack, Mussett, and Rigney 1973, 1981; Jenkins and Parrington 1976; Crompton and Luo 1993). Their small body size and sharp, tricuspid teeth are suggestive of an insectivorous diet. Various skeletal elements of *Morganucodon watsoni* (Kühne [1949], sometimes also referred to as *Eozostrodon;* see Clemens [1979]) from the Early Jurassic fissure fillings of the Pant Quarry, Wales, were sectioned in the study by Chinsamy and Hurum (2006).

Although a majority of the *Morganucodon* sections showed poorly preserved bone microstructure, an ulna showed woven bone tissue with large, random oriented osteocyte lacunae and several primary osteons (Chinsamy and Hurum 2006). Several enlarged erosion cavities are evident in the compacta but no secondary osteons were present. Substantial endosteal resorption had also occurred. Parallel-fibered bone tissue is evident in a localized area and has several rest lines, which indicate pauses in the rate of bone formation, and, hence, arrests in growth. A section of a humerus preserves only a distinct region of parallel-fibered bone tissue. The RBT of the femur measured 26 percent and although microstructural details of the mid-cortical area were diagenetically altered, a fairly thick layer of endosteal bone was visible around the medullary cavity. Thus, it is apparent that *Morganucodon* formed fibrolamellar bone during early ontogeny, but later in life, formed parallel-fibered bone tissue episodically. Although a well-defined outer circumferential lamellae (OCL) is lacking, the presence of parallel-fibered bone with lines of arrested growth (LAGs) suggest that growth was slow and intermittent later in ontogeny.

MULTITUBERCULATES

The multituberculates, which are superficially rodent-like, are the largest group of Mesozoic mammals. They appeared in the fossil record from about the Middle Jurassic (Kielan-Jaworowska and Hurum 2001, Butler and Hooker 2005), but they are best known from the Late Cretaceous and Early Tertiary of North America and Central Asia, though recent discoveries have extended their range to Africa, South America, Scandinavia, and Australia. Chinsamy and Hurum (2006) sampled femora from diagnostic articulated specimens of two multituberculates from the Gobi Desert, Mongolia: *Kryptobaatar dashzevegi* from the Early Campanian,

Djadoktha Formation; and *Nemegtbaatar gobiensis* from the Late Campanian, Baruungoyot Formation (for occurrences and discussion see Kielan-Jaworowska, Hurum, and Badamgarav [2003)]).

The midshaft region of the *Kryptobaatar* femur revealed a fairly thin-walled compact bone region with a hollow medullary cavity. The RBT of the bone is 10 percent. The bone appears to be a mixture of some parallel-fibered tissue (in some areas along the margin of the medullary cavity) and a woven-textured type of bone tissue (Fig. 10.3A). Several channels occur within the bone, which appears to be at an early stage of primary osteon formation. The peripheral and medullary margins of the bone are uneven and are distinctly resorptive in nature.

Kryptobaatar is distinctive in having almost the entire cortical bone predominantly consisting of a woven bone tissue with primary osteons, which suggest a rapid rate of bone formation. The uneven peripheral and perimedullary margins of the bone wall and the early stages of osteonal development suggest that the bone is from an immature individual. As such, it is not known whether bone formation slowed down in later stages of ontogeny, or whether they had determinate growth.

Four serial sections of the femur of *Nemegtbaatar* provided a good assessment of the microstructural changes associated with the growth in length and girth of the proximal part of the femur. The compacta is predominantly composed of compacted, fine cancellous bone (Enlow 1962). Superficially this tissue looks like woven fibrolamellar bone; however, on closer inspection, it is evident that this bone lies below a distinct tide line (= erosion line), which indicates its endosteal nature (Fig. 10.3B). The endosteal bone varies from the compacted fine cancellous bone tissue described to a type with more convoluted lamellae, and a more lamellated type of bone tissue. The periosteal bone deposits mainly consist of parallel-fibered bone with a distinct LAG interrupting the bone deposition. The RBT is about 15 percent.

The four thin sections all showed complex cortical stratifications that are associated with growth and remodeling of the growing femur. The compacted, fine-cancellous bone in the cortex indicates that at an earlier stage of ontogeny this region was close to the epiphyseal growth plate (Enlow 1962). As the femur grew in length, this region was relocated from its original location into the cortex of the growing bone. Such mixed bone tissues have been reported in the bones of young, rapidly growing individuals and are known to occur in the bones of white rats of different ages (Enlow 1962).

A small patch of periosteally formed woven bone tissue observed in *Nemegtbaatar* suggests that during early ontogeny, the initial periosteal cortical bone tissue was deposited rapidly. However, in the sections Chinsamy and Hurum (2006) studied, most of this early formed bone tissue was already resorbed, and mainly the later formed, periosteal lamellar bone with longitudinally and radially oriented channels is evident in the bone wall. Thus, given the overall histological structure of the bone (the occurrence of compacted fine cancellous bone, the presence of a localized

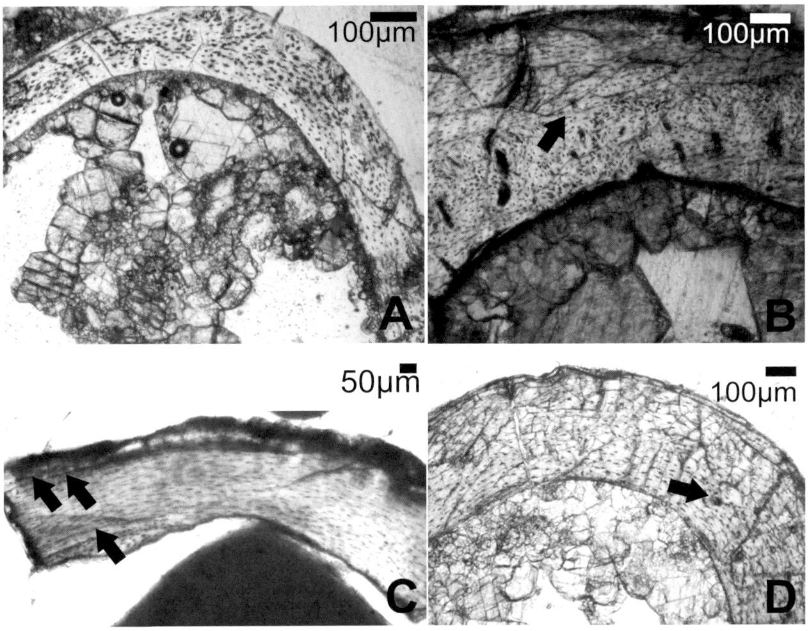

10.3. (A) Femur of *Kryptobaatar dashzevegi* (ZPALMgM-I/18) showing the overall bone structure of the thin bone wall. (B) Femoral section of *Nemegtbaatar gobiensis* (ZPALMgM-I/81). Arrow indicates the tide line below which lies endosteally formed bone tissue. (C) Femoral section of *Barunlestes butleri* (ZPALMgM-I/77). Arrows indicate lines of arrested growth in the compacta. (D) Femoral section of *Zalambdalestes lechei* (ZPALMgM -I/43). Arrow indicates a primary osteon.

region of woven bone, as well as vascularized lamellar bone), it appears that *Nemegtbaatar* had a fast rate of growth during early ontogeny, which slowed down during later stages of development.

The well-vascularized compacta of the Mesozoic multituberculates compares well with that of the mandible of the Paleocene multituberculate *Ptilodus*, which has a distinct, triple-layer structure of bone with an OCL and an inner endosteal circumferential layer that sandwiched a well-vascularized region of woven bone (Enlow and Brown 1958). Our *Kryptobaatar* sections appear to be from a juvenile, since only the woven bone region is preserved, whereas in *Nemegtbaatar*, which appears to have been an older individual, traces of all three regions as described for *Ptilodus* are present.

EUTHERIANS

Two eutherian species *Barunlestes butleri* and *Zalambdalestes lechei*, both from the Late Cretaceous of the Gobi Desert, were also studied by Chinsamy and Hurum (2006). Both are in the family Zalambdalestidae. *Zalambdalestes* is considered to have been a specialized jumping, hedgehog-sized animal, with a long snout. *Barunlestes* is similar in size but is more heavily built, and has a more robust skull. A femur of *Zalambdalestes*, and a rib and femur of *Barunlestes* from identifiable postcranial skeletons were thin-sectioned in the analysis (Fig. 10.3C). This small, rare sample of *Barunlestes* also permitted the assessment of histological variation between two skeletal elements in a single Mesozoic mammal.

In *Barunlestes butleri*, a complete transverse section of a rib showed a fairly thick bone wall, with a RBT of about 31 percent. It is largely composed of parallel-fibered bone tissue with several rest lines. The bone is generally poorly vascularized, except for a localized area of drift where a few channels for blood vessels occur and the osteocytes tend to have a random orientation. The sectioned fragment of a femur shows an outer parallel-fibered bone tissue with several LAGs (Fig. 10.3C). In a localized region, a bone tissue with more globular, haphazardly oriented osteocyte lacunae occurs, which may be indicative of a faster rate of bone deposition.

In *Zalambdalestes lechei*, the RBT of the femoral cross section is about 17 percent. The compacta consists of parallel-fibered bone tissue which is rather poorly vascularized, except in a localized area where several simple blood vessels are located (Fig. 10.3D). A few poorly defined primary osteons and isolated secondary osteons are visible in the compacta. In the peripheral region, the osteocytes tend to be more flattened and less abundant and a single LAG is visible. The medullary margin is irregular, which suggests that endosteal resorption was underway when the animal died.

The overall nature of the Mesozoic eutherian bone tissue suggests a slow rate of bone deposition, with *Barunlestes* having an even slower growth and hence depositional rate than *Zalambdalestes*. They both experienced a relatively slow rate of growth with periodic pauses indicated by rest lines. The *Zalambdalestes* specimen appears to have been a young individual since only parallel-fibered bone tissue with simple blood vessels are evident in its compacta. The *Barunlestes* rib preserves a fairly thick cortical region having parallel-fibered bone tissue with rest lines. However, in the femoral fragment studied, the more haphazardly organized osteocytes suggest a faster depositional rate during early ontogeny.

Bone Histology of Extant Mammals

MONOTREME BONE HISTOLOGY

Currently there are only two studies of the bone histology of the extant platypus, *Ornithorhynchus anatinus* (Enlow and Brown 1958; Chinsamy and Hurum 2006), and both examined a rather limited number of bones. These studies reported a few similarities, but also some different characteristics in the bone microstructure of *Ornithorhynchus*. Both studies reported that the bone compacta was richly vascularized and that a lamellar bone layer was present. However, the Enlow and Brown (1958) study described this layer as located between two layers of nonlamellar bone, whereas Chinsamy and Hurum (2006) found this lamellar layer to be peripherally located and suggested that perhaps these lamellar deposits are annuli.

Enlow and Brown (1958) examined sections of a femur and a rib and reported the absence of Haversian reconstruction and the presence of a vacant medullary cavity. They also mentioned that the *Ornithorhynchus* specimens had unusually small lumina that were surrounded by a ring of acellular, nonlamellar bone tissue. Chinsamy and Hurum

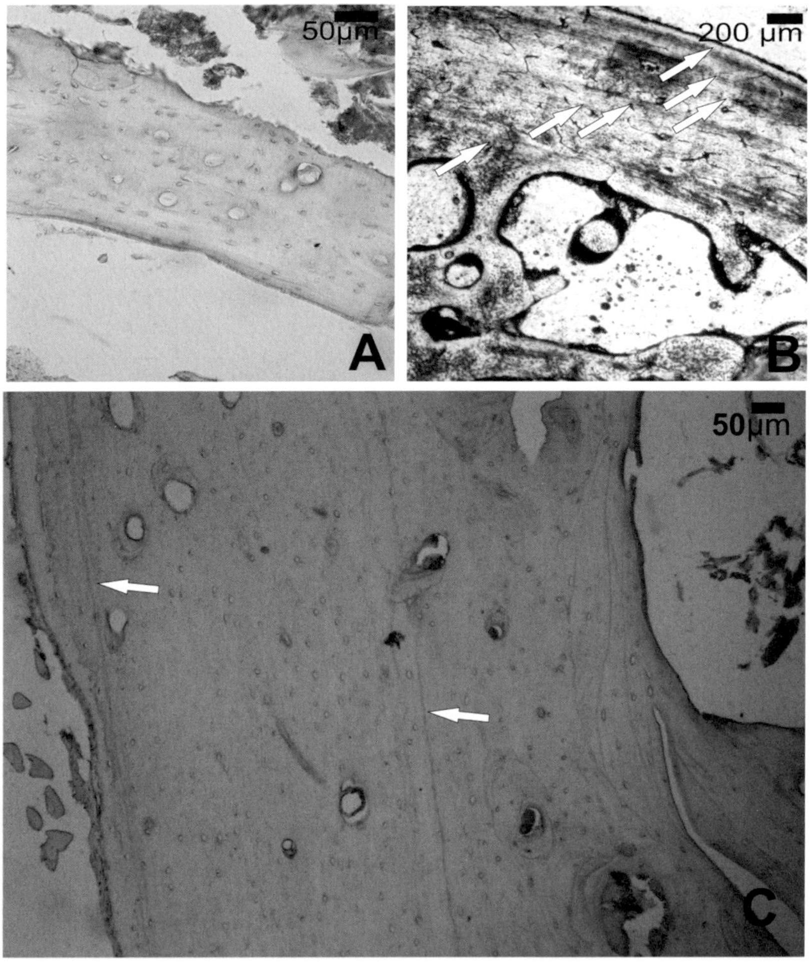

10.4. Bone histology of extant monotremes. (A) Juvenile platypus (*Ornithorhynchus anatinus*) femur, demineralized and stained with hematoxylin showing rapidly deposited bone tissue and a number of canals. (B) Adult platypus humerus (showing several thin layers of lamellar bone (arrows). (C) Adult echidna (*Tachyglossus aculeatus*) femur showing at least two growth rings.

(2006) also reported a richly vascularized compacta but found that a large number of secondary osteons were located in the bone wall and that an extensive development of trabeculae extended into the medullary cavity. This bone did not show the small lumina and acellular tissue that Enlow and Brown (1958) reported. Chinsamy and Hurum (2006) found that the *Ornithorhynchus* bone was generally formed fairly rapidly, although it was interrupted periodically by more slowly formed lamellar bone tissue.

In the current analysis of *Ornithorhynchus*, we examined the bone microstructure of a femur of a juvenile and an adult, and that of a humerus of an adult (Fig. 10.4). The sections were decalcified and stained with hematoxylin.

Ornithorhynchus Bone Microstructure

FEMORA The bone wall of the juvenile is thin (average thickness of 237 µm) and surrounds a large medullary cavity. The peripheral margin of the bone is uneven. A large number of channels in the bone are evident, and numerous globular-shaped osteocyte lacunae are visible (Fig. 10.4A). A LAG occurs near the margin of the medullary cavity. It is possible that this might represent a "hatching" line, as observed in extant lepidosaurs (Castanet et al. 1993). The bone wall of the adult femur of the platypus has large peri-medullary excavations into the bone wall. The actual bone wall thickness across the cross section is highly variable and measures from as little as 44 μm to about 225 μm. Many of the channels in the bone are secondarily enlarged.

HUMERI Two humeri were sectioned, one of which was from an individual of uncertain age, while the other was from an adult. The average bone wall thickness of the cross section of the former individual is 1,091 μm. The compacta is stratified into different regions—the innermost region (nearest the medullary cavity) comprises a relatively avascular region, above which is a relatively well-vascularized region mainly comprising primary osteons with some slightly enlarged. The osteocyte lacunae in this region appear to be disorganized and more rounded in shape. An uneven growth mark is visible. This is followed by a poorly vascularized region with flattened osteocyte lacunae that appear to be deposited in a more organized manner. Three to four growth marks interrupt the deposition of this bone tissue. The outermost two are closely spaced and are most likely rest lines indicating the attainment of adult body size.

The uneven growth mark visible within the compacta represents an erosion line, and the bone deposited below this line is endosteally formed. The periosteally formed bone lies above the uneven line, and this appears to be a more slowly formed lamellar type of bone tissue with few channels for blood vessels. The peripheral closely spaced lines suggest that the animal has attained adult body size and has virtually stopped appositional growth.

The (mineralized) cross section of the adult platypus humerus shows a fairly thick layer of endosteally formed bone under an erosion line. Just underneath the erosion line is a region of secondary osteons and compacted bone tissue. This is followed by a layer of poorly vascularized lamellar bone tissue that lines the medullary cavity. The periosteally formed bone in the compacta also comprises a poorly vascularized lamellar bone. Several rest lines are visible peripherally (Fig. 10.4B), which suggests that this individual had reached its adult body size.

Echidna Bone Histology

A thin section of an adult humerus was sampled for histological analysis. The bone was sectioned using a microtome and thereafter stained with hematoxylin.

The average bone wall thickness was about 886 μm. Several secondarily enlarged erosion cavities are visible in the compacta (Fig. 10.4C). The cement line or erosion line is distinctive in all of them, and several of the cavities show well-developed secondary bone deposits that result in secondary osteons. Around the medullary cavity, a poorly vascularized lamellar bone tissue is present. This tissue is interrupted by several growth lines which suggest that medullary expansion and redeposition had occurred several times. About midway across the bone wall, two LAGs are evident. These are closely associated, and it is likely that they were formed in the same season and do not represent two consecutive unfavorable seasons. The outer peripheral region also shows a pair of closely spaced LAGs (Fig. 10.4C). It is possible that these again represent double LAGs formed during one season (Castanet et al. 1993).

MARSUPIAL BONE HISTOLOGY

Macropus (Kangaroo) Bone Histology

There have been a few brief reports of kangaroo bone microstructure (Leahy 1991; Chinsamy-Turan 2005), but unfortunately none of these have provided a thorough assessment. Leahy (1991) and Chinsamy-Turan (2005) both reported the presence of growth rings in kangaroo bones. Here, we provide a more detailed description of the specimen sectioned by Chinsamy-Turan (2005). The specimen shows a fairly thick bone wall. The bone formed during early ontogeny appears to be a richly vascularized fibrolamellar bone tissue with a reticular orientation of the vascular channels (Fig. 10.5A). About halfway around the compacta, a distinct change in the bone microstructure occurs: A LAG is present that may be related to the attainment of sexual maturity. Thereafter, a narrow band of fibrolamellar bone tissue is deposited, and following this, another LAG is deposited. Subsequent bone deposited is a relatively slowly formed bone tissue with simple, radially organized blood vessels. This cycle of a LAG and slowly formed bone tissue is repeated about fourteen times (Fig. 10.5A).

From the preceding description, it is apparent that the kangaroo studied has a initial rapid growth (indicated by the fibrolamellar bone) and thereafter (possibly following sexual maturity), appositional growth slowed down substantially (as indicated by the change to simple blood vessels in a lamellar bone matrix). It seems that this slow increase in girth occurs for the next fourteen years and the bone wall thickens. It is evident that more closely constrained studies of histological changes during ontogenetic growth in kangaroos and other marsupials are needed. Nevertheless, this cursory inspection of marsupial bone has demonstrated that cyclical patterns of bone formation are evident in a kangaroo.

North American Opossum (*Didelphis virginiana*)

The fairly small (60–90 cm) *Didelphis virginiana* is the only marsupial found in North America and is commonly known as the North American opossum.

Thin sections of the opossum reveal a fairly thick bone wall with a RBT of about 33 percent. Thick bone walls are generally considered to

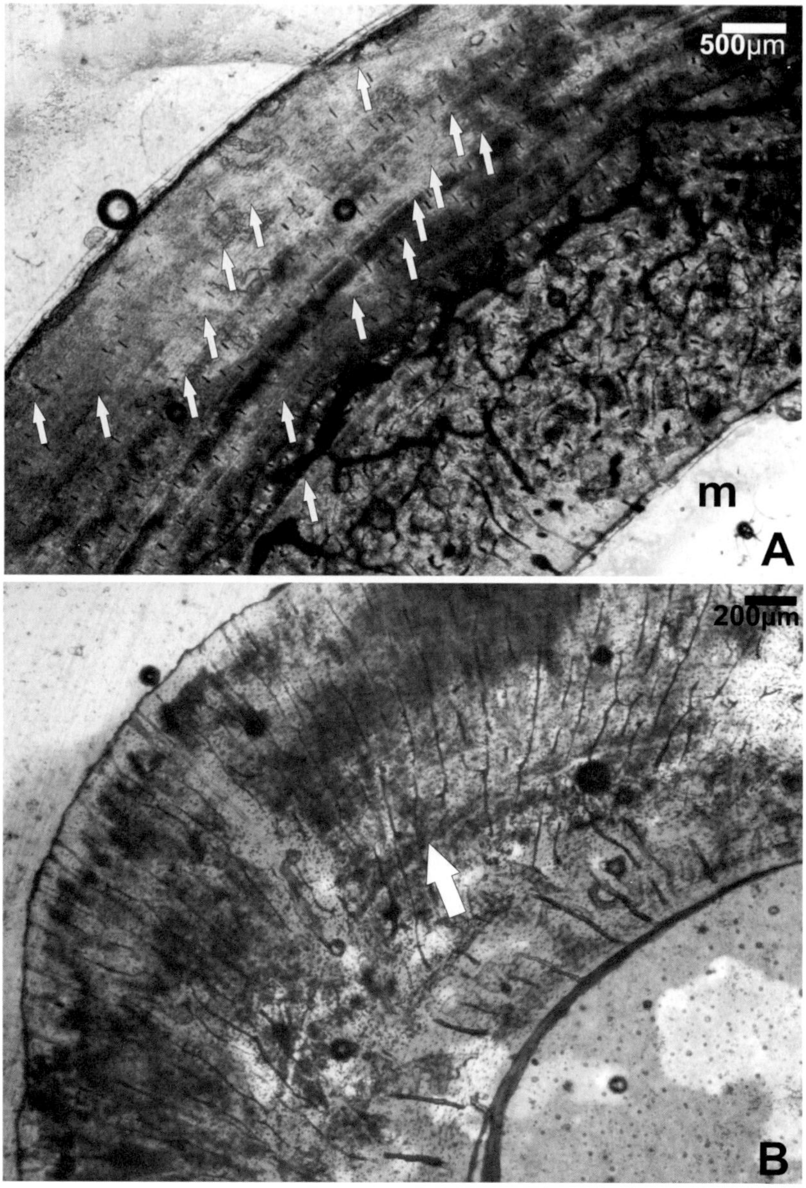

10.5. Marsupial bone tissues. (A) Kangaroo femur showing rapidly deposited fibrolamellar bone tissue nearest the medullary cavity (m), followed by a lamellar zonal bone tissue. Arrows indicate the periods of slowed growth. (B) North American opossum (*Didelphis virginiana*) showing a well vascularized compacta with radially arranged channels, interrupted by a single LAG (arrow).

be features of fossorial or amphibious animals (see, e.g., Wall 1983; Botha and Chinsamy 2004; Chinsamy and Hurum 2006). However, *Didelphis* is not amphibious, and it is generally thought that it does not make its own burrows; instead it uses burrows that have been excavated and abandoned by other animals. The bone wall is richly vascularized by a large number of radial channels (Fig. 10.5B). Even the inner circumferential layer that lines the medullary cavity has several radially oriented simple channels. The compacta has a distinctive LAG that is visible around the cross section. No secondary osteons or resorptive channels are visible in the cortical bone. The outermost part of the bone wall has a distinct layer of lamellar bone which most likely marks the attainment of adult body size. Since only one bone was examined, it is uncertain whether there would have been other rest lines forming during later stages of the animal's life (as observed in the kangaroo) (Fig. 10.5A).

Although opossums do not hibernate, during the cold North American winters, they tend to den in their burrows for several days at a time, only emerging to feed. Thus, it is likely that the distinctive LAG around the middle of the bone wall corresponds to a particularly severe winter.

EXTANT EUTHERIAN (PLACENTAL) BONE MICROSTRUCTURE

A host of studies have provided insight into the kinds of bone tissues present among extant placentals (e.g., Klevezal and Kleinenberg 1969; Enlow and Brown 1958; and Enlow 1962), and how their bone microstructure changes during growth and remodeling processes. Furthermore, since pigs and rats are often used as models for human bone, there is an extensive experimental literature on bone microstructure of these animals, especially pertaining to osteoporosis and biomechanical aspects (e.g., Morey and Baylink 1978).

Placental mammals generally grow rapidly during early ontogeny, and once sexual maturity is attained, their growth slows down drastically (e.g., Klevezal 1996). These growth patterns are reflected in the kind of bone tissue that forms, that is, during the early rapid phase of growth, azonal, fibrolamellar bone is deposited, and once growth has virtually ceased, the more slowly formed OCL, usually consisting of lamellar bone or parallel-fibered bone, is deposited (see, e.g., Chinsamy and Dodson 1995; Chinsamy-Turan 2005; Köhler and Moyà-Solà 2009).

Although not a common occurrence, placental mammals occasionally form periodically interrupted fibrolamellar bone, in other words, zonal bone. For example, this type of bone tissue has been documented in marine mammals (e.g., seals and cetaceans [Klevezal and Kleinenberg 1969]), mammals of temperate regions and hibernators (Klevezal and Kleinenberg 1969), as well as in polar bears (Chinsamy-Turan 2005). Recently, Köhler and Moyà-Solà (2009) described lamellar-zonal bone tissue in a growth series the extinct Plio-Pleistocene bovid *Myotragus*. They suggested that the flexible growth pattern observed in this mammal was probably an adaptive response to fluctuating resource levels in the environment. An isolated occurrence of a single LAG has been documented in an elk (Horner, de Ricqlès, and Padian 1999), although no environmental or life history data are known to permit an explanation of the possible cause of this interruption in its growth. Sander and Andrassy (2006) described LAGs in Late Pleistocene large mammals (e.g., cervids, bovids, a rhinocerotid, *Coelodonta antiquitatis*, and an elephantid, *Mammuthus primigenius*) that experienced glacial conditions.

The compacta of placental mammals generally shows extensive secondary reconstruction and a high concentration of secondary osteons which often reach dense Haversian proportions (Fig. 10.6A). In many mammals, secondary reconstruction begins fairly early in ontogeny.

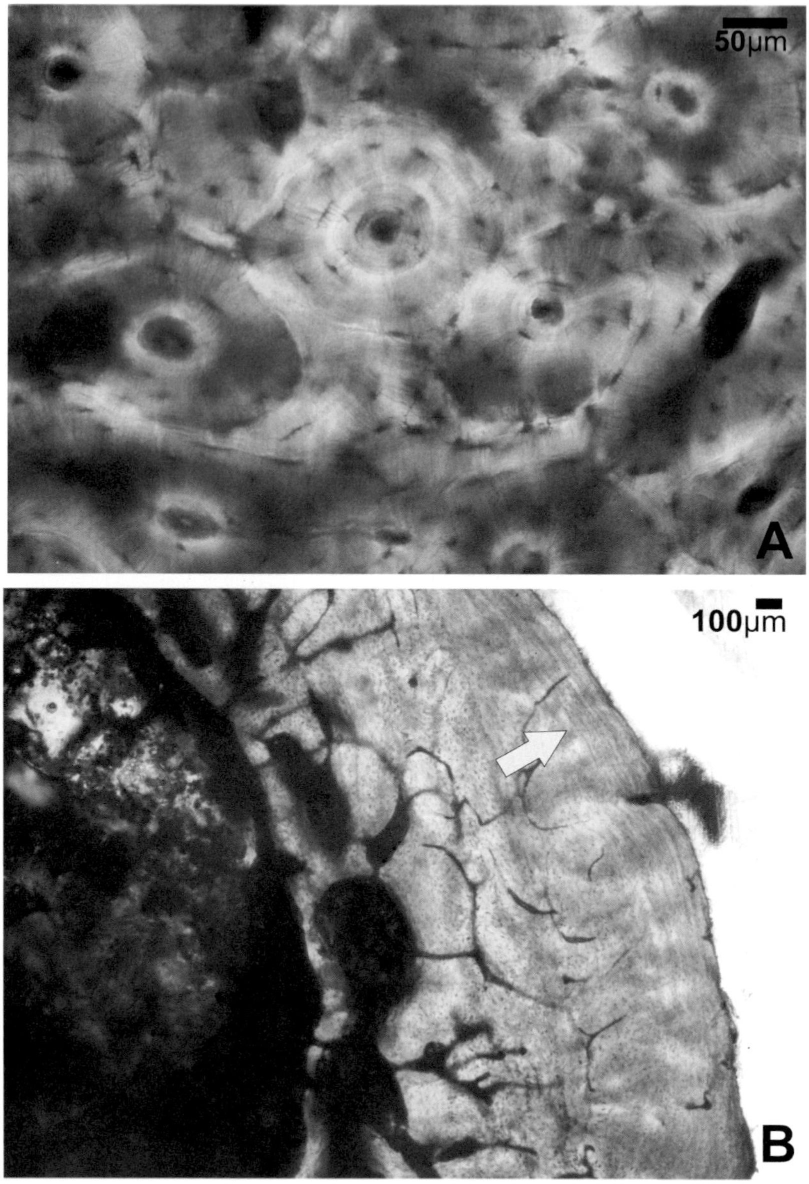

10.6. Examples of modern mammal bone. (A) Dense Haversian bone in a sloth. (B) *Marmot monax* femur showing several closely spaced peripheral rest lines (arrow).

During pregnancy and lactation, females often mobilize phosphates and calcium from their bones, which results in a high concentration of erosion cavities in their bones. Generally, these are later infilled with a centripetal deposition of lamellar bone, but often there is a disproportional infilling and bones of reproductively active females tend to be more osteoporotic. Although osteoporosis is quite frequent among female eutherians, it also occurs among elderly males.

Measuring about 76 cm and weighing up to 6.5 kg, the marmot (*Marmot monax)* is considered to be among the largest members of the Sciuridae (the squirrel family). They have fairly long life spans (about 10 years), and they hibernate annually inside their burrows. A femur of a marmot was thin-sectioned and studied under light microscopy. The average bone wall thickness is 1,261 μm and RBT is 16 percent. The compacta of the bone wall consists of a richly vascularized reticular fibrolamellar bone tissue (Fig. 10.6B). Around the medullary cavity, there is a distinctive layer of secondarily enlarged erosion cavities. A distinct LAG is evident about midway in the bone wall. This most likely corresponds to the animal's first hibernation. In the peripheral region of the bone wall, there are a series of closely spaced rest lines that appear to interrupt a lamellar bone tissue. The change in bone tissue from a fibrolamellar type to a more lamellar type with rest lines probably marks the attainment of sexual maturity. The more slowly formed lamellar bone tissue indicates that subsequent appositional growth was minimal and was punctuated by rest lines (Fig. 10.6B). It is possible that each of these lines corresponds to a hibernation season. Further studies are warranted to confirm this supposition.

Comparison of the Jurassic-Cretaceous Mammalian Bone with that of Their Nonmammmalian Therapsid Ancestors and Extant Mammals

Except for the *Kryptobaatar* and *Zalambdalestes* bones, which are likely from young individuals that have not yet attained adult body size, all other Mesozoic mammalian bone studied by Chinsamy and Hurum (2006) showed a change to a slower rate of bone deposition later in ontogeny. This suggests determinate growth patterns for all mammalian clades studied, the Mesozoic multituberculates, the fossil and extant eutherians, as well as the extant monotremes and marsupials. Since several nonmammalian therapsids, including dicynodonts, such as *Placerias* (Green, Schweitzer, and Lamm 2010) and *Kannemeyeria* (chapter 5 of this book), and more basal theriodonts, such as a scylacosaurid therocephalian (Ray, Botha, and Chinsamy 2004; chapter 8 of this book), as well as the nonmammaliaform cynodonts *Tritylodon* (Chinsamy and Hurum 2006), all show a determinate growth strategy, it is reasonable to suggest that a determinate body size evolved fairly early among the nonmammaliaform therapsids.

Interestingly, as compared to most of the nonmammaliaform cynodonts, the Mesozoic mammals appear to have slower bone depositional rates and hence growth rates. However, it is worth noting that among the nonmammaliaform cynodonts, varying growth strategies are evident: from rapid uninterrupted rates of growth (as seen, for example, in *Thrinaxodon* [Botha and Chinsamy 2005] [see Plate M, bottom] and *Cynognathus* [Botha and Chinsamy 2000] [see Plate O and Plate 10B]), to cyclical growth patterns (as observed in *Diademodon* [Botha and Chinsamy 2000; see Plate 10C]) and *Trirachodon* [Botha and Chinsamy 2004]), to slow rates of bone deposition as suggested by the bone histology of *Andescynodon* and *Massetognathus* (Chinsamy and Abdala 2008). Thus, it appears that the nonmammaliaform cynodonts display much variability in terms of growth and developmental strategies.

Although the Mesozoic mammalian sample comprised only five different taxa, observable differences in their bone histology can be directly linked to differences in the bone depositional rates and overall growth rates. Moreover, the multituberculates and the Mesozoic eutherians appear to have slower growth rates than modern monotremes, marsupials, and placentals. Thus, our study of a rather small sample of diverse Mesozoic mammals suggests that the rapid rates of growth most likely evolved during the Paleocene evolution of the eutherians, metatherians, and monotremes. With the already dominant position of eutherians in the Late Cretaceous in relation to metatherians and monotremes, this finding supports the hypothesis for an explosive evolution of Placentalia (Wible et al. 2007) in the Paleocene.

One of the striking similarities between the Mesozoic eutherians and extant monotremes, marsupials, and extant placentals examined here is the distinct occurrence of LAGs in the compacta of their bone walls, which suggests developmental plasticity—flexible growth rates. Indeed, such plasticity in growth rates is well represented among various synapsid representatives (Ray, Botha, and Chinsamy 2004; chapters 8 and 9 of this book), although there appears to be a trend toward more rapid sustained rates of growth among the Cynodontia (see, e.g., Botha and Chinsamy 2000, 2004; chapter 9 of this book). In the echidna (Fig. 10.4C) and the marmot (Fig. 10.5B), the growth marks in their bones (LAGs) appear to be linked to their hibernation strategies, and although it has not yet been proven, it is likely that the periodic deposition of lamellar bone (following a fibrolamellar layer of bone) in the polar bear also correlates to their annual torpor. Interestingly, the wide ability for hibernation among the Mammalia seems to involve the differential expression of genes that appear to be common to all mammals, in other words, hibernators do not have unique genes that give them the ability to hibernate (Carey, Andrews, and Martin 2003). Indeed the fact that hibernation/torpor are found in the deepest branches of extant Mammalia, for example, among the monotremes, supports the hypothesis that hibernation, which is possibly the most remarkable example of phenotypic plasticity, is plesiomorphic for all mammals (e.g., Carey, Andrews, and Martin 2003; Grigg, Beard, and Augee 2004). Studies of the echidna have shown that even within a single species, depending on where they live, individuals can be facultative or obligate hibernators (Grigg, Beard, and Augee 2004). Recently, Liow et al. (2009) documented that extant mammals that hibernate or that hide in burrows are less likely to turn up on an endangered species list and suggested that their ability to "sleep-or-hide" buffers them from changing environments and enables them to avert extinction. Could flexible growth rates and hibernation have been an adaptive strategy for the eutherians after the K/T event?

In our study, although we cannot ascertain whether the growth marks present in the bones of the Mesozoic mammals signal hibernation/torpor, we can assert that they directly indicate periodic arrests in growth. Thus, the growth rings or growth marks in the form of LAGs and/or annuli that are evident among the pelycosaurs (chapter 4 of this book) and a diverse array of nonmammalian therapsids (e.g., chapters 5, 8, and 9 of this book) suggest that flexible growth patterns are deeply rooted in the Synapsida and are most likely plesiomorphic for the clade. We propose that following the K/T extinction event, the ability to invoke different growth strategies in response to environmental conditions coupled with their small size and possible sleep-or-hide strategies might have played a major role in the success of survival for the eutherians, monotremes, and marsupials. Once conditions stabilized in the Eocene, the Eutherians were able to invoke their ancestral ability to deposit fibrolamellar bone in a sustained uninterrupted manner that enabled them to attain larger body sizes. Recent findings of specialized overwintering strategies, such as seasonal fluctuating metabolic rates in the high latitude, red deer (*Cervus elaphus*) (Arnold et al. 2004), as well as the documentation of zonal bone tissue in the dwarf bovid, *Myotragus* (Köhler and Moyà-Solà 2009), from a seasonally resource-limiting environment, and the report of growth rings in a polar bear (Chinsamy-Turan 2005) and in other hibernating mammals (Klevezal 1996) suggest that under certain conditions, placental mammals are capable of reverting to the plesiomorphic trait of flexible growth rates as seen among monotremes, marsupials, Mesozoic eutherians, and a variety of nonmammaliaform synapsids (Starck and Chinsamy 2002). Our rather limited analysis of extant marsupial bone microstructure suggests that they also evolved faster growth strategies and a wide range of body sizes, but like extant monotremes, they retained the plesiomorphic trait of variable growth dynamics.

Concluding Remarks

Our findings can be summarized as follows:

- Determinate growth strategy evolved early in the evolutionary history of the nonmammaliaform therapsids.
- The overall nature of the bone tissue of the two Mesozoic eutherians suggests a much slower rate of bone deposition than that of extant monotremes, marsupials, and eutherians; the extinct multituberculates; *Morganucodon*; and the nonmammaliaform therapsids.
- A flexible growth strategy is plesiomorphic for the mammalian lineage. It appears that the multituberculates grew at a faster rate than the contemporary eutherians which retained the plesiomorphic condition of flexible growth rates (Chinsamy and Hurum 2006).
- The fast growth rate of multituberculates, as suggested by their bone microstructure, provided an advantage in a changing environment such as the semi-desert of the Late Cretaceous Gobi region. Being of similar size to the Eutheria found in the same locations, but with a faster growth rate, they may have

reached sexual maturity earlier than the eutherians and, therefore, reproduced at a faster rate.

- The inherent flexible growth strategies of the Mesozoic eutherians may have given them an adaptive advantage at the K/T extinction, and after the K/T event, by enabling them to invoke sustained (energetically costly) or cyclical (energetically less costly) growth regimes in response to prevailing environmental conditions, thus enabling them to radiate and become the dominant mammals.
- We suggest that the rapid, uninterrupted growth rates of eutherians developed in the Paleocene, perhaps in response to favorable conditions, spurred the increase in body size. This seems to supports the explosion hypothesis for the evolution of eutherians (Wible et al. 2007), but more work on Paleocene mammals is needed.
- Extant monotremes (platypus and echidna) and marsupials (kangaroo) appear to have grown rapidly to sexual maturity, and then continued growing at lower (and flexible) growth rates to adult body size. More thorough, carefully constrained investigations into the bone microstructure of extant monotremes and marsupials are needed to provide a better understanding of how their life history is recorded in their bone microstructure.
- We suggest that extant mammals, such as hibernators (e.g., marmots and polar bears), as well as extinct mammals that lived in periodically stressful environments (e.g., *Myotragus*) reverted to the plesiomorphic flexible growth patterns of their ancestors.

Acknowledgments

We are grateful to Zofia Kielan-Jaworowska for providing us with the multituberculates and Mesozoic eutherians specimens for histological study. Susan Evans of University College of London provided multiple specimens of *Morganucodon*, and Jenny Jarvis of the University of Cape Town contributed a platypus femur for our original study. The late Tim Hamley (University of Queensland) is acknowledged for having provided the demineralized thin sections of the monotreme specimens. Kholeka Sidinile and Malcolm Lynn prepared the thin sections of the early mammals. Tom Kemp and Jennifer Botha-Brink are thanked for having reviewed this manuscript.

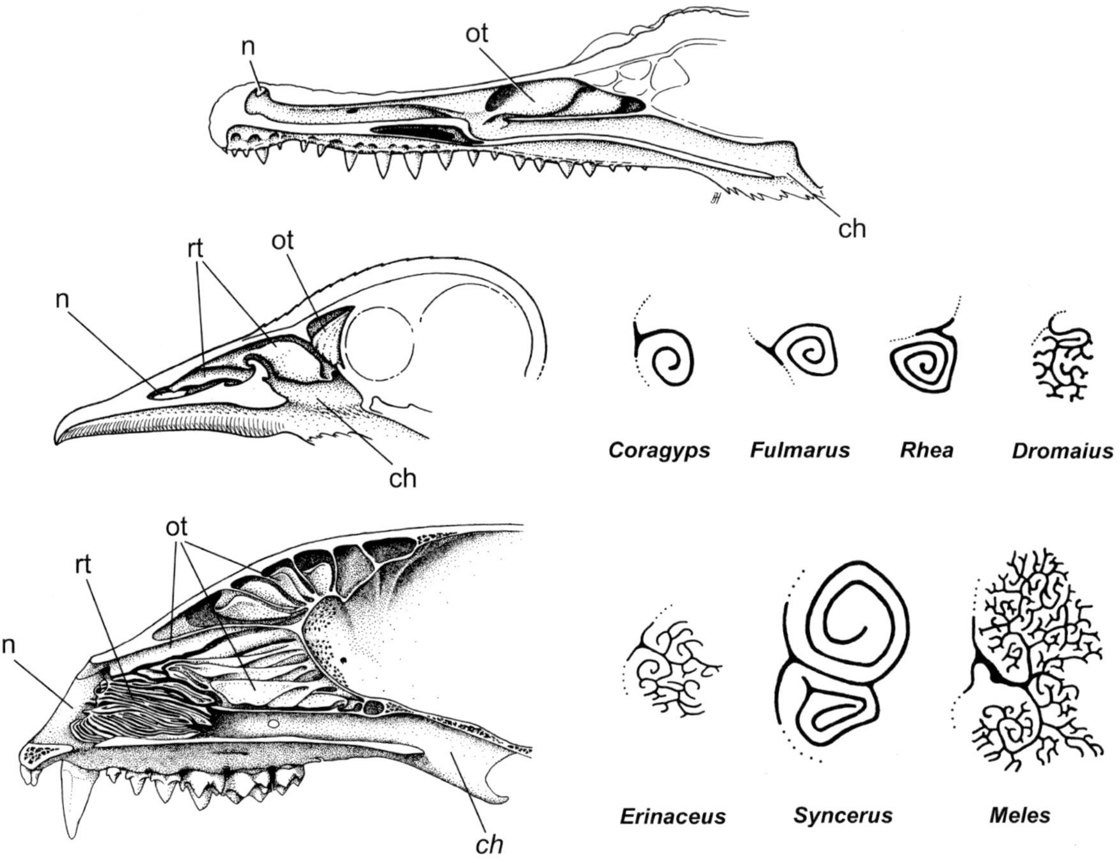

11.1. Nasal passages of extant amniotes. Reptiles, such as crocodilians (top), have up to three simple nasal conchae, all of which are exclusively covered with olfactory epithelium. In contrast, birds (center left) and mammals (bottom left) typically possess complex respiratory turbinates in addition to olfactory conchae. Also shown are cross sections of the respiratory turbinates of representative birds (center right) and mammals (bottom right). *Abbreviations:* n = external nostril; rt = respiratory turbinate; ot = olfactory turbinate; ch = internal nostril. Modified from Hillenius (1994).

The Evolution of Mammalian Endothermy

11

John A. Ruben, Willem J. Hillenius, Tom S. Kemp, and Devon E. Quick

Introduction

Whole body endothermy, or "warm-bloodedness," is a major specialization that has profoundly affected the biology of both mammals and birds and sharply distinguishes them physiologically from reptiles and all other vertebrates. It provides distinct physiological and ecological benefits, and thus may well be largely responsible for the present success of mammals and birds in a wide range of aerial, aquatic, and terrestrial environments (McNab 2003; Ruben 1995). Elevated rates of lung ventilation, oxygen consumption, and internal heat production (through aerobic metabolism), all hallmarks of endothermy, permit the maintenance of thermal homeostasis over a wide range of ambient temperatures. This enables mammals and birds to thrive in habitats with cold or highly variable thermal conditions and in nocturnal niches generally unavailable to ectothermic vertebrates (McNab 2003).

In addition, endothermy provides other important benefits as well. The expanded aerobic capacity associated with endothermy allows most mammals and birds to sustain levels of activity that are well beyond the capacity of almost all ectotherms (Bennett 1991; Ruben 1995). With some noteworthy exceptions, ectotherms, such as reptiles, are generally limited to nonsustainable, anaerobic metabolism for all activities beyond relatively slow movements. Although capable of spectacular bursts of intense exercise, ectotherms generally fatigue rapidly as a result of lactic acid accumulation (Bennett 1991; Ruben 1991). In contrast, the elevated rates of aerobiosis of endotherms enable these animals to sustain even relatively high levels of activity for extended periods of time, while simultaneously maintaining a reserve capacity for anaerobically fueled intense bursts of activity as well. Thus mammals and birds are able to maintain extensive territories or home ranges, to forage widely, and to migrate over large distances (Ruben 1995).

Many anatomical and physiological attributes of endothermy in mammals and birds bear striking similarity to one another (e.g., cardiovascular and enzymatic modifications [see Ruben 1995]). Nevertheless, given that elevated metabolic rates in these groups surely evolved independently of one another (Ruben 1995), selective factors linked to the origins of mammalian and avian endothermy must necessarily be considered independent of one another. Here, we concentrate on the origin of mammalian endothermy.

Given the significance of endothermic status to the lifestyle and ecological success of extant mammals, it is not surprising that the evolutionary origins of this attribute have received considerable attention (e.g., Bennett 1991; Else, Turner, and Hulbert 2004; Ruben 1995). Endothermy is the cumulative result of a cascade of physiological adjustments and morphological modifications (Kemp 2006b), and it is clearly a derived attribute that evolved from an ectothermic ancestral state. However, until recently it has remained largely unclear when early mammals or their ancestors began to approach, or attained, endothermic metabolic status. In this chapter, we critically examine evidence for when and in which groups of the therapsid-mammal continuum endothermy first appeared. We also evaluate data bearing on possible selective factors that may have driven early phases in the evolution of mammalian endothermy.

Metabolism and Thermoregulation in Mammals, Living and Extinct

Mammals maintain constant deep body temperatures largely by (i) minimizing thermal conductance with fur and, in some cases, by storing significant amounts of fat subcutaneously (reptiles maintain fat stores internally), and (ii) by sustaining metabolic heat production equivalent to the rate at which body heat is lost to the environment. To compensate for accelerated rates of heat loss, oxygen consumption increases at cold ambient temperatures. There is also a range of ambient temperatures over which heat production rates are minimal (i.e., at basal metabolic rate), the thermoneutral zone.

Variation in metabolic rate, especially during periods of rest or routine activity, comprises the core physiological difference between extant ectotherms (e.g., reptiles) and mammals. Mammals routinely have much higher rates of aerobiosis, or cellular oxygen consumption: at rest, mammalian metabolic rates are typically about 5–15 times greater than those of reptiles of the same body mass and temperature (Bennett 1973a, 1973b, 1982, 1991; Bennett and Dawson 1976; Else and Hulbert 1981; Schmidt-Nielsen 1984, 1990).

Thermoregulatory mechanisms in extant endothermic and ectothermic tetrapods also contrast sharply with one another. During periods of either rest and/or routine or accelerated levels of activity in endotherms, internal heat production rates are usually sufficient to enable them to maintain a constant body temperature over a wide range of ambient temperatures. In contrast, without access to a substantive external heat source (e.g., solar energy), ectotherms, especially those of temperate latitudes, are generally unable to achieve and maintain optimal body temperatures. In such cases, Q_{10} effects often cause ectotherms to appear sluggish, even in moderate temperatures. These markedly different thermoregulatory attributes formed much of the basis for the early reconstructions of "cold-blooded," brutish, slow-moving dinosaurs, but they were also the impetus for notions of endotherm superiority and the popular appeal of "warm-blooded" dinosaurs (e.g., Bakker 1968, 1975, 1980).

But in most circumstances, ectotherms are not merely passive "victims" of their external thermal environment. In spite of their low resting metabolic rates, extant ectotherms possess sophisticated temperature control mechanisms, and, whether tropical or temperate, they generally maintain high body temperatures and near homeothermy during periods of diurnal activity. As in endotherms, the reptilian hypothalamus is the primary thermostatic control center (Bartholomew 1982), and many lizards thermoregulate behaviorally during extended periods of diurnal activity at body temperatures that overlap those of endotherms (Avery 1982). In addition, a number of extant reptiles also exhibit an adaptive, mammal-like fever response in which the thermoregulatory set point is elevated significantly (e.g., from 37.4°C to 39.6°C in the lizard *Dipsosaurus*) through the action of endogenous and exogenous pyrogens on the hypothalamus (Firth and Turner 1982; Kluger, Ringler, and Anver 1975).

Given normal field conditions, therefore, many extant reptiles, whether tropical or temperate, are able to sustain marked and consistent variation between internal and external temperatures (Pearson 1954; Schmidt-Nielsen 1990). Even small lizards of cool environments at temperate latitudes are often sufficiently adept at solar basking that they sustain diurnal body temperatures that overlap, or in some cases exceed, those of many endotherms (Avery 1982; Greenberg 1980). For example, the small (25 g) Andean lizard *Liolaemus multiformis* maintains diurnal body temperatures some 31°–34° above the subfreezing/freezing ambient temperatures common at those altitudes (Pearson 1954).

Furthermore, the aforementioned thermal lability is especially characteristic of small ectotherms but diminishes with increasing body size. Some particularly large ectotherms (e.g., Komodo dragons, *Varanus komodoensis*) living in warm, equable climates are able to remain virtually homeothermic for long periods in their natural environments. In such cases, these animals achieve "inertial homeothermy" by virtue of their low surface area-to-volume ratios and correspondingly low heat loss rates (Frair, Ackman, and Mrosovsky 1972; McNab and Auffenberg 1976; Paladino, O'Connor, and Spotila 1990; Standora et al. 1984).

Given the generally equable, subtropical-tropical climates of much of the Late Paleozoic Era and almost all of the Mesozoic Era, it appears unlikely that thermoregulation posed a particular challenge to most therapsids and early mammals. Combinations of behavioral thermoregulation and thermal inertia likely enabled most nonmammalian therapsids/early mammals, whether ecto- or endothermic, or very small or very large, to maintain relatively high and stable body temperatures for extended periods of time. Mathematical models indicate that large Paleozoic/Mesozoic tetrapods (>100 kg) were likely to have been inertial homeotherms, relatively unaffected by diurnal temperature fluctuations (O'Connor and Dodson 1999; Paladino, Spotila, and Dodson 1997; Spotila et al. 1991). In such cases, even small fluctuations in body temperatures might have occurred only on a weekly or monthly scale. Moreover, as noted, given an external heat source, virtually all reptiles control body temperature (T_b)

to some extent (Cowles and Bogert 1944) and many species, especially diurnal lizards, maintain such high and precisely regulated temperatures when active (Brattstrom 1965) that some approach the active body temperatures of mammals and birds (Bowker, Wright, and Bowker 2010). Smaller therapsids and mammals (<75 kg), regardless of their metabolic status, might easily have utilized behavioral thermoregulation to achieve high and stable body temperatures during periods of diurnal activity. Accordingly, attainment of mammalian metabolic status need not have involved modification of thermostatic sensors or temperature regimes maintained behaviorally by the ectothermic ancestors of mammals.

These models indicate that, at least when active, nonmammalian therapsids and early mammals were probably capable of maintaining high and stable body temperatures regardless of their metabolic status. Thus, even if fully ectothermic, had therapsids possessed aerobic metabolic capacities and predatory habits equivalent to those of some modern tropical latitude lizards (e.g., *Varanus*), they may well have maintained large home ranges and actively pursued and killed prey (Bennett 1973b).

In all organisms, body mass affects not only thermoregulation but also influences resting or routine metabolic rate. Because of their greater mass, large ectotherms and endotherms maintain higher total metabolic rates than do small ectotherms and endotherms, respectively. However, because of the allometric relationship between animal mass and metabolic rate (where total metabolic rate scales at approximately $mass^{0.75}$), each gram of tissue in a small animal actually sustains a higher rate of oxygen consumption than does each gram of tissue in a larger animal. Thus, while a resting elephant obviously maintains a much higher rate of total heat production than does a resting mouse, the oxygen consumption rate of each gram of mouse is considerably greater than that of each gram of elephant. The physiological mechanisms and selective factors governing the allometry of metabolism have long puzzled biologists (Darveau et al. 2002; Randall, Burggren, and French 2002; Schmidt-Nielsen 1990), but the relationship between body mass and metabolic rate is well documented for all vertebrate and invertebrate classes (Hemmingsen 1960). Similar allometry almost surely occurred in nonmammalian therapsids/early mammals as well (Reid 1997). However, by itself, this metabolic scaling has no a priori implication for interpreting the total metabolic rates of these animals or the ectothermic or endothermic status of these taxa, whether living or extinct (Reid 1997). All small ectotherms maintain higher mass-specific metabolic rates than do larger individuals of related species. All other factors being equal, smaller ectotherms are no closer to achieving endothermic metabolic rates than are larger individuals. Similarly, large extant mammals are not less endothermic than are small ones.

Given the significance of metabolic status to the lifestyles of extant vertebrates, it is hardly surprising that identification and elaboration of physiological parameters associated with vertebrate ecto- and endothermy have received so much attention since the 1970s (Bennett 1991). Although by itself the physiology of extant taxa usually provides little, if any, historical perspective into the metabolic status of long extinct taxa, correlations between physiology and morphology can be extremely valuable in this regard.

Endothermy itself is virtually impossible to demonstrate clearly in extinct forms. Endothermy is almost exclusively an attribute of the "soft anatomy," which leaves a poor, usually nonexistent, fossil record. To support their high oxygen consumption levels, endotherms possess profound structural and functional modifications to facilitate oxygen uptake, transport, and delivery. Compared with ectotherms of similar mass, mammals have greatly expanded rates of lung ventilation, fully separated pulmonary and systemic circulatory systems, and expanded cardiac output. They also have greatly increased blood volume and blood oxygen–carrying capacities, as well as increased tissue aerobic enzymatic activities (Ruben 1995). Unfortunately, these key features of endothermic physiology are unlikely to ever be preserved in fossils—mammalian, avian, or otherwise.

Consequently, many hypotheses concerning possible endothermy in a variety of extinct vertebrates, especially dinosaurs and therapsids, have relied primarily on supposed correlations of metabolic rate with a variety of weakly supported criteria (see discussion that follows) (Bakker 1971, 1980, 1986; Padian and Horner 2004), which are, at best, equivocal (Bennett and Dalzell 1973; Farlow, Dodson, and Chinsamy 1995; Chinsamy and Hillenius 2004).

The Physiology of Endothermy in Extant Mammals

It has long been recognized that ~70 percent of heat production at rest in mammals is generated by the internal organs (i.e., liver, kidneys, brain, heart, intestines), even though they comprise only about 6–8 percent of total body mass (Turner, Hulbert, and Else 2006). There is increasing evidence that two related factors largely account for the increment in mammalian versus reptilian resting oxygen consumption rates: (i) in mammals, the internal organs are more massive than those structures in reptiles, and (ii) mass-specific metabolic rates of mammalian internal organs are markedly greater than in reptiles.

The major internal organs (liver, kidney, heart, brain) of the bearded dragon lizard (*Pogona*) comprise 5.3 percent of total body mass compared with 10 percent in the laboratory mouse (*Mus*). Mammalian internal organ mass scales according to the equation $0.18\ M^{0.83}$, while similar structures in reptiles scale at $0.14\ M^{0.75}$ (M = grams body mass) (Else and Hulbert 1985). Consequently, not only are internal organ mass elevations, or "*a*" constants, greater in mammals, but doubling of reptilian body mass results in an average increment of 54 percent in internal organ mass, while doubling of mammalian body mass is accompanied by a 68 percent increase in internal organ mass. These values correspond well with mass doubling–related increments in total resting metabolic rate in these taxa (i.e., 54 percent and 71 percent, respectively [Daan, Dirkjan, and Groenewald 1990]).

Unlike the minor contribution of striated musculature to resting metabolism, energy production (both aerobic and anaerobic) during exercise is primarily a product of the skeletal muscles, which compose about 40 percent of total body mass in mammals (Lindstedt and Schaeffer 2002; Turner, Hulbert, and Else 2006). At the organ and tissue level, factors underlying high metabolic capacity of the striated musculature in mammals are similar to those associated with elevated resting metabolic rate of the viscera. In mammals, each unit mass of skeletal muscle has a relatively high capacity for active metabolic rate, and there is proportionately more total mass of skeletal muscle in most endotherms than is present in ectotherms. Mammalian skeletal muscle mass (in grams) scales $0.42\ M^{1.01}$, compared to $0.19\ M^{1.09}$ in reptiles, and potential maximum oxygen consumption rates (as indexed by cytochrome oxidase activity) in skeletal muscle of the lab rat and the lizards *Sauromalas* and *Varanus* approximate variation in maximal rates of aerobiosis in these taxa (Else and Hulbert 1981).

As already noted, endothermy in extant mammals (and birds) is defined by elevated rates of resting (or routine) aerobic metabolism, which, in laboratory settings, are usually about five to fifteen times greater than that in ectotherms of similar mass and body temperature (e.g., Bennett 1991; Bennett, Hicks, and Cullum 2000). Routine metabolism is the primary source of heat that terrestrial endotherms use to maintain stable body temperatures in their thermoneutral zones. Additionally, although activity metabolism is not a requisite component of endothermy, most mammals have aerobic capacities and stamina that far exceed those in most ectotherms: Mammalian field metabolic rates generally exceed reptilian rates by twenty- to thirtyfold (e.g., Nagy 1987).

Neontological Evidence for the Origin of Mammalian Endothermy

Hypotheses explaining the transition to endothermy (e.g., Clarke and Pörtner 2010; Kemp 2006b) from ectothermic ancestors have traditionally centered on selection for either (i) increased thermoregulatory capacity via elevated resting metabolic rate or (ii) increased stamina via elevated levels of aerobically supported activity. In the latter scenario (the "aerobic capacity" model), resting metabolic rate is thought to have increased as a by-product of continuous selection for elevated aerobic capacity (Bennett and Ruben 1979).

Support for the aerobic capacity model for the origin of endothermy is provided by empirical evidence for a link between resting and maximal rates of aerobic metabolism in a number of extant vertebrates (e.g., Boily 2002; Dutenhoffer and Swanson 1996). In contrast, even substantially elevated resting metabolic rates would provide little thermoregulatory benefit to lizards (Bennett, Hicks, and Cullum 2000). Additionally, incipient elevation of resting metabolic rates seems to have first appeared in Permian Era (ca. 265 Ma), lion- to bear-sized predators (pristerognathid therocephalian synapsids) that lived in tropical or near-tropical climates. These taxa were likely to have been, due to their size, inertial homeotherms, even without the benefit of endothermy. Therefore, there would seem to be little initial selective advantage in elevation of resting metabolic rate for thermoregulatory purposes in such animals.

Certain attributes of plasma membrane physiology in extant tetrapods might provide further insight into the origin of endothermy. If the initial function of increased mammalian metabolic rate was largely to facilitate endogenous heat production, then a greater proportion of total cellular metabolism should be associated with these processes in mammals than in reptiles. Vertebrate tissues clearly specialized for thermogenesis dedicate unusually high fractions of total cellular metabolism (and heat production) to membrane-ion gradient maintenance (e.g., mammalian brown adipose tissue [Himms-Hagen 1990] and billfish brain heater organs [Block 1991]). However, in vitro incubation of liver and kidney slices demonstrates that reptiles and mammals devote similar fractions of total tissue oxygen consumption (25–35 percent) to maintenance of plasma- and mitochondrial-ion homeostasis (Brand et al. 1991; Hulbert and Else 1990). Consequently, on the basis of membrane physiology, there is no compelling reason to assume that resting metabolic rates have more to do with a necessity for heat production in mammals than in reptiles.

Evidence for Endothermy in the Fossil Record

In the absence of evidence from fossils for the cellular processes and soft tissue correlates, such as high pulmonary capacities and expanded cardiac outputs, hypotheses concerning possible endothermy in extinct vertebrates have typically relied primarily on supposed correlations of metabolic rate with a variety of more weakly supported possible correlates, such as predator-prey ratios, relative brain size, correlations with mammalian posture, estimated growth rates, bone histology, and oxygen isotope (O^{16}/O^{18}) composition of fossilized bone. Close scrutiny has revealed that when considered individually, virtually all of these correlations are not well supported. (For reviews, see Bennett and Ruben [1986]; Farlow, Dodson, and Chinsamy [1995]; Chinsamy and Hillenius [2004]).

Evolution of Endothermy in the Therapsid-Mammal Lineage

Unequivocal evidence for the presence of a fully developed insulative pelage in mammalian ancestors would constitute a strong signal for elevated rates of routine heat production and at least near-endothermic metabolic status. However, the evolution of integumentary appendages remains largely unresolved, but it is unlikely that a fully developed coat of fur evolved all at once. Possible intermediate morphologies, such as short, stubby protohairs or longer but isolated protohairs, would have contributed little to insulation but could have had mechanosensory or protective roles (Maderson 2003; Ruben and Jones 2000). In that case, sensory vibrissae, which are fully functional even as single hairs, may have evolved long before insulatory fur.

The oldest fossilized impression of fur is from the Early Cretaceous eutherian *Eomaia* (Ji et al. 2002) and from Late Paleocene multituberculates (Meng and Wyss 1997). Hillenius (2000) infers that a complete insulatory fur coat may have been present in mammals as early as the Late Triassic. However, these inferences may well have more significance for the advent of metabolically based thermoregulation than for the initial elevation of aerobic metabolic rates that likely preceded it.

Mammalian metabolic status is a highly derived physiological condition. Nevertheless, given the antiquity of the group, both initial selection for, and the eventual attainment of endothermy, were likely to have occurred ~200+ Mya. Thus, questions of when and why mammalian metabolic status was acquired can probably only be answered with positive evidence primarily from fossilized osteological features that were likely to have been functionally linked to the derived physiological processes and attributes that underlie endothermy. One line of evidence, the presence or absence of respiratory turbinates in extinct taxa, has shed light on the antiquity of endothermy in both birds and in the therapsid-mammal lineage. In all extant terrestrial mammals (and birds), respiratory turbinates are tightly and functionally linked to the high rates of lung ventilation that are requisite for maintenance of endothermic metabolic rates.

Evidence for Elevated Routine Metabolic Rates: Respiratory Turbinates

Nasal turbinates, or "conchae," are convoluted bony or cartilaginous projections from the walls of the nasal cavity. These structures are found in most modern amniotes and range in complexity from comparatively simple curls in reptiles to elaborately scrolled or branched structures in birds and mammals (Fig. 11.1). The turbinates are covered with moist epithelium and principally serve to increase the surface area available to these membranes.

There are two distinct types of turbinates, olfactory and respiratory, which differ in the type of epithelium they bear and in their location within the nasal cavity (Hillenius 1994). Olfactory turbinates are nearly ubiquitous among amniotes. These are lined with sensory (olfactory) epithelium and are generally situated outside the main path of respiratory air. Olfactory turbinates function specifically to enhance the sense of smell and are unrelated to metabolic physiology. In contrast, respiratory turbinates are found only among birds and mammals; they are completely absent in all modern reptiles. In both mammals and birds, the respiratory turbinates project directly into the main nasal air stream, are covered exclusively with non-sensory "respiratory" epithelium, and have a distinct functional association with endothermy (Geist 2000; Hillenius 1992, 1994).

The chronically elevated rates of lung ventilation of birds and mammals—a consequence of their elevated rates of metabolism even while resting, and thus a key attribute of endothermy—could potentially result

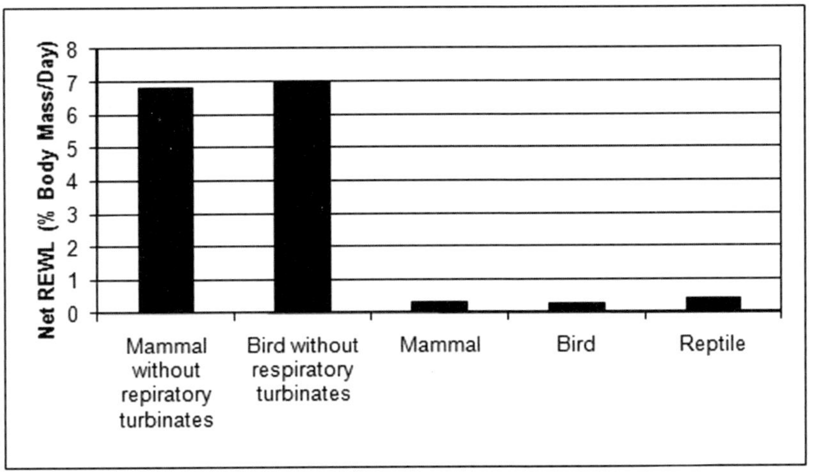

11.2. Daily net respiratory evaporative water loss (REWL) rates (i.e., REWL—metabolic water production) for a free-living 1-kg mammal, reptile, and bird, and probable REWL for a free-living mammal and bird lacking the use of respiratory turbinates (i.e., with a reptile-like nasal anatomy and net REWL rates/cm^3 of O_2 consumed). Without the water-conserving function of respiratory turbinates, daily water flux rates in many mammals and birds would be out of balance by about 30 percent. (Modified from Hillenius [1992]; Geist [2000]).

in significant desiccation and heat loss rates as large volumes of warm, humid air are expelled from the lungs (Fig. 11.2). However, respiratory turbinates provide a compensatory mechanism by absorbing a substantial portion of the heat and moisture contained in respiratory air through a mechanism of intermittent countercurrent exchange (cf. Geist 2000; Hillenius 1992; Jackson and Schmidt-Nielsen 1964). In the absence of such a recovery mechanism, respiratory heat and water loss rates associated with continuously high rates of oxidative metabolism and lung ventilation would likely exceed tolerable levels, and endothermy itself might well be unsustainable (Geist 2000; Hillenius 1992; Ruben 1995) (Fig. 11.2). Significantly, total surface area of the respiratory turbinates in both mammals and birds scales proportionately with mass-specific resting metabolic rates—both are proportional to body mass$^{0.75}$ (Owerkowicz and Crompton 2001).

Respiratory turbinates are present in 100 percent of extant terrestrial birds and mammals and are absent only in some plunge-diving marine birds that lack functional nostrils (e.g., cormorants, pelicans) and whales (cf. Hillenius 1994; Ruben et al. 1996). The widespread presence of respiratory turbinates among extant endotherms suggests that it is difficult for them to circumvent the need for these structures through alternative sites or mechanisms for water and heat recovery. It should also be pointed out that, energetically, respiratory turbinates are comparatively inexpensive to operate; their countercurrent heat exchange mechanism incurs no physiological energy expense (Collins, Pilkington, and Schmidt-Nielsen 1971). This contrasts sharply with potential alternative water recovery mechanisms, such as salt glands or enhanced kidney function, which require significant, and continuous, investments of cellular energy. Furthermore, maintenance of a countercurrent exchange mechanism in any portion of the respiratory tree other than the nasal cavity would be untenable. Such exchange sites in the body cavity would necessarily preclude deep-body homeothermy, while the presence of such a system in the trachea would likely interfere with the stability of brain temperatures because of the proximity to the brain-bound carotid arteries (e.g., Jones and Ruben 2001). The fact that deep nasopharyngeal temperature virtually equals core body temperature in extant mammals (e.g., Jackson and Schmidt-Nielsen 1964) and birds (Geist 2000) confirms that little or no significant heat exchange takes place in the trachea. Finally, the widespread distribution of respiratory turbinates among extant mammals and birds also indicates that these structures are homologous within each of these groups (although evolved independently in each clade [cf. Geist 2000; Hillenius 1992, 1994]). The rare cases of reduction or absence of respiratory turbinates among these taxa (e.g., in cetaceans, pelecaniforms) all clearly represent secondary, highly specialized developments.

In contrast to the nearly ubiquitous presence of respiratory turbinates in endotherms, these structures are entirely absent in all extant ectotherms (cf. Witmer 1995), nor are there other modifications of the nasal cavity designed specifically for the recovery of respiratory water vapor. Lung ventilation rates of these animals are sufficiently low that respiratory heat and water loss rates seldom create problems (Geist 2000; Randall, Burggren, and French 2002). Nasal heat exchange and modest recovery of respiratory moisture do occur in thermoregulating reptiles (e.g., Murrish and Schmidt-Neilsen 1970). However, without complex respiratory turbinates, the "simple" nasal passages of ectotherms are effective only at low ventilation rates. For the chronically higher rates of metabolism and ventilation characteristic of mammalian endothermy, specialized respiratory turbinate systems are necessary for complete heat and water exchange (Geist 2000; Hillenius 1992, 1994; Jackson and Schmidt-Nielsen 1964).

Finally, embryological and anatomical studies confirm that although the olfactory turbinates are a primitive attribute of amniotes, the respiratory turbinates are specialized structures in mammals and birds that arose independently in each lineage (Witmer 1995). The most parsimonious conclusion is that the respiratory turbinates evolved in concert with expanding chronic metabolic and lung ventilation rates and the origin of endothermy (Hillenius 1994). Consequently, the presence or absence of respiratory turbinates among fossil amniotes may be a good indicator of the minimal routine (i.e., resting) ventilation rates and, by extension, metabolic rates of these taxa (Hillenius 1994; Jones and Ruben 2001; Ruben et al. 1996).

RESPIRATORY TURBINATES IN SYNAPSIDS AND EARLY MAMMALS

Although turbinates are occasionally found preserved in fossils, their fragile nature usually precludes their preservation and recovery. Among fossil mammals and non-mammalian synapsids, the presence of nasal turbinates, as well as their function, is most readily revealed by characteristic patterns of bony ridges or turbinate attachment scars left behind on the walls of the nasal cavity. Attachment ridges for olfactory turbinates are

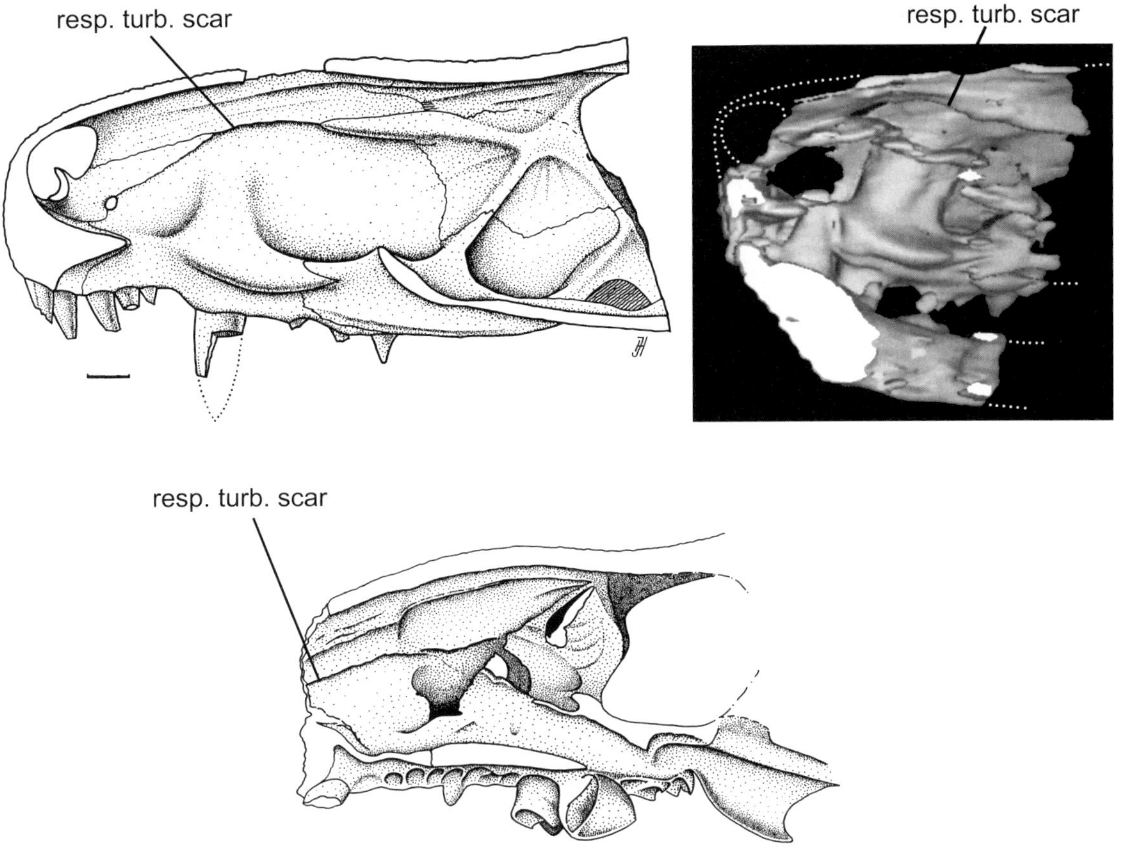

located posterodorsally, outside the main path of airflow, whereas those of the respiratory turbinates are located in the anterolateral rostrum in close proximity, and parallel, to the nasolachrymal duct, directly in the path of respired air. Ridges for olfactory turbinates are found throughout the synapsids, including ancestral forms such as pelycosaurs. However, ridges most likely associated with respiratory turbinates occur in only two lineages of advanced theriodonts, the therocephalians and cynodonts (Hillenius 1994), which suggests that the routine ventilation rates of these animals had risen sufficiently above ancestral levels to warrant the presence of a compensatory mechanism to limit respiratory water loss rates. The earliest appearance of these is in the Late Permian therocephalians, such as the bear-sized pristerognathid *Glanosuchus* (Fig. 11.3). However, pristerognathids retain the primary palatal condition of ancestral amniotes, where the anterior location of the internal nostrils limits the size of the nasal respiratory chamber. Consequently, their capacity to recover respiratory water vapor was probably limited as well, suggesting that the chronic ventilation rates of these early therocephalians likely had increased only relatively modestly beyond ancestral rates (Hillenius 1994). Significantly, in later therocephalians, the nasal respiratory chamber became proportionately larger, and the internal nostrils migrated posteriorly (Fig. 11.4). This posterior migration of the choanae preceded the

11.3. Respiratory turbinates (resp. turb.) in nonmammalian therapsids. Attachment scars for respiratory turbinates in Late Permian therocephalians, such as the pristerognathid *Glanosuchus* (top left) and an unnamed pristerognathid, SAM-PK-K226 (top right) (three-dimensional reconstructions from computed tomography scan). Respiratory turbinate scars as also found in Triassic cynodonts, such as *Massetognathus* (bottom). After Hillenius (1994).

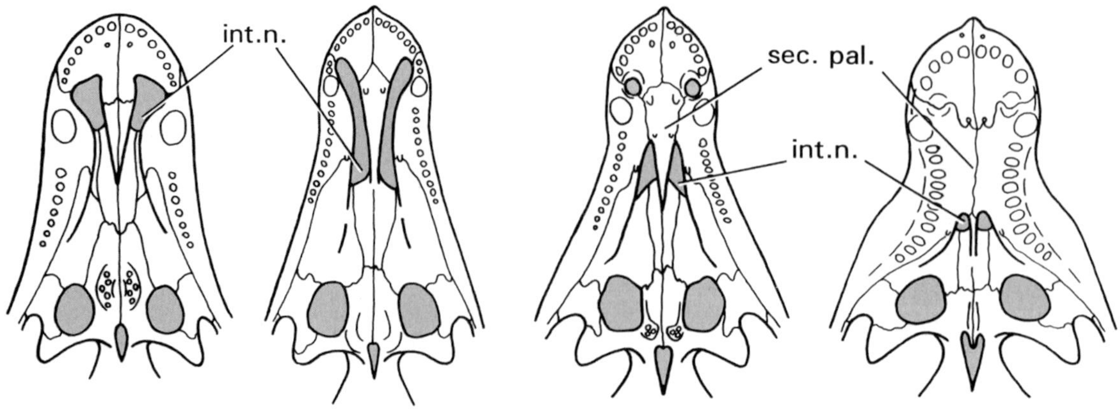

11.4. Evolution of the secondary palate (sec. pal.) in therocephalians. Schematic diagrams of the palatal configurations of representative therocephalians, scaled to a common size. The secondary palate itself probably evolved for mechanical reasons associated with jaw function (Thomason and Russell 1986) but was preceded by an expansion of the respiratory portion of the nasal cavity, as indicated by the posterior migration of the internal nostrils (int. n.). This allowed more room for nasal conditioning of respiratory air and recovery of water vapor and suggests that routine ventilation rates and metabolic rates of therocephalians were increasing incrementally. After Hillenius (1994).

development of the therocephalian bony secondary palate (cf. Hillenius 1994), which most likely evolved subsequently for mechanical reasons relating to bite strength (Thomason and Russell 1986). As the respiratory portion of the nasal cavity became larger, more space became available for the conditioning of respiratory air and recovery of water vapor. This association suggests that therocephalian ventilation rates were gradually expanding as well, which is consistent with the increasingly mammal-like features of these groups (Kemp 2005, 2007b; Thomason and Russell 1986).

Attachment scars of respiratory turbinates also occur in several Triassic nonmammaliaform cynodonts, including the gomphodont *Massetognathus* (Fig. 11.3), indicating that some cynodonts likely also possessed elevated resting lung ventilation rates and metabolic rates. The fact that the secondary palate of nonmammaliaform cynodonts, as well as other mammal-like features such as postural and masticatory modifications, evolved independently from that of therocephalians (Kemp 2005; Thomason and Russell 1986) suggests that the expansion of metabolic rates in therocephalians and cynodonts occurred largely independently in these two lineages (Hillenius 1994). Nonmammaliaform cynodonts gave rise to the earliest mammals toward the end of the Triassic Period (Hopson 1994). The nasal turbinate ridges of the earliest mammals (e.g., Fig. 10.2) are essentially identical to those of extant mammals (Hillenius 1994), suggesting that the evolution of "mammalian" routine respiration rates and endothermy was largely complete by this time.

Independent confirmation of the trend toward mammalian metabolic status within the theriodont therapsids is provided by concomitant increase in the relative cross-sectional area of the main nasal passage in these taxa. Expansion of the nasal cavity is necessary to accommodate the presence of respiratory turbinates, as well as to facilitate increased lung

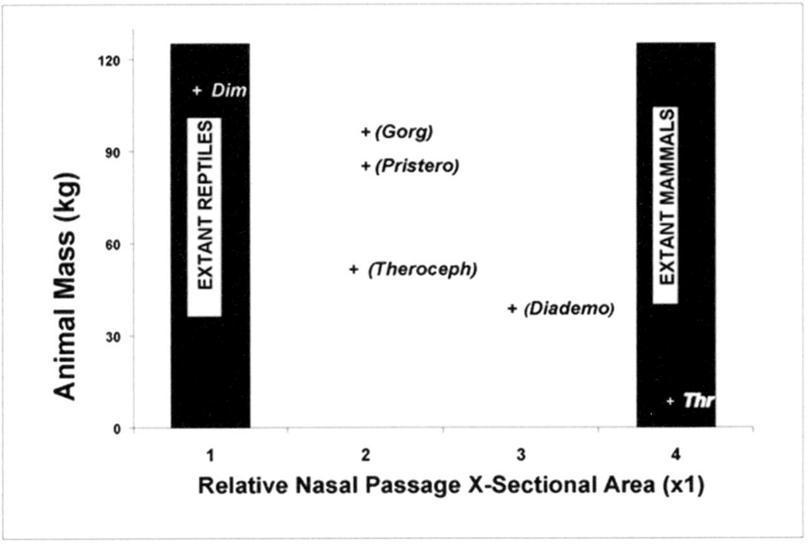

11.5. Nasal passage relative cross-sectional area in extant ecto- and endotherms and in Early Permian–Early Triassic synapsids. Early, larger taxa (e.g., *Dimetrodon*) possessed relatively narrow nasal passages unlikely to have housed fully developed respiratory turbinates. Later forms such as *Diademodon* and *Thrinaxodon* possessed wider, mammal-like nasal passages consistent with the presence of fully functional, mammal-like respiratory turbinates. Transitional forms (i.e., therocephalian and gorgonopsian therapsids) seem likely to have possessed incipient turbinates, perhaps consistent with intermediate levels of lung ventilation and metabolic rates. Relative nasal cross-sectional areas for extant taxa based on regressions from Ruben et al. (1996): *Dimetrodon* (Dim), Early Perm. (Museum of Comparative Zoology, #1347); *Diademodon* (Diademo), Early Triassic (South African Museum, QR#1296); gorgonopsian (Gorg), Late Permian (South African Museum, #10054); *Pristerognathus* (Pristero), Middle Permian (South African Museum, #K226); therocephalian (Theroceph), M. Permian (South African Museum, #K312); *Thrinaxodon* (Thr), Early Triassic (Univ. of California Mus. Paleontology, #40466).

ventilation rates in endotherms without unduly increasing the resistance to airflow. Consequently, most extant mammals and birds possess nasal passageways with cross-sectional areas about fourfold that of reptiles of similar masses (Ruben et al. 1996). Significantly, as Permo-Triassic theriodont therapsids become smaller and more mammal-like through time, nasal passage cross-sectional area becomes proportionately larger until that parameter is fully endotherm-like as in, for example, the small Early Triassic cynodont *Thrinaxodon* (Fig. 11.5; see Plate M, bottom). Together, enlarged nasal cavities and longitudinal rostral scars (accompanied by the nasolachrymal canal) strongly suggest that lung ventilation rates and the necessity to deal with potentially excessive rates of respiratory water and heat loss was steadily increasing in these taxa. It is also reasonable to infer that by this stage, resting metabolic rates in nonmammaliaform cynodonts were likely sufficient to have sustained metabolically based thermoregulatory status.

The Transition to Mammalian Endothermy

Endothermy in extant taxa is a complex, multi-functional attribute affecting virtually every aspect of mammalian biology, from elevated levels of routine activity and expanded habitat availability, to the reliable functioning of an enlarged brain and complex behavior patterns. We have argued that during the earliest stages in the evolution of mammalian endothermy, selective factors favored increased levels of sustainable aerobic activity and stamina. However, during this interim, levels of aerobic metabolism must, at some point, have increased sufficiently that it became advantageous for transitional forms to retain internally produced heat as a supplement to pre-existing, ectothermically based thermoregulatory strategies. Initially this probably had little effect beyond extending the length of the daily activity by allowing an earlier start and later finish to the active day. Ultimately, however, the outcome was the more or less constant internal body temperature of most mammals for most of their life.

To attain endothermically based constant and elevated body temperature, a number of morphological and physiological changes were necessary. Initial necessities must have included development of an insulatory pelt linked to a capacity to vary heat conductance by altering subcutaneous blood flow patterns and degree of piloerection of the hairs. There was also the need of mechanisms for both evaporative cooling, presumably by panting initially, and elevation of the metabolic rate in order to extend temporarily the range of ambient temperatures outside the thermoneutral zone.

Kemp (2006b, 2007b) proposed a model for the origin of full mammalian endothermy based on the concept of correlated progression. Changes in all these and many other characters were necessary, yet the organism's phenotype at all times had to remain fully integrated. The several functions of endothermy can similarly be viewed as evolving step by step in correlation with one another—ever-increasing aerobic activity levels, ecological thermoregulation expansion of activity periods and climatic range, and later physiological thermoregulation allowing the complex brain and associated behavior, including parental care and provision (Hayes 2010).

Concluding Remarks

Endothermy in extant vertebrates is a complex system, and accounting for its origin is correspondingly difficult. The most likely explanation for the initial selective advantage of an elevated metabolic rate is the aerobic capacity model, with thermoregulatory functions evolving subsequently. The question of when endothermic status was actually achieved depends on recognizing unequivocal anatomical correlates of endothermy, and only fully developed respiratory turbinates are likely to fulfill this criterion. Attachment scars for these structures are first recognizable in two lineages of Late Permian/Early Triassic therapsid taxa—therocephalians and nonmammaliaform cynodonts—suggesting that resting and/or routine metabolic rates had been gradually increasing in these taxa over a span of some 40 My.

The acquisition of fully mammalian endothermy (i.e., high, constant body temperature linked to a complete complement of mammal-like morphological structures and physiological processes) was probably a gradual, correlated acquisition of characters, many of which appeared well before the evolution of the earliest true mammals. Virtually modern endothermic physiology and thermoregulation was probably present in the earliest Mammaliaformes of the Late Triassic.

Acknowledgments

Jim Farlow and Al Bennett are thanked for having reviewed this manuscript.

References

R

Abdala, F. 1996. Redescripión del cráneo y reconsideración del la validez de *Cynognathus minor* (Eucynodontia-Cynognathidae) del Triásico Inferior de Mendoza. *Ameghiniana* 33:115–126.

———. 2007. Redescription of *Platycraniellus elegans* (Therapsida, Cynodontia) from the Lower Triassic of South Africa, and the cladistic relationships of eutheriodonts. *Palaeontology* 50:591–618.

Abdala, F., and M. Allinson. 2005. The taxonomic status of *Parathrinaxodon proops* (Therapsida: Cynodontia), with comments on the morphology of the palate in basal cynodonts. *Palaeontologia Africana* 41:45–52.

Abdala, F., J. C. Cisneros, and R. M. H. Smith. 2006. Faunal aggregation in the Early Triassic Karoo Basin: Earliest evidence of shelter-sharing behavior among tetrapods? *Palaios* 21:507–512.

Abdala, F., and N. P. Giannini. 2002a. Chiniquodontid cynodonts: Systematic and morphometric considerations. *Palaeontology* 45:1151–1170.

———. 2002b. Gomphodont cynodonts of the Chañares Formation: The analysis of an ontogenetic sequence. *Journal of Vertebrate Paleontology* 20:501–506.

Abdala, F., P. J. Hancox, and J. Neveling. 2005. Cynodonts from the uppermost Burgersdorp Formation, South Africa, and their bearing on the biostratigraphic subdivision and correlation of the Triassic *Cynognathus* Assemblage Zone. *Journal of Vertebrate Paleontology* 25:192–199.

Abdala, F., and M. C. Malabarba. 2007. Enamel microstructure in *Exaeretodon*, a Late Triassic traversodontid (Therapsida Cynodontia). *Revista Brasileira de Paleontologia* 10(2):71–78.

Abdala, F., J. Neveling, and J. Welman. 2006. A new trirachodontid cynodont from the lower levels of the Burgersdorp Formation (Lower Triassic) of the Beaufort Group, South Africa and the cladistic relationships of Gondwanan gomphodonts. *Zoological Journal of the Linnean Society* 147:383–413.

Abdala, F., and A. M. Ribeiro. 2003. A new traversodontid cynodont from the Santa Maria Formation (Ladinian-Carnian) of southern Brazil, with a phylogenetic analysis of Gondwanan traversodontids. *Zoological Journal of the Linnean Society* 139:529–545.

———. 2010. Distribution and diversity patterns of Triassic cynodonts (Therapsida, Cynodontia) in Gondwana. *Palaeogeography, Palaeoclimatology, Palaeoecology* 286:202–217.

Abdala, F., B. S. Rubidge, and J. van den Heever. 2008. The oldest therocephalians (Therapsida, Eutheriodonta) and the early diversification of Therapsida. *Palaeontology* 51:1011–1024.

Allin, E. F. 1975. The evolution of the mammalian middle ear. *Journal of Morphology* 147:403–438.

Alroy, J. 1999. The fossil record of North American mammals: Evidence for a Paleocene evolutionary radiation. *Systematic Biology* 48(1):107–118.

Amprino, R. 1947. La structure du tissu osseux envisagée comme expression de différences dans la vitesse de l'accroissement. *Archives de Biologie* 58:315–330.

Anderson, J. S., and R. R. Reisz. 2004. *Pyozia mesenensis*, a new small varanopid (Synapsida, Eupelycosauria) from Russia: "Pelycosaur" diversity in the Middle Permian. *Journal of Vertebrate Paleontology* 24:173–179.

Angielczyk, K. D. 2001. Preliminary phylogenetic analysis and stratigraphic congruence of the dicynodont anomodonts (Synapsida: Therapsida). *Palaeontologia Africana* 37:53–79.

———. 2004. Phylogenetic evidence for and implications of a dual origin of propaliny in anomodont therapsids (Synapsida). *Paleobiology* 30:268–296.

———. 2007. New specimens of the Tanzanian dicynodont *"Cryptocynodon" parringtoni* Von Huene, 1942 (Therapsida, Anomodontia), with an expanded analysis of Permian dicynodont phylogeny. *Journal of Vertebrate Paleontology* 27:116–131.

Angielczyk, K. D., P. D. Roopnarine, and S. C. Wang. 2005. Modeling the role of primary productivity disruption in End Permian extinctions, Karoo Basin, South Africa. *New Mexico Museum of Natural History and Science Bulletin* 3:16–23.

Angilletta, M. J., Jr., T. D. Steury, and M. W. Sears. 2004. Temperature, growth rate, and body size in ectotherms: Fitting pieces of a life-history puzzle. *Integrative and Comparative Biology* 44:498–509.

Archibald, J. D. 1996. *Dinosaur Extinction and the End of an Era: What the Fossils Say*. New York: Columbia University Press.

Archibald, J. D., A. O. Averianov, and E. G. Ekdale. 2001. Late Cretaceous relatives of rabbits, rodents, and other extant eutherian mammals. *Nature* 414:62–65.

Archibald, J. D., and D. H. Deutschman. 2001. Quantitative analysis of the timing of the origin and diversification of extant placental orders. *Journal of Mammalian Evolution* 8:107–124.

Arnold, W., T. Ruf, S. Reimoser, F. Tataruch, K. Onderscheka, and F. Schober. 2004. Nocturnal hypometabolism as an overwintering strategy of red deer (*Cervus elaphus*). *American Journal of Physiology, Regulatory, Integrative and Comparative Physiology* 286:174–181.

Atayman, S., B. S. Rubidge, and F. Abdala. 2009. Taxonomic evaluation of tapinocephalid dinocephalians. *Palaeontologia Africana* 44:87–90.

Averianov, A. O., and P. P. Skutschas. 2000. [The oldest placental mammal]. *Doklady Akademii Nauk* 374:130–132 (in Russian).

Avery, R. A. 1982. Field studies of body temperatures and thermoregulation. In *Biology of the Reptilia*, edited by C. Gans and H. F. Pough, 93–166. London: Academic Press.

Bain, A. G. 1845. On the discovery of the fossil remains of bidentals and other reptiles in South Africa. *Proceedings of the Geological Society London* 4:499–500.

———. 1856. On the geology of Southern Africa. *Transactions of the Geological Society of London* 7:175–192.

Baird, D., and O. F. Patterson, III. 1967. Dicynodont-Archosaur fauna in the Pekin Formation (Upper Triassic) of North Carolina. *Geological Society of America Special Paper* 115:11.

Bakker, R. T. 1968. The superiority of dinosaurs. *Discovery* 3:11–22.

———. 1971. Dinosaur physiology and the origin of mammals. *Evolution* 25:636–658.

——— 1975. Dinosaur renaissance. *Scientific American* 232:58–78.

———. 1980. Dinosaur heresy—Dinosaur renaissance. In *A Cold Look at Warmblooded Dinosaurs*, edited by R. D. K. Thomas and E. C. Olson, 351–462. Boulder, Colo.: Westview Press.

———. 1986. *The Dinosaur Heresies*. New York: Morrow and Co.

Bandyopadhyay, S. 1988. A kannemeyeriid dicynodont from the Middle Triassic Yerrapalli Formation. *Philosophical Transactions of the Royal Society of London, B* 320:185–233.

Bandyopadhyay, S., T. K. Roy Chowdhury, and D. P. Sengupta. 2002. Taphonomy of some Gondwana vertebrate assemblages of India. *Sedimentary Geology* 147:219–245.

Barbarena, M. C., and D. C. Araujo. 1975. Tetrapodes fósiles de Sudamerica y deriva continental. *Actas I Congreso Argentino de Paleontología y Bioestratigrafía (Tucumán)* 1:497–504.

Barghusen, H. R. 1968. The lower jaw of cynodonts (Reptilia, Therapsida) and the evolutionary origin of mammal-like adductor musculature. *Postilla* 116:1–49.

Barry, T. S. 1974. A new dicynodont ancestor from the Upper Ecca. *Annals of the South African Museum* 64:117–136.

Bartholomew, G. A. 1982. Physiological control of body temperature. In *Biology of the Reptilia*, edited by C. Gans and H. F. Pough, 167–211. London: Academic Press.

Battail, B. 1993. On the biostratigraphy of Triassic therapsid-bearing formations. *New Mexico Museum of Natural History and Science Bulletin* 3:31–35.

———. 2000. A comparison of Late Permian Gondwanan and Laurasian amniote faunas. *Journal of African Earth Sciences* 31:165–174.

———. 2005. Late Triassic traversodontids (Synapsida: Cynodontia) in southern Africa. *Palaeontologia Africana* 41:67–80.

Battail, B., and M. Surkov. 2000. Mammal-like reptiles from Russia. In *The Age of Dinosaurs in Russia and Mongolia*, edited by M. J. Benton, M. A. Shishkin, D. M. Unwin, and E. N. Kurochkin, 86–119. Cambridge: Cambridge University Press.

Beer, de G. R. 1937. *The Development of the Vertebrate Skull*. London: Oxford University Press.

Bennett, A. F. 1973a. Blood physiology and oxygen-transport during activity in two lizards, *Varanus gouldii* and *Sauromalushispidus*. *Comparative Biochemistry and Physiology* 46:673–690.

———. 1973b. Ventilation in two species of lizards during rest and activity. *Comparative Biochemistry and Physiology* 46:653–671.

———. 1982. The energetics of reptilian activity. In *Biology of the Reptilia*, edited by C. Gans and H. F. Pough, 155–199. New York: Academic Press.

———. 1991. The evolution of active capacity. *Journal of Experimental Biology* 160:1–23.

———. 1996. Aerodynamics and thermoregulatory function of the dorsal sail of Edaphosaurus. *Paleobiology* 22:496–506.

Bennett, A. F., and B. Dalzell. 1973. Dinosaur physiology: A critique. *Evolution* 27:170–174.

Bennett, A. F., and W. R. Dawson. 1976. Metabolism. In *Biology of the Reptilia*, edited by C. Gans and W. R. Dawson, 127–223. New York: Academic Press.

Bennett, A. F., J. W. Hicks, and A. J. Cullum. 2000. An experimental test of the thermoregulatory hypothesis for the evolution of endothermy. *Evolution* 54:1768–1773.

Bennett, A. F., and J. A. Ruben. 1979. Endothermy and activity in vertebrates. *Science* 206:649–654.

———. 1986. The metabolic and thermoregulatory status of therapsids. In *The Ecology and Biology of Mammal-Like Reptiles*, edited by N. Hotton, III, P. D. MacLean, J. J. Roth, and E. C. Roth, 207–218. Washington: Smithsonian Institution Press.

Berman, D. S. 1979. *Edaphosaurus* (Reptilia, Pelycosauria) from the Lower Permian of northeastern United States, with description of a new species. *Annals of Carnegie Museum* 48:185–202.

Berman, D. S., R. R. Reisz, J. R. Bolt, and D. Scott. 1995. The cranial anatomy and relationships of the synapsid *Varanosaurus* (Eupelycosauria: Ophiacodontidae) from the Early Permian of Texas and Oklahoma. *Annals of Carnegie Museum* 64:99–133.

Berman, D. S., S. S. Sumida, and R. E. Lombard. 1997. Biogeography of primitive amniotes. In *Amniote Origins: Completing the Transition to Land*, edited by S. S. Sumida and K. L. M. Martin, 85–140. London: Academic Press.

Berrigan, D., and E. L. Charnov. 1994. Reaction norms for age and size at maturity in response to temperature: a puzzle for life historians. *Oikos* 70:474–478.

Bertram, J. E. A., and A. A. Biewener. 1988. Bone curvature: Sacrificing strength for load predictability. *Journal of Theoretical Biology* 131:75–92.

Bininda-Emonds, O. R. P., M. Cardillo, K. E. Jones, R. D. E. MacPhee, R. M. D. Beck, R. Grenyer, S. A. Price, R. A. Vos, J. L. Gittleman, and A. Purvis. 2007. The delayed rise of present-day mammals. *Nature* 446:507–512.

Block, B. A. 1991. Endothermy in fish: Thermogenesis, ecology and evolution. In *Biochemistry and Molecular Biology of Fishes*, edited by P. W. Hochachka and T. Mommsen, 269–311. Amsterdam: Elsevier.

Boily, P. 2002. Individual variation in metabolic traits of wild nine-banded armadillos (*Dasypus novemcinctus*), and the aerobic capacity model for the evolution of endothermy. *Journal of Experimental Biology* 205:3207–3214.

Bonaparte, J. F. 1966. Una nueva "fauna" Triásicade Argentina (Therapsida: Cynodontia Dicynodontia). Consideraciones filogenéticas y paleobiogeográficas. *Ameghiniana* 4:243–296.

———. 1980. El primer ictidosaurio (Reptilia–Therapsida) de América del Sur, *Chaliminia musteloides*, del Triásico Superior de La Rioja, República Argentina. *Actas II Congreso Argentino de Paleontologia y Bioestratigrafia y I Congreso Latinoamericano de Paleontologia. Buenos Aires Tomo* 1:123–133.

Bonaparte, J. F., J. Ferigolo, and A. M. Ribeiro. 2001. A primitive Late Triassic "ictidosaur" from Rio Grande do Sul, Brazil. *Palaeontology* 44:623–635.

Bonaparte, J. F., A. G. Martinelli, and C. L. Schultz. 2005. New information on *Brasilodon* and *Brasilitherium* (Cynodontia, Probainognathia) from the Late Triassic of southern Brazil. *Revista Brasileira de Paleontologia* 8:25–46.

Bonaparte, J. F., A. G. Martinelli, C. L. Schultz, and R. Rubert. 2003. The sister group of mammals: Small cynodonts from the Late Triassic of southern Brazil. *Revista Brasileira de Paleontologia* 5:5–27.

Bond, G. 1969. The fauna of the *Tapinocephalus* Zone (Beaufort Beds of the Karoo). *Annals of the South African Museum* 56:1–73

———. 1973. The palaeontology of Rhodesia. *Bulletin of the Geological Survey of Rhodesia* 70:1–121.

Boonstra, L. D. 1946. Report on some fossil reptiles from Gunyanka's Kraal, Busi Valley. *Proceedings and Transactions of the Rhodesian Scientific Association* 41:46–49.

———. 1963. Early dichotomies in the therapsids. *South African Journal of Science* 59:176–195.

———. 1964. The girdles and limbs of the pristerognathid Therocephalia. *Annals of the South African Museum* 48:121–165.

———. 1969. The fauna of the *Tapinocephalus* Zone (Beaufort Beds of the Karoo). *Annals of the South African Museum* 56:1–73.

———. 1971. The early therapsids. *Annals of the South African Museum* 59:17–46.

Bordy, E. M., P. J. Hancox, and B. S. Rubidge. 2004. Main Karoo Basin development during the deposition of the Elliot Formation (Late Triassic-Early Jurassic), Karoo Supergroup, South Africa. *South African Journal of Geology* 107:395–410.

Botha, J. 2002. *The Palaeobiology of the Non-Mammalian Cynodonts Deduced from Bone Microstructure and Stable Isotopes.* PhD diss., University of Cape Town.

———. 2003. Biological aspects of the Permian dicynodont *Oudenodon* (Therapsida: Dicynodontia) deduced from bone histology and cross-sectional geometry. *Palaeontologia Africana* 39:37–44.

Botha, J., F. Abdala, and R. M. H. Smith. 2007. The oldest cynodont: new clues on the origin and diversification of Cynodontia. *Zoological Journal of the Linnean Society* 149:477–492.

Botha, J., and K. Angielczyk. 2007. An integrative approach to distinguishing the Late Permian dicynodont species *Oudenodon bainii* and *Tropidostoma microtrema* (Therapsida: Anomodontia). *Palaeontology* 50:1175–1209.

Botha, J., and A. Chinsamy. 2000. Growth patterns deduced from the histology of the cynodonts

Diademodon and *Cynognathus*. *Journal of Vertebrate Paleontology* 20:705–711.
———. 2004. Growth and life habits of the Triassic cynodont *Trirachodon*, inferred from bone histology. *Acta Palaeontologica Polonica* 49:619–627.
———. 2005. Growth patterns of *Thrinaxodon*, a non-mammalian cynodont from the Early Triassic of South Africa. *Palaeontology* 48:385–394.
Botha, J., J. Lee-Thorp, and A. Chinsamy. 2005. The palaeoecology of the non-mammalian cynodonts *Diademodon* and *Cynognathus* from the Karoo Basin of South Africa, using stable light isotope analysis. *Palaeogeography Palaeoclimatology Palaeoecology* 223:303–316.
Botha, J., and R. M. H. Smith. 2006. Rapid vertebrate recuperation in the Karoo Basin of South Africa following the end-Permian extinction. *Journal of African Earth Sciences* 45(4–5):502–514.
Botha-Brink, J., and F. Abdala. 2008. A new cynodont record from the *Tropidostoma* Assemblage Zone of the Beaufort Group: Implications for the early diversity of cynodonts in South Africa. *Palaeontologia Africana* 43:1–6.
Botha-Brink, J., and K. D. Angielczyk. 2010. Do extraordinarily high growth rates in Permo-Triassic dicynodonts (Therapsida, Anomodontia) explain their success before and after the end-Permian extinction? *Zoological Journal of the Linnean Society* 160:341–365.
Botha-Brink, J., and S. P. Modesto. 2009. Anatomy and relationships of the Middle Permian varanopid *Heleosaurus scholtzi* based on a social aggregation from the Karoo Basin of South Africa. *Journal of Vertebrate Paleontology* 29:389–400.
———. 2011. A new specimen of the therocephalian synapsid *Olivierosuchus parringtoni* from the Lower Triassic South African Karoo Basin. *Palaeontology* 54:591–606.
Bou, J., M. J. Castiella, J. Ocana, and A. Casinos. 1990. Multivariate analysis and locomotor morphology in insectivores and rodents. *Zoologischer Anzeiger* 225:287–294.
Bourdon, E., J. Castanet, de A. Ricqlès, H. Lamrous, and J. Cubo. 2009. Bone growth marks reveal protracted growth in New Zealand kiwi (Aves, Apterygidae). *Biology Letters*. doi:10.1098/rsbl.2009.0310.
Bouvier, M., and W. L. Hylander. 1981. Effect of bone strain on cortical bone structure in macaques (*Macaca mulatta*). *Journal of Morphology* 167:1–12.
———. 1996. The mechanical or metabolic function of secondary osteonal bone in the monkey *Macaca fascicularis*. *Archives of Oral Biology* 41:941–950.
Bowker, R. G., C. L. Wright, and G. E. Bowker. 2010. Patterns of body temperatures: Is lizard thermoregulation chaotic? *Journal of Thermal Biology* 35:1–5.
Bowring, S. A., D. H. Erwin, Y. G. Jin, M. W. Martin, K. Davidek, and W. Wang. 1998. U/Pb zircon geochronology and tempo of the end-Permian mass extinction. *Science* 280:1039–1045.
Bramwell, C. D., and P. B. Fellgett. 1973. Thermal regulation in sail lizards. *Nature* 242:203–205.
Brand, M. D., P. Couture, P. L. Else, K. W. Withers, and A. J. Hulbert. 1991. Evolution of energy-metabolism—Proton permeability of the inner membrane of liver-mitochondria is greater in a mammal than in a reptile. *Biochemical Journal* 275:81–86.
Brattstrom, B. H. 1965. Body temperature in reptiles. *American Midland Naturalist* 73:376–422.
Brink, A. S. 1951. On the genus *Lystrosaurus* Cope. *Transactions of the Royal Society of South Africa* 33:107–120.
———. 1954. *Thrinaxodon* and some other *Lystrosaurus* Zone cynodonts in the collection of the National Museum, Bloemfontein. *Researches of the National Museum, Bloemfontein* 1:115–126.
———. 1955. A study on the skeleton of *Diademodon*. *Palaeontologia Africana* 3:3–39.
———. 1960. A new type of primitive cynodont. *Palaeontologia Africana* 7:119–154.
———. 1963. On *Bauria cynops* Broom. *Palaeontologia Africana* 9:129–138.
———. 1965. A new ictidosauchid (Scaloposauria) from the *Lystrosaurus* Zone. *Palaeontologia Africana* 9:129–138.
Brinkman, D. 1988. Size-independent criteria for estimating relative age in *Ophiacodon* and *Dimetrodon* (Reptilia, Pelycosauria) from the Admiral and Lower Belle Plains Formations of west-central Texas. *Journal of Vertebrate Paleontology* 8:172–180.
Bristowe, M. A. 2004. *The Reconstruction of a Juvenile Coelophysoid Theropod Dinosaur from the Forest Sandstone Formation (Karoo Sequence) of Zimbabwe, and Its Significance in Identifying the Taxon Concerned*. MSc diss., University of the Witwatersrand, Johannesburg.
Bristowe, M. A., and M. A. Raath. 2004. A juvenile coelophysoid skull from the Early Jurassic of Zimbabwe, and the synonomy of *Coelophysis* and *Syntarsus*. *Palaeontologia Africana* 40:31–42.
Broom, R. 1903. On the remains of *Lystrosaurus* in the Albany Museum. *Records of the Albany Museum* 1:3–8.
———. 1906a. The classification of the Karroo Beds of South Africa. *Geological Magazine* 5:3–36.
———. 1906b. On the Permian and Triassic faunas of South Africa. *Geological Magazine* 5:29–30.

———. 1906c. On the remains of *Erythrosuchus africanus* Broom. *Annals of the South African Museum* 5:187–197.

———. 1906d. On the South African diaptosaurian reptile *Howesia*. *Proceedings of the Zoological Society London* 2:591–600.

———. 1907. On the geological horizons of the vertebrate genera of the Karroo Formation. *Records of the Albany Museum* 2:156–163.

———. 1909. An attempt to determine the horizons of the fossil vertebrates of the Karroo. *Annals of the South African Museum* 7:285–289.

———. 1932. *The Mammal-like Reptiles of South Africa and the Origin of Mammals.* London: H. F and G. Witherby.

Buffrénil, V. de. 1980. Mise en évidence de l'incidence des conditions de milieu sur la croissance de *Crocodylus siamensis* (Schneider, 1801) et valeur des marques de croissance squelettiques pour l'évaluation de l'âge individuel. *Archives de Zoologie Expérimentale Générale* 121:63–76.

Buffrénil, V. de, A. de Ricqlès, C. E. Ray, and D. P. Domning. 1990. Bone histology of the ribs of the *Archaeocetes* (Mammalia: Cetacea). *Journal of Vertebrate Paleontology* 10(4):455–466.

Bühler, P. 1986. Das Vogelskellat—hochentwickelter knochen-leicht-bau. *Arcus* 5:221–22.

Busbey, A. B., and C. Gow. 1984. A new protosuchian crocodile from the Upper Triassic Elliot Formation of South Africa. *Palaeontologia Africana* 25:127–149.

Butler, E. 2010. *The Postcranial Skeleton of the Early Triassic Non-Mammalian Cynodont Galesaurus planiceps: Implications for Biology and Lifestyle.* MSc diss., University of the Free State.

Butler, P. M., and J. J. Hooker. 2005. New teeth of allotherian mammals from the English Bathonian, including the earliest multituberculates. *Acta Palaeontologica Polonica* 50:185–207.

Butler, R. J., R. M. H. Smith, and D. B. Norman. 2007. A primitive ornithischian dinosaur from the Late Triassic of South Africa, and the early evolution and diversification of Ornithischia. *Proceedings of the Royal Society B* 274:2021–2046.

Camp, C. L., and S. P. Welles. 1956. Triassic dicynodont reptiles. Part I: The North American genus *Placerias. Memoir of the University of California* 13:255–304.

Campione, N. E., and R. R. Reisz. 2010. *Varanops brevirostris* (Eupelycosauria: Varanopidae) from the Lower Permian of Texas, with discussion of varanopid morphology and interrelationships. *Journal of Vertebrate Paleontology* 30:724–746.

Canoville, A., and M. Laurin. 2010. Evolution of humeral microanatomy and lifestyle in amniotes, and some comments on paleobiological inferences. *Biological Journal of the Linnean Society* 100:384–406.

Carey, H. V., M. T. Andrews, and S. L. Martin. 2003. Mammalian hibernation: Cellular and molecular responses to depressed metabolism and low temperature. *Physiological Reviews* 83:1153–1181.

Carrier, D. R. 1987. The evolution of locomotor stamina in tetrapods: Circumventing a mechanical constraint. *Paleobiology* 13:326–341.

Carroll, R. L. 1964. The earliest reptiles. *Journal of the Linnean Society* (Zoology) 45:61–83.

———. 1988. *Vertebrate Paleontology and Evolution.* New York: Freeman.

———. 1991. The origins of reptiles. In *Origins of the Higher Groups of Tetrapods: Controversy and Consensus*, edited by H. P. Schultze and L. Trueb, 331–353. Ithaca, N.Y.: Cornell University Press.

Carter, D. R., W. C. Hayes, and D. J. Schurman. 1976. Fatigue life of compact bone—II. Effects of microstructure and density. *Journal of Biomechanics* 9:211–218.

Carter, D. R., and D. M. Spengler. 1978. Mechanical properties and composition of cortical bone. *Clinical Orthopaedics and Related Research* 135:192–217.

Casamiquela, R. M. 1964. *Estudios Icnológicos.* Buenos Aires: Imprenta del Colegio Industrial, Pío X.

Case, T. J. 1978. Speculations on the growth rate and reproduction of some dinosaurs. *Paleobiology* 4:320–328.

Castanet, J. 1994. Age estimation and longevity in reptiles. *Gerontology* 40:174–192.

———. 2006. Time recording in bone microstructures of endothermic animals; functional relationships. *C. R. Palevol* 5:629–636.

Castanet, J., and M. Baez. 1991. Adaptation and evolution in Gallotia lizard from the Canary Islands: Age, growth, maturity and longevity. *Amphibia-Reptilia* 12:81–102.

Castanet, J., S. Croci, F. Aujard, M. Perret, J. Cubo, and E. de Margerie. 2004. Lines of arrested growth in bone and age estimation in a small primate: *Microcebus murinus*. *Journal of Zoology* 263:31–39.

Castanet, J., H. Francillon-Vieillot, J. Meunier, and A. de Ricqlès. 1993. Bone and individual aging. In *Bone Growth*, edited by B. K. Hall. 245–283. Boca Raton, Fla.: CRC Press.

Castanet, J., A. Grandin, A. Arbourachid, and A. de Ricqlès. 1996. Expression of growth dynamic in the structure of the periosteal bone in the mallard, *Anas platyrynchos. Comptes Rendus de l'Académie des Sciences Paris, Life Sciences, Animal Biology and Pathology* 319:301–308.

Castanet, J., and E. M. Smirina. 1990. Introduction to the skeletochronological method in amphibians and reptiles. *Annales des Sciences Naturelles, Zoologie* 11:191–96.
Catuneanu, O., P. J. Hancox, and B. S. Rubidge. 1998. Reciprocal flexural behaviour and contrasting stratigraphies, a new Basin development model for the Karoo retroarc foreland system, South Africa. *Basin Research* 10:417–439.
Catuneanu O., H. Wopfner, P. G. Eriksson, B. Cairncross, B. S. Rubidge, R. M. H. Smith, and P. J. Hancox 2005 The Karoo basins of south-central Africa. *Journal of African Earth Sciences* 43:211–253.
Chatterjee, S. 1982. A new cynodont reptile from the Triassic of India. *Journal of Paleontology* 56:203–214.
Chinsamy, A. 1990. Physiological implications of the bone histology of *Syntarsus rhodesiensis* (Saurischia: Theropoda). *Palaeontologia Africana* 27:77–82.
———. 1993a. Bone histology and growth trajectory of the prosauropod dinosaur *Massospondylus carinatus* Owen. *Modern Geology* 18:319–329.
———. 1993b. Image analysis and the physiological implications of the vascularization of femora in archosaurs. *Modern Geology* 18:319–329.
———. 1997. Assessing the biology of the fossil vertebrates through bone histology. *Palaeontologia Africana* 33:29–35.
Chinsamy, A., and F. Abdala. 2008. Palaeobiological implications of the bone microstructure of South American traversodontids (Therapsida: Cynodontia). *South African Journal of Science* 104:225–230.
Chinsamy, A., L. Chiappe, and P. Dodson. 1994. Growth rings in Mesozoic birds. *Nature* 368:196–197.
Chinsamy, A., L. Codorniú, and L. Chiappe. 2008. Developmental growth patterns of the filter feeder pterosaur *Pterodaustro guiñazui. Biology Letters* 4:282–285.
———. 2009. Palaeobiological implications of the bone histology of *Pterodaustro guinazui*. *Anatomical Record* 292:1462–1477.
Chinsamy, A., and P. Dodson. 1995. Inside a dinosaur bone. *American Scientist* 83:174–180.
Chinsamy, A., S. A. Hanrahan, R. M. Neto, and M. Seeley. 1995. A skeletochronological assessment of age in *Angolosaurus skoogi*, a lizard living in an aseasonal environment. *Journal of Herpetology* 29(3):457–460.
Chinsamy, A., and W. J. Hillenius. 2004. Physiology of nonavian dinosaurs. In *The Dinosauria*, 2nd ed., edited by D. B. Weishampel, P. Dodson, and H. Osmólska, 643–659. Berkeley: University of California Press.
Chinsamy, A., and J. Hurum. 2006. Bone microstructure and growth patterns of early mammals. *Acta Palaeontologia Polonica* 51(2):325–338.
Chinsamy, A., and M. A. Raath. 1992. Preparation of fossil bone for histological study. *Palaeontologia Africana* 29:39–44.
Chinsamy, A., T. Rich, and P. Vickers-Rich. 1998. Polar dinosaur bone histology. *Journal of Vertebrate Paleontology* 18:385–390.
Chinsamy, A., and B. S. Rubidge. 1993. Dicynodont (Therapsida) bone histology: Phylogenetic and physiological implications. *Palaeontologia Africana* 30:97–102.
Chinsamy, A., and A. Tumarkin-Deratzian. 2009. Pathologic bone tissues in a turkey vulture and a nonavian dinosaur: Implications for interpreting endosteal bone and radial fibrolamellar bone in fossil dinosaurs. *Anatomical Record* 292:1478–1484.
Chinsamy-Turan, A. 2005. *The Microstructure of Dinosaur Bone: Deciphering Biology with Fine Scale Techniques.* Baltimore: Johns Hopkins University Press.
Cifelli, R. L. 1999. Tribosphenic mammal from the North American Early Cretaceous. *Nature* 401:363–366.
Cifelli, R. L., and C. L. Gordon. 2007. Re-crowning mammals. *Nature* 447:918–920.
Cisneros J. C., F. Abdala, B. S. Rubidge, P. C. Dentzien-Dias, and A. de O. Bueno. 2011. Dental occlusion in a 260-million-year-old therapsid with saber canines from the Permian of Brazil. *Science* 331:1603–1605.
Cisneros, J. C., and L. A. Tsuji. 2009. *Nycteroleter* affinities of a Permian parareptile from the South African Karoo Basin. *Acta Palaeontologica Polonica* 54:165–169.
Clark, J. M., M. Montellano, J. A. Hopson, R. Hernandez, and D. Fastovsky. 1997. An Early or Middle Jurassic tetrapod assemblage from the La Boca Formation, northeastern Mexico. In *In the Shadow of the Dinosaurs—Early Mesozoic Tetrapods,* edited by N. C. Fraser and H-D. Sues, 295–302. Cambridge: Cambridge University Press.
Clark, J. M., and H. D. Sues. 2002. Two new basal crocodylomorph archosaurs from the Lower Jurassic and the monophyly of the Sphenosuchia. *Zooogical Journal of the Linnean Society* 136:77–95.
Clarke, A., and H. O. Pörtner. 2010. Temperature, metabolic power and the evolution of endothermy. *Biological Reviews* 85:703–727.
Clemens, W. A. 1979. A problem in morganucodontid taxonomy. *Zoological Journal of the Linnean Society* 66:1–14.
———. 1986. On the Triassic and Jurassic mammals. In *The Beginning of the Age of Dinosaurs: Faunal*

Change acrss the Triassic-Jurassic Boundary, edited by K. Padian, 237–246. Cambridge: Cambridge University Press.

Cluver, M. A. 1971. Cranial morphology of the dicynodont genus *Lystrosaurus*. *Annals of the South African Museum* 56:155–274.

———. 1974. The skull and mandible of a new cistecephalid dicynodont. *Annals of the South African Museum* 64:137–155.

———. 1978. The skeleton of the mammal-like reptile *Cistecephalus* with evidence of a fossorial mode of life. *Annals of the South African Museum* 76:213–246.

Cluver, M. A., and N. H. Hotton. 1981. The genera *Dicynodon* and *Diictodon* and their bearing on the classification of the Dicynodontia (Reptilia, Therapsida). *Annals of the South African Museum* 83:99–146.

Cluver, M. A., and G. M. King. 1983. A reassessment of the relationships of Permian Dicynodontia (Reptilia, Therapsida) and a new classification of dicynodonts. *Annals of the South African Museum* 91:195–273.

Colbert, E. H. 1948. The mammal-like reptile *Lycaenops*. *Bulletin of the American Museum of Natural History* 89:357–404.

Colbert, E. H., and J. W. Kitching. 1977. Triassic cynodont reptiles from Antarctica. *American Museum Novitates* 2611:1–30.

———. 1981. Scaloposaurian reptiles from the Triassic of Antarctica. *American Museum Novitates* 2709:1–22.

Cole, D. I. 1992. Evolution and development of the Karoo Basin. In *Inversion Tectonics of the Cape Fold Belt, Karoo and Cretaceous Basins of Southern Africa*, edited by M. J. De Wit, and I. G. D. Ransome, 87–99. Rotterdam: A. A. Balkema.

Collins, J. C., T. C. Pilkington, and K. Schmidt-Nielsen. 1971. A model of respiratory heat transfer in a small mammal. *Biophysical Journal* 11:886–914.

Cooper, M. R. 1981. The prosauropod dinosaur *Massospondylus carinatus* from Zimbabwe: Its biology, mode of life and phylogenetic significance. *Occasional Papers of the National Museums of Rhodesia B, Natural Sciences* 6:689–840.

Cormack, D. 1987. *Ham's Histology*. New York: Lippencott.

Cowles, R. B., and C. M. Bogert. 1944. A preliminary study of the thermal requirements of desert reptiles. *Bulletin of the American Museum of Natural History* 83:261–296.

Cox, C. B. 1964. On the palate, dentition and classification of the fossil reptile *Endothiodon* and related genera. *American Museum Novitates* 2071:1–25.

———. 1972. The pectoral region and habits of a new digging dicynodont. In *Studies in Vertebrate Evolution*, edited by K. A. Joysey and T. S. Kemp, 173–190. Edinburgh: Oliver and Boyd.

———. 1998. The jaw function and adaptive radiation of the dicynodont mammal-like reptiles of the Karoo Basin of South Africa. *Zoological Journal of the Linnean Society* 122:349–384.

Crompton, A. W. 1972a. The evolution of the jaw articulation of cynodonts. In *Studies in Vertebrate Evolution*, edited by K. A. Joysey and T. S. Kemp, 231–251. Edinburgh: Oliver and Boyd.

———. 1972b. Postcanine occlusion in cynodonts and tritylodonts. *Bulletin of the British Museum of Natural History (Geology)* 21:29–71.

Crompton, A. W., and N. Hotton III. 1967. Functional morphology of the masticatory apparatus of two dicynodonts (Reptilia, Therapsida). *Postilla* 109:1–51.

Crompton, A. W., and Z. X. Luo. 1993. Relationships of the Liassic mammals *Sinoconodon*, *Morganucodon oehleri*, and *Dinnetherium*. In *Mammal Phylogeny: Mesozoic Differentiation, Multituberculates, Monotremes, Early Therians, and Marsupials*, edited by F. S. Szalay, M. J. Novacek, and M. C. McKenna, 30–44. New York: Springer-Verlag.

Cruickshank, A. R. I. 1965. On a specimen of the anomodont reptile *Kannemeyeria latifrons* (Broom) from the Manda Formation of Tanzania (Tangyanyika). *Proceedings of the Linnean Society London* 176:149–157

———. 1978. Feeding adaptations in Triassic dicynodonts. *Palaeontologia Africana* 21:121–132.

Cubo, J., F. Ponton, M. Laurin, E. de Margerie, and J. Castanet. 2005. Phylogenetic signal in bone microstructure of sauropsids. *Systematic Biology* 54:562–574.

Currey, J. D. 1987. The evolution of the mechanical properties of amniote bone. *Journal of Biomechanics* 20:1035–1044.

———. 2002. *Bones: Structure and Mechanics*, 2nd ed. Princeton, N.J.: Princeton University Press.

———. 2003. The many adaptations of bone. *Journal of Biomechanics* 36:1487–1495.

Curry, K. A. 1999. Ontogenetic histology of *Apatosaurus* (Dinosauria: Sauropoda): New insights on growth rates and longevity. *Journal of Vertebrate Paleontology* 19:654–665.

Daan, S., M. Dirkjan, and A. Groenewald. 1990. Avian basal metabolite rate: Their association with body composition and energy expenditure in nature. *American Journal of Physiology R.* 259(2):333–340.

Damiani, R. J. 2001a. A systematic revision and phylogenetic analysis of Triassic mastodonsauroids

(Temnospondyli, Stereospondyli). *Zoological Journal Linnaean Society* 133:379–482.
———. 2001b. *Parotosuchus* (Amphibia, Temnospondyli) from the Cynognathus Assemblage Zone of South Africa, cranial morphology and relationships. *Alcheringa* 25:351–379.
———. 2004. Temnospondyls from the Beaufort Group (Karoo Basin) of South Africa and their biostratigraphy. *Gondwana Research* 7:165–173.
———. 2008. A giant skull of the temnospondyl *Xenotosuchus africanus* from the Middle Triassic of South Africa and its ontogenetic implications. *Acta Palaeontologica Polonica* 53:75–84.
Damiani, R. J., S. P. Modesto, A. M. Yates, and J. Neveling. 2003. Earliest evidence of cynodont burrowing. *Proceedings of the Royal Society London* B 270:1747–1751.
Damiani, R. J., J. Neveling, P. J. Hancox, and B. S. Rubidge. 2000. First trematosaurid temnospondyl from the *Lystrosaurus* Assemblage Zone of South Africa and its biostratigraphic implications. *Geological Magazine* 137(6):659–665.
Damiani, R. J., C. Vasconcelos, A. Renaut, J. Hancox, and A. Yates. 2007. *Dolichuranus primaevus* (Therapsida: Anomodontia) from the Middle Triassic of Namibia and its phylogenetic relationships. *Palaeontology* 50:1531–1546.
Darveau, C. A., R. K. Suarez, R. D. Andrews, and P. W. Hochachka. 2002. Allometric cascade as a unifying principle of body mass effects on metabolism. *Nature* 417:166–170.
Davies, K. C. 1981. Heulandite deposition in fossil reptile bones from the Karoo of Zambia. *Journal of Paleontology* 55(1):47–61.
De Beer, F. C., R. Prevec, J. Cisneros, and F. Abdala. 2008. Hidden structure of fossils revealed by neutron and X-ray tomography. In *Neutron Radiography*: Proceedings of the 8th World Conference on Neutron Radiography WCNR-8, edited by M. Arif and R. G. Downing, 452–461. Lancaster, Pa.: DEStech Publications.
Depew, M., and C. A. Simpson. 2006. 21st century neontology and comparative develoment of the vertebrate skull. *Journal of Experimental Zoology (Molecular Development and Evolution)* 235:1256–1291.
Depew, M., C.A. Simpson, M. Morasso, and L. R. Rubenstein. 2005. Reassesing the Dlx code: The genetic regulation of branchial arch skeletal pattern and develoment. *Journal of Anatomy* 207:501–561.
Dilkes, D. W., and R. R. Reisz. 1996. First record of a basal synapsid ("mammal-like reptile") in Gondwana. *Proceedings of the Royal Society London B* 263:1165–1170.
Dunn, E. J. 1873. *Report on the Stormberg Coal-fields.* Parliamentary Report G31. Cape of Good Hope.
———. 1878. *Geological report on the Stormberg Coal-fields.* Parliamentary Report G4. Cape of Good Hope.
Dutenhoffer, M. S., and D. L. Swanson. 1996. Relationship of basal to summit metabolic rate in passerine birds and the aerobic capacity model for the evolution of endothermy. *Physiological Zoology* 69:1232–1254.
Du Toit, A. L. 1954. *Geology of South Africa,* 3rd ed. Edinburgh and London: Oliver and Boyd.
Eckhart, L., L. D. Valle, K. Jaeger, C. Ballaun, S. Szabo, A. Nardi, M. Buchberger, M. Hermann, L. Alibardi, and E. Tschachler. 2008. Identification of reptilian genes encoding hair keratin-like proteins suggests a new scenario for the evolutionary origin of hair. *Proceedings of the National Academy of Sciences* 105:18419–18423.
Edinger, T. 1951. The brains of the Odontognathae. *Evolution* 5:6–24.
Egberts, S., T. Rowe, H. D. Sues, Z. Luo, and F. Jenkins. 2009. The first semi-aquatic synapsid from the Early Jurassic of Arizona. *Journal of Vertebrate Paleontology* 29:91A.
Else, P. L., and A. J. Hulbert. 1981. Comparison of the "mammal machine" and the "reptile machine": Energy production. *American Journal of Physiology* 240:R3–R9.
———. 1985. An allometric comparison of the mitochondria of mammalian and reptilian tissue: The implications for the evolution of endothermy. *Journal of Comparative Physiology, Series B* 156:3–11.
Else, P. L., N. Turner, and A. J. Hulbert. 2004. The evolution of endothermy: Role for membranes and molecular activity. *Physiological and Biochemical Zoology* 77:950–958.
Enlow, D. H. 1962. A study of the post-natal growth and remodelling of bone. *American Journal of Anatomy* 110(2):79–102.
———. 1963. *Principles of Bone Remodeling: an Account of Post-Natal Growth and Remodeling Processes in Long Bones and the Mandible.* Springfield, Ill.: Charles C. Thomas Publishers.
———. 1969. The bone of reptiles. In *Biology of the Reptilia, Volume 1: Morphology A,* edited by C. Gans, A. Bellairs, and T. Parsons, 45–77. London: Academic Press.
———. 1990. *Facial Growth,* 3rd ed. Philadelphia: W. B. Saunders Company.
Enlow, D. H., and S. O. Brown. 1956. A comparative histological study of fossil and recent bone tissues, Part I. *Texas Journal of Science* 8:405–443.

———. 1957. A comparative histological study of fossil and recent bone tissues, Part II. *Texas Journal of Science* 9:186–214.
———. 1958. A comparative histological study of fossil and recent bone tissues, Part III. *Texas Journal of Science* 10:187–230.
Erickson, G. M. 1996a. Daily deposition of dentine in juvenile *Alligator* and assessment of tooth replacement rates using incremental line counts. *Journal of Morphology* 228:189–194.
———. 1996b. Incremental lines of von Ebner in dinosaurs and the assessment of tooth replacement rates using growth line counts. *Proceedings of the National Academy of Sciences* USA 93:14623–14627.
———. 2005. Assessing dinosaur growth patterns: A microscopic revolution. *Trends in Ecology and Evolution* 20:677–684.
Erickson, G. M., P. J. Makovicky, P. J. Currie, M. A. Norell, S. A. Yerby, and C. A. Brochu. 2004. Gigantism and comparative life-history parameters of tyrannosaurid dinosaurs. *Nature* 430:772–775.
Erickson, G. M., P. J. Makovicky, B. D. Inouye, Z. Chang-Fu, and G. Ke-Qin. 2009. A life table for *Psittacosaurus*: Initial insight into Ornithiscian population biology. *Anatomical Record* 292:1514–1521.
Erickson, G. M., O. W. M Rauhut, Z. Zhou, A. H. Turner, B. D. Inouye, D. Hu, and M. A. Norell. 2009. Was dinosaurian physiology inherited by birds? Reconciling slow growth in *Archaeopteryx*. *PLoS ONE* 4(10):e7390. doi:10.1371/journal.pone.0007390.
Erismis, U. C., and A. Chinsamy. 2010. Ontogenetic changes in the epiphyseal cartilage of *Rana (Pelophylax) caralitana* (Anura: Ranidae). *Anatomical Record* 293:1825–1837.
Erwin, D. 1994. The Permo-Triassic extinction. *Nature* 367:231–236.
Etnier, S. A. 2001. Flexural and torsional stiffness in multi-jointed biological beams. *Biological Bulletin* 200:1–8.
Evans, S. E., and K. A. Kermack. 1997. Assemblages of small tetrapods from the Early Jurassic of Britain. In *In the Shadow of the Dinosaurs—Early Mesozoic Tetrapods*, edited by N. C. Fraser and H-D. Sues, 271–283. Cambridge: Cambridge University Press.
Falcon-Lang, H. J. 2003. Late Carboniferous tropical dryland vegetation in an alluvial-plain setting, Joggins, Nova Scotia, Canada. *Palaios* 18:197–211.
Falcon-Lang, H. J., W. J. Nelson, S. Elrick, C. V. Looy, P. R. Ames, and W. A. DiMichele. 2009. Incised channel fills containing conifers indicate that seasonally dry vegetation dominated Pennsylvanian tropical lowlands. *Geology* 37:923–926.
Falkowski, P. G., M. E. Katz, A. Milligan, K. Fennel, B. S. Cramer, M. P. Aubry, R. A. Berner, M. Novacek, and W. M. Zapole. 2005. The rise of oxygen over the past 205 million years and the evolution of large placental mammals. *Science* 309:2202–2204.
Farlow, J. O., P. Dodson, and A. Chinsamy. 1995. Dinosaur biology. *Annual Review of Ecology and Systematics* 26:445–471.
Fiorillo, A. R., K. Padian, and C. Musikasinthorn. 2000. Taphonomy and depositional setting of the *Placerias* Quarry (Chinle Formation: Late Triassic, Arizona). *Palaios* 15:373–386.
Firth, B. J., and J. S. Turner. 1982. Sensory, neural and hormonal aspects of thermoregulation. In *Biology of the Reptilia*, edited by C. Gans and H. F. Pough, 213–274. London: Academic Press.
Fish, F. E. 1993. Comparison of swimming kinematics between terrestrial and semiaquatic opossums. *Journal of Mammalogy* 74:275–284.
Fisher, D. C. 1987. Mastodon procurement by paleoindians of the Great Lakes Region: Hunting or scavenging? In *The Evolution of Human Hunting*, edited by M. N. Nitecki and D. V. Nitecki, 309–421. New York: Plenum Press.
———. 1988. Season of death of the Hiscock mastodonts. *Bulletin of the Buffalo Society of Natural Sciences* 33:115–125.
———. 1990. Age, sex, and season of death of the Grandville mastodont. *The Michigan Archeologist* 36:141–160.
———. 1996. Extinction of proboscideans in North America. In *The Proboscidea: Evolution and Palaeoecology of Elephants and Their Relatives*, edited by J. Shoshani and P. Tassy, 296–315. New York: Oxford University Press.
Fisher, D. C., J. Trapani, J. Shoshani, and M. S. Woodford. 1998. Schreger angles in mammoth and mastodon tusk dentin. *Current Research in the Pleistocene* 15:105–107.
Flannery, T. F., M. Archer, T. H. Rich., and R. Jones. 1995. A new family of monotremes from the Cretaceous of Australia. *Nature* 377:418–420.
Florides, G. A., S. A. Kalogirou, S. A. Tassou, and L. Wrobel. 2001. Natural environment and thermal behaviour of *Dimetrodon limbatus*. *Journal of Thermal Biology* 26:15–20.
Flynn, J. J., J. M. Parrish, B. Rakotosamimanana, L. Ranivoharimanana, W. F. Simpson, and A. R. Wyss. 2000. New traversodontids (Synapsida: Eucynodontia) from the Triassic of Madagascar. *Journal of Vertebrate Paleontology* 20:422–427.
Foote, M., J. P. Hunter, C. M. Janis, and J. J. Sepkoski Jr.1999. Evolutionary and preservational constraints on origins of biologic groups: Divergence times of eutherian mammals. *Science* 283:1310–1314.

Foote, M., and A. I. Miller. 2007. *Principles of Paleontology*. New York: W. H. Freeman and Co.
Foote, M., and R. R. Reisz. 2008. A new species of *Emydops* (Synapsida, Anomodontia) and a discussion of dental variability and pathology in dicynodonts. *Journal of Vertebrate Paleontology* 28:770–787.
Fordyce, N. 2010. *Cold case files: Palaeoforensics and taphonomic study of the Mamafura dicynodont from the end-Permian*. BSc honors thesis. University of Cape Town.
Fourie, H., and B. S. Rubidge. 2009. The postcranial skeleton of the basal therocephalian *Glanosuchus macrops* (Scylacosauridae) and comparison of morphological and phylogenetic trends amongst the Theriodontia. *Palaeontologia Africana* 44:27–39.
Fox, D. L. 2000. Growth increments in *Gomphotherium* tusks and implications for late Miocene climate change in North America. *Palaeogeography, Palaeoclimatology, Palaeoecology* 156:327–48.
Fox, R. C., G. P. Youzwyshyn, and D. W. Krause. 1992. Post-Jurassic mammal-like reptile from the Palaeocene. *Nature* 358:233–235.
Frair, W., R. G. Ackman, and N. Mrosovsky. 1972. Body temperature of *Dermochelys coriacea:* Warm turtle from cold water. *Science* 177:791–793.
Francillon-Vieillot, H., V. de Buffrénil, J. Castanet, J. Géraudie, F. J. Meunier, J. Y. Sire, L. Zylberberg, and A. de Ricqlès. 1990. Microstructure and mineralisation of vertebrate skeletal tissues. In *Skeletal Biomineralisation: Patterns, Processes and Evolutionary Trends*, edited by J. G. Carter, 471–530. New York: Van Nostrand Reinhold.
Fraser, N. 2006. *Dawn of the dinosaurs: Life in the Triassic*. Bloomington: Indiana University Press.
Fröbisch, J. 2007. The cranial anatomy of *Kombuisia frerensis* Hotton (Synapsida, Dicynodontia) and a new phylogeny of anomodont therapsids. *Zoological Journal of the Linnean Society* 150:117–144.
———. 2008. Global taxonomic diversity of anomodonts (Tetrapoda, Therapsida) and the terrestrial rock record across the Permo-Triassic boundary. *PLoS One* 3:1–14.
———. 2009. Composition and similarity of global anomodont-bearing tetrapod faunas. *Earth-Science Reviews* 95:119–157.
Fröbisch, J., K. D. Angielczyk, and C. A. Sidor. 2010. The Triassic dicynodont *Kombuisia* (Synapsida, Anomodontia) from Antarctica, a refuge from the terrestrial Permian-Triassic mass extinction. *Naturwissenschaften* 97:187–196.
Fröbisch, J., and R. R. Reisz. 2008. A new species of *Emydops* (Synapsida, Anomodontia) and a discussion of dental variability and pathology in dicynodonts. *Journal of Vertebrate Paleontology* 28:770–787.
———. 2009. The Late Permian herbivore *Suminia* and the early evolution of arboreality in terrestrial vertebrate ecosystems. *Proceedings of the Royal Society B* 276:3611–3618.
Gaffney, E. S., and J. W. Kitching. 1994. The most ancient African turtle. *Nature* 369:55–58.
Galton, P. M., and J. van Heerden. 1985. Partial hindlimb of *Blikanosaurus cromptoni* n. gen. and n. sp. representing a new family of prosauropod dinosaurs from the Upper Triassic of South Africa. *Geobios* 18:509–516.
Gao, K. Q., R. C. Fox, C. F. Zhou, and D. Q. Li. 2010. A new nonmammalian eucynodont (Synapsida: Therapsida) from the Triassic of northern Gansu Province, China, and its biostratigraphic and biogeographic implications. *American Museum Novitates* 3685:1–24.
Gastaldo, R. A., J. Neveling, C. K. Clark, and S. S. Newberry. 2009. The terrestrial Permian-Triassic boundary event bed is a nonevent. *Geology* 37(3):199–202.
Gauthier, J. A., J. A. Kluge, and T. Rowe. 1988. Amniote phylogeny and the importance of fossils. *Cladistics* 4:105–209.
Gay, S. A., and A. R. I. Cruickshank. 1999. Biostratigraphy of the Permian tetrapod faunas from the Ruhuhu Valley, Tanzania. *Journal of African Earth Science* 29:195–210.
Gebauer, E. V. I. 2007. Phylogeny and evolution of the Gorgonopsia with a special reference to the skull and skeleton of GPIT/RE/7113 ("*Aelurognathus?*" *parringtoni*). Geowissenschaftlichen Fakultät, Eberhard-Karls Universität, Tübingen.
Geist, N. R. 2000. Nasal respiratory turbinate function in birds. *Physiological and Biochemical Zoology* 73:581–589.
Germain, D. and M. Laurin. 2005. Microanatomy of the radius and lifestyle in amniotes (Vertebrata, Tetrapoda). *Zoologica Scripta* 34:335–350.
Girondot, M., and M. Laurin. 2003. Bone Profiler: A tool to quantify, model and statistically compare bone section compactness profiles. *Journal of Vertebrate Paleontology* 23:458–461.
Godefroit, P., and B. Battail. 1997. Late Triassic cynodonts from Saint Nicolas de Port (northeastern France). *Geodiversitas* 19:567–631.
Golonka, J. 2007. Late Triassic and Early Jurassic palaeogeography of the world. *Palaeogeography, Palaeoclimatology, Palaeoecology* 244:297–307.
Goodwin, M. B., and J. R. Horner. 2004. Cranial histology of pachycephalosaurs (Ornithischia: Marginocephalia) reveals transitory structures

inconsistent with head-butting behavior. *Paleobiology* 30:253–267.
Gow, C. E. 2000. The skull of *Protosuchus haughtoni*, an Early Jurassic crocodyliform from southern Africa. *Journal of Vertebrate Paleontology* 20:49–56.
———. 2001. A partial skeleton of the tritheledontid *Pachygenelus* (Therapsida: Cynodontia). *Palaeontologia Africana* 37:93–97.
Gow, C. E., and J. W. Kitching. 1988. Early Jurassic crocodilomorphs from the Stormberg of South Africa. *Neues Jahrbuch für Geologie und Paläontologie, Monatshefte* 1988:517–536.
Gow, C. E., and B. S. Rubidge. 1997. The oldest procolophonoid (Amniota, Parareptilia)—New discovery from the Lower Beaufort of South Africa. *Palaeontologia Africana* 34:49–53.
Graham G. T., E. R. Pianka, and J. D. McEachran. 2001. Allometry of clutch and neonate sizes in monitor lizards (Varanidae: *Varanus*). *Copeia* 2001(2):443–458.
Green, J., V. Schneider, M. Schweitzer, and J. Clarke. 2005. New evidence for non-*Placerias* dicynodonts in the Late Triassic (Carnian-Norian) of North America. *Journal of Vertebrate Paleontology* 25(3) supplement:65A–66A.
Green, J. L., M. H. Schweitzer, and E. T. Lamm. 2010. Limb bone histology and growth in *Placerias hesternus* (Therapsida: Anomodontia) from the Upper Triassic of North America. *Palaeontology* 53:347–364.
Greenberg, N. 1980. Physiological and behavioral thermoregulation in living reptiles. In *A Cold Look at the Warmblooded Dinosaurs*, edited by R. D. K. Thomas and E. C. Olson, 141–166. Boulder, Colo.: Westview Press.
Grigg, G. C., L. A. Beard, and M. L. Augee. 2004. The evolution of endothermy and its diversity in mammals and birds. *Physiological and Biochemical Zoology* 77(6):982–997.
Grine, F. E., C. A. Forster, M. A. Cluver, and J. A. Georgi, 2006. Cranial variability, ontogeny and taxonomy of *Lystrosaurus* from the Karoo Basin of South Africa: Amniote paleobiology. In *Perspectives on the Evolution of Mammals, Birds, and Reptiles*, edited by M. Carrano, R. W. Blob, T. J. Gaudin, and J. R. Wible, 432–503. Chicago: University of Chicago Press.
Grine, F. E., B. D. Hahn, and C. E. Gow. 1978. Aspects of relative growth and variability in *Diademodon* (Reptilia: Therapsida). *South African Journal of Science* 74:50–58.
Grine, F. E., E. S. Vrba, and A. R. I. Cruickshank. 1979. Enamel prisms and diphyodonty: Linked apomorphies of Mammalia. *South African Journal of Science* 75:114–120.
Groenewald, G. H., and J. W. Kitching. 1995. Biostratigraphy of the *Lystrosaurus* Assemblage Zone. In *Biostratigraphy of the Beaufort Group (Karoo Supergroup)*, edited by B. S. Rubidge, 35–39. SACS Biostraigraphic Series No. 1. Pretoria, South Africa: Council for Geoscience.
Groenewald, G. H., J. Welman, and J. A. MacEachern. 2001. Vertebrate burrow complexes from the Early Triassic *Cynognathus* Zone (Driekoppen Formation, Beaufort Group) of the Karoo Basin, South Africa. *Palaios* 16:148–160.
Gross, W. 1934. Die Typen des mikroskopischen Knochenbaues bei foslen Stegocephalen und Reptilien. *Zeitschrift für Anatomie und Entwicklungsgeschichte*103:731–764.
Gubin, Y. M., and S. M. Sinitza. 1993. Triassic terrestrial tetrapods of Mongolia and the geological structure of the Sain-Sar-Bulak locality. In *The Nonmarine Triassic*, edited by S. G. Lucas and M. Morales, 169–170. Albuquerque: The New Mexico Museum of Natural History and Science Bulletins.
Haack, S. C. 1986. A thermal model of the sailback pelycosaur. *Paleobiology* 12:450–458.
Ham, A. W. 1953. *Histology*. 2nd ed. New York: Lippencott.
Hammer, W. R. 1990. *Thrinaxodon* from Graphite Peak, central Transantarctic Mountains. *US Antarctic Journal* 25:26–27.
———. 1995. New therapsids from the Upper Fremouw Formation (Triassic) of Antarctica. *Journal of Vertebrate Paleontology* 15:105–112.
Hancox, P. J. 1998. *A Stratigraphic, Sedimentological and Palaeonvironmental Synthesis of the Beaufort-Molteno Contact in the Karoo Basin*. PhD diss., University of the Witwatersrand, Johannesburg.
Hancox, P. J., R. J. Damiani, and B. S. Rubidge. 2000. First occurrence of *Paracyclotosaurus* (Temnospondyli, Capitosauridae) in the Karoo Basin of South Africa and its biostratigraphic significance. *South African Journal of Science* 96:135–137.
Hancox, P. J., and B. S. Rubidge. 1994. A new dicynodont therapsid from South Africa, implications for the biostratigraphy of the Upper Beaufort (*Cynognathus* Assemblage Zone). *South African Journal of Science* 90:98–99.
———.1996 The first specimen of the Mid Triassic dicynodont *Angonisaurus* from the Karoo of South Africa: Implications for the dating and biostratigraphy of the *Cynognathus* Assemblage Zone, Upper Beaufort Group. *South African Journal of Science* 92:391–392.
———. 1997. The role of fossils in interpreting the development of the Karoo Basin. *Palaeontologia Africana* 33:41–54.

———. 1998. *A Stratigraphic, Sedimentological and Palaeonvironmental Synthesis of the Beaufort-Molteno Contact in the Karoo Basin*. PhD diss., University of the Witwatersrand, Johannesburg.

———. 2000. The continental Triassic of South Africa. *Zentralblatt für Geologie und Paläontologie, Teil I* 1998:1285–1324.

Hancox, P. J., M. A. Shishkin, B. S. Rubidge, and J. W. Kitching. 1995. A threefold subdivision of the *Cynognathus* Assemblage Zone (Beaufort Group, South Africa) and its palaeogeographic implications. *South African Journal of Science* 91:143–144.

Haughton, S. H. 1924. The fauna and stratigraphy of the Stormberg Series. *Annals of the South African Museum* 12(8):323–497.

———. 1926. On Karoo vertebrates from Nyasaland. *Transactions of the Geological Society of South Africa* 29:69–83.

Hayes, J. P., and T. Garland. 1995. The evolution of endothermy: Testing the aerobic capacity model. *Evolution* 49(5):836–847.

———. 2010. Metabolic rates, genetic constraints, and the evolution of endothermy. *Journal of Evolutionary Biology* 23:1868–1877.

Heath, J. E. 1968. The origins of thermoregulation. In *Evolution and the Environment*, edited by E. T. Drake, 259–278. New Haven, Conn.: Yale University Press.

Heerden, J. van. 1979. The morphology and taxonomy of *Euskelosaurus* (Reptilia: Saurischia; Late Triassic) from South Africa. *Navorsinge van die Nasionale Museum* 4:21–84.

Heever, J. van den. 1980. On the validity of the therocephalian family Lycosuchidae (Reptilia, Therapsida). *Annals of the South African Museum* 81:111–125.

———. 1994. The cranial anatomy of the early Therocephalia (Amniota: Therapsida). *Annals of the University of Stellenbosch* 1:1–59.

Heever, J. van den, F. Abdala, and B. S. Rubidge. 2009. Southern African Bauriidae (Therocephalia, Eutherapsida): Implications for Early-Middle Triassic biostratigraphy. *Journal of Vertebrate Paleontology* 29:194A.

Hemmingsen, A. M. 1960. Energy metabolism as related to body size and respiratory surfaces, and its evolution. *Reports of the Steno Memorial Hospital* 9:7–110.

Herring, S. W., and P. Ochareon. 2005. Bone—Special problems of the craniofacial region. *Orthodontics & Craniofacial Research* 8:174–182.

Herring, S. W., and S. Teng. 2000. Strain in the braincase and its sutures during function. *American Journal of Physical Anthropology* 112:575–593.

Herring, S. W., S. Teng, X. Huang, R. J. Mucci, and J. Freeman. 1996. Patterns of bone strain in the zygomatic arch. *Anatomical Record Part A: Discoveries in Molecular, Cellular, and Evolutionary Biology* 246:446–457.

Hieronymus, T. L. 2006. Quantitative microanatomy of jaw muscle attachment in extant diapsids. *Journal of Morphology* 267:954–967.

Hieronymus, T. L., L. M. Witmer, D. H. Tanke, and P. J. Currie. 2009. The facial integument of centrosaurine ceratopsids: morphological and histological correlates of novel skin structures. *Anatomical Record* 292:1370–1396.

Hillenius, W. J. 1992. The evolution of nasal turbinates and mammalian endothermy. *Paleobiology* 18:17–29.

———. 1994. Turbinates in therapsids: Evidence for Late Permian origins of mammalian endothermy. *Evolution* 48:207–229.

———. 2000. Septomaxilla of nonmammalian synapsids: Soft-tissue correlates and a new functional interpretation. *Journal of Morphology* 245:29–50.

Hillenius, W. J., and J. A. Ruben 2004. The evolution of endothermy in terrestrial vertebrates: Who? When? Why? *Physiological and Biochemical Zoology* 77:1019–1042.

Hiller, N., and N. Stavrakis. 1984. Permo-Triassic fluvial systems in the southeastern Karoo Basin, South Africa. *Palaeogeography, Palaeoclimatology, Palaeoecology* 45:1–21.

Himms-Hagen, J. 1990. Brown adipose tissue thermogenesis: Interdisciplinary studies. *FASEB Journal* 4:2890–2898.

Hoffman, M. M., and I. Schour. 1940. Quantitative studies in the development of the rat molar I. The growth pattern of primary and secondary dentin (from birth to 500 days of age). *Anatomical Record* 78:233–251.

Hohn, A. A. 1980. Analysis of growth layers in the teeth of *Tursiops truncates* using light microscopy, microradiography, and SEM. In *Age Determination of Toothed Whales and Sirenians*, edited by W. F. Perrin and A. C. Myrick Jr., 155–160. Cambridge: Report to the International Whaling Commission, Special Report.

Hölzforster, F., H. Stollhofen, and I. G. Stanistreet. 1999. Lithostratigraphy and depositional environments in the Waterberg-Erongo area, central Namibia, and correlation with the main Karoo Basin, South Africa. *Journal of African Earth Sciences* 29:105–123.

———. 2000. Lower Permian deposits of the Huab area, NW Namibia, a continental to marine transition. *Communications of the Geological Survey of Namibia* 12:247–257.

Hook, R. W., and N. Hotton III. 1991. A new sphenacodontid pelycosaur (Synapsida) from the Wichita Group, Lower Permian of north-central Texas. *Journal of Vertebrate Paleontology* 11:37–44.

Hopson, J. A. 1973. Endothermy, small size, and the origin of mammalian reproduction. *The American Naturalist* 107:446–462.

———. 1979. Paleoneurology. In *Biology of the Reptilia,* edited by C. A. Gans, R. G. Northcutt, and P. Ulinsky, 39–146. London: Academic Press.

———. 1980. Relative brain size in dinosaurs. In *A Cold Look at the Warm-Blooded Dinosaurs,* edited by R. D. K. Thomas and E. C. Olson, 287–310. Boulder, Colo.: Westview Press.

———. 1985. Morphology and relationships of *Gomphodontosuchus brasiliensis* von Huene (Synapsida, Cynodontia, Tritylodontoidea) from the Triassic of Brazil. *Neues Jahrbuch für Geologie und Paläontologie, Monatshefte* 1985:285–299.

Hopson, J. A. and H. R. Barghusen. 1986. An analysis of therapsid relationships. In *The Ecology and Biology of Mammal-Like Reptiles,* edited by N. Notton III, P. D. McLean, J. J. Roth, and E. C. Roth, 83–106. Washington: Smithsonian Institution Press.

———. 1991. Systematics of nonmammalian Synapsida and implications for patterns of evolution in synapsids. In *Origins of the Higher Groups of Tetrapods: Controversy and Consensus,* edited by H. P. Schultze and L. Trueb, 635–694. Ithaca, N.Y.: Cornell University Press.

———. 1994. Synapsid evolution and the radiation of non-eutherian mammals. In *Major Features of Vertebrate Evolution,* edited by D. R. P. Prothero and R. M. Schoch, 190–219. Knoxville: Paleontological Society.

Hopson, J. A., and J. W. Kitching. 2001. A probainognathian cynodont from South Africa and the phylogeny of nonmammalian cynodonts. *Bulletin of the Museum of Comparative Zoology Harvard* 156:5–35.

Hopson, J. A., and H. D. Sues. 2006. A traversodont cynodont from the Middle Triassic (Ladinian) of Baden-Wurtemberg (Germany). *Palaeontologische Zeitschrift* 80:124–129.

Horner, J. R., and K. Padian. 2004. Age and growth dynamics of *Tyrannosaurus rex. Proceedings of the Royal Society B* 271:1875–1880.

Horner, J. R., A. de Ricqlès, and K. Padian. 1999. Variation in dinosaur skeletochronology indicators: Implications for age assessment and physiology. *Paleobiology* 25:295–304.

———. 2000. Long bone histology of the hadrosaurid dinosaur *Maiasaura peeblesorum:* Growth dynamics and physiology based on an ontogenetic series of skeletal elements. *Journal of Vertebrate Paleontology* 20:115–129.

Hotton, N., and J. W. Kitching. 1963. Speculation on upper Beaufort deposition. *South African Journal of Science* 59:254–258.

———. 1986. Dicynodonts and their role as primary consumers. In *The Ecology and Biology of Mammal-Like Reptiles,* edited by P. D. MacLean, J. J. Roth, and E. C. Roth, 71–82. Washington: Smithsonian Institution Press.

Hotton N., P. D. MacLean, J. J. Roth, and E. C. Roth. 1986. *The Ecology and Biology of Mammal-Like Reptiles.* Washington: Smithsonian Institution Press.

Ho Tun, E. 1979. Volcaniclastic material in the Lower Beaufort Group, Karoo rocks. *Abstracts of Geological Society of South Africa Geocongress '79,* 1:197–199.

Hu, Y., J. Meng, Y. Wang, and C. Li. 2005. Large Mesozoic mammals fed on young dinosaurs. *Nature* 433:149–152.

Hua, S., and V. de Buffrénil. 1996. Bone histology as a clue in the interpretation of functional adaptations in the Thalattosuchia (Reptilia, Crocodylia). *Journal of Vertebrate Paleontology* 16:703–717.

Huber, P., S. G. Lucas, and A. P. Hunt. 1993. Revised age and correlation of the Upper Triassic Chatham Group (Deep River Basin, Newark Supergroup), North Carolina. *Southeastern Geology* 33:171–193.

Hulbert, A. J., and P. L. Else. 1990. The cellular basis of endothermic metabolism—A role for leaky membranes. *News in Physiological Sciences* 5:25–28.

Hunt, A. P., and S. G. Lucas. 1998. Vertebrate tracks and the myth of the belly-dragging, tail-dragging tetrapods of the Late Paleozoic. *New Mexico Museum of Natural History and Science Bulletin* 12:67–69.

Hunter, J. P., and C. M. Janis. 2006. Spiny Norman in the Garden of Eden? *Ausktribosphenos* and the biogeography of eutherian origins. *Journal of Mammalian Evolution* 13:89–123.

Hurum, J. H. 1994. The snout and orbit of Mongolian multituberculates studied by serial sections. *Acta Palaeontologica Polonica* 39:181–221.

Hurum, J. H., Z. X. Luo, and Z. Kielan-Jaworowska. 2006. Were mammals originally venomous? *Acta Palaeontologica Polonica* 51:1–11.

Huttenlocker, A. K., K. Angielczyk, and A. Lee. 2006. Osteohistology of *Sphenacodon* (Synapsida: Sphenacodontidae) and the hidden diversity of growth patterns in basal synapsids. *Journal of Vertebrate Paleontology* 26(3) supplement:79–80A.

———. 2008. *Comparative Osteohistology of Hyperelongate Neural Spines in Basal Synapsids (Vertebrata: Amniota): Growth and Mechanical*

Considerations. MSc thesis. California State University, San Bernardino.
———. 2009. An investigation into the cladistic relationships and monophyly of therocephalian therapsids (Amniota: Synapsida). *Zoological Journal of the Linnean Society* 157:865–981.
Huttenlocker, A. K., D. Mazierski, and R. Reisz. 2011. Comparative osteohistology of hyperelongate neural spines in the Edaphosauridae (Amniota: Synapsida). *Palaeontology* 54:573–590.
Huttenlocker, A. K., E. Rega, and S. Sumida. 2010. Comparative anatomy and osteohistology of hyperelongate neural spines in the sphenacodontids *Sphenacodon* and *Dimetrodon* (Amniota: Synapsida). *Journal of Morphology* 271:1407–1421.
Huttenlocker, A. K., C. A. Sidor, and R. M. H. Smith. 2011. A new specimen of *Promoschorhynchus* (Therapsida: Therocephalia: Akidnognathidae) from the Early Triassic of South Africa and its implications for theriodont survivorship across the Permo-Triassic boundary. *Journal of Vertebrate Paleontology* 31:405–421.
Hutton, J. M. 1986. Age determination of living Nile crocodiles from the cortical stratification of bone. *Copeia* 2:332–341.
Ivakhnenko, M. F. 1990. The late Paleozoic faunal assemblage of tetrapods from the deposits of the Mezen River. *Paleontological Journal* 23:81–90.
———. 1996. New primitive therapsids from the Permian of eastern Europe. *Paleontological Journal* 30:337–343.
———. 1999. Biarmosuches from the Ocher Faunal Assemblage of Eastern Europe. *Paleontological Journal* 33:289–296.
———. 2000a. Estemmenosuches and primitive theriodonts from the Late Permian. *Paleontological Journal* 34:189–197.
———. 2000b. The Nikkasauridae-problematic primitive therapsids from the Late Permian of the Mezen localities. *Paleontological Journal* 34:S179–S186.
———. 2003. Eotherapsids from the east European placket (Late Permian). *Paleontological Journal* 37:S339–S465.
Jacobs, L. L., D. A. Winkler, K. D. Newman, E. M. Gomani, and A. Deino. 2005. Therapsids from the Permian Chiweta Beds and the Age of the Karoo Supergroup in Malawi. *Palaeontologia Electronica* 8(1):28A:23p. Available at: http://palaeo-electronica.org/2005_1/jacobs28/jacobs28.pdf. Accessed March 19, 2011.
Jackson, D. C., and K. Schmidt-Nielsen. 1964. Countercurrent heat exchange in the respiratory passages. *Proceedings of the National Academy of Sciences* 51:1192–1197.
Jans, M. M. E. 2005. *Histological Characterization of the Degradation of Archaeological Bone.* PhD thesis, Vrije Universiteit Amsterdam.
Jans, M. M. E., C. M. Nielsen-Marsh, C. I. Smith, M. J. Collins, and H. Kars. 2004. Characterisation of microbial attack on archaeological bone. *Journal of Archaeological Science* 31:87–95.
Jasinoski, S. C. 2008. *Comparative Cranial Biomechanics of Lystrosaurus and the Generalised Dicynodont Oudenodon.* PhD diss., University of Bristol.
Jasinoski, S. C., E. J. Rayfield, and A. Chinsamy. 2009. Comparative feeding biomechanics of *Lystrosaurus* and the generalized dicynodont *Oudenodon. Anatomical Record* 292:862–874.
———. 2010a. Functional implications of dicynodont cranial suture morphology. *Journal of Morphology* 271:705–728.
———. 2010b. Mechanics of the scarf premaxilla-nasal suture in the snout of *Lystrosaurus. Journal of Vertebrate Paleontology* 30:1283–1288.
Jenkins, F. A., Jr. 1971. The postcranial skeleton of African cynodonts. *Bulletin of the Peabody Museum of Natural History* 36:1–216.
Jenkins, F. A., Jr., and F. R. Parrington. 1976. The postcranial skeletons of the Triassic mammals *Eozostrodon, Megazostrodon* and *Erythrotherium. Philosophical Transactions of the Royal Society of London B* 273:387–431.
———. 1990. Monotremes and the biology of Mesozoic mammals. *Netherlands Journal of Zoology* 40:5–31.
Jerison, H. J. 1973. *Evolution of the Brain and Intelligence.* New York: Academic Press.
Ji, Q., Z. X. Luo, and S. Ji. 1999. A Chinese triconodont mammal and mosaic evolution of the mammalian skeleton. *Nature* 398:326–330.
Ji, Q., Z. X. Luo, C. X. Yuan, A. R. Tabrum. 2006. A swimming mammaliaform from the Middle Jurassic and ecomorphological diversification of early mammals. *Science* 311:1123–1127.
Ji, Q., Z. X. Luo, C. X. Yuan, J. R. Wible, J. P. Zhang, and J. A. Georgi. 2002. The earliest known eutherian mammal. *Nature* 416:816–822.
Jin, M., Y. Hu, Y. Li, and Y. Wang. 2006. The mammal fauna in the Early Cretaceous Jehol Biota: Implications for diversity and biology of Mesozoic mammals. *Geological Journal* 41: 439–463.
Johnson, M. R. 1976. *Stratigraphy and Sedimentology of the Cape and Karoo Systems in the Eastern Cape Province.* MSc diss., Rhodes University, Grahamstown.
———. 1991. Sandstone petrography, provenance and plate tectonic setting in Gondwana context of the south-eastern Cape Karoo Basin. *South African Journal of Geology* 94:137–154.

Jones, S. J., and A. Boyde. 1974. The organization and gross mineralization patterns of the collagen fibres in Sharpey fibre bone. *Cell and Tissue Research* 148:83–96.

Jones, T. D., and J. A. Ruben. 2001. Respiratory structure and function in theropod dinosaurs and some related taxa. In *Proceedings of the International Symposium in Honor of John H. Ostrom (February 13–14, 1999): New Perspectives on the Origins and Evolution of Birds.*, edited by J. Gauthier and L. F. Gall, 443–461. New Haven, Conn.: Yale University Press.

Jun, L., B. S. Rubidge, and L. Jinling. 2009a. New basal synapsid supports Laurasian origin for therapsids. *Acta Palaeontologica Polonica* 54(3):393–400.

Jun, L., B. S. Rubidge, and L. Jinling. 2009b. A new specimen of *Biseridens qilianicus* indicates its phylogenetic position as the most basal anomodont. *Proceedings of the Royal Society of London.* doi:10.1098/rspb.2009.0883. Published online.

Kamiya, H., T. Yoshida, N. Kusuhashi, and H. Matsuoka. 2006. Enamel texture of the tritylodontid mammal-like reptile, occurred from the Lower Cretaceous of central Japan. *Materials Science and Engineering C* 26:707–709.

Kammerer, C. F., and F. Abdala. 2009. *Procynosuchus* Broom, 1937 (Therapsida, Cynodontia): Proposed precedence over *Cyrbasiodon* Broom, 1931 and *Parathrinaxodon* Parrington, 1936. *Bulletin of Zoological Nomenclature* 66:64–69.

Kammerer, C. F., and K. D. Angielczyk. 2009. A proposed higher taxonomy of anomodont therapsids. *Zootaxa* 2018:1–24.

Kammerer, C. F., J. J. Flynn, L. Ranivoarimana, and A. R. Wyss. 2008. New material of *Menadon besairiei* (Cynodontia: Traversodontidae) from the Triassic of Madagascar. *Journal of Vertebrate Paleontology* 28:445–464.

Kammerer, C. F., J. J. Flynn, L. Ranivoharimanana, and A. R. Wyss. 2010. The first record of a probainognathian (Cynodontia: Chiniquodontidae) from the Triassic of Madagascar. *Journal of Vertebrate Paleontology* 30:1889–1894.

Kemp, T. S. 1969. Functional morphology of the gorgonopsid skull. *Philosophical Transactions of the Royal Society B* 256:1–83.

———. 1972. Whaitsiid Therocephalia and the origin of cynodonts. *Philosophical Transactions of the Royal Society B* 264:1–54.

———. 1978. Stance and gait in the hindlimb of a therocephalian mammal-like reptile. *Journal of Zoology* 186:143–161.

———. 1979. The primitive cyndont *Procynosuchus:* Functional anatomy of the skull and relationships. *Philosophical Transactions of the Royal Society B* 285:73–122.

———. 1980a. Aspects of the structure and functional anatomy of the Middle Triassic cynodont *Luangwa. Journal of Zoology* 191:193–239.

———. 1980b. The primitive cynodont *Procynosuchus:* Structure, function and evolution of the postcranial skeleton. *Philosophical Transactions of the Royal Society B* 288:217–258.

———. 1982. *Mammal-Like Reptiles and the Origin of Mammals.* London: Academic Press.

———. 1983. The relationships of mammals. *Zoological Journal of the Linnean Society* 77:353–384.

———. 1986. The skeleton of a baurioid therocephalian therapsid from the Lower Triassic (*Lystrosaurus* Zone) of South Africa. *Journal of Vertebrate Paleontology* 6:215–232.

———. 1988. Interrelationships of the Synapsida. In *The Phylogeny and Classification of the Tetrapods,* edited by M. J. Benton, 2:1–22. Oxford: Oxford University Press.

———. 2005. *The Origin and Evolution of Mammals.* Oxford: Oxford University Press.

———. 2006a. The origin and evolution of the therapsid mammal-like reptiles: A palaeobiological hypothesis. *Journal of Evolutionary Biology* 19:1231–1248.

———. 2006b. The origin of mammalian endothermy: A paradigm for the evolution of complex biological structure. *Zoological Journal of the Linnean Society* 147:473–488.

———. 2007a. One step at a time: Review of Amniote paleobiology: Perspectives on the evolution of mammals, birds, and reptiles. *Bioscience* 57:452–454.

———. 2007b. The origin of higher taxa: macroevolutionary processes, and the case of the mammals. *Acta Zoologica* 88:3–22.

———. 2009a. The endocranial cavity of a nonmammalian cynodonts *Chiniquodon theotenicus* and its implications for the origin of the mammalian brain. *Journal of Vertebrate Paleontology* 29:1188–1198.

———. 2009b. Phylogenetic interrelationships and the pattern of evolution of the therapsids: Testing for polytomy. *Palaeontologia Africana* 44:1–12.

———. 2010. New perspectives on the evolution of Late Palaeozoic and Mesozoic terrestrial tetrapods. In *Lecture Notes in Earth Science,* edited by S. Bandyopadhyay, 1–26. Berlin: Springer.

Kermack, K. A., F. Mussett, and H. W. Rigney. 1973. The lower jaw of *Morganucodon. Zoological Journal of the Linnean Society* 53:87–175.

———. 1981. The skull of *Morganucodon. Zoological Journal of the Linnean Society* 71:1–158.

Ketcham, R. A., and W. D. Carson. 2001. Acquisition, optimization and interpretation of X-ray computed tomographic imagery: Application to geosciences. *Computers and Geosciences* 21:381–400.

Keyser, A. W. 1966. Some indications of arid climate during the deposition of the Beaufort Series. *Annals of the Geological Survey of South Africa* 5:77–79.

———. 1973. A new Triassic vertebrate fauna from South West Africa. *Palaeontologia Africana* 16:1–15.

———. 1975. A re-evaluation of the cranial morphology and systematics of some tuskless Anomodontia. *Memoirs of the Geological Survey of South Africa* 67:1–110.

———. 1979. A review of the biostratigraphy of the Beaufort Group in the Karoo Basin of South Africa. *Geological Society of South Africa Geocongress '79* 2:13–31.

Keyser, A. W., and R. M. H. Smith. 1978. Vertebrate biozonation of the Beaufort Group with special reference to the Western Karoo Basin. *Annals of the Geological Survey of South Africa* 12:1–36.

Kielan-Jaworowska, Z., R. L. Cifelli, and Z.-X. Luo. 2004. *Mammals from the Age of Dinosaurs: Origins, Evolution, and Structure.* New York: Columbia University Press.

Kielan-Jaworowska, Z., and D. Dashzeveg. 1989. Eutherian mammals from the Early Cretaceous of Mongolia. *Zoologica Scripta* 18:347–355.

Kielan-Jaworowska, Z., and J. H. Hurum. 2001. Phylogeny and systematics of multituberculate mammals. *Palaeontology* 44:389–429.

———. 2006. Limb posture in early mammals: Sprawling or parasagittal. *Acta Palaeontologica Polonica* 51:393–406.

Kielan-Jaworowska, Z., J. H. Hurum, and D. Badamgarav. 2003. An extended range of multituberculate *Kryptobaatar* and distribution of mammals in the Upper Cretaceous of the Gobi Desert. *Acta Palaeontologica Polonica* 48:273–278.

Kielan-Jaworowska, Z., and B. A. Trofimov. 1980. Cranial morphology of the Cretaceous eutherian *Barunlestes. Acta Palaeontologica Polonica* 25:167–185.

———. 1986. Brain evolution in Mesozoic mammals. In *Vertebrates, Phylogeny, and Philosophy. Contributions to Geology Special Paper* 3, edited by K. M. Flanagan and J. A. Lillegraven, 21–34. Laramie: University of Wyoming.

King, G. M. 1981a. The functional anatomy of a Permian dicynodont. *Philosophical Transactions of the Royal Society, London B* 291:244–321.

———. 1981b. The postcranial skeleton of *Robertia broomiana*, an early dicynodont (Reptilia, Therapsida) from the South African Karoo. *Annals of the South African Museum,* 84:203–231.

———. 1988. Anomodontia. In *Handbuch der Paläoherpetologie (Encyclopedia of Paleoherpetology)*, edited by P. Wellnhofer, 17C:1–174. Stuttgart: Gustav Fisher.

———. 1990. *The Dicynodonts: A Study in Paleobiology.* London: Chapman and Hall.

———. 1991. The aquatic *Lystrosaurus:* a palaeontological myth. *Historical Biology* 4:285–322.

———. 1993. Species longevity and generic diversity in dicynodont mammal-like reptiles. *Palaeogeography, Palaeoclimatology, Palaeoecology* 102:321–332.

King, G. M., and M. A. Cluver. 1991. The aquatic *Lystrosaurus:* An alternative lifestyle. *Historical Biology* 4:323–341.

King, G. M., and I. Jenkins. 1997. The dicynodont *Lystrosaurus* from the Upper Permian of Zambia: Evolutionary and stratigraphical implications. *Palaeontology* 40:149–156.

King, G. M., B. W. Oelofsen, and B. S. Rubidge. 1989. The evolution of the dicynodont feeding system. *Zoological Journal of the Linnean Society* 96:185–211.

Kingsolver, J. G., and D. W. Pfennig. 2004. Individual-level selection as a cause of Cope's rule of phyletic size increase. *Evolution* 58:1608–1612.

Kissel, R. A., and R. R. Reisz. 2004. Synapsid fauna of the Upper Rock Lake Shale near Garnett, Kansas and the diversity pattern of early amniotes. In *Recent Advances in the Origin and Early Radiation of Vertebrates,* edited by G. Arratia, M. V. H. Wilson, and R. Cloutier, 409–428. München: Verlag Dr. Friedrich Pfeil.

Kitazoe, Y., H. Kishino, P. J. Waddell, N. Nakajima, T. Okabayashi, T. Watabe, and Y. Okuhara. 2007. Robust time estimation reconciles views of the antiquity of placental mammals. *PLoS ONE* 2(4):e384. doi:10.1371/journal.pone.0000384.

Kitching, J. W. 1970. A short review on the Beaufort zoning in South Africa. In *I.U.G.S. 2nd Symposium on Gondwana Stratigraphy and Palaeontology,* edited by S. H. Haughton, 309–312. Pretoria, South Africa: Council for Scientific and Industrial Research.

———. 1977. The distribution of the Karroo vertebrate fauna. *Memoir Bernard Price Institute Palaeontological Research, University Witswatersrand* 1:1–131.

———. 1995. Biostratigraphy of the *Dicynodon* Assemblage Zone. In *SACS Biostratigraphic Series No. 1,* edited by B. S. Rubidge, 29–34. Pretoria, South Africa: Council for Geoscience.

Kitching, J. W., and M. A. Raath. 1984. Fossils from the Elliot and Clarens formations (Karoo sequence) of the northeastern Cape, Orange Free State and

Lesotho, and suggested biozonation based on tetrapods. *Palaeontologia Africana* 25:111–125.

Klevezal, G. A. 1996. *Recording Structures of Mammals: Determination of Age and Reconstruction of Life History.* Translated by M. V. O. Mina. A. V. Rotterdam: A. A. Balkema.

Klevezal, G. A., and S. E. Kleinenberg. 1969. *Age Determination of Mammals from Annual Layers in Teeth and Bones.* Translated by J. Salkind. Jerusalem: Israel Program of Scientific Translations.

Kluger, M. J., D. H. Ringler, and M. R. Anver. 1975. Fever and survival. *Science* 188:166–168.

Koch, P. L., D. C. Fisher, and D. Dettman. 1989. Oxygen isotope variation in the tusks of extinct proboscideans: A measure of season of death and seasonality. *Geology* 17:515–519.

Köhler, M. and S. Moyà-Solà. 2009. Physiological and life history strategies of a fossil large mammal in a resource-limited environment. *Proceedings of the National Academy of Sciences* 106:48:20354–20358.

Kriloff, A., D. Germain, A. Canoville, P. Vincent, M. Sache, and M. Laurin. 2008. Evolution of bone microanatomy of the tetrapod tibia and its use in paleobiological inference. *Journal of Evolutionary Biology* 21:807–826.

Kühne, W. G. 1949. On a triconodont tooth of a new pattern from a fissure-filling in South Glamorgan. *Proceedings of the Zoological Society of London* 119:345–350.

———. 1956. *The Liassic Therapsid Oligokyphus.* London: British Museum of Natural History.

Kutty, T. S., S. L. Jain, and T. Roy Chowdhury. 1987. Gondwana sequence of the Northern Pranhita-Godavari valley: Its stratigraphy and vertebrate faunas. *Palaeobotanist* 36:214–229.

Langer, M. C. 2000. The first record of dinocephalians in South America: Late Permian (Rio do Rasto Formation) of the Parana Basin. *Neues Jahrbuch für Geologie und Paläontologie Abhandlungen* 215:69–95.

———. 2005. Studies of continental late Triassic tetrapod biochronology II. The Ischigualastian and a Carnian global correlation. *Journal of South American Earth Sciences* 19:219–239.

Lanyon, L. E. 1992. Control of bone architecture by functional load bearing. *Journal of Bone and Mineral Research* 7:8369–8375.

Latimer, E. M., C. E Gow, and B. S. Rubidge. 1995. Dentition and feeding niche of *Endothiodon* (Synapsida; Anomodontia). *Palaeontologia Africana* 32:75–82.

Laurin, M. 1993. The osteology and relationships of *Haptodus garnettensis*, a Pennsylvanian synapsid from Kansas. *Journal of Vertebrate Paleontology* 13:200–229.

Laurin, M., A. Canoville, and A. Quilhac. 2009. Use of paleontological and molecular data in supertrees for comparative studies: The example of lissamphibian femoral microanatomy. *Journal of Anatomy.* doi: 10.1111/j.1469-7580.2009.01104.x.

Laurin, M., and M. Girondot. 1999. Embryo retention in sarcopterygians, and the origin of the extra-embryonic membranes of the amniotic egg. *Annales des Sciences Naturelles, Zoologie* 20:99–104.

———. 2004. The evolution of body size, Cope's rule and the evolution of amniotes. *Systematic Biology* 53:594–622.

Laurin, M., M. Girondot, and M. Loth. 2004. The evolution of long bone microstructure and lifestyle in lissamphibians. *Paleobiology* 30:589–613.

Laurin, M., and R. R. Reisz. 1995. A reevaluation of early amniote phylogeny. *Zoological Journal of the Linnean Society* 113:165–223.

———. 1996. The osteology and relationships of *Tetraceratops insignis*, the oldest known therapsid. *Journal of Vertebrate Paleontology* 16:95–102.

———. 1998. New data on the cranial anatomy of *Lycaenops* (Synapsida, Gorgonopsidae), and reflections of the possible presence of streptostyly in gorgonopsians. *Journal of Vertebrate Palaeontology* 18:765–776.

Laws, R. M. 1962. Age determination of pinnipeds with special reference to growth layers in the teeth. *Zeitschrift für Sauetierkunde* 27:129–146.

Leahy, G. D. 1991. Lamellar-zonal bone in fossil mammal: Implications for dinosaur and therapsid paleophysiology. *Journal of Vertebrate Paleontology* 11:42A.

Lee, A. H. 2004. Histological organization and its relationship to function in the femur of *Alligator mississippiensis*. *Journal of Anatomy* 204:197–207.

Lee, M. S. Y., C. E. Gow, and J. W. Kitching. 1997. Anatomy and relationships of the pareiasaur *Pareiasuchus nasicornis* from the Upper Permian of Zambia. *Palaeontology* 40:307–335.

Lepper, J., M. A. Raath, and B. S. Rubidge 2000. A diverse dinocephalian fauna from Zimbabwe. *South African Journal of Science* 96:403–405.

Li, G., and Z. X. Luo. 2006. A Cretaceous symmetrodont therian with some monotreme-like postcranial features. *Nature* 439:195–200.

Li, J., and Z. Cheng 1995. A new Late Permian vertebrate fauna from Dashankou, Gansu with comments on Permian and Triassic vertebrate assemblage zones of China. In *Short Papers of Sixth Symposium on Mesozoic Terrestrial Ecosystems and Biota*, edited by A. L. Sun, and Y. Q. Wang, 33–37. Beijing: China Ocean Press.

Li, J., B. S. Rubidge, and Z. Cheng. 1996. A primitive anteosaurid dinocephalian from

China—Implications for the distribution of earliest therapsid faunas. *South African Journal of Science* 92:252–253.

Lillegraven, J. A., and G. Krusat. 1991. Cranio-mandibular anatomy of *Haldanodon exspectatus* (Docodonta; Mammalia) from the Late Jurassic of Portugal and its implications to the evolution of mammalian characters. *Contributions to Geology, University of Wyoming* 28:39–138.

Lindstedt, S. L., and P. J. Schaeffer. 2002. Use of allometry in predicting anatomical and physiological parameters of mammals. *Laboratory Animals* 36:1–19.

Liow, L. H., M. Fortelius, K. Lintulaakso, H. Mannila, and N. C. Stenseth. 2009. Lower extinction risk in sleep or hide mammals. *The American Naturalist* 173(2):264–272.

Liu, J., and P. E. Olsen. 2010. The phylogenetic relationships of Eucynodontia (Amniota: Synapsida). *Journal of Mammalian Evolution.* doi: 10.1007/s10914-010-9136-8.

Liu, J., and J. Powell. 2009. Osteology of *Andescynodon* (Cynodontia: Traversodontidae) from the Middle Triassic of Argentina. *American Museum Novitates* 3674:1–19.

Liu, J., B. S. Rubidge, and J. Li. 2010. A new specimen of *Biseridens qilianicus* indicates its phylogenetic position as the most basal anomodont. *Proceedings of the Royal Society B* 277:285–292.

———. 2009. New basal synapsid supports Laurasian origin for therapsids. *Acta Palaeontologica Polonica* 54:393–400.

Liu, J., and H. D. Sues. 2010. Dentition and tooth replacement of *Boreogomphodon* (Cynodontia: Traversodontidae) from the Upper Triassic of North Carolina, USA. *Vertebrata Palasiatica,* 48(3):169–184.

Long, R. A., and P. A. Murry. 1995. Late Triassic (Carnian and Norian) tetrapods from the southwestern United States. *New Mexico Museum of Natural History and Science Bulletin* 4:1–254.

Lucas, F. A. 1904. A new batrachian and a new reptile from the Triassic of Arizona. *Proceedings of the United States National Museum* 27:193–195.

Lucas, S. G. 1998. *Placerias* (Reptilia, Dicynodontia) from the Upper Triassic of the Newark Supergroup, North Carolina, USA, and its biochronological significance. *Neues Jahrbuch für Geologie und Paläontologie Monatshefte* 7:432–448.

Lucas, S. G., and A. B. Heckert. 2002a. Skull of the dicynodont *Placerias* from the Upper Triassic of Arizona. *New Mexico Museum of Natural History and Science Bulletin* 21:127–130.

———. 2002b. Tetrapods and the subdivision of Permian time. In *Carboniferous and Permian of the World,* edited by L. V. Hills, C. M. Henderson, and E. W. Bamber, 479–491. Memoir 19. Calgary: Canadian Society of Petroleum Geologists.

Lucas, S. G., and A. P. Hunt. 1993. A dicynodont from the Upper Triassic of New Mexico and its biochronological significance. *New Mexico Museum of Natural History and Science Bulletin* 3:321–325.

Lucas, S. G., and Z. X. Luo. 1993. *Adelobasileus* from the Upper Triassic of western Texas: The oldest mammal. *Journal of Vertebrate Paleontology* 13:309–334.

Luo, Z. X. 1994. Sister-group relationships of mammals and transformations of diagnostic mammalian characters. In *In the Shadow of the Dinosaurs—Early Mesozoic Tetrapods,* edited by N. C. Fraser and H-D. Sues, 98–128. Cambridge: Cambridge University Press.

Luo, Z. X., P. J. Chen, G. Li, and M. Chen. 2007. A new eutriconodont mammal and evolutionary development of early mammals. *Nature* 446:288–293.

Luo, Z. X., R. L. Cifelli, and Z. Kielan-Jaworowska. 2001. Dual origin of tribosphenic mammals. *Nature* 409:53–57.

Luo, Z. X., W. Crompton, and A. Sun. 2001. A new mammaliaform from the Early Jurassic and evolution of mammalian characteristics. *Science* 292:1535–1540.

Luo, Z. X., Q. Ji, J. R. Wible, and C. X. Yuan. 2003. An Early Cretaceous tribosphenic mammal and metatherian evolution. *Science* 302:1934–1939.

Luo, Z. X., Z. Kielan-Jaworowska, and R. L. Cifelli. 2004. Evolution of dental replacement in mammals. *The Carnegie Museum of Natural History Bulletin* 36:159–175.

———. 2007. Transformation and diversification in early mammal evolution. *Nature* 450:1011–1019.

Luo, Z. X., and X. C. Wu. 1997. The small tetrapods of the Lower Lufeng Formation, Yunnan, China. In *In the Shadow of the Dinosaurs—Early Mesozoic Tetrapods,* edited by N. C. Fraser and H-D. Sues, 251–270. Cambridge: Cambridge University Press.

Macleod, G. K., R. M. H. Smith, P. L. Koch, and P. D. Ward. 2000. Timing of mammal-like reptile extinctions across the Permian Triassic boundary in South Africa. *Geology* 28:227–230.

MacRae, C. 1999. *Life Etched in Stone—Fossils of South Africa.* Johannesburg: The Geological Society of South Africa.

Macrini, T. E., T. Rowe, and J. L. van de Berg. 2007. Cranial endocasts from a growth series of *Monodelphis domestica* (Didelphidae, Marsupialia): A study of individual and ontogenetic variation. *Journal of Morphology* 268:844–865.

Maddin, H. C. 2008. Cranial anatomy of *Ennatosaurus tecton* (Synapsida: Caseidae) from the Middle Permian of Russia and the evolutionary relationships of Caseidae. *Journal of Vertebrate Paleontology* 28:160–180.
Maddin, H. C., D. C. Evans, and R. R. Reisz. 2006. An Early Permian varanodontine varanopid (Synapsida: Eupelycosauria) from the Richards Spur locality, Oklahoma. *Journal of Vertebrate Paleontology* 26:957–966.
Maddin, H. C., C. A. Sidor, and R. R. Reisz. 2008. Cranial anatomy of Ennatosaurus tecton (Synapsida: Caseidae) from the Middle Permian of Russia and the evolutionary relationships of Caseidae. *Journal of Vertebrate Paleontology* 28:160–180.
Maderson, P. F. A. 2003. Mammalian skin evolution: A reevaluation. *Experimental Dermatology* 12:233–236.
Magwene, P. M. 1993. *What's Bred in the Bone: Histology and Cross-Sectional Geometry of Mammal-Like Reptile Long Bones-Evidence of Changing Physiological and Biomechanical Demands.* MSc thesis, Harvard University.
Maisch, M. W. 2001. Observations on Karoo and Gondwana vertebrates. Part 2: A new skull reconstruction of *Stahleckeria potens* von Huene, 1935 (Dicynodontia, Middle Triassic) and a reconsideration of kannemeyeriiform phylogeny. *Neues Jahrbuch für Geologie und Paläontologie Abhandlungen* 220:127–152.
———. 2002. A new basal lystrosaurid dicynodont from the Upper Permian of South Africa. *Palaeontology* 45:343–359.
Mansfield, A. W. 1958. The biology of the Atlantic walrus *Odobenus rosmarus rosmarus* (Linnaeus) in the Eastern Canadian Arctic. *Fisheries Research Board of Canada Manuscript Report Series* 653:1–146.
Marchiafava, V., E. Bonucci, and A. Ascenzi. 1974. Fungal osteoclasia: A model of dead bone resorption. *Calcified Tissue Research* 14:195–210.
Margerie, E. de. 2002. Laminar bone as an adaptation to torsional loads in flapping flight. *Journal of Anatomy* 201:521–526.
Margerie, E. de., J. Cubo, and J. Castanet. 2002. Bone typology and growth rate: Testing and quantifying "Amprino's rule" in the mallard (*Anas platyrhynchos*). *Comptes Rendus de l'Académie des sciences, Série Biologies* 325:221–230.
Margerie, E. de., J-P. Robin, D. Verrier, J. Cubo, R. Groscolas, and J. Castanet. 2004. Assessing a relationship between bone microstructure and growth rates: A fluorescent labelling study in the king penguin chick (*Aptenodytes patagonicus*). *Journal of Experimental Biology* 207:869–879.
Margerie, E. de., S. Sanchez, J. Cubo, and J. Castanet. 2005. Torsional resistance as a principal component of the structural design of long bones: comparative multivariate evidence in birds. *Anatomical Record Part A: Discoveries in Molecular, Cellular, and Evolutionary Biology* 282:49–66.
Markey, M. J., R. P. Main, and C. D. Marshall. 2006. In vivo cranial suture function and suture morphology in the extant fish *Polypterus*: Implications for inferring skull function in living and fossil fish. *Journal of Experimental Biology* 209:2085–2102.
Markey, M. J., and C. R. Marshall. 2007. Linking form and function of the fibrous joints in the skull: A new quantification scheme for cranial sutures using the extant fish *Polypterus endlicherii*. *Journal of Morphology* 268:82–102.
Marsh, H. 1980. Age determination of the *Dugong* (*Dugong dugon* (Muller)) in Northern Australia and its biological implications. In *Age Determination of Toothed Whales and Sirenians*, edited by W. F. Perrin and A. C. Myrick Jr., 181–201. Cambridge: Report to the International Whaling Commission, Special Report.
Martin, R. B., D. B. Burr, and N. A. Sharkey. 1998. *Skeletal Tissue Mechanics*. New York: Springer-Verlag.
Martin, T., and Z. X. Luo. 2005. Paleontology: Homoplasy in the mammalian ear. *Science* 307:861–862.
Martinelli, A. G., J. F. Bonaparte, C. L. Schultz, and R. Rubert. 2005. A new tritheledontid (Therapsida, Eucynodontia) from the Late Triassic of Rio Grande do Sul (Brazil) and its phylogenetic relationships amongst carnivorous non-mammalian eucynodonts. *Ameghiniana* 42:91–208.
Martinelli, A. G., and G. W. Rougier. 2007. On *Chaliminia musteloides* (Eucynodontia: Tritheledontidae) from the Late Triassic of Argentina, and a phylogeny of Ictidosauria. *Journal of Vertebrate Paleontology* 27:442–460.
Martinez, R. N., C. L. May, and C. Forster. 1996. A new carnivorous cynodont from the Ischigualasto Formation (Late Triassic, Argentina), with comments on eucynodont phylogeny. *Journal of Vertebrate Paleontology* 16:271–284.
Maxwell, W. D., and M. J. Benton. 1990. Historical tests for the absolute completeness of the fossil record of tetrapods. *Paleobiology* 16:322–335.
Mazierski, D., and R. R. Reisz. 2006. Adult skeletal remains of *Ianthasaurus*, the oldest known edaphosaur (Synapsida: Eupelycosauria) from the Upper Pennsylvanian of Kansas. *Journal of Vertebrate Paleontology* 26(3) supplement:97–98A.
———. 2010. Description of a new specimen of *Ianthasaurus hardestiorum* (Eupelycosauria: Edaphosauridae) and a re-evaluation of its phylogenetic

relationships. *Canadian Journal of Earth Sciences.* 47:901–912.
Mazin, J. M., and G. M. King. 1991. The first dicynodont from the Late Permian of Malagasy. *Palaeontology* 34:837–842.
McNab, B. K. 2003. Metabolism—Ecology shapes bird bioenergetics. *Nature* 426:620–621.
McNab, B. K., and W. A. Auffenberg. 1976. The effect of large body size on temperature regulation of the komodo dragon, *Varanus Komodoensis. Comparative Biochemistry and Physiology* 55A:345–350.
McShea, D. W. 1994. Mechanisms of large-scale evolutionary trends. *Evolution* 48:1747–1763.
Mendrez, C. H. 1975. Principales variations du palais chez les thérocéphales sud-Africains (Pristerosauria et Scaloposauria) au cours du Permien Supérieur et du Trias Inférieur. Colloque Internationale C.N.R.S. No. 218. *Problèmes Actuels de Paléontologie-Évolution des Vertèbres:*379–408.
Meng, J., Y. Hu, Y. Wang, X. Wang, and C. Li. 2006. A Mesozoic gliding mammal from northeastern China. *Nature* 444:889–893.
Meng, J., and A. R. Wyss. 1997. Multituberculate and other mammal hair recovered from *Palaeogene Excreta. Nature* 385:712–714.
Milner, A. R. 1993. The Westphalian tetrapod fauna: Some aspects of its geography and ecology. *Journal of the Geological Society of London* 144:495–506.
Modesto, S. P. 1994. The Lower Permian synapsid *Glaucosaurus* from Texas. *Palaeontology* 37:51–60.
———. 1995. The skull of the herbivorous synapsid *Edaphosaurus boanerges* from the Lower Permian of Texas. *Palaeontology* 38:213–239.
Modesto, S. P., and J. Botha-Brink. 2010. Burrow cast with *Lystrosaurus* skeletal remains from the Lower Triassic of South Africa. *Palaios* 25:274–281.
Modesto, S. P., and R. J. Damiani. 2003. Taxonomic status of *Thelegnathus browni* Broom, a procolophonid reptile from the South African Triassic. *Annals of Carnegie Museum* 72:53–64.
Modesto, S. P., and R. R. Reisz. 1990a. A new skeleton of *Ianthasaurus hardestii,* a primitive edaphosaur (Synapsida: Pelycosauria) from the Upper Pennsylvanian of Kansas. *Canadian Journal of Earth Sciences* 27:834–844.
———. 1990b. Taxonomic status of *Edaphosaurus raymondi* Case. *Journal of Paleontology* 64:1049–1051.
———. 1992. Restudy of Permo-Carboniferous synapsid *Edaphosaurus novomexicanus* Williston and Case, the oldest known herbivorous amniote. *Canadian Journal of Earth Sciences* 29:2653–2662.
Modesto, S. P., and B. S. Rubidge. 2000. A basal anomodont therapsid from the lower Beaufort Group, Upper Permian of South Africa. *Journal of Vertebrate Paleontology* 20:515–521.
Modesto, S. P., B. S. Rubidge, I. Visser, and J. Welman. 2003. A new basal dicynodont from the Upper Permian of South Africa. *Palaeontology* 46:211–223.
Modesto, S. P., B. S. Rubidge, and J. Welman. 1999. The most basal anomodont therapsid and the primacy of Gondwana in the evolution of the anomodonts. *Proceedings of the Royal Society of London* B266:331–337.
Modesto, S. P., and N. Rybczynski. 2000. The amniote faunas of the Russian Permian, implications for Late Permian terrestrial vertebrate biogeography. In *The Age of Dinosaurs in Russia and Mongolia,* edited by M. J. Benton, E. N. Kurochkin, M. A. Shishkin, and D. M. Unwin, 17–34. New York: Cambridge University Press.
Modesto, S. P., C. Sidor, B. S. Rubidge, and J. Welman. 2001. A second varanopseid skull from the Upper Permian of South Africa: implications for Late Permian "pelycosaur" evolution. *Lethaia* 34:249–259.
Montes, L., J. Castanet, and J. Cubo. 2010. Relationship between bone growth rate and bone tissue organization in amniotes: First test of Amprino's rule in a phylogenetic context. *Animal Biology* 60:25–41.
Morey, E. R. and D. J. Baylink. 1978. Inhibition of bone formation during space flight. *Science* 201:1138–1141.
Moss, M. L. 1969. Evolution of mammalian dental enamel. *American Museum Novitates* 2360:1–39.
Mueller, B., and S. Chatterjee. 2007. Dicynodonts (Synapsida: Therapsida) from the Late Triassic Dockum Group of Texas. *Journal of Vertebrate Paleontology* 24 (3) supplement: 121A.
Munyikwa, D. 2001. *Cranial Morphology of a Primitive Dinocephalian (Amniota, Therapsida) from the Madumabisa Mudstone Formation, Zimbabwe.* MSc diss., University of the Witwatersrand, Johannesburg.
Murrish, D. E., and K. Schmidt-Neilsen. 1970. Exhaled air temperature and water conservation in lizards. *Respiration Physiology* 10:151–158.
Nagy, K. A. 1987. Field metabolic rates and food requirement scaling in mammals and birds. *Ecological Monographs* 57:111–128.
Nash, D. S. 1968. A crocodile from the Upper Triassic of Lesotho. *Journal of Zoology* 156:163–179.
———. 1975. The morphology and relationships of a crocodilian, *Orthosuchus stormbergi,* from the Upper Triassic of Lesotho. *Annals of the South African Museum* 67:227–329.
Nesbitt S. 2003. *Arizonasaurus* and its implications for archosaur divergence. *Proceedings of the Royal Society London B* 270:S234–S237.

——— 2005. Osteology of the Middle Triassic pseudosuchian archosaur *Arizonasaurus babbitti*. *Historical Biology* 17:19–47.

Neveling, J. 2002. *The Biostratigraphy and Sedimentology of the Contact Area between the Lystrosaurus and Cynognathus Assemblage Zones (Beaufort Group, Karoo Supergroup)*. PhD thesis, University of the Witwatersrand, Johannesburg.

———. 2004. Stratigraphic and sedimentological investigation of the contact between the *Lystrosaurus* and *Cynognathus* Assemblage Zones (Beaufort Group: Karoo Supergroup). *Bulletin of the Council for Geoscience of South Africa* 137:1–165.

Neveling, J., B. S. Rubidge, and P. J. Hancox. 1999. A lower *Cynognathus* Assemblage Zone fossil from the Katberg Formation (Beaufort Group, South Africa). *South African Journal of Science* 95:555–556.

Nicolas, M., and B. S. Rubidge. 2009. Changes in Permo-Triassic Terrestrial tetrapod ecological representation in the Beaufort Group (Karoo Supergroup) of South Africa. *Lethaia* 43:45–59.

Nicolas, M. V. M. 2007. *Tetrapod Biodiversity through the Permo-Triassic Beaufort Group (Karoo Supergroup) of South Africa*. PhD diss., University of the Witwatersrand, Johannesburg.

North, F. W. 1878. Colonial mining engineers report on the coalfield of the Stormbergen. *Parliamentary Report of the Cape of Good Hope*, G47.

Ochev, V. G., and M. A. Shishkin, 1989. On the principles of global correlation of the continental Triassic on the tetrapods. *Acta Palaeontologica Polonica* 34:149–73.

O'Connor, M. P., and P. Dodson. 1999. Biophysical constraints on the thermal ecology of dinosaurs. *Paleobiology* 25:341–368.

Oliveira, T. V. 2010. *Postura e locomoção em cinodontes do Triássico Sul-Americano: um estudo de caso baseado em Trucidocynodon riograndensis Oliveira, Soares & Schultz, 2010 (Triássico Superior, Formação Santa Maria, Bacia do Paraná; Rio Grande do Sul, Brasil)*. PhD diss. Universidade Federal do Rio Grande do Sul.

Oliveira, T. V., M. B. Soares, and C. L. Schultz. 2010. *Trucidocynodon riograndensis* gen. nov. et sp. nov. (Eucynodontia), a new cynodont from the Brazilian Upper Triassic (Santa Maria Formation). *Zootaxa* 2382:1–71.

Olson, E. C. 1975. Permo-Carboniferous paleoecology and morphotypic series. *American Zoologist* 15:371–389.

Olson, P. E., and P. Galton. 1984. A review of the reptile and amphibian assemblages from the Upper Triassic of Lesotho. *Annals of the South African Museum* 67:227–329.

———. 1986. Relationships and ecology of the early therapsids and their predecessors. In *The Ecology and Biology of Mammal-Like Reptiles*, edited by N. I. Hotton, P. D. McLean, J. J. Roth, and E. C. Roth, 47–60. Washington: Smithsonian Institution Press.

Orlov, Y. A. 1958. The carnivorous dinocephalians of the Isheevo fauna (titanosuchians). *Trudi Paleontological Institute of the Academy of Sciences, U.S.S.R.* 72:3–113.

Owen, R. 1845. Description of certain fossil crania discovered by A. G. Bain Esq., in the sandstone rocks in the southwestern extremity of Africa, referable to different species of an extinct genus of Reptilia (*Dicynodon*), and indicative of a new tribe or suborder of Sauria. *Transactions of the Geological Society of London* 2:59–84.

———. 1859. On the orders of fossil and recent Reptilia and their distribution in time. *Report of British Association for the Advancement of Science* 1859:153–166.

Owerkowicz, T., and A. W. Crompton. 2001. Allometric scaling of respiratory turbinates and trachea in mammals and birds. *Journal of Vertebrate Paleontology* 21 (supplement):86A.

Packard, M. J., and R. S. Seymour. 1997. Evolution of the amniote egg. In *Amniote Origins: Completing the Transition to Land*, edited by S. S. Sumida and K. L. M. Martin, 265–291. London: Academic Press.

Padian, K., and J. R. Horner. 2004. Dinosaur physiology. In *The Dinosauria*, edited by D. B. Weishampel, P. Dodson, and H. Osmólska, 660–671. Berkeley: University of California Press.

Padian, K., J. R. Horner, and A. de Ricqlès. 2004. Growth in small dinosaurs and pterosaurs: The evolution of Archosaurian growth strategies. *Journal of Vertebrate Paleontology* 24(3):555–571.

Paladino, F. V., M. P. O'Connor, and J. R. Spotila. 1990. Metabolism of leatherback turtles, giganto thermy, and thermoregulation of dinosaurs. *Nature* 344:858–860.

Paladino, F. V., J. R. Spotila, and P. Dodson. 1997. A blueprint for giants: Modeling the physiology of large dinosaurs. In *The Complete Dinosaur*, edited by J. O. Farlow and M. K. Brett-Surnam, 491–504. Bloomington: Indiana University Press.

Parrington, F. R. 1946. On the cranial anatomy of cynodonts. *Proceedings of the Zoological Society of London* 116:181–197.

Parrish, J. T. 1993. Climate of the supercontinent Pangea. *The Journal of Geology* 101:215–233.

Peabody, F. E. 1961. Annual growth zones in living and fossil vertebrates. *Journal of Morphology* 108:11–62.

Pearson, H. S. 1924a. A dicynodont reptile reconstructed. *Proceedings of Zoological Society of London* 1924:827–855.

———. 1924b. The skull of the dicynodont reptile *Kannemeyeria. Proceedings of Zoological Society of London* 1924:793–826.

Pearson, O. P. 1954. Habits of the lizard *Liolaemus Multiformis Multiformis* at high latitudes in Southern Peru. *Copeia* 1954:111–116.

Piantoni, C., N. R. Ibargüengoytía, and V. E. Cussac. 2006a. Age and growth of the Patagonian lizard *Phymaturus patagonicus. Amphibia-Reptilia* 27:385–392.

———. 2006b. Growth and age of the southernmost distributed gecko of the world (*Homonota darwini*) studied by skeletochronology. *Amphibia-Reptilia* 393–400.

Piñeiro, G., M. Verde, M. Ubilla, and J. Ferigolo. 2003. First basal synapsids ("pelycosaurs") from the Upper Permian-?Lower Triassic of Uruguay, South America. *Journal of Paleontology* 77:389–392.

Pivorunas, A. 1970. *Allometry in the Limbs and Sail of Dimetrodon.* MSc thesis, University of Illinois, Chicago.

Plochocki, J. H., J. P. Rivera, C. Zhang, and S. A. Ebba. 2008. Bone modeling response to voluntary exercise in the hindlimb of mice. *Journal of Morphology* 269:313–318.

Poole, D. F. G. 1956. The structure of the teeth of some mammal-like reptiles. *Quarterly Journal of Microscopical Science* 97(2):303–312.

Pörtner, H. O. 2004. Climate variability and the energetic pathways of evolution: The origin of endothermy in mammals and birds. *Physiological and Biochemical Zoology* 77:959–981.

Pough, F. H., J. B. Heiser, and W. N. McFarland. 1996. *Vertebrate Life.* New Jersey: Prentice-Hall.

Prasad, G. V. R. 2009. Divergence time estimates of mammals from molecular clocks and fossils: Relevance of new fossil finds from India. *Journal of Biosciences* 34:649–659.

Quiroga, J. C. 1979. The brain of two mammal-like reptiles (Cynodontia-Therapsida). *Journal für Hirnforschung* 20:341–350.

———. 1980. The brain of the mammal-like reptile *Probainognathus jenseni* Therapsida, Cynodontia). A correlative paleoneurological approach to the neocortex at the reptile-mammal transition. *Journal für Hirnforschung* 21:99–336.

———. 1984. The endocranial cast of the advanced mammallike reptile *Therioherpeton cargnini* (Therapsida-Cynodontia) from the Middle Triassic of Brazil. *Journal für Hirnforschung* 25:285–290.

Raath, M. A. 1969. A new coelurosaurian dinosaur from the Forest Sandstone of Rhodesia. *Arnoldia* 4:1–25.

———. 1977. *The Anatomy of the Triassic Theropod Syntarsus rhodesiensis (Saurischia: Podokosauridae) and a Consideration of Its Biology.* PhD diss., Rhodes University, Grahamstown.

Rafferty, K. L., and S. W. Herring. 1999. Craniofacial sutures: Morphology, growth, and in vivo masticatory strains. *Journal of Morphology* 242:167–179.

Randall, D., W. W. Burggren, and K. French. 2002. *Eckert's Animal Physiology: Mechanisms and Adaptations,* 5th ed. New York: W. H. Freeman.

Ray, S. 1999. Permian reptilian fauna from the Kundaram Formation, Pranhita-Godavari Valley, India. *Journal of African Earth Sciences* 29:211–218.

———. 2000. Endothiodont dicynodonts from the Late Permian Kundaram Formation, India. *Palaeontology* 43:375–404.

———. 2001. Small Permian dicynodonts from India. *Paleontological Research* 5:177–191.

———. 2005. *Lystrosaurus* (Therapsida, Dicynodontia) from India: Taxonomy, relative growth and cranial dimorphism. *Journal of Systematic Palaeontology* 3:203–221.

———. 2006. Functional and evolutionary aspects of the postcranial anatomy of dicynodonts (Synapsida, Therapsida). *Palaeontology* 49:1263–1286.

Ray, S., and S. Bandyopadhyay. 2003. Late Permian vertebrate community of the Pranhita-Godavari valley, India. *Journal of Asian Earth Sciences* 21:643–654.

Ray, S., S. Bandyopadhyay, and R. Appana. 2010. Bone histology of a kannemeyeriid dicynodont *Wadiasaurus:* Paleobiological implications. In *New Aspects of Mesozoic Biodiversity,* edited by S. Bandyopadhyay, 73–89. Lecture Note Series in Earth Sciences 132. Berlin Hiedelberg: Springer-Verlag.

Ray, S., S. Bandyopadhyay, and D. Bhawal. 2009. Growth patterns as deduced from bone microstructure of some selected neotherapsids with special emphasis on dicynodonts: Phylogenetic implications. *Palaeoworld* 18:53–66.

Ray, S., J. Botha, and A. Chinsamy. 2004 Bone histology and growth patterns of some nonmammalian therapsids. *Journal of Vertebrate Palaeontology* 24:634–648.

Ray, S., and A. Chinsamy. 2003. Functional aspects of the postcranial anatomy of the Permian dicynodont *Diictodon* and their ecological implications. *Palaeontology* 46:151–183.

———. 2004. *Diictodon feliceps* (Therapsida, Dicynodontia): Bone histology, growth and biomechanics. *Journal of Vertebrate Palaeontology* 24:180–194.

Ray, S., A. Chinsamy, and S. Bandyopadhyay. 2005. *Lystrosaurus murrayi* (Therapsida, Dicyndontia): Bone histology, growth and lifestyle adaptations. *Palaeontology* 48:1169–1185.

Ray, S., D. Mukherjee, and S. Bandyopadhyay. 2009. Growth patterns of fossil vertebrates as deduced from bone microstructure: Case studies from India. *Journal of Biosciences* 34 (5):661–672.

Rayfield, E. J. 2007. Finite element analysis and understanding the biomechanics and evolution of living and fossil organisms. *Annual Review of Earth and Planetary Sciences* 35:541–576.

Rega, E., S. S. Sumida, K. Noriega, C. Pell, and A. Lee. 2005. Evidence-based paleopathology I: Ontogenetic and functional implications of dorsal sails in *Dimetrodon. Journal of Vertebrate Paleontology* 25(3):103A.

Rega, E., N. Wideman, and C. Brochu. 2002. Paleopathology of amniote specimens from the Late Paleozoic and Mesozoic of North America: Comparison of gross morphological and histological analyses. *Journal of Vertebrate Paleontology* 22(3) supplement:98A.

Reichel, M., C. L. Schultz, and M. B. Soares. 2009. A new traversodontid cynodont (Therapsida, Eucynodontia) from the Middle Triassic Formation of Rio Grande do Sul, Brazil. *Palaeontology* 52:229–250.

Reid, R. E. H. 1987. Bone and dinosaurian "endothermy." *Modern Geology* 11:133–154.

———. 1990. Zonal "growth rings" in dinosaurs. *Modern Geology* 15:19–48.

———. 1993. Apparent zonation and slowed growth in a small Cretaceous theropod. *Modern Geology* 18:391–406.

———. 1996. Bone histology of the Cleveland-Lloyd Dinosaurs and of dinosaurs in general, Part I: Introduction: Introduction to bone tissues. *Brigham Young University Geology Studies* 41:25–71.

———. 1997. Dinosaurian physiology: The case for "intermediate" physiology. In *The Complete Dinosaur,* edited by J. O. Farlow and M. K. Brett-Surnam, 449–473. Bloomington: Indiana University Press.

Reilly, D. T., and A. H. Burstein. 1975. The elastic and ultimate properties of compact bone tissue. *Journal of Biomechanics* 8:393–405.

Reinemund, J. A. 1955. Geology of the Deep River Coal Field North Carolina. *Geological Survey Professional Paper* 246:1–159.

Reisz, R. R. 1972. Pelycosaurian reptiles from the middle Pennsylvanian of North America. *Bulletin of the Museum of Comparative Zoology, Harvard* 144:27–62.

———. 1975. Pennsylvanian pelycosaurs from Linton, Ohio and Nýřany, Czechoslovakia. *Journal of Paleontology* 49:522–527.

———. 1986. Pelycosauria. In *Encyclopedia of Paleoherpetology*, 17A, edited by P. Wellnhofer, 1–102. Stuttgart: Gustav Fischer Verlag.

Reisz, R. R., and D. S Berman. 1986. *Ianthasaurus hardestii n. sp.*, a primitive edaphosaur (Reptila, Pelycosauria) from the Upper Pennsylvanian Rock Lake Shale near Garnett, Kansas. *Canadian Journal of Earth Sciences* 23:77–91.

Reisz, R. R., D. Berman, and D. Scott. 1992. The cranial anatomy and relationships of *Secodontosaurus,* an unusual mammal-like reptile (Synapsida: Sphenacodontidae) from the early Permian of Texas. *Zoological Journal of the Linnean Society* 104:127–184.

Reisz, R. R., D. W. Dilkes, and D. S. Berman. 1998. Anatomy and relationships of *Elliotsmithia longiceps* Broom, a small synapsid (Eupelycosauria: Varanopseidae) from the Late Permian of South Africa. *Journal of Vertebrate Paleontology* 18:602–611.

Reisz, R. R., S. J. Godfrey, and D. Scott. 2009. *Eothyris* and *Oedaleops:* Do these Early Permian synapsids form a clade? *Journal of Vertebrate Paleontology* 29:39–47.

Reisz, R. R., and P. J. Hancox. 2001. Cranial description and taxonomic re-evaluation of *Kannemeyeria argentinensis* (Therapsida: Dicynodontia). *Palaeontologia Africana* 37:81–91.

Reisz, R. R., and S. P. Modesto. 2007. *Heleosaurus scholtzi* from the Permian of South Africa: A varanopid synapsid, not a diapsid reptile. *Journal of Vertebrate Paleontology* 27:234–239.

Reisz, R. R., D. Scott, H-D. Sues, D. C. Evans, and M. A. Raath. 2005. Embryos of an Early Jurassic prosauropod dinosaur and their evolutionary significance. *Science* 309:761–764.

Reisz, R. R., and H-D. Sues. 2000. Herbivory in late Paleozoic and Triassic terrestrial vertebrates. In *Evolution of Herbivory in Terrestrial Vertebrates Perspectives from the Fossil Record,* edited by H-D. Sues, 9–42. Cambridge: Cambridge University Press.

Reisz, R. R., H. Wilson, and D. Scott. 1997. Varanopseid synapsid skeletal elements from Richards Spur, a Lower Permian fissure fill located near Fort Sill, Oklahoma. *Oklahoma Geology Notes* 57:160–170.

Renaut, A. J. 2001. *A Re-Evaluation of the Cranial Morphology and Taxonomy of the Triassic Dicynodont Genus Kannemeyeria.* PhD thesis, University of the Witwatersrand, Johannesburg.

Renaut, A. J., and P. J. Hancox. 2001. Cranial description and taxonomic re-evaluation of *Kannemeyeria argentinensis* (Therapsida: Dicynodontia). *Palaeontologia Africana* 37:81–91.

Retallack, G. J., and E. S. Krull. 1999. Landscape ecological shift at the Permian-Triassic boundary

in Antarctica. *Australian Journal of Earth Sciences* 46:785–812.
Retallack, G. J., R. M. H. Smith, and P. D. Ward. 2003. Vertebrate extinction across Permian-Triassic boundary in Karoo Basin, South Africa. *Geological Society of America Bulletin* 115:1133–1152.
Rich, T. H., T. F. Flannery, P. Trusler, and P. Vickers-Rich. 2001. Corroboration of the Garden of Eden hypothesis. In *Faunal and Floral Migrations and Evolution in SE Asia-Australia*, edited by I. Metcalfe, J. M. B. Smith, M. Morwood, and I. Davidson, 323–352. Rotterdam: A. A. Balkema Publishers.
Rich, T. H., P. Vickers-Rich, A. Constantine, T. F. Flannery, L. Kool, and N. van Klaveren. 1997. A tribosphenic mammal from the Mesozoic of Australia. *Science* 278:1438–1442.
Ricqlès, A. de. 1969. Recherches paléohistologiques sur les os longs des Tétrapodes. II, Quelques observations sur la structure des longs des Thériodontes. *Annales de Paléontologie (Vertébrés)* 55:1–52.
———. 1972. Recherches paléohistologiques sur les os longs des Tétrapodes. III, Titanosuchiens, Dinocéphales et Dicynodontes. *Annales de Paléontologie (Vertébrés)* 58:17–60.
———. 1974a. Evolution of endothermy: Histological evidence. *Evolutionary Theory* 1:51–80.
———. 1974b. [Paleohistological research on the long bones of tetrapods IV: Eotheriodonts and pelycosaurs]. *Annales de Paléontologie (Vertébrés)* 60:3–39, 8 plates (in French).
———. 1975. Recherches paléohistologiques sur les os longs des Tétrapodes. VII, Sur la classification, la signification fonctionelle et l'histoire des tissues osseux des Tétrapodes (première partie: structures). *Annales de Paléontologie* (Vertébrés) 61:49–129.
———.1976. On bone histology of fossil and living reptiles with comments on its functional and evolutionary significance. In *Morphology and Biology of Reptiles*, edited by A. d'A. Bellairs and C. B. Cox, Linnean Society Symposium Series 3, 123–150. London: Academic Press.
Ricqlès, A. de., J. Castanet, H. Francillon-Vieillot. 2004. The "message" of bone tissue in paleoherpetology. *Italian Journal of Zoology* 1:3–12.
Ricqlès, A. de., F. J. Meunier, J. Castanet, and H. Francillon-Vieillot. 1991. Comparative microstructure of bone. In *Bone. Volume 3: Bone Matrix and Bone Specific Products*, edited by B. K. Hall, 1–77. Boca Raton: CRC Press.
Ricqlès, A. de., K. Padian, F. Knoll, and J. R. Horner. 2008. On the origin of high growth rates in archosaurs and their ancient relatives: Complementary histological studies on Triassic archosauriforms and the problem of a "phylogenetic signal" in bone histology. *Annales de Paléontologie* 94:57–76.
Robinson, P. L. 1970. The Indian Gondwana formations—A review. In *First IUGS International Symposium on Gondwana stratigraphy, I.U.G S. Buenos Aires, South America*, edited by A. J. Amos, 201–268. Paris:United Nations Educational, Scientific, and Cultural Organization.
Rogers, A. W. 1903. Early lithostratigraphic subdivision of the Karoo. Listed in Rogers (1905).
———. 1905. *An Introduction to the Geology of the Cape Colony*. London: Longmans Green.
Romer, A. S. 1927. Notes on the Permo-Carboniferous reptile *Dimetrodon*. *The Journal of Geology* 35:673–689.
———. 1948. Relative growth in pelycosaurian reptiles. *Special Publication of the Royal Society of South Africa, Robert Broom Commemorative Volume*:45–55.
———. 1957. Origin of the amniote egg. *Scientific Monthly* 85:57–63.
———. 1958. Tetrapod limbs and early tetrapod life. *Evolution* 12:365–369.
———. 1961. Synapsid evolution and dentition. *International Colloquium on the Evolution of Mammals* 1:9–56.
———. 1969. The Chañares (Argentina) Triassic reptile fauna V. A new chiniquodontid cynodont, *Probelesodon lewisi*—Cynodont ancestry. *Breviora* 333:1–24.
———. 1970. The Chañares (Argentina) Triassic reptile fauna VI. A chiniquodontid cynodont with an incipient squamosal-dentary jaw articulation. *Breviora* 344:1–18.
Romer, A. S., and L. W. Price. 1940. Review of the Pelycosauria. *Geological Society of America Special Papers* 28:1–538.
Roopnarine, P. D, K. D. Angielczyk, S. C. Wang, and R. Hertog. 2007. Food web models explain instability of Early Triassic terrestrial communities. *Proceedings of the Royal Society B.* 274:2077–2086.
Rose K., and J. D. Archibald. 2005. *The Rise of Placental Mammals: Origins and Relationships of the Major Extant Clades*. Baltimore: The Johns Hopkins University Press.
———. 2006. *The Beginning of the Age of Mammals*. Baltimore: The Johns Hopkins University Press.
Roussouw, P. J., and J. De Villiers. 1952. *The geology of the Merweville area, Cape Province*. Geological Survey of South Africa Explanation to sheet 189.
Roussouw, P. J., E. I. Meyer, M. P. Mulder, and C. G. Stocken. 1964. Die geologie van die Swartberge, die Kangovallei en die omgewing van Prins Albert, K.P. *Geological Survey of South Africa*.

Rowe, T. 1989. A new species of the theropod dinosaur *Syntarsus* from the Early Jurassic Kayenta Formation of Arizona. *Journal of Vertebrate Paleontology* 9:125–136.

———. 1996 Coevolution of the mammalian middle ear and neocortex. *Science* 273:651–654.

Rowe, T., T. H. Rich, P. Vickers-Rich, M. Springer, and M. O. Woodburne. 2008. The oldest platypus and its bearing on divergence timing of the platypus and echidna clades. *Proceedings of the National Academy of Science* 105:1238–1242.

Roy Chowdhury, T. 1970. Two new dicynodonts from the Triassic Yerrapalli Formation of Central India. *Palaeontology* 13:132–144.

Ruben, J. A. 1991. Reptilian physiology and the flight capacity of *Archaeopteryx*. *Evolution* 45:1–17.

———. 1995. The evolution of endothermy in mammals and birds: From physiology to fossils. *Annual Review of Physiology* 57:69–95.

Ruben, J. A., W. J. Hillenius, N. R. Geist, A. Leitch, T. D. Jones, P. J. Currie, J. R. Horner, and G. Espe III. 1996. The metabolic status of some Late Cretaceous dinosaurs. *Science* 273:1204–1207.

Ruben, J. A., and T. D. Jones. 2000. Selective factors associated with the origin of fur and feathers. *American Zoologist* 40:585–596.

Rubidge, B. S. 1988. *A Palaeontological and Palaeoenvironmental Synthesis of the Permian Ecca-Beaufort Contact between Prince Albert and Rietbron, Cape Province, South Africa*. PhD diss., University of Port Elizabeth, South Africa.

———. 1990a. The cranial morphology of a new species of the genus *Eodicynodon* (Therapsida; Dicynodontia). *Navorsinge van die Nasionale Museum Bloemfontein* 7:1–25.

———. 1990b. A new vertebrate biozone at the base of the Beaufort Group Karoo sequence (South Africa). *Palaeontologia Africana* 27:17–20.

———. 1990c. Redescription of the cranial morphology of *Eodicynodon oosthuizeni* (Therapsida, Dicynodontia). *Navorsinge van die Nasionale Museum Bloemfontein* 7:29–42.

———. 1991. A new primitive dinocephalian mammal-like reptile from the Permian of southern Africa. *Palaeontology* 34:547–559.

———. 1994. *Australosyodon*, the first primitive anteosaurid dinocephalian from the Upper Permian of Gondwana. *Palaeontology* 37:579–594.

———. 1995a. Biostratigraphy of the Beaufort Group (Karoo Supergroup). *South African Committee for Stratigraphy Biostratigraphic Series* 1:1–46. Pretoria: South African Committee for Stratigraphy and Biostratigraphy.

———. 1995b. Biostratigraphy of the *Eodicynodon* Assemblage Zone. In *Biostratigraphy of the Beaufort Group (Karoo Supergroup)*, edited by B. S. Rubidge. *SACS Biostratigraphic Series* 1:3–7. Pretoria: South African Committee for Stratigraphy and Biostratigraphy.

———. 1995c. Did mammals originate in Africa? South African fossils and the Russian connection. *Sidney Haughton Memorial Lecture* 4. Cape Town: Royal Society of South Africa.

Rubidge, B. S., K. Angielczyk, W. De Klerk, and F. Abdala. 2009. Gondwanan Middle Permian Synapsid Radiations. *Journal of Vertebrate Paleontology* 29:173A.

Rubidge, B. S., D. H. Erwin, J. Ramezani, S. A. Bowring, and W. J. De Klerk. 2010. The first radiometric dates for the Beaufort Group, Karoo Supergroup of South Africa. *Proceedings of the 16th Conference of the Palaeontological Society of Southern Africa, Howick* 82–83.

Rubidge, B. S., and J. A. van den Heever. 1997. Morphology and systematic position of the dinocephalian *Styracocephalus platyrhynchus*. *Lethaia* 30:157–168.

Rubidge, B. S., and J. A. Hopson. 1990. A new anomodont therapsid from South Africa, and its bearing on the ancestry of Dicynodontia. *South African Journal of Science* 86:43–45.

———. 1996. A primitive anomodont Therapsid from the base of the Beaufort Group (Upper Permian) of South Africa. *Zoological Journal of the Linnaean Society* 117:117–139.

Rubidge, B. S., and J. W. Kitching. 2003. A new burnetiamorph (Therapsida, Biarmosuchia) from the Lower Beaufort of South Africa, and a phylogenetic analysis of the Burnetiidae. *Palaeontology* 46:199–210.

———. 2005. Re-uniting lost continents—Fossil reptiles from the ancient Karoo and their wanderlust. *South African Journal of Geology* 108:135–172.

Rubidge, B. S., J. W. Kitching, and J. A. van den Heever. 1983. First record of a therocephalian (Therapsida: Pristerognathidae) from the Ecca of South Africa. *Navorsing Nasionale Museum, Bloemfontein* 4:229–235.

Rubidge, B. S., and C. A. Sidor. 2001. Evolutionary patterns among Permo-Triassic therapsids. *Annual Review of Ecology and Systematics* 32:49–480.

———. 2002. On the cranial morphology of the basal therapsids *Burnetia* and *Proburnetia* (Therapsida: Burnetiidae). *Journal of Vertebrate Paleontology* 22:257–267.

Rubidge, B. S., C. A. Sidor, and S. P., Modesto. 2006. A new burnetiamorph (Therapsida: Biarmosuchia) from the Middle Permian of South Africa. *Journal of Paleontology* 80:740–749.

Sander, P .M. 1997. Nonmammalian synapsid enamel and the origin of mammalian enamel prisms: the bottoms-up perspective. In *Tooth Enamel Microstructure,* edited by W. V. Koenigswald and P. M. Sander, 41–62. Rotterdam: A. A. Balkema.

———. 1999. The microstructure of reptilian tooth enamel: Terminology, function, and phylogeny. *Münchner Geowissenschaftliche Abhandlungen Reihe A: Geologie und Paläontologie* 38:1–102.

———. 2000. Long bone histology of the Tendaguru sauropods: Implications for growth and biology. *Paleobiology* 26(3):466–488.

Sander, P .M., and P. Andrássy. 2006. Lines of arrested growth and long bone histology in Pleistocene large mammals from Germany: What do they tell us about dinosaur physiology? *Palaeontographica Abt. A* 277:143–159.

Satsangi, P. P. 1987. The vertebrate faunas from the Permian and the Lower Triassic Gondwana sequences of India. *Geological Survey of India Special Publication* 11:165–178.

Scheyer, T. M., N. Klein, and P. M. Sander. 2010. Developmental palaeontology of Reptilia as revealed by histological studies. *Seminars in Cell and Developmental Biology* 21:462–470.

Scheyer, T. M., and P. M. Sander. 2009. Bone microstructure and mode of skeletogenesis in osteoderms of three pareiasaur taxa from the Permian of South Africa. *Journal of Evolutionary Biology* 22(6):1153–1162.

Schmidt-Nielsen, K. 1984. *Scaling.* Cambridge: Cambridge University Press.

———. 1990. *Animal Physiology: Adaptation and Environment,* 4th ed. Cambridge: Cambridge University Press.

Schoch, R. R., and A. R. Milner. 2000. Stereospondyli. Stem-stereospondyli, Rhinesuchidae, Rhytidostea, Trematosauroidea, Capitosauroidea. In *Encyclopedia of Paleoherpetology* edited by P. Wellnhofer, 3B. Stuttgart: Fischer.

Schour, I., and M. M. Hoffman. 1939a. Studies in tooth development I. The 16 microns calcification rhythm in the enamel and dentin from fish to man. *Journal of Dental Research* 18:91–102.

———. 1939b. Studies in tooth development II. The rate of apposition of enamel and dentin in man and other mammals. *Journal of Dental Research* 18:161–175.

Schultz, C. L. 2000. Biostratigraphy of southern Brazilian Middle-Upper Triassic. *Revista Brasileira de Geociências* 30:495–498.

Schultze, H. P., and L. Trueb. 1991. *Origins of the Higher Groups of Tetrapods: Controversy and Consensus.* Ithaca, N.Y.: Cornell University Press.

Seeley, F. R. S. 1898. On the skull of *Mochlorhinus platyceps* from Bethulie, Orange Free State, preserved in the Albany Museum, Grahamstown. *Annals and Magazine of Natural History* 1:164–176.

Seeley, H. G. 1892. Researches on the structure, organisation and classification of the fossil Reptilia. *Philosophical Transactions of the Royal Society* 179:59–109.

Seki, Y., M. S. Schneider, and M. A. Meyers. 2005. Structure and mechanical behavior of a toucan beak. *Acta Materialia* 53:5281–5296.

Sengupta, D. P. 1995. Chigutisaurid temnospondyls from the late Triassic of India and a review of the Chigutisauridae. *Palaeontology* 38:313–339.

Sereno, P. C. 2006. Shoulder girdle and forelimb in a Cretaceous multituberculate: Form, functional evolution, and a proposal for basal mammalian taxonomy. In *Amniote Paleobiology: Perspectives on the Evolution of Mammals, Birds, and Reptiles,* edited by M. T. Carrano, T. J. Gaudin, R. W. Blob, and J. R. Wible, 315–370. Chicago: University of Chicago Press.

Setoguchi, T., H. Matsuoka, and M. Matsuda. 1999. New discovery of an Early Cretaceous tritylodontid (Reptilia, Therapsida) from Japan and the phylogenetic reconstruction of Tritylodontidae based on the dental characters. In *Proceedings of the Seventh Annual Meeting of the Chinese Society of Vertebrate Paleontology,* edited by Y.–Q. Wang and T. Deng, 117–124. Beijing: China Ocean Press.

Sharov, A. G. 1971. [New flying reptiles from the Mesozoic of Kazakhstan and Kirgizia]. *Trudy Academia Nauk SSSR, Palaeontoloitscheki Institut* 130:104–113 (in Russian).

Shishkin, M. A., and V. G. Ochev. 1993. Permo-Triassic transition and the early Triassic history of the Euramerican tetrapod fauna. *New Mexico Museum of Natural History and Science Bulletin* 3:435–437.

Shishkin, M. A., V. G. Ochev, V. R. Lovzovsky, and I. V. Novikov. 2000. Tetrapod biostratigraphy of the Triassic of Eastern Europe. In *The Age of Dinosaurs in Russia and Mongolia,* edited by M. J. Benton, M. A. Shishkin, D. M. Unwin, and E. N. Kurochkin, 120–139. Cambridge: Cambridge University Press.

Shishkin, M. A., and B. S. Rubidge. 2000. A relict rhinesuchid (Amphibia, Temnospondyli) from the Lower Triassic of South Africa. *Palaeontology* 43:653–670.

Shishkin, M. A., B. S. Rubidge, and P. J. Hancox. 1995. Vertebrate biozonation of the Upper Beaufort Series of South Africa—A new look on correlation of the Triassic biotic events in Euramerica and southern Gondwana. *Sixth Symposium on Mesozoic Terrestrial Ecosystems*, 39–41. Beijing: China Press.

Shishkin, M. A., B. S. Rubidge, and J. W. Kitching. 1996. A new Lydekkerinid (Amphibia, Temnospondyli) from the Lower Triassic of South Africa—Implications for evolution of early capitosauroid cranial pattern. *Philosophical Transactions of the Royal Society of London B* 351:1635–1659.

Shishkin, M. A., and J. Welman. 1994. A new find of *Trematosuchus* (Amphia, Temnospondyli) from the *Cynognathus* Assemblage Zone of South Africa. *Palaeontologia Africana* 31:39–49.

Shubin, N. H., P. E. Olsen, and H-D. Sues. 1997. Early Jurassic small tetrapods from the Mc Coy Brook Formation of Nova Scotia, Canada. In *In the Shadow of the Dinosaurs—Early Mesozoic Tetrapods*, edited by N. C. Fraser and H-D. Sues, 242–250. Cambridge: Cambridge University Press.

Sidor, C. A. 2003. Evolutionary trends and the origin of the mammalian lower jaw. *Paleobiology* 29:605–640.

Sidor, C. A., and P. J. Hancox. 2006. *Elliotherium kersteni*, a new tritheledontid from the Lower Elliot Formation (Upper Triassic) of South Africa. *Journal of Paleontology* 80:333–342.

Sidor, C. A., and J. A. Hopson. 1998. Ghost lineages and "mammalness": Assessing the temporal pattern of character acquisition in the Synapsida. *Paleobiology* 24:254–273.

Sidor, C. A., J. A. Hopson, and A. W. Keyser. 2004. A new burnetiamorph therapsid from the Teekloof Formation, Permian of South Africa. *Journal of Vertebrate Paleontology* 24:938–950.

Sidor, C. A., and B. S. Rubidge. 2006. *Herpetoskylax hopsoni*, a new biarmosuchian (Therapsida: Biarmosuchia) from the Beaufort Group of South Africa. In *Amniote Paleobiology: Perspectives on the Evolution of Mammals, Birds, and Reptiles*, edited by M. T. Carrano, T. J. Gaudin, R. W. Blob and J. R. Wible, 76–113. Chicago: Chicago University Press.

Sidor, C. A., and R. M. H. Smith. 2004. A new galesaurid (Therapsida: Cynodontia) from the Lower Triassic of South Africa. *Palaeontology* 47:535–556.

———. 2007. A second burnetiamorph therapsid from the Permian Teekloof Formation of South Arica and its associated fauna. *Journal of Vertebrate Paleontology* 27:420–430.

Sidor, C. A., and J. Welman. 2003. A second specimen of *Lemurosaurus pricei* (Therapsida: Burnetiamorpha). *Journal of Vertebrate Paleontology* 23:631–642.

Sigogneau, D. 1970. *Révision systématique des gorgonopsiens sud-africaines., Cahiers de Paleontologie*. Paris: Centre National de la Recherche Scientifique.

Sigogneau, D., and P. Tchudinov. 1972. Reflections on some Russian eotheriodonts. *Palaeovertebrata* 5:79–109.

Sigogneau-Russell, D. 1989. Theriodontia 1. Phthinosuchia, Biarmosuchia, Eotitanosuchia, Gorgonopsia. Vol. 17B/1, *Encyclopedia of Paleoherpetologie*. Stuttgart: Gustav Fischer Verlag.

Simms, M. J., and A. H. Ruffell. 1990. Climate and biotic change in the late Triassic. *Journal of the Geological Society, London* 147:321–327.

Simon, R. V., C. A. Sidor, K. D. Angielczyk, and R. M. H. Smith. 2010. First record of a tapinocephalid (Therapsida: Dinocephalia) from the Ruhuhu Formation (Songea Group) of southern Tanzania. *Journal Vertebrate Palaeontology* 30:1289–1293.

Singh, I. J., E. A. Tonna, and C. P. Gandel. 1974. A comparative histological study of mammalian bone. *Journal of Morphology* 144:421–438.

Skedros, J. G., and K. J. Hunt. 2004. Does the degree of laminarity correlate with site-specific differences in collagen fibre orientation in primary bone? An evaluation in the turkey ulna diaphysis. *Journal of Anatomy* 205:121–134.

Skedros, J. G., M. W. Mason, and R. D. Bloebaum. 1994. Differences in osteonal micromorphology between tensile and compressive cortices of a bending skeletal system: Indications of potential strain-specific differences in bone microstructure. *Anatomical Record* 239:405–413.

Skedros, J. G., M. W. Mason, M. C. Nelson, and R. D. Bloebaum. 1996. Evidence of structural and material adaptation to specific strain features in cortical bone. *Anatomical Record Part A: Discoveries in Molecular, Cellular, and Evolutionary Biology* 246:47–63.

Skedros, J. G., S. M. Sorenson, and N. H. Jenson. 2007. Are distributions of secondary osteon variants useful for interpreting load history in mammalian bones? *Cells Tissues Organs* 185:285–307.

Skedros, J. G., S. C. Su, and R. D. Bloebaum. 1997. Biomechanical implications of mineral content and microstructural variations in cortical bone of horse, elk, and sheep calcanei. *Anatomical Record Part A: Discoveries in Molecular, Cellular, and Evolutionary Biology* 249:297–316.

Smiley, T., C. A. Sidor, A. Maga, and O. Ide. 2008. The vertebrate fauna of the Upper Permian of Niger. VI. First evidence of a gorgonopsian therapsid. *Journal of Vertebrate Paleontology* 28:543–547.

Smith, R. M. H. 1980. The lithology, sedimentology and taphonomy of flood-plain deposits of the Lower Beaufort (Adelaide Subgroup) strata near Beaufort West. *Transactions Geological Society South Africa* 83:399–413.

———. 1987a. Helical burrow casts of therapsid origin from the Beaufort Group (Permian) of South Africa. *Palaeogeography, Palaeoclimatology, Palaeoecology* 60:155–170.

———. 1987b. Morphology and depositional history of exhumed Permian point bars in the southwestern Karoo, South Africa. *Journal of Sedimentary Petrology* 60:258–276.

———. 1990a. Alluvial palaeosols and pedofacies sequences in the Permian Lower Beaufort (Late Permian), Karoo Basin, South Africa. *Journal of Sedimentary Petrology* 60:258–276.

———. 1990b. A review of stratigraphy and sedimentary environments in the Karoo Basin of South Africa. *Journal of African Earth Sciences* 10:117–137.

———. 1993a. Sedimentology and ichnology of floodplain palaeosurfaces in the Beaufort Group (Late Permian), Karoo Sequence, South Africa. *Journal of Sedimentary Petrology* 57:19–29.

———. 1993b. Vertebrate taphonomy of Late Permian floodplain deposits in the southwestern Karoo Basin of South Africa. *Palaios* 8:45–67.

———. 1995. Changing fluvial environments across the Permian-Triassic boundary in the Karoo Basin, South Africa and possible causes of tetrapod extinctions. *Palaeogeography, Palaeoclimatology, Palaeoecology* 117:81–104.

———. 2000. Sedimentology and taphonomy of Late Permian vertebrate fossil localities in southwestern Madagascar. *Palaeontologia Africana* 36:25–41.

Smith, R. M. H., and J. Botha. 2005. The recovery of terrestrial vertebrate diversity in South African Karoo Basin after the End-Permian extinction. *Comptes Rendus Palevol* 4:555–568.

Smith, R. M. H., P. G. Eriksson, and W. J. Botha. 1993. A review of the stratigraphy and sedimentary environments of the Karoo-aged Basins of southern Africa. *Journal of African Earth Sciences* 16:143–69.

Smith, R. M. H., and A. W. Keyser. 1995. Biostratigraphy of the *Tapinocephalus* Assemblage Zone. In *SACS Biostratigraphic Series,* edited by B. S. Rubidge, 8–17. Pretoria, South Africa: Council for Geoscience.

Smith, R. M. H., and J. W. Kitching. 1997. Sedimentology and vertebrate taphonomy of the *Tritylodon* Acme Zone, a reworked palaeosol in the Lower Jurassic Elliot Formation, Karoo Supergroup, South Africa. *Palaeogeography, Palaeoclimatology, Palaeoecology.* 131:29–50.

Smith, R. M. H., C. A. Marsicano, and J. A. Wilson. 2009. Sedimentology and paleoecology of a diverse Early Jurassic tetrapod tracksite in Lesotho, Southern Africa. *Palaios* 24:672–684.

Smith, R. M. H., B. S. Rubidge, and C. A. Sidor. 2006. A new burnetiid (Therapsida: Biarmosuchia) from the Upper Permian of South Africa and its biogeographic implications. *Journal of Vertebrate Paleontology* 26:331–343.

Smith, R. M. H., and P. D. Ward. 2001. Pattern of vertebrate extinctions across an event bed at the Permian-Triassic boundary in the Karoo Basin of South Africa. *Geology* 29:1147–1150.

Smith-Gill, S. J. 1983. Developmental plasticity: Developmental conversion versus phenotypic modulation. *American Zoologist* 23:47–55.

Sollas, I. B. J., and W. J. Sollas. 1914. A study of a skull of *Dicynodon* by means of serial sections. *Philosophical Transactions of the Royal Society of London B* 204:201–225.

———. 1916. On the structure of the dicynodont skull. *Philosophical Transactions of the Royal Society of London B* 207:531–539.

South African Committee for Stratigraphy (SACS). 1980. *Stratigraphy of South Africa, Part 1 Lithostratigraphy of the Republic of South Africa, South West Africa/Namibia, and the Republic of Bophuthatswana, Transkei and Nenda.* Handbook of the Geological Survey of South Africa 8. Pretoria, South Africa: South African Committee for Stratigraphy.

Spotila, J. R., M. P. O'Connor, P. Dodson, and F. V. Paladino. 1991. Hot and cold running dinosaurs: Body size, metabolism and migration. *Modern Geology* 16:203–227.

Springer M. S, W. J. Murphy, E. Eizirik, and S. J. O'Brien. 2003. Placental mammal diversification and the Cretaceous–Tertiary boundary. *Proceedings of the National Academy of Science* 100:1056–1061.

Standora, E. A., J. R. Spotila, J. A. Keinath, and C. R. Shoop. 1984. Body temperatures, diving cycles, and movement in the subadult leatherback turtle, *Dermochelys Coriacea. Herpetologica* 40:169–176.

Stanley, S. M. 1973. An explanation for Cope's Rule. *Evolution* 27:1–25.

Starck, J. M., and A. Chinsamy. 2002. Bone microstructure and developmental plasticity in birds and other dinosaurs. *Journal of Morphology* 254:232–246.

Stein, B. R. 1989. Bone density and adaptation in semi-aquatic mammals. *Journal of Mammalogy* 70:467–476.

Sues, H-D. 1985a. First record of the tritylodontid *Oligokyphus* (Synapsida) from the lower Jurassic of Western North America. *Journal of Vertebrate Paleontology* 5:328–335.

———. 1985b. The relationships of the Tritylodontidae (Synapsida). *Zoological Journal of the Linnean Society* 85:205–217.

———. 1992. No Palaeocene "mammal-like reptile." *Nature* 359:278.

———. 2001. On *Microconodon*, a Late Triassic cynodont from the Newark Supergroup of eastern North America. *Bulletin of the Museum of Comparative Zoology* 156:37–48.

Sues, H-D., and J. A. Boy. 1988. A procynosuchid cynodont from central Europe. *Nature* 331:523–524.

Sues, H-D., J. M. Clark, and F. A. Jenkins. 1997. A review of the Early Jurassic tetrapods from the Glen Canyon Group of the American Southwest. In *In the Shadow of the Dinosaurs—Early Mesozoic Tetrapods*, edited by N. C. Fraser and H-D. Sues, 284–294. Cambridge: Cambridge University Press.

Sues, H-D., and J. A. Hopson. 2010. Anatomy and phylogenetic relationships of *Boreogomphodon jeffersoni* (Cynodontia: Gomphodontia) from the Upper Triassic of Virginia. *Journal of Vertebrate Paleontology* 30:1202–1220.

Sues, H-D., and F. A. Jenkins Jr. 2006. The postcranial skeleton of *Kayentatherium wellesi* from the Lower Jurassic Kayenta Formation of Arizona and the phylogenetic significance of postcranial features in tritylodontid cynodonts. In *Amniote Paleobiology: Perspectives on the Evolution of Mammals, Birds, and Reptiles.*, edited by M. T. Carrano, T. J. Gaudin, R. W. Blob, and J. R. Wible, 114–152. Chicago: Chicago University Press.

Sues, H-D., and P. E. Olsen. 1990. Triassic vertebrates of Gondwanan aspect from the Richmond Basin of Virginia. *Science* 249:1020–1023.

Sullivan, C., R. R. Reisz, and R. M. H. Smith. 2003. The Permian mammal-like herbivore *Diictodon*, the oldest known example of sexually dimorphic armament. *Proceedings of the Royal Society of London. Series B* 270:173–178.

Sumida, S. S. 1989. New information on the pectoral girdle and vertebral column in *Lupeosaurus* (Reptilia, Pelycosauria). *Canadian Journal of Earth Sciences* 26:134–1349.

Sumida, S. S., E. Rega, and K. Noriega. 2005. Evidence-based paleopathology II: Impact on phylogenetic analysis of the genus *Dimetrodon*. *Journal of Vertebrate Paleontology* 25(3) supplement:120A.

Sun, A. 1991. A review of Chinese therocephalian reptiles. *Vertebrata Palasiatica* 29:85–94.

Sun, A-L. 1988. Additional study on *Sinognathus gracilis* (Cynodontia; Reptilia). *Vertebrata Palasiatica* 26:173–180.

Surkov, M. V., and M. J. Benton. 2004. The basicranium of dicynodonts (Synapsida) and its use in phylogenetic analysis. *Palaeontology* 47:619–638.

———. 2008. Head kinematics and feeding adaptations of the Permian and Triassic dicynodonts. *Journal of Vertebrate Paleontology* 28:1120–1129.

Surkov, M. V., N. N. Kalandadze, and M. J. Benton. 2005. *Lystrosaurus georgi*, a dicynodont from the Lower Triassic of Russia. *Journal of Vertebrate Paleontology* 25:402–413.

Sutton, M. 2008. Tomographic techniques for the study of exceptionally preserved fossils. *Proceedings of the Royal Society B* 275:1587–1593.

———. 1968. Morphology and systematics of the North Dvina cynodonts (Reptilia, Therapsida: Upper Permian. *Postilla* 126:1–15.

Tatarinov, L. P. 1960. Otkrytiye psevdosukhiy v verkhney Permi SSSR. *Paleontologicheskii Zhurnal* 1961:117–132.

———. 1963. New Late Permian therocephalian. *Paleontologicheskii Zhurnal* 1963(4):76–94.

———. 1968. Morphology and systematics of the Northern Dvina cynodonts (Reptilia, Therapsida; Upper Permian). *Postilla* 126:1–51.

———. 1994. On an unusual mammalian tooth from the Mongolian Jurassic. *Paleontologicheskii Zhurnal* 1994:97–105.

Tatarinov, L. P., and E. N. Matchenko. 1999. A find of an aberrant tritylodont (Reptilia, Cynodontia) in the Lower Cretaceous of the Kemerovo region. *Paleontological Journal* 33:422–428.

Thackeray, J. F. 1991. Growth increments in teeth of *Diictodon* (Therapsida). *Koedoe* 34(1):7–11.

Thomason, J. J., and A. P. Russell. 1986. Mechanical factors in the evolution of the mammalian secondary palate: A theoretical analysis. *Journal of Morphology* 189:199–213.

Thulborn, T., and S. Turner. 2003. The last dicynodonts: an Australian relic. *Proceedings of the Royal Society B* 270:985–993.

Tollman, S. M., F. E. Grine, and B. D. Hahn. 1980. Ontogeny and sexual dimorphism in *Aulacephalodon* (Reptilia, Anomodontia). *Annals of the South African Museum* 81:159–186.

Tomkins, J. L., N. R. Le Bas, M. P. Witton, D. M. Martill, and S. Humphries. 2010. Positive allometry and the prehistory of sexual selection. *The American Naturalist* 176:141–148.

Tracy, C. R., J. S. Turner, and R. B. Huey. 1986. A biophysical analysis of possible thermoregulatory adaptations in sailed pelycosaurs. In *The Ecology and Biology of Mammal-like Reptiles*, edited by N. Hotton III, P. D. MacLean, J. J. Roth, and E. C. Roth, 195–206. Washington: Smithsonian Institution Press.

Tripathi, C., and P. P. Satsangi. 1963. *Lystrosaurus* fauna of the Panchet series of the Raniganj coalfield. *Palaeontologia Indica* 37:1–53.

Tucker, M. E., and M. J. Benton. 1982. Triassic environments, climates and reptile evolution.

Palaeogeography Palaeoclimatology Palaeoecology 40:361–379.

Tumarkin-Deratzian, A. R. 2007. Fibrolamellar bone in wild adult *Alligator mississippiensis*. *Journal of Herpetology* 41(2):341–345.

Turner, B. R. 1979. *Geology of the uraniferous Beaufort Group near Beaufort West, Excursions Guide Book, Geocongress '79*. Johannesburg: Geological Society of South Africa.

Turner, J. S., and C. R. Tracy. 1986. Body size, homeothermy and the control of heat exchange in mammal-like reptiles. In *The Ecology and Biology of Mammal-Like Reptiles*, edited by N. Hotton, III, P. D. MacLean, J. J. Roth, and E. C. Roth, 185–194. Washington: Smithsonian Institution Press.

Turner, N., A. J. Hulbert, and P. L. Else. 2006. Limits to physical performance and metabolism across species. *Current Opinion in Clinical Nutrition and Metabolic Care* 9:691–696.

Turvey, S. T., O. R. Green, and R. N. Holdaway. 2005. Cortical growth marks reveal extended juvenile development in New Zealand moa. *Nature* 435:940–943.

Unwin, D. M., and N. N. Bakhurina. 1994. *Sordes pilosus* and the nature of the pterosaur flight apparatus. *Nature* 371:62–64

Van der Walt, M., M. Day, B. Rubidge, A. K. Cooper, and I. Netterberg. 2010. A new GIS-based biozone map of the Beaufort Group (Karoo Supergroup), South Africa. *Palaeontologia Africana* 45:1–5.

Varricchio, D. J., P. C. Sereno, Z. Xijin, T. Lin, J. A. Wilson, and G. H. Lyon. 2008. Mud-trapped herd captures evidence of distinctive dinosaur sociality. *Acta Palaeontologica Polonica* 53:567–578.

Vega-Dias, C., M. W. Maisch, and C. L. Schultz. 2004. A new phylogenetic analysis of Triassic dicynodonts (Therapsida) and the systematic position of *Jachaleria candelariensis* from the Upper Triassic of Brazil. *Neues Jahrbuch für Geologie und Paläontologie, Abhandlungen* 231:145–166.

Vogel, S. 1988. *Life's Devices: the Physical World of Animals and Plants*. Princeton, N.J.: Princeton University Press.

———. 2003. *Comparative Biomechanics: Life's Physical World*. Princeton, N.J.: Princeton University Press.

Voigt, S., and M. Ganzelewski. 2010. Towards the origin of amniotes: Diadectomorph and synapsid footprints from the early Late Carboniferous of Germany. *Acta Palaeontologica Polonica* 55:57–72.

Von Huene, F. 1925. Die Südafrikanische Karroo-Formation als geologisches und faunitisches Lebensbild. *Fortschritte der Geologie und Palaeontologie* 12:1–24.

———. 1950. Die Theriodontiere des ostafrikanischen Ruhuhu-Gebietes in der Tübinger Sammlung. *Neues Jahrbuch fur Geologie und Paläontologie* 92:47–136.

Von Huene, F., D. R. Montgomery, and R. M. H. Smith. 2000. Altered river morphology in South Africa related to the Permian Triassic extinction. *Science* 289:1740–1743.

Vullo, R., V. Girard, D. Azar, and D. Néraudeau. 2010. Mammalian hairs in Early Cretaceous amber. *Naturwissenschaften*. doi 10.1007/s00114-010-0677-8.

Wall, P. W. 1983. The correlation between high limb-bone density and aquatic habits in recent mammals. *Journal of Paleontology* 57(2):197–207.

Walter, L. R. 1988. The limb posture of kannemeyeriid dicynodonts: Functional and ecological considerations. In *The Beginnings of the Age of Dinosaurs*, edited by K. Padian, 89–98. Cambridge: Cambridge University Press.

Wang, X., Z. Zhou, F. Zhang, and X. Xu. 2002. A nearly completely articulated rhamphorhynchoid pterosaur with exceptionally well-preserved wing membranes and "hairs" from Inner Mongolia, northeast China. *Chinese Science Bulletin* 47:226–232.

Ward, P. D., J. Botha, R. Buick, M. O. de Kock, D. H. Erwin, G. H.Garrison, J. I. Kirschvink, and R. M. H. Smith. 2005. Abrupt and gradual extinction among Late Permian land vertebrates in the Karoo Basin, South Africa. *Science* 307:709–714.

Ward, P. D., D. R. Montgomery, and R. M. H. Smith. 2000. Altered river morphology in South Africa related to the Permian Triassic extinction. *Science* 289:1740–1743.

Warren, A. A., and R. Damiani. 1999. Stereospondyl amphibians from the Elliot Formation of South Africa. *Palaeontologia Africana* 35:171–179.

Warren, A. A., and M. N. Hutchinson. 1983. The last labyrinthodont? A new brachyopid (Amphibia, temnospondyli) from the early Jurassic Evergreen Formation of Queensland, Australia. *Philosophical Transactions of the Royal Society of London B* 303:1–62.

Warren, J. W. 1963. *Growth Zones in the Skeleton of Recent and Fossil Vertebrates*. PhD diss., University of California, Los Angeles.

Warren, W. C., L. W. Hillier, J. A. Marshall Graves, E. Birney, C. P. Ponting, F. Grützner, K. Belov, W. Miller, L. Clarke, A.T. Chinwalla, S. Yang, A. Heger, D.P. Locke, P. Miethke, P.D. Waters, F. Veyrunes, L. Fulton, B. Fulton, T. Graves, J. Wallis, X. S. Puente, C. López-Otín, G. R. Ordóñez, E. E. Eichler, L. Chen, Z. Cheng, J. E. Deakin, A. Alsop, K. Thompson, P. Kirby, A. T. Papenfuss, M. J. Wakefield, T. Olender, D. Lancet,

G. A. Huttley, A. F. Smit, A. Pask, P. Temple-Smith, M. A. Batzer, J. A. Walker, M. K. Konkel, R. S. Harris, C. M. Whittington, E. S. Wong, N. J. Gemmell, E. Buschiazzo, I. M. Vargas Jentzsch, A. Merkel, J. Schmitz, A. Zemann, G. Churakov, J. Ole Kriegs, J. Brosius, E. P. Murchison, R. Sachidanandam, C. Smith, G. J. Hannon, E. Tsend-Ayush, D. McMillan, R. Attenborough, W. Rens, M. Ferguson-Smith, C. M. Lefèvre, J. A. Sharp, K. R. Nicholas, D. A. Ray, M. Kube, R. Reinhardt, T. H. Pringle, J. Taylor, R. C. Jones, B. Nixon, J. Dacheux, H. Niwa, Y. Sekita, X. Huang, A. Stark, P. Kheradpour, M. Kellis, P. Flicek, Y. Chen, C. Webber, R. Hardison, J. Nelson, K. Hallsworth-Pepin, K. Delehaunty, C. Markovic, P. Minx, Y. Feng, C. Kremitzki, M. Mitreva, J. Glasscock, T. Wylie, P. Wohldmann, P. Thiru, M. N. Nhan, C. S. Pohl, S. M. Smith, S. Hou, M. B. Renfree, E. R. Mardis, and R. K. Wilson. 2008. Genome analysis of the platypus reveals unique signatures of evolution. *Nature* 453:175–183.

Watson, D. M. S. 1913. The Beaufort Beds of the Karroo System of South Africa. *Geological Magazine* 10:388–393.

———.1914 The zones of the Beaufort Beds of the Karroo System of South Africa. *Geological Magazine* 11:203–208.

———. 1931. On the skeleton of a bauriamorph reptile. *Proceedings of the Zoological Society, London* 1163–1205.

———. 1948. *Dicynodon* and its allies. *Proceedings of the Zoological Society* 118:823–877.

———. 1960. The anomodont skeleton. *Transactions of the Zoological Society of London* 29:131–208.

Watson, D. M. S., and A. S. Romer. 1956. A classification of therapsid reptiles. *Bulletin of the Museum of Comparative Zoology Harvard* 111:37–89.

Weide, D. M., C. A. Sidor, K. D. Angielczyk, and R. M. H. Smith. 2009. A new record of *Procynosuchus delaharpeae* (Therapsida: Cynodontia) from the Upper Permian Usili Formation, Tanzania. *Palaeontologia Africana* 44:21–26.

Weiner, S., and W. Traub. 1992. Bone structure: From angstroms to microns. *FASEB Journal* 6:879–885.

White, R. V. 2002. Earth's biggest "whodunnit": Unravelling the clues in the case of the end-Permian mass extinction. *Philosophical Transactions of the Royal Society of London B* 360:2963–2985.

Wible, J. R., G. W. Rougier, M. J. Novacek, and R. J. Asher. 2007. Cretaceous eutherians and Laurasian origin for placental mammals near the K/T boundary. *Nature* 447:1003–1006.

Williston, S. W. 1904. Notice of some new Reptiles from the Upper Triassic of Wyoming. *Journal of Geology* 12:688–694.

Witmer, L. M. 1995. Homology of facial structures in extant archosaurs (birds and crocodilians), with special reference to paranasal pneumaticity and nasal conchae. *Journal of Morphology* 225:269–327.

Wood, C. B., and D. N. Stern. 1997. The earliest prisms in mammalian and reptilian enamel. In *Tooth Enamel Microstructure*, edited by W. V. Koenigswald and P. M. Sander, 63–83. Rotterdam: A. A. Balkema.

Woodburne, M. O., T. H. Rich, and M. S. Springer. 2003. The evolution of tribospheny and the antiquity of mammalian clades. *Molecular Phylogeny and Evolution* 28:360–385.

Wyley, A. 1859. *Notes on a Journey in Two Directions across the Cape Colony, Made in the Years 1857–8, with a View To Determine the Character and Order of the Various Geological Formations*, Appendix to Parliamentary Report G, Cape Town: Cape of Good Hope.

Yates, A.M. 2003. The first definite dinosaur from the Lower Elliot Formation (Norian: Upper Triassic). *Palaeontologia Africana* 39:63–68.

Yates, A. M., and J. W. Kitching. 2003. The earliest known sauropod dinosaur and the first steps towards sauropod locomotion. *Proceedings of the Royal Society of London B* 270:1753–1758.

Yates, A. M., M. F. Bonnan, J. Neveling, A. Chinsamy, and M. Blackbeard. 2009. A new transitional sauropodomorph dinosaur from the Early Jurassic of South Africa and the evolution of sauropod feeding and quadrupedalism. *Proceedings of the Royal Society B* 277:787–794.

Index

I

Forerunners of Mammals was designed by Jamison Cockerham at Indiana University Press, set in type by Jamie McKee at MacKey Composition, and printed by Thomson-Shore, Inc.

The text type is Electra, designed by William A. Dwiggins (circa 1935) and the headings and captions are set in Frutiger, designed by Adrian Frutiger (circa 1975), both issued by Adobe Systems.